MOS INTEGRATED CIRCUITS

Theory, Fabrication, Design, and Systems Applications of MOS LSI

MOS INTEGRATED CIRCUITS

Theory, Fabrication, Design, and Systems Applications of MOS LSI

Prepared by the Engineering Staff of American Micro-systems, Inc.

Edited by

William M. Penney
Senior Technical Staff

Lillian Lau
Corporate Staff

ROBERT E. KRIEGER PUBLISHING COMPANY
HUNTINGTON, NEW YORK
1979

Original Edition 1972
Reprint Edition 1979

Printed and Published by
ROBERT E. KRIEGER PUBLISHING COMPANY, INC.
645 NEW YORK AVENUE
HUNTINGTON, NEW YORK 11743

Printed in the United States of America

Library of Congress Cataloging in Publication Data

American Micro-systems, inc.
MOS integrated circuits.

Reprint of the edition published by Van Nostrand Reinhold Co., New York, in series: Microelectronics series.
Includes bibliographical references and index.
1. Integrated circuits—Large scale integration.
2. Metal oxide semiconductors. I. Penney, William M.
II. Lau, Lillian. III. Title.
[TK7874.A53 1979] 621.381'73 79-1039
ISBN 0-88275-897-7

CONTRIBUTORS

G. S. ASHCRAFT
J. L. FARLEY
J. N. FORDEMWALT
J. A. HORTON
S. JASPER
W. M. PENNEY
R. J. SADOWSKI
L. A. TAYLOR
P. VUTZ
G. P. WALKER
G. A. WETLESEN

Preface

The objective of this book is to present the techniques that are necessary to implement systems using Metal Oxide Semiconductor (MOS) Large-Scale Integration (LSI). These techniques can be used to partition a system, prepare the detailed logic designs of the partitioned system, perform detailed circuit design using a method based on a simplified model of the MOS device and, finally, obtain near-optimum integrated circuits with respect to size, cost, and functional capability. Also included are basic physical theory of the MOS transistor and description of several processes that may be used to manufacture MOS large-scale arrays.

The book is primarily written for practicing systems and logic designers in the digital equipment business, although circuit development engineers should find the circuit design data and topological considerations useful. The level of presentation assumes an electrical or electronic engineering background, a general understanding of basic semiconductor theory and a familiarity with Boolean algebra as applied to digital systems design. In the mathematical treatment, only those equations that are necessary to ensure clarity have been included.

Much of the material in this book has evolved from an engineering course in MOS technology. AMI has presented this course to engineers and designers in the United States since 1966, and more recently to engineers in Europe and Japan. The wealth of material in the original course book and the updated editions, as well as the comments of attendees of the courses throughout the years, has been invaluable.

In the first chapter, W. M. Penney discusses the advantages and limitations of MOS large-scale integrated circuits with respect to performance and cost plus the effect of this new technology on supplier/vendor business relationships. Included is a listing of major areas of application for which MOS arrays are well-suited. These areas of application are divided into categories; the first is where MOS has already established a firm foothold, and the second includes new markets or products in which MOS arrays are likely to gain significant acceptance in the near future. Also presented in this chapter is a brief review of the evolution of the MOS device from early concepts of the field effect principle through the development of MOS integrated circuits.

Chapter 2, principally written by P. Vutz, contains an analysis of the physical theory of the MOS device followed by the derivation of the basic design equations necessary for proper application of the MOS transistor in large-scale arrays. Next, is a description of the device parameters and characteristics that are of importance in circuit design and analyses. Also included in this chapter is a review of semiconductor basic physical theory by J. N. Fordemwalt. This material has been purposely limited to the amount necessary for an understanding of the principles of operation of the MOS transistor and the development of the device design equations.

In Chapter 3, G. A. Wetlesen presents a review of basic silicon processing and assembly techniques applicable to the manufacture of MOS large-scale integrated circuits. A detailed description of the typical *P*-channel MOS process is followed by descriptions of other currently popular MOS processes. The features and characteristics of these processes are compared with those of the high-voltage *P*-channel process.

The design of complex digital integrated circuits that employ the MOS device is presented by J. A. Horton in Chapter 4. The basic MOS inverter is examined in detail using both graphical and analytical techniques. The dc transfer characteristics of the inverter are derived and the transient response is examined. A number of design examples are presented to demonstrate the application of the design concepts to practical MOS large-scale circuits.

R. J. Sadowski and G. S. Ashcraft discuss, in Chapter 5, the various logic design techniques that are useful as a result of the unique characteristics of the MOS device. After reviewing the basic types of logic that can be implemented with MOS, the advantages and limitations of each are examined. Next, typical MOS logic configurations are analyzed in detail, followed by design examples that illustrate the application of these concepts to typical design problems. Following this, the effects of particular logic designs on testing the final MOS products are considered.

The factors that must be analyzed and the trade-offs that must be evaluated in the design of systems in which MOS LSI circuits are used are covered in Chapter 6. In this chapter, L. A. Taylor and R. J. Sadowski discuss such system design considerations as system partitioning and provide guidelines for evaluating the trade-offs necessary to achieve the most efficient (MOS) system in terms of cost, performance, and reliability. Preceding this, W. M. Penney reviews some of the important interface points between the system user and the MOS supplier and compares the advantages against the limitations of each point. Finally, logic simulation is discussed and G. S. Ashcraft presents the philosophies and schemes employed to test the complex MOS arrays.

The application of the MOS device in memory types of products is presented by S. Jasper in Chapter 7. Basic memory cell concepts are discussed followed by descriptions of the various types of MOS memory systems that employ these cells. At the conclusion of this chapter is a review of typical system applications of MOS memory products, including their use as logic functions in addition to the normal data storage function.

In Chapter 8, J. A. Horton discusses MOS circuit layout (topological) philosophy and circuit design rules. A typical product development flow chart is presented that serves to tie together much of the material discussed in previous chapters. Considerations that are important in deciding whether circuit layouts should be performed manually or by computer aided design schemes are also examined. Following a description of the steps in preparing MOS circuit layout drawings, typical circuit design rules and tolerances are presented, together with their effect on the yield of the manufacturing process.

System reliability of the MOS large-scale arrays is discussed in Chapter 9 by J. L. Farley. Included is a description of significant MOS array failure modes and mechanisms, followed by techniques that can be used by circuit designers and process engineers to minimize their occurrence. Also discussed are screening techniques to eliminate potential reliability risks and a reliability test vehicle that permits continuous monitoring and measurement of the process reliability at a very low cost to the manufacturer.

The appendix contains a discussion of some concepts of semiconductor surface physics that are applicable in the analysis of MOS device characteristics. This presentation was written by G. P. Walker and is based on original work that he performed at the Raytheon Company with P. Lin and W. Ford.

The editors gratefully acknowledge the whole-hearted cooperation of all the authors who contributed to this book. We also express our sincere appreciation to the authors of the original MOS course book and to the many individuals at AMI who gave their enthusiastic support during the various stages of manuscript preparation. We are all especially grateful to the management of AMI for their wise counsel and encouragement. We also thank them for the opportunity and resources that made this book possible.

W. M. Penney
L. Lau

Contents

MOS INTEGRATED CIRCUITS

Theory, Fabrication, Design, and Systems Applications of MOS LSI

1 MOS LSI Technology

1.1 INTRODUCTION

Metal oxide semiconductor (MOS) large-scale integration (LSI) is spearheading what may prove to be the most significant development in the electronic industry since the introduction of the bipolar transistor. When combined with good product planning and imaginative system design, MOS large-scale arrays result in high performance, low cost, reliable systems, and equipment of all types. With MOS technology, it is possible to build medium-speed, complex logic arrays and long delay shift registers with minimum power dissipation and low cost per gate, or bit of memory. This process is also ideally suited for the implementation of medium-size memory systems, both random access read-write and read-only, where memory speed requirements preclude the use of existing magnetic devices. Multiplexing and analog switching applications are an MOS "natural" because of the high off-resistance and zero offset voltage of the MOS transistor. However, the greatest advantage of MOS technology is that highly complex logic functions that operate at moderate speeds can be produced at a lower cost than possible with any other existing technology. Even though bipolar technology will probably, in the foreseeable future, continue to outperform MOS circuits when compared on a switching speed basis, such limitations will be largely offset by the ease with which custom LSI arrays can be designed and fabricated with MOS technology.

This chapter will explain the meaning of MOS LSI, why it is of vital interest to system and logic designers, what its advantages are, as well as its limitations, and where it can be advantageously applied.

Although MOS LSI may be defined in several ways, in this text we will mean a silicon semiconductor die (also called chip or array) that contains many interconnected logic functions and, in some instances, memory elements. These functions may comprise as much as an entire system by themselves or as little as a portion of subsystem within a complex system. This definition deliberately avoids the temptation to stipulate that an MOS LSI array must contain a predetermined number of equivalent logic gates or MOS transistors. However, most MOS LSI chips presently contain more than 100 logic gates and over 400 MOS transistors. These values are likely to double in the near future. The definition of an MOS array in this text is based more on the capability of the array to perform some function that heretofore required the interconnection of many separate integrated circuits, discrete parts or modules, and printed circuit boards. This definition of LSI focuses attention onto the importance of the interconnection of the logic functions, as well as the functions themselves. A major factor that added impetus to the development of LSI was the need to reduce the number of costly, unreliable interconnections that were required by the previously available packaged logic functions. Because MOS LSI accomplishes both jobs—interconnect and logic—new products and systems that were previously uneconomical to manufacture are now possible.

1.2 EVOLUTION OF MOS LSI

Although MOS LSI is considered a new technology in the 20-year-old semiconductor world, the basic concepts of the MOS insulated gate field effect transistor (MOSFET), also sometimes called IGFET, predate the bipolar transistor. Only through the comparatively recent development of a stable, predictable, high-yield manufacturing process has MOS LSI become a reality today. Prior to this development, poor control and lack of complete understanding of solid state surface conditions resulted in low yields and unstable electrical characteristics.

According to the literature, initial attempts to realize active solid-state transistors that operate on a field effect principle were undertaken by J. E. Lilienfield in the early 1930s. Oskar Heil disclosed his findings in a British patent issued in 1935. The first working model of the unipolar or field effect transistor was developed at Bell Telephone Laboratories where the transistor effect was discovered in 1948 by Bardeen and Brattain.[1] These scientists observed the effect during a series of experiments in which they attempted to modulate the current through point contacts on a germanium block. Their findings thus paved the way for the development of point contact transistors and the invention of the bipolar junction transistors, in which the basic principle of operation is one of minority injection rather than field effect (majority current flow). Attention was then focused on these two types of bipolar transistors and the development of the FET was virtually halted.

In 1952, Shockley[2] described an FET with a control electrode which consisted of a reverse-biased junction. Such junction FETs were subsequently built and tested according to G. C. Dacey and I. M. Ross, who also formulated an analytical treatment of the performance limits in 1955. However, early attempts to fabricate MOSFETs were unsuccessful because of the lack of a controllable and stable surface.

A major breakthrough in semiconductor processing occurred in the early 1960s with the development of the silicon planar process.[3] The growing, etching, and regrowing af an insulating layer of oxide on top of the silicon substrate provided a means of obtaining a much more stable surface and controlling the physical geometry far more accurately than was previously possible. It then became practical to consider the development of MOSFETs to handle functions which at that time were being handled either by vacuum tubes or, in a more costly manner, by bipolar transistors. Attempts at MOSFET fabrication were hindered by charge migration problems, which led to deterioration of their electrical characteristics. With further process refinements, the first reliable and reproducible discrete MOSFET resulted. Simple integrated circuits were produced in 1964, ten years after the first conventional silicon transistors were fabricated and more than thirty years after the FET concept was discovered.

Process control problems continued to plague the MOS manufacturer until about 1967. At that time, the yield of MOS circuits with stable performance characteristics was increased through refinements in the basic MOS process in conjunction with tightened process control and the institution of strict, clean room environments at critical processing steps. Since then, yields have continued to rise as a result of manufacturing experience and development of better production equipment, tools, and raw materials.

The very nature of the MOS process coupled with the size and complexity of LSI chips demands high-quality silicon wafers. The silicon suppliers have not only satisfied this need but have also produced the larger diameter wafers (2 to 3 inches) to reduce manufacturing costs. Photolithographic equipment suppliers have developed new equipment to produce the complex photomasks with the high resolution and dimensional control necessary for larger arrays. Better wafer fabrication equipment, such as that needed to provide the high accuracy required in the sequence of photomasking operations, has also become available. Improvements have been made too in many other production equipment and materials essential to high-yield semiconductor manufacturing.

In addition to the general upgrading in the quality and capability of the manufacturing materials and equipment, knowledge of semiconductor surface phenomena has increased greatly, resulting in improvements at various critical process points. Another factor that helped sustain the rise in yield was the development of circuit design techniques to take better advantage of the unique characteristics of MOSFETs. These techniques generally tended to reduce circuit and array size with no loss in performance. Then in the late 1960s, computers were used to control the generation of artwork and photomasks which provided more uniform tooling and increased usage of proven, predesigned basic circuits within complex arrays.

Of the basic MOS technologies now available, the metal gate *P*-channel process is most widely used. A cross section of an MOS transistor is shown in Figure 1-1. A thin slice (8 to 10 mils) of lightly doped *N*-type silicon material, called a wafer, serves as the supporting substrate or body of the MOS transistor. Two closely spaced, heavily doped *P*-type regions, the source and the drain, are formed within the substrate by selective diffusion of a substance that provides "holes" as majority electrical carriers. A thin deposited layer of metal, called the gate, covers the area between the source and drain regions, but is electrically insulated from them and the substrate by a thin layer (1000-1500 Å) of silicon dioxide. The gate serves as the control element of the MOS transistor creating, when properly biased, a conductive path or channel between the source and drain regions. If a conducting channel exists with zero gate bias, the transistor operates in the "depletion mode." However, if the MOSFET is normally nonconducting without gate bias, and a channel can be formed only by applying a sufficiently negative voltage to the gate, the transistor operates in the "enhancement mode." This mode is preferable for LSI digital circuits and is easiest to produce with the *P*-channel process.

The minimum value of gate voltage just sufficient to cause channel formation is known as the gate threshold voltage and is dependent upon the particular manufacturing process and the crystal lattice orientation of the silicon ingot from which the wafer is sliced. Voltages equal to or more negative than the threshold level cause the surface of the *N*-type silicon beneath the gate to invert to essentially *P*-type; hence, the name *P*-channel.

Since the gate is electrically insulated from the substrate, drain, and source regions, the input resistance of the MOSFET is extremely high

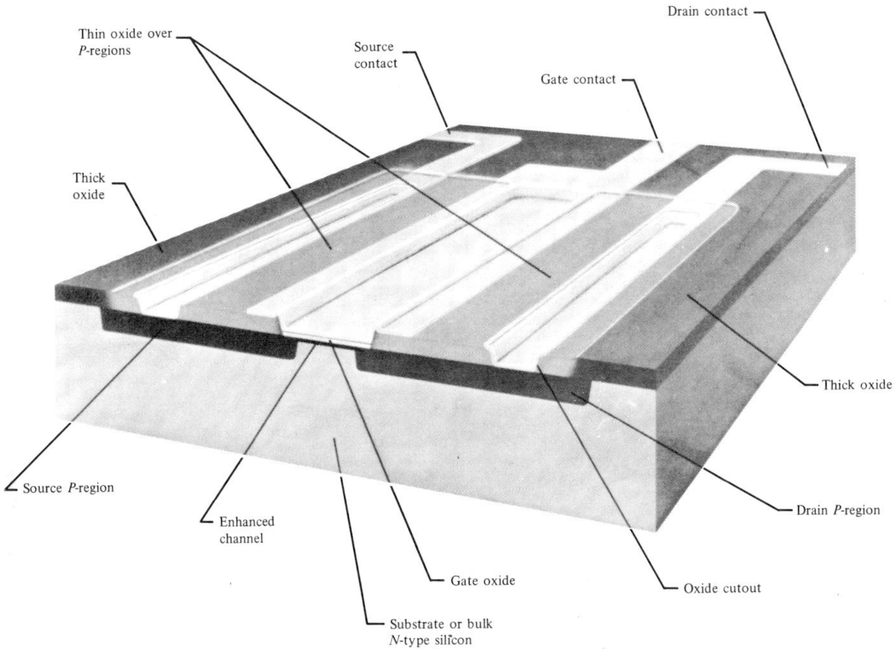

Figure 1–1 Cross section of an MOS device in conducting state.

(in the range of 10^{15} Ω at room temperature) and degrades very little even at elevated temperatures. A thorough analysis of the physical principles of operation and a detailed examination of the characteristics of *P*-channel MOSFET are provided in Chapter 2. Chapter 3 contains a detailed explanation of other MOS manufacturing technologies, as well as the basic *P*-channel process.

Symbols often used to schematically represent an MOS transistor are shown in Figure 1-2. Since the circuit design techniques developed in this book are applicable to large-scale arrays with either *P*- or *N*-channel MOSFETs, the symbol in Figure 1-2c will generally be used. It is understood that the body or substrate is connected to the most positive circuit potential for arrays fabricated with the *P*-channel process. Any circuit with a *P*-channel MOSFET can be made suitable for an *N*-channel transistor by simply reversing the power supply leads and polarity of external signals. Throughout this book the source is used as the reference terminal,

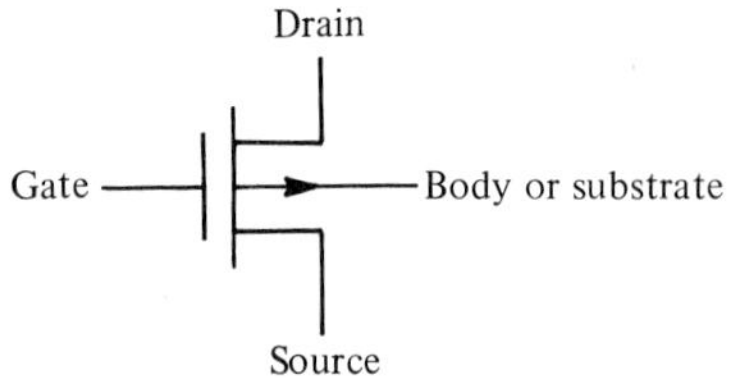

(a) *P*-channel MOSFET

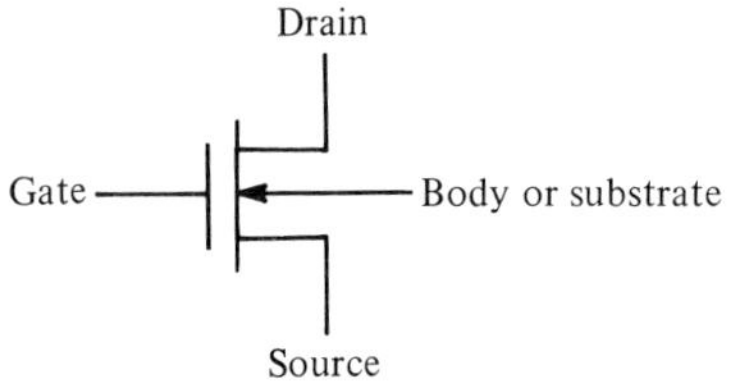

(b) *N*-channel MOSFET

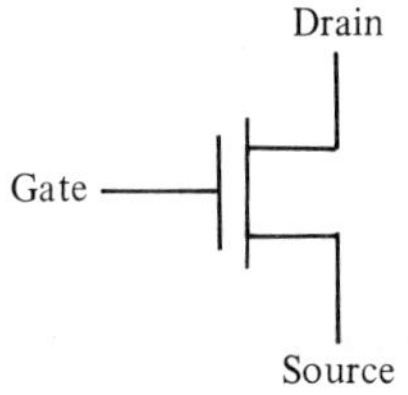

(c) MOSFET (*N*- or *P*-channel) in an array

Figure 1–2 Symbols for a MOSFET.

generally ground unless otherwise specified. The input signal is usually applied to the gate electrode and the output signal is taken from the drain terminal.

An important characteristic of the MOSFET is that the source and drain regions are interchangeable. As shown in Chapter 4, "MOS Circuit

Design Theory," this bilateral operation feature often is used to advantage in complex logic circuits.

1.3 ADVANTAGES OF MOS LSI

For many system manufacturers there are two basic advantages in using MOS LSI, viz., economy and performance. Because of the successful application of MOS LSI in numerous systems, the economic advantage is generally conceded, particularly where production volumes are high. However, the claim for performance advantages is almost certain to raise controversy, especially among advocates of bipolar semiconductor technology. Therefore, it is important to distinguish between superior system performance and transistor circuit performance per se. Indeed, the principal limitations of MOS arrays are a result of the slower speed characteristics of present MOSFETs compared to their bipolar counterparts. The two basic advantages are further subdivided into the following categories and are examined in detail in subsequent sections.

- ECONOMIC FACTORS
  a. lower system cost of ownership due to higher reliability and better maintainability
  b. lower initial cost for system
  c. low cost complex arrays due to simpler process and small size
  d. automation possible because of the design process

- PERFORMANCE FACTORS
  a. higher performance system can be designed at same cost
  b. better system reliability and maintainability
  c. performance of MOS LSI is readily predictable

1.3.1 Economic Advantages

- Reduced System Cost of Ownership

Lower cost of ownership results from increased reliability and reduced maintenance costs when MOS LSI is used in the system. Although reliability and maintainability are not usually considered as cost factors in the initial purchase price of a system, they are major factors in the analysis of the actual costs of a system during its useful lifetime. Since most of the reasons for the improvements in reliability and maintainability are also pertinent to economy and performance, they will be examined in detail in the discussion of performance factors.

- Lower Initial Cost

The initial cost of a system that is designed with MOS LSI can be much less than one that uses standard, low complexity integrated circuits (IC) and conventional interconnection, packaging, and assembly techniques. The low cost is primarily a result of the large decrease in the number of semiconductor circuits, which in turn causes a significant reduction in the number of interconnections, printed circuit (pc) boards, connectors, mother boards, enclosures, etc. These reductions provide large savings in labor costs because the assembly and checkout of the system are simplified. The combined effect of these savings is the principal advantage of MOS, viz., low cost per function.

The smaller number of interconnections within the MOS LSI system is probably the most important single factor responsible for both lower initial cost and increased system reliability. This is a result of the ability of a single MOS complex array to replace many conventional ICs and their associated interconnections. The cost of interconnections remains substantial, even though the cost per active circuit function has decreased drastically since semiconductors were first widely used in systems designed in the late 1950s. Chang[4] points out that studies indicate that the cost of permanently connecting two terminals electrically varies from a minimum of \$0.05 using highly automated techniques in large production volumes to more than \$1.00 for low volume, high reliability equipment. In contrast, the cost of interconnecting two points on an integrated circuit using photolithographic techniques can be as little as a few thousandths of a cent. With this magnitude difference in interconnection costs, even if reliability were not a consideration, there is sufficient economic incentive for using complex arrays.

Figure 1-3 from Chang[4] points out the reduction in total manufacturing costs of systems as they progressed from the use of transistors (and other discrete parts) to conventional ICs and finally to complex LSI circuits. Note that as more complex semiconductors are used in systems, the interconnection costs decrease substantially. However, no matter which generation semiconductor technology is used in the system, as time progresses, the cost of the semiconductors reduces to but a small fraction of the total system cost. In the past this cost decrease could be attributed to "product maturity" as the usage of each generation of semiconductors reached high volumes. Since the vast majority of MOS arrays are custom rather than standard products, as has been the case for the two prior generations, this cost trend may not be valid for LSI. Memory systems, however, which consist mostly of MOS LSI standard products, are likely to follow the trends indicated by past history.

It is important to understand that the mere substitution of MOS arrays

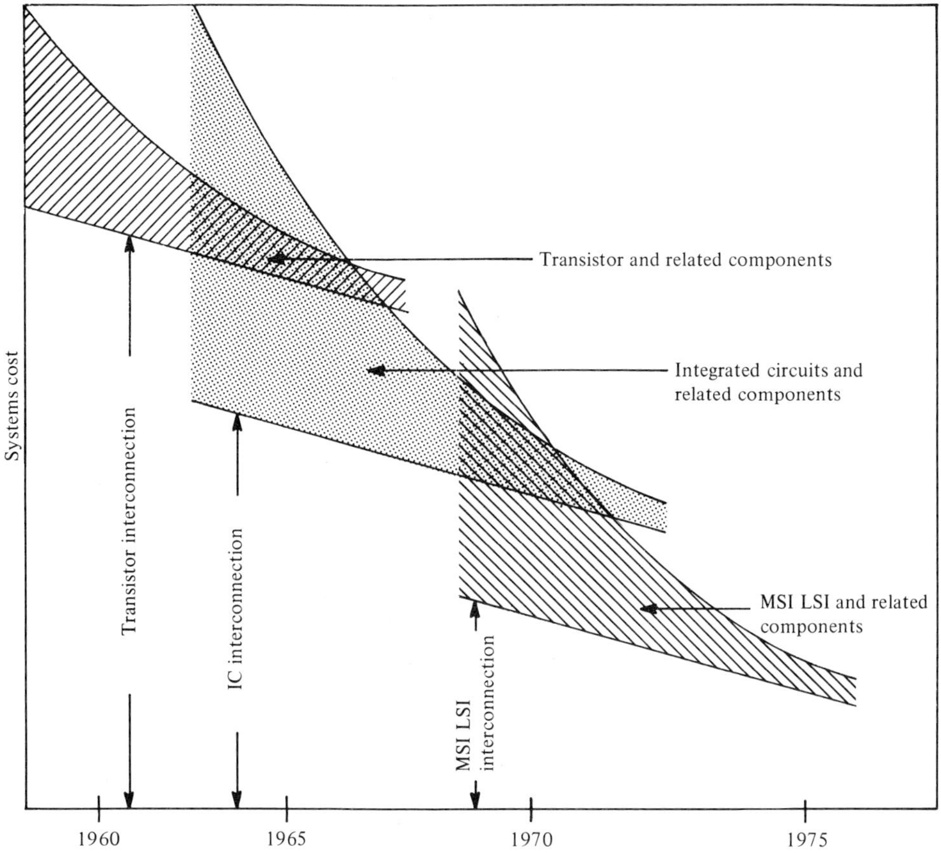

Figure 1–3 System cost as a function of level of integration (after Chang[1]).

for conventional ICs and discrete semiconductor circuits in existing systems will not necessarily afford a reduction in the cost of that system. As is discussed later in Chapter 6, "System Design with MOS Arrays," few existing systems can make efficient use of MOS LSI without major redesign.

- Simpler Process

The relatively low cost of the arrays is a consequence of the simpler, standardized manufacturing process as compared to bipolar IC technology. Table 1-1 shows a comparison of the two basic manufacturing processes.

A similar comparison by Warner[5] suggests that the bipolar process is about 30 percent more difficult to manufacture than the MOS process when all factors are considered. However, when the MOS process is compared with the bipolar process that is required for LSI, the differences are considerably exaggerated. Although bipolar LSI circuits are likely to require two levels (layers) of interconnect as do MOS LSI circuits, addi-

Table 1–1 Comparison of Metal Gate *P*-Channel MOS and Double-Diffused Epitaxial Bipolar Fabrication Processes

	MOS	*Double Diffused*
Diffusions	1	4 + Epitaxial
Process Steps	38 to 45	130
High-Temperature Process Steps	2	10
Masking Steps	4 or 5	8

tional complications arise with the bipolar process. Disregarding the diffused resistor which is sometimes oriented to serve as interconnect, bipolar circuits require at least two more masking steps plus a number of additional operations such as oxide deposition, etching, alloying, etc. Unlike the bipolar process, the MOS process requires no additional steps to provide two layers of interconect, as will be shown shortly.

As shown in Table 1-1, the number of critical process steps, particularly the high-temperature diffusions plus the masking operations, is greatly reduced in the MOS process. Consequently, the yields of MOS arrays greatly exceed those of bipolar chips of comparable functional complexity. Of even greater importance is the fact that the MOS process can be held constant, regardless of the array that is being fabricated. With the MOS process, circuit performance is varied by the design of the MOSFET geometry, not by varying the process (diffusion doping, time, and temperatures) as is often done in bipolar technology. Each array, even though it may be a custom design, benefits from the supplier's accumulated manufacturing experience. More realistically, the bipolar process is approximately 50 percent more difficult than the standard V_T metal gate MOS.

• Smaller Circuit Size

Another important factor that gives MOS LSI a cost per function advantage is the relatively small size of very complex MOS arrays. Since yield varies inversely with die area, this means lower die costs and, ultimately, lower selling price. The factors that contribute to the small size of MOS arrays are the small area (typically about 1 sq mil) required by a single MOSFET, the use of active MOS load transistors in place of large diffused load resistors, and the simpler process.

A study by Warner[5] revealed that the area of the MOS circuits was typically less than 20 percent of the area required by a bipolar circuit for similar functions. Warner attributes the size advantage of MOS arrays in part to the fact that fewer elements are needed to achieve a given circuit function and fewer metal to silicon contacts are required. With increase in array complexity, the advantage of MOS is increased.

The simpler MOS process provides at least three size benefits. First, since there are fewer masking, diffusion, and etching steps, there is less tolerance buildup throughout the total process. Consequently, there is less area wasted in an MOS array to allow for accumulation of tolerances. Second, MOSFETs require no isolation junctions or "wells," as do bipolar transistors. These isolation junctions can consume as much as 30 percent of the active area of a bipolar IC chip because of the wide lateral diffusion that occurs during the long diffusion step needed to form the deep isolation wells.

The third benefit results from the provision of two layers or levels of interconnect, although some restrictions are imposed upon one of the two layers. This second layer is formed during the normal MOSFET processing by diffusing *P*-doped interconnection lines or "tunnels." These lines typically have a resistance of about 100 Ω/□. However, unlike bipolar ICs, which use much higher currents, this amount of series resistance is tolerable for many circuit connections because of the low-magnitude currents that occur in MOS arrays.

As will be seen later, it is even possible to provide a third level of interconnect in some MOS processes, such as silicon gate. Again, this layer of interconnect is basically "free", i.e., it requires no additional processing steps beyond those normally required to fabricate the actual MOSFETs. This additional layer of interconnect allows crossing under conductive paths and considerably reduces the area required by MOS circuits.

- Amenable to Design Automation

Since the design of the MOSFET for specific performance characteristics is, unlike that of a bipolar transistor, accomplished by straightforward mathematical techniques, it is easily automated with a computer program.[6] Frequently used standard logic funtions, commonly called cells, can be designed, checked, and stored on magnetic tape for use in designing new MOS arrays. Several sophisticated computer programs are available for the generation of a layout drawing of an array, including the interconnection of logic functions. These computer programs can be used either to prepare the various masters for the required masks or to cut the artwork for the mask masters that are prepared in a separate nonautomated operation. The automated design process is particularly well-suited for MOS LSI memory products, such as random access memories (RAMs), read only memories (ROMs), and variable bit length shift registers. For nonrepetitive functions, such as random logic, the chips are often larger than if they were designed manually. The design automation process is applicable to other development functions, including circuit design, test generation, logic simulation, etc., as well as layout. Therefore, the process is described in greater detail in subsequent chapters where applicable. Obviously, design automation tools can provide cost savings and shorten the development cycle of certain types of MOS LSI circuits.

1.3.2 Performance Advantages

- Improved System Performance

Performance advantages of MOS LSI must be carefully qualified. Because of the unique features of MOS LSI, systems that were previously impractical can now be economically designed and manufactured. Systems are now feasible where size, power consumption, or weight restrictions are beyond the state of the art of conventional ICs, assembly, and packaging techniques. The use of MOS LSI can also improve system reliability and maintainability. However, the realization of these advantages requires considerable changes in the disciplines of circuit design through system design to overcome the inherent limitations of present MOS technologies. These limitations are explored later in this chapter. Chapter 4, "MOS Circuit Design Theory," and Chapter 6, "System Design with MOS Arrays," detail how these limitations can be circumvented.

One of the most significant advantages of MOS LSI, as pointed out by Notz *et al.*,[7] is that the system designer has a new freedom of choice. He can design either a higher-performance system with no increase in system cost or a low-cost system with no sacrifice in performance. Because of the small size of the typical MOSFET in an array, active devices may be employed as needed, with little penalty in chip area. Also, its extremely high input impedance permits high fanout within the chip. These two features allow the system designer to provide much more logic within a given die size. Whereas in the past, system economics dictated the minimization of logic, MOS LSI reduces the cost per logic function so drastically that it is no longer a valid design criterion. This increased logic capability can be used, as noted by Notz *et al.*, to provide greater computing or processing power, replace high cost software with hardware, or simplify system interfaces.

Several techniques have been developed to overcome the inherent speed limitations of the MOSFET in many system applications. The use of dynamic logic with multiphase logic clocks[8,9] and ratioless circuits are examples (see Chapters 4 and 5). Complex logic gates can be implemented with as many as eight logic levels per clock period, as discussed in Chapter 5. Dynamic logic offers other benefits to the system, viz., reduced size and power consumption.

Higher system noise immunity is still another advantage provided by

MOS technology. The MOSFET is well-suited for use in dynamic logic circuits because its gate, which acts as a virtually "perfect" capacitor, serves as a temporary memory for input data. The MOSFET can also conduct current in either direction. This bilateral feature provides a means for transferring the data by charging and discharging the capacitive storage nodes in an array. The ability to store and transfer charge permits the use of clocked load devices and ratioless circuits which greatly reduce the dc power as well as the size of MOS arrays. The relatively high turn-on or threshold voltage of the standard *P*-channel process, coupled with the large logic level swings, results in typical noise margins of at least 1 V over the normal commercial equipment temperature range.

- Improved Reliability and Maintainability

The improvement in system reliability with MOS LSI has been mentioned during the discussion of the economic benefits, but it must also be considered as a performance factor, especially when systems must meet contractual reliability requirements. The mere reduction in parts count and interconnect are obvious contributors to better system reliability. In addition, because of the small size and minimal power consumption of even complex MOS arrays, the system designer may find it attractive to employ functional redundancy or majority voting logic to further increase the system reliability.

However, the principal reason for the improvement in system reliability is that the number of interconnections within the system is greatly reduced. Even though the total number of interconnections in the system is not drastically decreased, when the number of batch-produced connections on the MOS chips is considered, extensive test and usage data have established the reliability of these mass fabricated connections. The quality of the interconnections within an MOS chip is relatively easy to control and detect compared to the difficulty of controlling and detecting the wide variety of connections from and external to the MOS chip. Usually, a terminal, commonly called a "pad," on the chip is connected to a package lead or pin with a thermal compression ball bond or an ultrasonic wedge bond. Since these connections are significant IC failure modes, these bonds should be minimized. Therefore, to achieve better system reliability, an MOS array is designed to have a high ratio of functions (logic gates) to package pins.

The following example illustrates how this ratio affects the reliability of a system. Consider a typical system in which most of the electronics function consisted of approximately 100 bipolar ICs in 14- and 16-lead packages. These were replaced with six MOS LSI circuits in 40-lead packages. Assuming full use of all pins in all packages, 240 leads replaced approximately 1500. Since the original premise was that the connections (bonds) to these leads were the major cause of IC failures, then the use of MOS arrays provided a decrease in the failure rate attributed to this particular function by a factor of about 6 to 1. Of course, the exact effect of this large decrease on the system reliability can be determined only after all other components and factors that contribute to the total system failure rate are examined.

Figure 1-4 shows the reduction in total interconnect for a 10,000 circuit system as the number of circuits per chip increases from one (early IC era) to 100 (typical present day LSI chip) and, eventually, to 1000 (a level of integration not yet attained). Note the large increase in circuit (gates) to pin ratios as the complexity of the IC increases.

An increase in system reliability, through reduction in system interconnect and quantity of parts, inherently tends to improve system maintainability. Additional maintenance benefits can be achieved by taking advantage of these reductions in designing the layout and packaging of the system. Faster, simplified checkout routines can be devised and easier access can

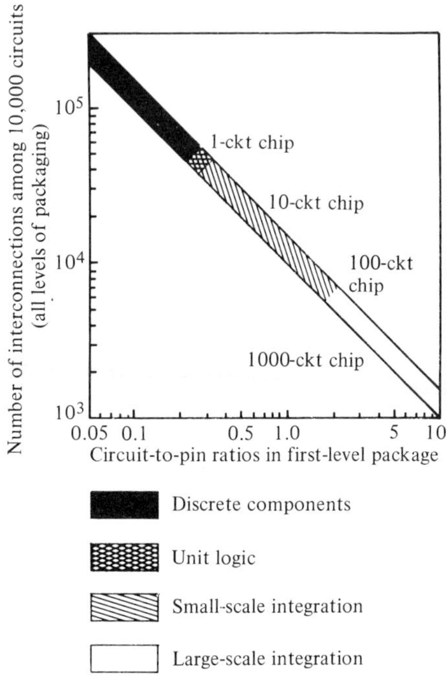

Figure 1–4 Effects of chip complexity on total system interconnect and circuit-to-pin ratio (Electronics, Feb. 20, 1967; copyright McGraw-Hill, Inc., 1967).

be provided to less reliable components of the system to reduce down time. Details on the reliability of MOS LSI are covered in Chapter 9.

- Predictable Design

Another feature of MOS that leads to performance advantages as well as economic advantages is the relative ease with which MOS circuits can be designed to meet specific performance requirements. This means that there is a high probability that an array will meet specifications in the first design effort. The key to successful MOS LSI designs is a stable process which has been thoroughly characterized so that a designer need only to vary the topology of the MOSFET to achieve a given set of performance characteristics. Once a process is characterized and the design limits are established, the average circuit designer has little trouble determining proper MOSFET dimensions. He has no concern for diffusion times, temperatures, doping concentrations, etc.; parasitic diode effects are minimal with MOS technology. MOS design automation further guarantees consistent performance.

The high probability that a properly designed MOS array will meet its performance specifications without redesign is seldom appreciated until the user is faced with the cost of a redesign. Inasmuch as complex MOS LSI logic circuits are mostly custom designs, this is an especially valuable feature. Undoubtedly, the design process will be refined and automated to a greater extent in the near future to reduce further the development cost and time span and increase the probability of the initial design to successfully meet specification requirements.

1.4 MOS LSI—ITS PROBLEMS AND LIMITATIONS

Although it is impossible to anticipate all problems that may arise with any relatively new technology, the principal performance limitations of MOS LSI are generally well defined and, in many system applications, any limitations that may exist can be circumvented. The same two broad categories used in the preceding section—economy and performance—can be used to segregate the basic limitations of MOS LSI quite satisfactorily, but a third category will contribute clarity. This category includes the business management problems associated with MOS LSI.

- ECONOMIC FACTORS
  a. high development costs and limited production runs
  b. high cost of changing an array
  c. high cost of testing complex logic arrays
  d. high cost of suitable packages
  e. effect on system cost may be minimal

- PERFORMANCE FACTORS
  a. slower speed of MOS versus bipolar
  b. techniques used to circumvent MOS performance limitations complicate other parts of system
  c. high voltage MOS incompatible with existing bipolar systems

- BUSINESS-MANAGEMENT FACTORS
  a. supplier-customer interface point no longer clearly defined
  b. custom versus standard products
  c. long lead-time required for development of custom MOS LSI

1.4.1 Economic Problems and Considerations

- High Development Costs

The development costs of MOS LSI chips can be quite high and vary considerably depending on the required performance, type of logic, and circuit complexity. The use of computer programs to automate much of the routine tasks helps reduce the direct costs associated with the design of an array. However, the costs of developing and maintaining the design automation programs plus the cost of the equipment to perform the actual tasks of layout, drafting, and mask making are very high. In addition, if a design change is required in a chip after the tooling is either complete or well on its way to completion, it is quite costly. Unlike other systems that use conventional ICs, where a change might involve only adding or changing a few packages and some connections, the entire tooling may have to be modified or redone. Because of the high cost of design error, it is important that the designer take precautions to minimize the chance of such errors. Therefore, it is becoming common practice to provide computer aids to assist the designer and reduce the risk of human error.[10] The computer is used by logicians to verify the correctness of his logic design by means of a simulation program, as described in Chapter 5. Various computer programs have been developed to help the circuit engineer determine proper device sizes and aid in analyzing his design for proper performance. The conclusion is that custom MOS arrays can be designed properly the first time, but it is a rather costly procedure.

- High Testing Costs

The cost of preparing a test pattern or program for the MOS array must be included in the development costs. Frequently, this is generated during the design phase when the system logic design is simulated with the aid of a computer, but it also can be prepared manually. Although either method of preparation is costly, it is especially so if performed manually, since it is so time consuming. The magnitude of the testing problem can be appreciated when one considers the number of possible inputs on an array plus the number of internal storage elements. With n inputs (each of which can be either a logical 1 or 0) and m internal states (also binary in nature) as many as 2^{n+m} test words may be required to thoroughly check out the array. If, for example, an array has 20 inputs and 20 internal states, 2^{40} or more than 1×10^{12} test words might be required! Even at the rate of a million tests per second, approximately 278 hours would be required to complete all tests. It is also possible to design complex arrays that are impossible to test without providing extra circuits or pads that are not normally required when the array is used in the system. These testing difficulties are thoroughly discussed in Chapters 5 and 6, but the designer should be aware of the testing cost implications when designing his system.

- Limited Production Quantities

The relatively small production volume requirements of many systems that would benefit from the use of MOS LSI creates a dilemma. Because of the small volume, the high costs of design and development, testing, and any modifications must be amortized over fewer units than before. The very fact that a single MOS array replaces numerous discrete parts and conventional ICs limits its volume of usage to several orders of magnitude below that of simpler semiconductors. Hopefully, the lower cost per function provided by MOS LSI will allow the system manufacturer to expand his market by reducing the cost of his product and, consequently, his selling price.

- Minimal Effect on System Cost

An important economic consideration is the percentage of total system costs that is affected with MOS LSI in place of the existing circuits, pc boards, connectors, etc. Clearly, when the cost of the electronics and related items that can be replaced with MOS LSI is but a small fraction of the total system cost, even free MOS LSI circuits offer little savings in the initial cost of a system. The modern digital computer is a prime example. As pointed out by Palevsky,[11] electronic component costs typically amount to only about 30 percent of the manufacturing costs of this type of machine.

- High Packaging Costs

The MOS LSI chip must be packaged to protect it from contaminants, and yet the package must be convenient for the system manufacturer to handle. Multilead, hermetic packages suitable for MOS LSI are very costly relative to the 14- and 16-lead dual in-line packages used for most bipolar ICs. For minimum costs, packages with the least number of pins should be used, but this would most likely result in more MOS LSI chips per system. As we will see in the discussion of system partitioning in Chapter 6, the selection of a package is but one of the many decisions that involve complex trade-offs for optimum system costs. Attempts to use less expensive packages such as plastic for large MOS arrays have been moderately successful, but the use of plastic for simpler MOS logic functions is becoming more commonplace. It is almost a certainty that the high cost of ceramic and metal hermetic packages will ultimately force the development of suitable plastic packages.

The present methods for connecting the pads on the chip to the package leads are costly. Unlike the batch fabrication processes by which many individual MOS arrays are formed simultaneously on a single silicon wafer, both the die attach and the lead bonding operations are performed individually. Although several methods have been developed that allow an operator to form all the connections from chip to package leads simultaneously, these generally complicate the wafer fabrication process. The beam lead and the flip chip techniques, which are well described in the literature,[12,13] both require additional process steps. The spider bond and the beam lead laminate[14-16] do not require extra process steps during wafer fabrication. However, they presently appear unsuitable for chips that require a large number of leads because it is difficult economically to produce multilead frames that have sufficient lead rigidity and placement precision for lead bonding.

1.4.2 Performance Limitations

- Slow Speed

The principal performance limitation of MOS LSI is the comparatively slow switching speed of the MOSFET. Typical pair delays of less than 150 ns within a complex logic array are not easy to achieve with standard commercial temperature variations and power supply tolerances. Output stages of an MOS LSI chip with typical pc board capacitance loads are even slower. The slow switching speed of the MOSFET is due to its limited ability to charge and discharge capacitive loads. These loads consist of the self or intrinsic capacitance of the device which is a function of the device dimensions and process parameters plus the stray circuit capacitance which is dependent upon the physical layout of the array. This limited ability to charge and discharge capacitance is a consequence of the comparatively low transconductance (g_m) or current carrying capability of typical MOSFETs used in LSI. Any increase in the size of the device to enhance its g_m and, therefore, its speed, is of little benefit since the self capacitance of the MOSFET also increases proportionately.

However, as stated by Warner,[5] since the g_m of a MOSFET is highly dependent upon the state of the art of MOS technology, we can expect improvement in the future. Indeed, this is already the case as exemplified by developments such as the *N*-channel process which has a higher inherent mobility by a factor of 2 to 3, and the self-aligned gate techniques such as the silicon gate and some ion implantation processes which reduce the self capacitance of the MOSFET. Although present day *P*-channel MOS can operate comfortably at clock rates up to 2 MHz, variations of the basic process such as silicon gate and ion implantation with self-aligned gates work well at 10 MHz or better. More complex MOS processes such as complementary symmetry (CMOS) offer operational speeds of 20 MHz with good likelihood of soon doubling that figure. However, these improvements are achieved only with more complex, hence more costly, processes as compared to the metal gate *P*-channel MOS process.

Despite these improvements in MOS processes, as pointed out by Crawford,[17] bipolar technology has a speed advantage of at least 10 to 1 due to its higher g_m per unit area.

An often quoted figure of merit for a particular IC logic family is the product of propagation delay (T_{PD} in ns) and static power (P_D in mW) of a simple gate or inverter. A low speed-power product is, of course, desirable. The speed-power figure of merit of a typical standard cell *P*-channel MOS dynamic logic gate is compared with two widely used bipolar logic families, series 54/74 TTL and 900 DTμL, in Table 1–2.

Table 1–2 Comparison of MOS and Bipolar Speed-Power Product of a Simple Gate

Process	*Logic Family*	T_{PD} *(ns)*	P_D *(mW)*	*Fig. of Merit (pJ)*
MOS	STD V_T, 2-Phase Dynamic	53	4.0	212
MOS	Low V_T, 2-Phase Dynamic	69	0.75	52
Bipolar	900 DTμL	30	8.5	255
Bipolar	54/74 TTL	10	10	100
Bipolar	54H/74H TTL	6	22	132
Bipolar	54L/74L TTL	33	1.0	33

In the interest of providing a fair comparison, the T_{PD} for bipolar is the typical value at 25°C, whereas for MOS it is at a junction temperature of 90°C, since the T_{PD} of MOS unlike bipolar increases by about one-third going from 25° to 90°C. The P_D for both bipolar and MOS is the typical value at 25°C, assuming a 50 percent duty cycle, which is equivalent to averaging the power dissipated for an output logic 0 with that for an output logic 1 level. This tends to favor bipolar, since the P_D of MOS decreases with increasing temperature.

As can be seen from this table, the speed-power product of standard threshold voltage (V_T) MOS is approximately two times more (worse) than that of the standard 54/74 TTL family; but is actually better than that of the 900 series DTL. On the other hand, the figure of merit for low V_T MOS is better than DTL and standard TTL, but it is about 60 percent worse than the low power TTL family.

- System Problems

Whereas dynamic logic provides many advantages, as pointed out in the previous section, it has a limitation in some systems applications. Because dynamic logic relies on the ability to store data on the gate capacitance of an MOSFET for proper operation, it requires a minimum system clock rate to assure that the stored charges do not deteriorate below usable levels. Also, since the charge leakage is increased at elevated temperatures, a higher minimum system clock rate is needed to assure proper operation at high temperatures.

Dynamic logic also requires the use of at least two and perhaps four clock phases. Although it is possible to generate the two phase clocks on the MOS chip, they consume considerable area and power to provide the large clock signal amplitudes required for most MOS circuits. When either 4- or 2-phase ratioless MOS logic is employed, it is usually necessary to generate the clock phases external to the MOS LSI chips because of the inability of the MOSFET to drive, at normal clock rates, the large capacitive loads presented by ratioless circuits.

Whereas MOS LSI achieves optimum speed and power performance when 4-phase logic is used, the speed is attained by shifting the design burden to the equipment that generates the 4-phase clocks. It is not an easy task to generate the fast, large amplitude clock levels with minimal overshoot and precise timing that 4-phase logic requires for proper operation.

- Incompatibility With Bipolar Circuits

The standard threshold voltage channel MOS process operates from power supply levels that are quite different from those required by bipolar ICs. Of even greater importance are the incompatibilities of the input and output logic (voltage) levels of MOS circuits and bipolar ICs. Existing systems that contain mostly bipolar ICs require buffers or level translators if they must interface with standard threshold voltage MOS arrays. The various low V_T MOS processes can minimize this inconvenience and cost, but they usually provide very limited fanout to bipolar current sinking logic due to their low g_m.

Because of the low g_m of the MOSFET, output stages of MOS LSI chips require large geometry devices to drive the pc board and interconnect capacitance. Consequently, MOS output stages are relatively slow compared to an internal stage. When interfacing with widely used TTL or DTL inputs, the MOS output device must be capable of sinking 1.6 mA per input. Since the current handling capability of the MOSFET is a direct function of the device size (width), output stages are not only slow but consume considerable chip area. This tends to increase die size and, ultimately, increase the cost of the final product.

1.4.3 Business-Management Considerations

- Customer-Supplier Interface

The advent of MOS LSI has created several problems of common concern to both the semiconductor supplier and the systems manufacturer. Perhaps foremost is the loss of the distinct dividing line that once separated the market of the system manufacturer from that of the semiconductor supplier. The primary cause of this problem is the increasing number of system functions that can be performed by a single MOS LSI chip. Understandably, the systems manufacturer is concerned as he sees his suppliers contributing more and more to the value of his end product, whatever the system may be. Some systems companies view this trend as a serious intrusion into their domain and have, therefore, decided to establish facilities to design and even manufacture their own complex MOS LSI arrays. On the other hand, most semiconductor vendors view the situation quite differently. Rather than constituting a threat to the survival of the systems companies, they see themselves as suppliers of better products at lower cost per function. Both parties realize that there are now many possible points of interface between supplier and customer in developing a system with MOS LSI. (These interfaces are described in detail in Chapter 6.) However, the already difficult interface problems are virtually impossible to resolve satisfactorily if the customer is concerned that his supplier is becoming his competitor. Although the optimum interface point that is agreed upon may require that the system manufacturer reveal proprietary information to his MOS LSI supplier, by carefully selecting reputable suppliers, he can minimize his risks of losing trade secrets or having a supplier becoming a competitor. As appropriately pointed out by Penisten,[18] the successful solution to this problem requires involvement by the top management of both customer and supplier. Only when both parties cooperate fully can an interface be established that effectively employs the strong points of each company to achieve mutually beneficial goals.

• Custom Design vs. Standard Products

Several problems arise in designing MOS LSI into a new system or equipment. First is the difficulty in finding useful MOS LSI off-the-shelf standard parts that are available from more than one source. Second is the lengthy (and costly) development period required to build the custom MOS LSI chips that are likely to be needed to give the new product any advantages over the competition. Furthermore, the decision to design with custom MOS LSI is almost certain to affect the lead-time of the system. Factors that influence this decision are: the anticipated system or equipment production volume and selling price, system specifications, allowable time span for the development and production of the system, and the development time and costs for custom MOS LSI circuits. If, by chance, the system manufacturer discovers reasonably priced off-the-shelf MOS LSI circuits that are available from more than one vendor, and, if all the systems requirements can be met by using these LSI circuits, the decision is quite simple.

In reality, this rather ideal situation will rarely, if ever, be encountered except for systems in which mostly memory products such as shift registers, ROMS and RAMS are used. More likely than not, for optimum performance, reliability and cost of the system or equipment, custom MOS LSI circuits will be required. The alternative of attempting to adapt the system design to the few existing standard products is almost certain to affect severely the size and reliability of the equipment or system such that the product will offer little, if any, competitive advantage. As pointed out by Suran,[19] custom designs are economically sensible for high-volume, simple equipment as well as for low-volume, complex equipment, since they greatly reduce parts count and simplify diagnostic procedures. Although there is some controversy regarding the relative merits of custom and standard MOS LSI products, most semiconductor suppliers[20] feel that, with the exception of the memory market, custom MOS LSI will prevail during the next decade because it best achieves the economies possible with MOS LSI. The standardization will not be at the circuit level, as previously noted, but at the logic cell and process level. Standardization at these points coupled with computer aids provide the best means of overcoming the problems of high-development costs and long-development lead-time as well as who, user or supplier, will do the custom design. Design automation permits the system manufacturer to retain control over the design of his product while drastically reducing the development time of the MOS LSI supplier. Numerous design automation programs exist, but the methods used to produce a custom designed MOS LSI chip vary drastically. It is conceivable, however, that design automation programs will soon improve to the point where they will shorten the custom

MOS LSI chip turnaround time and simultaneously reduce development costs, to completely eliminate their present disadvantages.

1.5 OTHER MOS PROCESSES

Because the standard threshold voltage *P*-channel is the most widely used MOS LSI process today, much of the discussion so far has dealt with its characteristics. There are several other important MOS processes that improve various characteristics of MOSFETs. Certainly one of the most important MOSFET electrical characteristics that requires improvement is speed; but it should not be achieved at the expense of other unique

Table 1–3 Comparison of Characteristics of MOS Processes to Standard V_T Metal Gate *P*-Channel

Type of MOS process → / *Factors*	P-channel 100 Low V_T	P-channel Si nitride 111 Low V_T	P-channel silicon gate	Ion implant counter-doped channel—low V_T	Complementary MOS	N-channel low V_T
1. Process						
a. Simplicity	≈	–	–	–	––	–
b. Reproducibility	≈	–	≈	–	–	–
2. Electrical						
a. Speed	–	≈	+	+	++	+
b. Power consumption	+	+	+	+	++	+
c. Noise immunity	–	–	–	–	≈	–
d. Output drive capability	–	+	≈	++	++	++
e. Bipolar compatible levels	++	++	++	++	++	++
f. Clock drive requirements	++	++	++	++	++	++
g. No. of power supplies required	≈	≈	≈	+	++	–
3. Layout						
a. Area of typical inverter	–	≈	+	+	––	≈
b. Levels of interconnect	≈	≈	+	≈	≈	≈
c. Field inversion problems	–	≈	–	≈	≈	–
4. Other						
a. Proven reliability	–	–	–	––	–	––
b. Cost	–	–	–	–	––	–
c. Availability	–	––	–	–	–	––

Legend: ++ Much better than STD V_T *P*-channel process.
+ Better than STD V_T *P*-channel process.
≈ Approximately equal to STD V_T *P*-channel process.
– Worse than STD V_T *P*-channel process.
–– Much worse than STD V_T *P*-channel process.

features of MOS that have led to its successful application in LSI circuits. Table 1–3 shows how each of the other important MOS processes compare with the standard V_T *P*-channel MOS with regard to characteristics that are usually considered important in the design, production, and application of MOS LSI. A detailed explanation of each of these processes is provided in Chapter 3. The comparison is based on four major factors, viz., process, electrical, layout, and other. Each of the six MOS processes listed is rated as being either equal to, better than, much better than or, conversely, worse than or much worse than, the standard V_T *P*-channel process. Admittedly, this comparison is quite subjective and qualitative, but it does provide the neophyte user with a basis for his investigation into which process best fits his requirements. It also should be realized that there are many variations of the major processes listed in the table and that these modifications could appreciably change the ratings of certain characteristics within the major groupings.

The two major process considerations listed are simplicity and reproducibility, although cost could be included here as well. Simplicity refers to the total number of process steps or operations, but special consideration is given to the number of masking and high-temperature operations. Process reproducibility is related to the ability to produce the same MOSFET characteristics over a long period of time or manufacturing history. It also includes the relative ease of controlling the process within predetermined bounds or limits over a period of time.

The second major group, electrical characteristics, includes parameters that are important in determining whether MOS LSI can be used in the design of a system and how. Speed, power consumption, noise immunity, and output drive capability have been discussed previously and require little clarification. The comparison for the first three parameters assumes an internal inverter of typical dimensions as used in a complex array. Output drive refers to the ability of a given output stage to drive (switch) a specified load capacitance, usually about 25 pF. Bipolar compatibility refers to the ability of output devices of reasonable size in a given MOS process to directly drive a bipolar IC without the aid of any intermediate buffer, external resistor, or level translator. It also includes the ability of the MOS process to operate directly from bipolar output levels. Logic levels are the main consideration here, with some weight given to compatibility of the MOS process with normal bipolar power supply voltages. Clock driven requirements were evaluated with respect to the amplitude of the clock swing required for proper operation which, of course, assumes the use of dynamic logic. Since large clock amplitudes are more difficult to generate, especially at high clock rates, the high V_T *P*-channel process is ranked lowest. The number of power supplies, which includes clock supplies as well as bias supplies, is important for two reasons. First, it is obviously more costly to provide several voltage levels in the system. Second, the additional supplies require extra package pins, or in the instance when a package is already pin-limited, it means that some input or output must be eliminated, which is not a trivial problem.

The layout factors selected for the comparison are those items that affect the chip area required to implement typical circuit functions in complex logic and memory MOS LSI chips. It assumes a system speed requirement such that the circuit speed can be achieved with any of the MOS processes. It also assumes that dynamic logic is employed, as typical of most MOS LSI circuits now being designed and produced. Field inversion, which is explained in detail in Section 2.3 of Chapter 2, refers to the likelihood of inadvertent formation of an undesirable parasitic MOS transistor when an array is being laid out.

The last group is the comparison of proven reliability, cost, and availability, factors of utmost concern to the potential user of MOS LSI. Because of the rather poor early history of MOS, proven reliability is weighed heavily. This does not imply that processes other than *P*-channel are inherently unreliable, but it does point out the lack of current accumulated data to establish their true relative reliability rankings. In time, one or more of these processes will prove to be as reliable, if not more so, than today's *P*-channel process. Those who are willing to project future developments feel that silicon gate technology offers inherently better reliability despite a more complex process. The cost factor evaluated is the actual manufacturing cost, not the sales price, which is obviously subject to wide fluctuations depending upon market conditions. Since some MOS suppliers are willing to sell their products at cost or even at a loss to capture what they consider a desirable customer or market, the cost factor is difficult for the user to evaluate realistically. The low relative cost of *P*-channel MOS results from extensive manufacturing history and large production volumes, factors that are traditionally required to achieve low unit costs. Assuming that technological developments continue near the present rate, it is very likely that several other MOS processes will, within a year or so, be as economical to produce as the present *P*-channel process. Of course, by that time, the cost of *P*-channel should also be less. But, because of the extensive improvements already incorporated into the *P*-channel process, further cost reductions are likely to occur at a much slower rate than with the other MOS processes. The last factor in the rating, availability, indicates to the potential customer the likelihood of his finding several sources for his MOS LSI design. Whereas the standard V_T *P*-channel process with basically similar characteristics is available from numerous vendors, the electrical characteristics of low V_T processes vary drastically, depending upon the particular semiconductor supplier selected.

1.6 APPLICATIONS FOR MOS LSI

The main objective of this section is to point out the business areas in which MOS LSI is now well-established. Another objective is to point out the broad markets and products in which MOS LSI is likely to gain significant acceptance in the near future.

It is important to bear in mind in this discussion that it is almost always necessary to redesign or repartition the system if MOS LSI is to provide any significant advantages in existing systems or products. As pointed out earlier, MOS LSI is best suited for highly complex logic functions that operate at moderate clock rates and slow speed, low power, complex logic, and memory functions. It can be optimally applied in these areas only when one has complete freedom to design the system from the inception around the unique characteristics offered by MOS LSI.

1.6.1 Present Applications of MOS LSI

Most of the MOS LSI products of today have been produced by the standard V_T *P*-channel process and, to a lesser extent, by low V_T variations of *P*-channel MOS, despite their relatively slow operating speeds.

MOS LSI is already firmly established in both the commercial-industrial and military-space markets. In the former area, MOS LSI has almost completely taken over the fast growing electronic desk calculator market. Calculators already exist that use one MOS LSI chip to perform all the logical and memory operations for 12-digit display machines that sell for less than $200. The computer time-shared remote terminal market is another large segment in which vast quantities of MOS LSI are being used.

The calculator and the remote terminal are both ideally suited for MOS LSI implementation with resultant lower cost equipment. MOS LSI is well-established in the medium-speed, small-capacity read-write and read-only random access memory market where it outperforms existing core memories and already is less costly. The memory product applications of MOS LSI are covered in detail in Chapter 7.

The jumbo jet entertainment and passenger services system and the engine malfunction monitoring system are other applications that incorporate MOS LSI. Here even the unique analog characteristics of MOSFETs are utilized in the multiplexing of control and audio signals. MOS LSI, especially serial shift registers, find considerable application in commercial equipment designed to test multi-pin large-scale arrays and printed circuit boards that contain complex logic functions. Vending machine controllers are just now appearing with MOS LSI replacing much of the former electromechanical equipment.

The military and space markets were one of the first to employ MOS LSI, although early products were not always too successful. Since then, however, reliability records have been excellent, especially for devices used in NASA space programs where the low power consumption of MOS LSI is very important. MOS arrays have been used as miniature timers for artillery fuses since the first complex MOS circuits were developed. Similarly, secure military communications systems continue to be among the largest consumers of complex MOS LSI products.

MOS LSI is just now beginning to appear in the consumer market products. Electronic organs were the first products in which significant quantities of MOS LSI were used for frequency dividers and tone generators. Home-appliance timing and control units are now beginning to appear on the market as manfacturers recognize the greater reliability with MOS LSI compared to the older mechanical timers.

1.6.2 New Markets for MOS LSI

Regarding the untapped markets for MOS LSI, perhaps a statement made in late 1969 best illustrates the magnitude of the market for electronics in the immediate future.[21] It was stated that "the most profound changes in the history of the home, the factory, the hospital, and the battlefield will occur in the next 10 years because of electronics."

According to the source, in the 1970s the average American will be living in a world controlled by and dependent upon semiconductor electronics. How much of this expanding market will be captured by MOS LSI is surely speculative at this time. In an attempt to point out these new markets for MOS LSI, the business areas are again categorized into three groups, viz., industrial-commercial, military-space, and consumer.

• Industrial-Commercial

The sheer size of the computer market invites the attention of almost all semiconductor suppliers. So far, except for memory products and the central processing units (CPU) of a few minicomputers, MOS LSI has not yet been a serious contender for the majority of the electronics in this huge market. However, Eckert[11] has stated that LSI will have a large impact on the way computers are designed. With LSI, more hardware logic will be built into the machines. He concludes that LSI will have a great effect on the character and efficiency of the computer in that LSI will focus attention on the problem of reducing the cost of all computer peripheral equipment such as printers, displays, tape units, etc. This is significant because these peripherals are beginning to become more costly than the computer itself. It is difficult to predict how much of this market is suitable for MOS LSI because of the increasing demands for higher processing speeds. However, the computer peripherals themselves are a natural application of MOS LSI, even if the computer CPU speed requirements eliminate MOS from consideration.

The minicomputer, especially those units in the lower price, slower processing speed category, will most certainly incorporate MOS LSI because of its low cost per function. Computer remote terminals will continue to consume large quantities and types of MOS LSI. Small, special-purpose computers and process control equipment will employ MOS LSI for functions ranging from arithmetic-logic units, displays, mutiplexers and analog to digital converters to complete memories. The list of applications could be extended almost indefinitely, but the foregoing should provide an insight into the possibilities that exist in the computer industry.

Differing only slightly from the pure computer field are the following potential markets for MOS LSI: traffic control including mass transit, public utilities, sports, education, and law enforcement. Present-day traffic controllers that operate from a general-purpose (GP) computer can use MOS LSI in a limited way in peripheral equipment. As special-purpose controllers are designed to handle the ever-increasing flow of traffic, the use of MOS LSI should increase significantly. The ever-growing transportation industry is already well on the way to automatic control of both mass-transit passenger trains and railroad freight shipments. In this industry, MOS LSI can be advantageously applied in peripheral gear, such as sensors, multiplexers, buffers, etc., and in the memory function of associated special-purpose data-processing equipment. The public utilities such as water, gas, and electricity will, with the use of low-cost MOS LSI, be able to automatically read, record, and perhaps bill their customers by transmitting the multiplexed data over the telephone lines. Spectator and participant sports events and games will be timed, scored, and results displayed more and more with electronic equipment. This application is ideal for MOS LSI with its rather modest speed requirements and need for lightweight, portable equipment. The field of education is another "natural" for MOS LSI applications, especially in programmed learning machines where small-size memories and rather slow logic speeds suffice. Character generators for display, multiplexers, and keyboards for each training unit are typical applications for MOS LSI.

The law enforcement and crime prevention market currently use a considerable amount of computers and computer controlled equipment; but there is a large segment of detection and monitoring equipment that could be economically designed and manufactured with MOS LSI. Law enforcement agencies have a need for private communications systems which could benefit from the use of MOS LSI in simple equipment such as voice scramblers.

The vast communications market offers potential business for MOS LSI in products such as data modems and telephone switching equipment, especially with the increasing use of remote terminals.

The medical field is just beginning to realize the functions and services that can be provided more economically by electronic equipment. The rapidly rising hospital care costs are certain to lead to electronic patient monitoring, diagnostic, and record keeping equipment, for which MOS LSI certainly is suitable. MOSFET sensors, analog to digital converters, multiplexers, comparators, etc., are all potential large-volume requirements. In the near future, doctors are even likely to equip their offices with either terminal equipment linking them to various hospitals or perhaps less sophisticated electronic equipment to perform diagnostic services on the spot. The potential for MOS LSI with its features of low power and small size, which lead to easy portability, are limited only by the imagination of the medical profession user and electronic designer. The medical laboratory is another place where much of the previous discussion regarding needs for new electronic medical equipment is also very applicable. Another phase of the medical field especially well-suited to the characteristics of MOS LSI is prosthetics. The control functions and sensors for artificial limbs and body organs are again almost ideal applications for MOS LSI.

• Military-Space

The military-space electronics market, yet untapped by MOS LSI, is huge. However, it is difficult to foresee much success for MOS LSI in a large portion of the market unless the many high-speed systems are redesigned entirely to partition them in a more optimum manner for MOS LSI. Most military system manufacturers prefer to do their own design and vendors must be willing to interface at those points that give the system manufacturer the most complete control of the design. Once this is accomplished, MOS LSI should find fairly widespread acceptance, especially in portable equipment where lightweight, low power consumption, and reliability are more important than speed. The ability to implement algorithms, such as Fourier analysis, fast Fourier transform, and digital filtering plus pipeline processing with MOS LSI should be attractive to designers of new systems. Navigational computers may be designed with algorithms that were not economically practical to implement until the advent of MOS LSI. Data reduction, compression, and error-correcting systems such as used on a satellite to handle the mass of data acquired also fit the characteristics of MOS LSI well.

• Consumer

New consumer markets for MOS LSI are restricted somewhat by the need for low cost. The automotive electronics field is potentially huge. Significant penetration into the market is likely only where MOS LSI products prove less costly than existing products and fulfill a need that cannot now be economically accomplished by any other means. MOS LSI is well-suited for applications in speedometers and speed controllers, malfunction detection, status monitors and displays, routine service reminders, fuel injection and antiskid braking systems, antitheft sensors and alarms, air conditioner-heater climate controls, electronic clocks, and many others. The high noise immunity of MOS is an asset in the severe electrical environment of the automobile.

In the home, there are several potential markets for MOS LSI products. A low-cost electronic calculator is the most promising prospect, although a micro-minicomputer would be welcomed by many households to monitor purchases, bills, and payments due. Home entertainment systems and environment controls to purify the air as well as maintain correct temperature and humidity are possible new product areas for MOS LSI. The use of electronic timing and control devices in consumer appliances, lighting, and tools will become more widespread as the price of MOS LSI is lowered.

In summary, MOS LSI will flourish along with growth of present applications; but its expansion into new markets for electronics surely offers its most exciting potential.

REFERENCES

1. J. Bardeen and W. H. Brattain, "The Transistor: A Semiconductor Triode," *Phys. Rev.*, **74,** 230 (July, 1948); also U.S. Patent 2524035, October, 1950
2. W. Shockley, "A Unipolar 'Field-Effect' Transistor," *Proc. I.R.E.*, **40,** 1365–1376 (Nov. 1952).
3. G. E. Moore, "Semiconductor Integrated Circuits," Chap. 5 of "Principle of Microelectronic Engineering," E. Keonjian, ed., McGraw-Hill Book Co., Inc., New York, 1962.
4. M. Chang "Developments in Digital Integrated Circuits," paper presented at WESCON New Solid State Devices Session, San Francisco, August 1969.
5. R. M. Warner, "Comparing MOS and Bipolar Integrated Circuits," *IEEE Spectrum* **4,** 50–58 (June 1967).

6. L. Curran, "Computers Make a Big Difference in MOS Designs," *Electronics,* **42,** 82–92 (October 13, 1969).
7. W. A. Notz et al., "Benefiting the System Designer" and "Organizing for Processing Power," *Electronics,* **40,** 130–133 and 139–141 (February 20, 1967).
8. Y. T. Yen, "Transient Analysis of Four-Phase MOS Switching Circuits," *IEEE J. Solid State Circuits,* **SC-3,** 1–5 (March 1968).
9. L. Cohen, R. Rubinstein, and F. Wanlass, "MTOS Four Phase Clock Systems," 1967 Northeast Electronics Research and Engineering Meeting Rec., **9,** 170–171 (November 1967).
10. D. Frohman-Bentchkowsky, and L. Vadasz, "Computer-Aided Design and Characterization of Digital MOS Integrated Circuits," *IEEE J. Solid State Circuits,* **SC-4,** no. 2, 57–64, (April 1969).
11. G. Hollander, J. P. Eckert, B. Pollard, and M. Palevsky, "Panel Discussion, Impact of LSI on the Next Generation of Computers," *Computer Design,* **8,** 48–59 (June 1969).
12. R. Speer, ed., "Chip Bonding: Promises and Perils," *Electronic Design,* **17,** 61–79 (October 25, 1969).
13. L. Curran, "Rival Preleading Schemes Head for a Market Showdown," *Electronics,* **41,** 88–96 (December 9, 1968).
14. L. Curran, "In Search of a Lasting Bond," *Electronics,* **41,** 72–80 (November 25, 1968).
15. J. Morley and G. Trolsen, "New Beam-Lead Connection Method Boosts Semiconductor Memory Yields," *Electronics,* **42,** 105–110 (December 22. 1969).
16. "Spider Approach," *Electronics,* **41,** 58–60 (September 1968).
17. R. H. Crawford, *MOS FET in Circuit Design,* McGraw-Hill, New York, 1967.
18. G. E. Penisten, "Impact of LSI Technology on the Electronics Market," paper presented at WESCON, San Francisco, August 1969
19. J. J. Suran, "A Perspective on Integrated Electronics," *IEEE Spectrum,* **7,** 67–79 (January 1970).
20. R. L. Petritz, "Current Status of Large Scale Integration Technology," *IEEE J. Solid State Circuits,* **SC-2,** no. 4, 130–147, (December 1967).
21. R. D. Speer, ed., "A Changing World with Microcircuits," *Electronic Design,* **18,** 84–89 (January 4, 1970).

GENERAL REFERENCES

R. M. Warner, Jr. and J. N. Fordemwalt, eds., *Integrated Circuits, Design Principles and Fabrications,* McGraw-Hill, New York, 1965.

A. A. Alaspa and G. F. Durguall, "COS/MOS Parallel Processor Array," *IEEE J. Solid State Circuits,* **SC-5,** 221–227 (October 1970).

J. R. Burns, "Switching Response of Complementary Symmetry MOS Transistor Logic Circuits," *RCA Rev.,* **25,** 627–661, (December 1964).

G. Cheroff, D. L. Critchlow, R. H. Dennard, and L. M. Terman, "IGFET Circuit Performance—*n*-channel versus *p*-channel," *IEEE J. Solid State Circuits,* **SC-4,** 267–271 (October 1969).

F. Faggen and T. Klein, "Silicon Gate Technology," *Solid State Electronics,* **13,** 1125–1144 (1970).

T. Klein, "Technology and Performance of Integrated Complementary MOS Circuits," *IEEE J. Solid State Circuits,* **SC-4,** 122–130, June 1969.

J. D. Macdougall, K. Manchester, and R. B. Palmer, "Ion Implantation Offers a Bagful of Benefits for MOS," *Electronics,* **43,** 86–90 (June 22, 1970).

J. M. Shannon, J. Stephen, and J. H. Freeman, "MOS Frequency Soars with Ion-Implanted Layers," *Electronics,* **42,** 96–100 (February 3, 1969).

L. L. Vadasz et al., "Silicon Gate Technology," *IEEE Spectrum,* **6,** 28–34 (October 1969).

2

Basic Theory and Characteristics of the MOS Device

2.1 INTRODUCTION

This chapter considers the theory and characteristics that are important for an understanding of the operation of the MOS transistor. The chapter is divided into three main sections. The first section presents a review of some of the basic semiconductor physical theory that is necessary for an understanding of the analyses which follow. This review is brief, and is intended mainly to refresh the memory of those whose work is not directly in these areas.

The second section analyzes the MOS transistor in terms of its physical structure and derives the basic MOS transistor equations. As much as possible, the analyses and derivations are kept in common electrical engineering terminology, but are sufficiently complete to provide a good model of the basic MOS transistor.

In the third section, the results of the previous analyses are rewritten into a form commonly used in LSI design work. The characteristics of real MOS transistors and circuit elements are also examined in detail. This section includes data that are useful in many circuit design problems and in discussions of related device characteristics that affect circuit operation. Important symbols are defined at the end of this chapter under "Nomenclature."

2.2 BASIC SEMICONDUCTOR THEORY

It is predicted by quantum mechanics that the electrons of an isolated atom may exist only in certain discrete energy levels or orbitals. These energy levels are characterized by specific values of four quantum numbers: these are *n,* which designates approximately the total energy of the electron; *l,* which designates the orbital angular momentum; *m,* which indicates the orientation in space of the orbital angular momentum; and *s,* which designates the electron's spin angular momentum. According to the Pauli exclusion principle, no two electrons in a system (the "system" here being the atom) may have the same four quantum numbers.

2.2.1 Simple Band Theory

When two atoms approach one another, as in bonding, the levels must split so that there will be energy levels to accommodate all of the electrons of the system. When the system consists of a large number of atoms bound together, as is the case in a crystalline material, the higher energy levels tend to merge and blend into two separate bands of allowed energies, which are often separated by an *energy gap* between the two allowed energy bands. This gap is often referred to as the forbidden band or forbidden gap. The lower band, called the *valence band,* is that band of energies occupied by the electrons that are bound between pairs of atoms, as part of a chemical bond between those atoms. The upper band is referred to as the *conduction band.* Electrons that occupy energy levels in the conduction band are free and, consequently, may take part in the conduction of electronic currents. The width of the energy gap is a sensitive measure of the energy needed by an electron to break a chemical bond, and is very closely related to the relative conductivity of various materials. The wider the gap, the lower the conductivity. As is shown in Figure 2-1a, crystalline quartz with a band gap of 6(eV), an insulator, shows a bulk resistivity on the order of 2×10^{20} Ωcm. On the other hand, in a metallic conductor like silver (Figure 2-1b), the valence and conduction bands actually overlap, indicating that it is possible for electrons in the chemical bonds to conduct current. Thus, silver shows a bulk resistivity of approximately 1×10^{-6} Ωcm. It is interesting to note that this resistivity spread of 26 orders of magnitude is one of the widest variations of any physical quantity ordinarily encountered.

In the region between these extremes is that class of materials known as "semiconductors." Unlike the conductors, these materials have a finite gap between the valence and conduction bands, but the gap is much

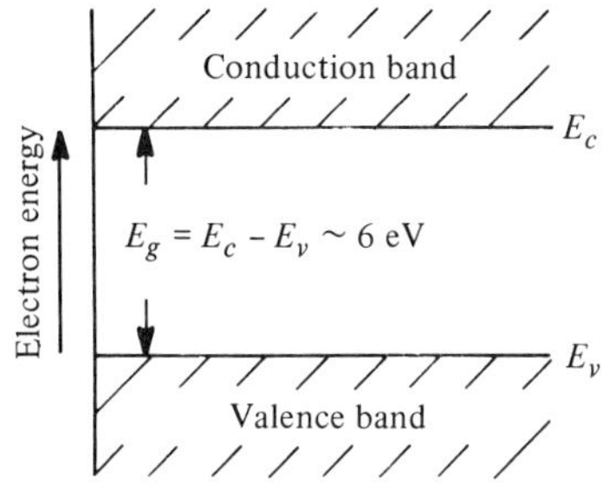

(a) Insulator (e.g., fused quartz)
$\rho \sim 2 \times 10^{20}$ Ωcm

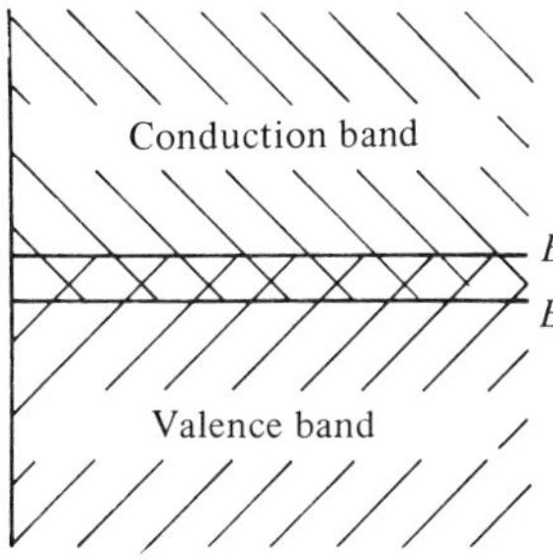

(b) Metal (e.g., silver)
$\rho \sim 10^{-6}$ Ωcm

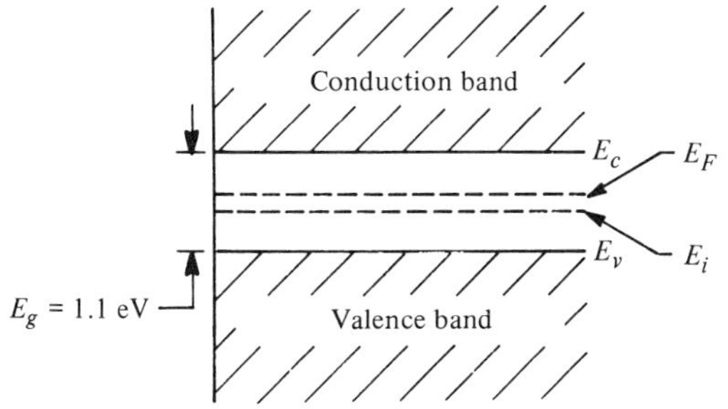

(c) Semiconductor (e.g., silicon)
$\rho \sim 2 \times 10^{5}$ Ωcm

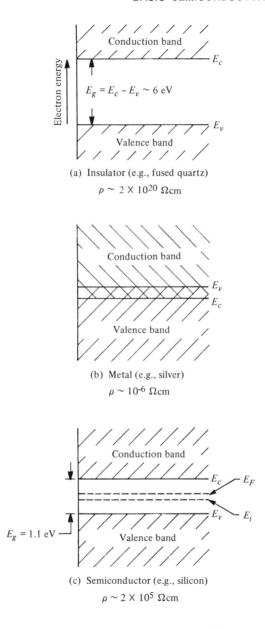

Figure 2–1 Element characteristics.

narrower than that of the insulators. For example, pure silicon, Figure 2-1c, with a band gap of 1.1 eV, has a resistivity of about 2×10^{5} Ωcm.

2.2.2 Fermi Level

The Fermi level is defined as that energy level at which the probability of finding an electron is just one-half. Since electrons are always to be found in the valence band, but only at times in the conduction band, the 0.5 probability level will usually fall in the forbidden gap. (Even though no electron may occupy an energy level in the forbidden gap, the *probability function* still exists there.) With a little thought, it can be seen intuitively that in pure intrinsic semiconductor materials this level will fall exactly in the middle of the forbidden gap. The Fermi level is shown as E_F in Figure 2-1c, and the intrinsic Fermi level is shown as E_i. As was pointed out above, in an absolutely pure semiconductor, the two levels are identical; the addition of an electrically active impurity to the semiconductor will shift the Fermi level from its intrinsic, mid-gap position.

2.2.3 Intrinsic Semiconductors

Silicon is the most widely used elemental semiconductor material. It is an *intrinsic semiconductor,* in that even in its pure state, it is still a semiconductor. Although extremely pure (undoped) silicon is seldom used in the manufacture of integrated circuits, consideration of its characteristics is important to an understanding of device theory.

The silicon atom, Figure 2-2a, consists of a nucleus of 14 protons and 14 neutrons which is surrounded by 14 electrons. Of these 14 orbital electrons, 10 occupy "closed" shells and are not involved in determining the chemical or common physical properties of silicon. The outermost 4 electrons, the *valence electrons,* are the ones that determine physical and chemical properties of silicon In the silicon crystal, each of these elec-

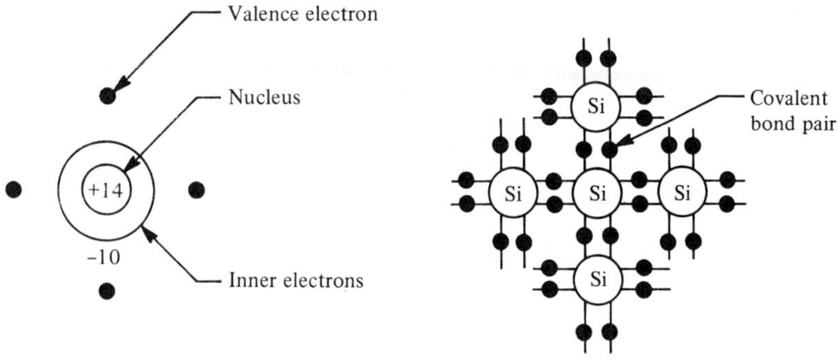

(a) Free silicon atom (b) Crystalline silicon

Figure 2–2 Silicon structure.

trons forms a covalent bond with another silicon atom by joining with a valence electron from the other atom. Thus, each silicon atom is bonded to four other silicon atoms by four *covalent bond pairs* (see Figure 2-2b). As with the carbon crystal (diamond), the resulting bond angles are 109° 28′, which is characteristic of the tetrahedral bonding structure. The basic unit cell of silicon may be described as that resulting from a pair of interpenetrating, face-centered cubic structures.

2.2.4 Intrinsic Carrier Generation

If a valence electron is given sufficient energy, more than 1.1 eV in the case of silicon, it can break out of the chemical bonding state and be free to move about in the lattice. It is then a *conduction electron.* A *hole* (or absence of an electron) is left where the electron was bonded. Since the atom was electrically neutral before the electron departed, a net positive charge is now associated with that atom. Thus, it may be considered that the hole has a positive charge.

Observe that in the case of intrinsic carrier generation, both an electron and a hole are generated simultaneously. Hence, in intrinsic silicon, the number of electrons will exactly equal the number of holes, or

$$n = p = n_i \tag{2-1}$$

where n is the free electron concentration, p is the free hole concentration, and n_i by definition is the free electron concentration in intrinsic silicon.

If we define N_c as the density of energy levels or states in the conduction band available to electrons, N_v as the density of energy levels in the valence band available to holes, and E_c and E_v as the lower edge of the conduction band and the upper edge of the valence band, respectively, then

$$n = N_c e^{-(E_c - E_F)/kT} \tag{2-2}$$

and

$$p = N_v e^{-(E_F - E_v)/kT} \tag{2-3}$$

where k is the Boltzmann constant and T is the absolute temperature. If we take the hole-electron concentration product, we find

$$np = n_i^2 = N_c N_v e^{-(E_c - E_v)/kT} \tag{2-4}$$

or

$$n_i = (N_c N_v)^{1/2} e^{-E_g/2kT} \tag{2-5}$$

where E_g is the band gap width. Note that the Fermi energy E_F dropped out when Equation 2-4 was formed. Experimentally, n_i has been found to be $1.5 \times 10^{10}/\text{cm}^3$ at 300°K.

2.2.5 Doped Semiconductors

As was stated earlier, silicon is rarely used in the pure state. Usually, some impurity called a *dopant* is added in small controlled amounts. Consider for a moment Figure 2-3.

The elemental semiconductors, particularly silicon and germanium, with four outermost valence electrons, are in Group IV. On either side of these, Groups III and V have three and five outermost electrons, respectively. If a boron atom were substituted for a silicon atom in the silicon lattice, the boron atom with only three available electrons would be able to form bonds to only three of the four adjacent silicon atoms and a hole would be formed (see Figure 2-4a). Initially, the hole, loosely bound to the boron atom, is electrically neutral. However, it is very easy for an electron from a nearby silicon to silicon bond to fall into this hole and effectively move the hole away from the boron atom. Since the boron atom will accept an electron, boron and the other elements of Group III are referred to as acceptors. Once the hole has left the boron atom, it is as free to conduct as a hole generated by any other means. Note, however, that the boron atom now has an *extra* electron and is an immobile negative ion.

If a Group V atom, such as phosphorus, is introduced into the silicon lattice, it will have an extra electron which may easily break away, becoming a conduction electron, and leaving behind a *positive* ion (see Figure 2-4b). The phosphorus may thus be said to be a donor, since it donates an electron to the conduction band.

The energy necessary to remove a hole from an acceptor atom or the

	Group	
III	IV	V
B	C	N
Al	Si	P
Ga	Ge	As
In	Sn	Sb

Figure 2–3 Portion of the periodic table.

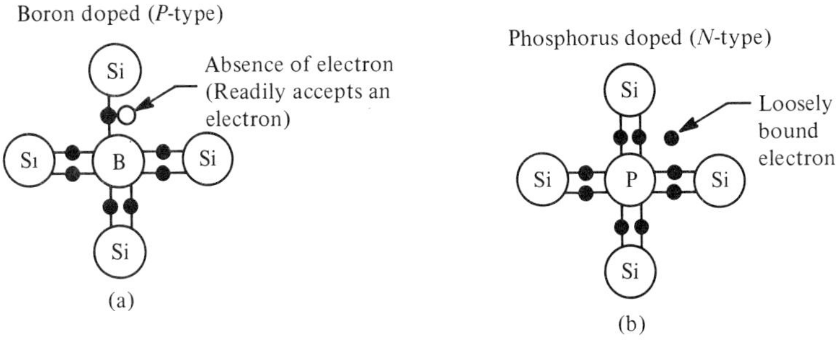

Figure 2–4 Doped silicon characteristics.

electron from a donor atom is called the *ionization energy*. Commonly accepted values for the ionization energy of boron or phosphorus in silicon are 0.039 and 0.045 eV, respectively.

Since these values are so small, nearly all acceptor and donor atoms will be ionized at room temperature. Hence, it is a very safe approximation to say that, in lightly doped silicon,

$$n \simeq N_D \tag{2-6}$$

and

$$p \simeq N_A \tag{2-7}$$

where N_D is the concentration of donor atoms and N_A is the acceptor atom concentration. Likewise, from $n_i^2 = np$, the electron concentration in acceptor doped silicon (n_p) will be

$$n_p \simeq \frac{n_i^2}{N_A} \tag{2-8}$$

and the hole concentration in donor doped silicon (p_n) will be

$$p_n \simeq \frac{n_i^2}{N_D} \tag{2-9}$$

In silicon doped with acceptor type atoms, conduction is predominantly by positive holes; thus, it is said to be *P*-type silicon. Similarly, since negative electron conduction predominates in donor doped silicon, it is said to be *N*-type. A handy "memory jogger" is to remember that acce*P*tor atoms give rise to *P*-type and do*N*or atoms to *N*-type.

As can be seen from Equation 2-9, in *N*-type material, where $N_D >> n_i$, with electrons in the majority, holes will be in the minority. Consequently, we often use the terminology of majority and minority carriers.

Similarly, for *P*-type material, with $N_A >> n_i$, holes will be the majority carrier and electrons the minority carrier.

2.2.6 Dependence of Fermi Potential on Doping Concentration

As pointed out earlier, the position of the Fermi level in the forbidden gap, exactly mid-gap for intrinsic material, varies up or down as dopants are added. Let us see how the Fermi level will depend on the doping concentrations of a donor element. If we combine the equations

$$n = N_c e^{-(E_c - E_F)/kT} \tag{2-2}$$

and

$$n_i = (N_c N_v)^{1/2}\, e^{-(E_c - E_v)/2kT} \tag{2-5}$$

using the assumptions that

$$N_c \simeq N_v \tag{2-7}$$

and

$$n \simeq N_D$$

with the definition of ϕ_F

$$q\phi_F = \left(E_F - \frac{E_c + E_v}{2}\right) = (E_F - E_i) \tag{2-10}$$

Noting that $(E_c + E_v)/2$ is E_i, we then find that

$$n = n_i e^{q\phi_F/kT} \tag{2-11}$$

Since we have assumed that $n \simeq N_D$, it is safe to say that

$$N_D = n_i e^{q\phi_F/kT} \tag{2-12}$$

Likewise, it can be shown for acceptor doping that

$$N_A = n_i e^{-q\phi_F/kT} \tag{2-13}$$

Thus, it can be seen that for any known donor or acceptor atom concentration, the corresponding value for the Fermi level or Fermi potential may be calculated. It might be noted in passing that for *P*-type material, the Fermi level will lie closer to the valence band edge, whereas for *N*-type, the Fermi level will be closer to the conduction band edge.

2.2.7 Carrier Mobility and Conductivity

With free electrons or holes in silicon, the conductivity of the material must be proportional to the number of free electrons or holes. However, another factor must be considered, and that is the mobility of the carriers under the influence of an electric field.

The mobility of holes is significantly less than electrons, as is shown in Figure 2-5, for the carrier mobility in bulk silicon.

If we define hole mobility (μ_p) for convenience as

$$\mu_p = \frac{1}{qp\rho} \tag{2-14}$$

where q is the electronic charge, p is the hole concentration and ρ is the resistivity, we can then say

$$\sigma = \frac{1}{\rho} \tag{2-15}$$

where σ stands for conductivity. Using our assumption that $p \simeq N_A$ we can say

$$\sigma = \frac{1}{\rho} \simeq q\mu_p N_A \tag{2-16}$$

Similarly for N-type material

$$\sigma = \frac{1}{\rho} = q\mu_n n \simeq q\mu_n N_D \tag{2-17}$$

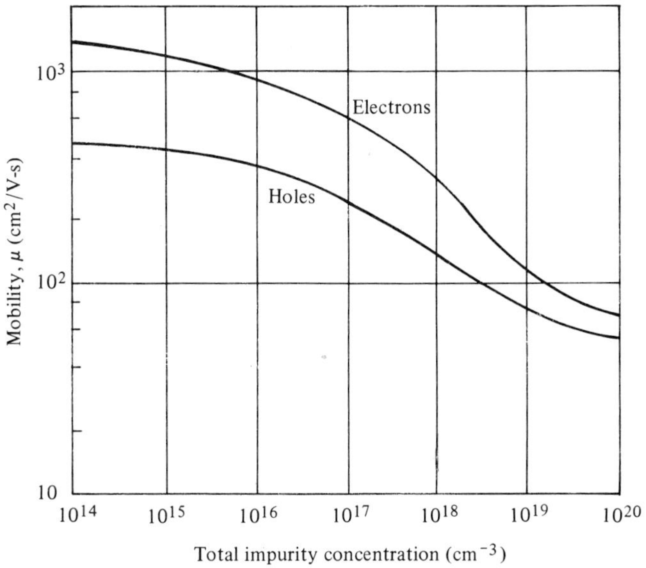

Figure 2–5 Bulk mobility in silicon as a function of dopant concentration (after Evans[1]).

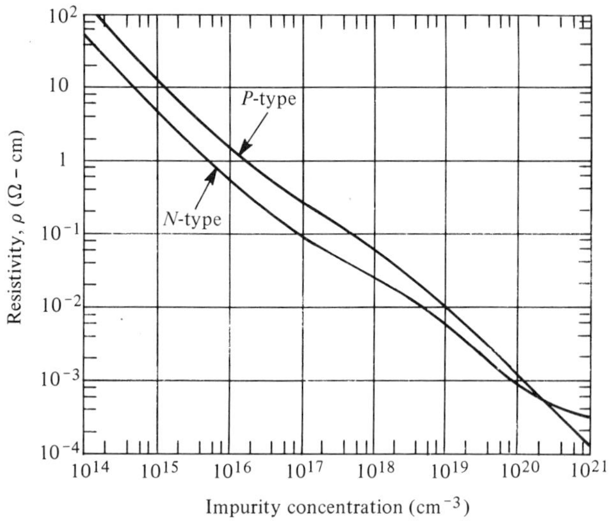

Figure 2–6 Silicon resistivity at 300°K as a function of dopant concentration (after Irvin[2]).

Figure 2-6 shows the resistivity of silicon at 300°K as a function of impurity concentration.

2.2.8 *P-N* Junction

A *P-N* junction is, by definition, the boundary between a *P*-region and an *N*-region in a single crystal. We frequently speak of it as the "metallurgical junction," i.e., the place in the crystal where the donor atom concentration and the acceptor atom concentration are just equal. This is illustrated (in one dimension vs. concentration) in Figure 2-7 for the case where an acceptor (e.g., boron) type impurity is diffused into a uniformly donor (phosphorus) doped crystal. The term X_j is used to denote the actual distance from the surface of the metallurgical junction. Note that X_j *is* defined as that distance where $N_A = N_D$.

2.2.9 Depletion Layer

A *P-N* junction at equilibrium (no external voltage applied) will have associated with it a "depletion layer" or depletion region, which is essentially devoid of mobile charge. This charge-depleted region is a result of the field due to the "built-in" voltage (ϕ_B) associated with the junction. To see how this built-in voltage arises, let us examine the band structure in

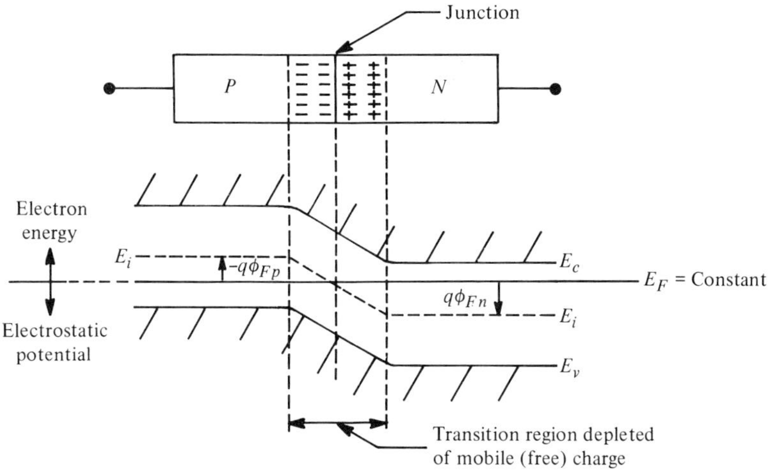

Figure 2–7 Metallurgical junction.

the region of the junction, as shown in Figure 2-8. Note that the Fermi level is constant throughout, and that the remainder of the energy level structure has shifted to accommodate that fact.

We can see that in the *P*-region, we will have the Fermi potential ϕ_{Fp}; from Equation 2-13

$$N_A = p = n_i e^{-q\phi_F/kT}$$

and solving for ϕ_{Fp}, we obtain for the *P*-region

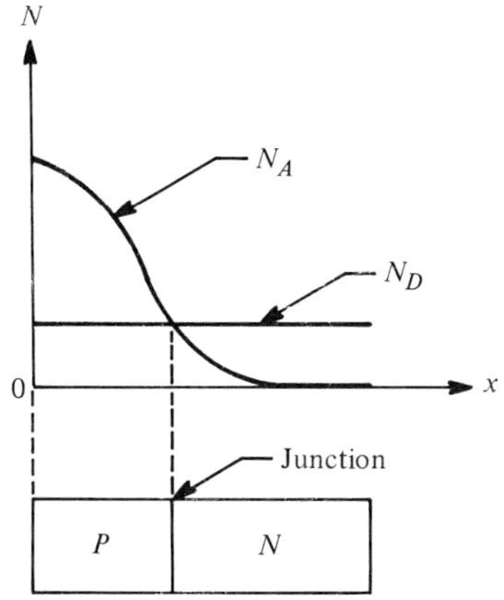

Figure 2–8 Band structure in the region of the junction.

$$\phi_{Fp} = \frac{kT}{q} \ln \frac{N_A}{n_i} \tag{2-18}$$

For the N-region

$$N_D = n = n_i e^{q\phi_F/kT} \tag{2-12}$$

therefore

$$\phi_{Fn} = \frac{kT}{q} \ln \frac{N_D}{n_i} \tag{2-19}$$

Since ϕ_B is the difference in the two ϕ_{Fp} values,

$$\phi_B = \phi_{Fn} - \phi_{Fp} = \frac{kT}{q} \ln \frac{N_A N_D}{n_i^2} \tag{2-20}$$

The mobile charge in the region of the junction reacts to the field due to ϕ_B by migrating out of the vicinity of the junction until sufficient fixed charge (ionized donor or acceptor atoms) is uncovered to just counter the field of ϕ_B. The effect of this depletion region may be determined by solving Poisson's equation

$$V^2\phi = q \frac{(N_A - N_D)}{\kappa\epsilon_0} \tag{2-21}$$

Usually, where the junction under consideration is in a single plane, the one-dimensional form of Poisson's equation is appropriate. Thus, the equation may be written in this form:

$$\frac{d^2\phi}{dx^2} = -\frac{\rho}{\kappa\epsilon_0} \tag{2-22}$$

where ρ is the charge per unit volume, κ is the dielectric constant, and ϵ_0 is the permittivity of free space. We assume the following boundary conditions,

$\rho = -qN_A$ on the P-side of the junction

$\rho = qN_D$ on the N-side of the junction

and $\rho = 0$ elsewhere outside the depletion region. Finally, for conservation of charge, there must be equal quantities of charge residing on the two sides of the junction, and the ratio of the depletion layer penetration into the two sides must be equal to the inverse of the doping ratios. Thus, if X_{dn} and X_{dp} are the depletion layer penetration distances into the N- and P-sides, respectively, then

$$N_A X_{dp} = N_D X_{dn} \tag{2-23}$$

If the boundary conditions of a one-sided step junction are now applied, either $N_D \gg N_A$ or $N_A \gg N_D$ (this condition holds true for most diffused junctions), the solutions are

$$X_{dn} = \left(\frac{2\kappa\epsilon_0\phi}{qN_D}\right)^{1/2} \tag{2-24}$$

for $N_D \gg N_A$ (*P*+ into lightly doped *N*)

or

$$X_{dp} = \left(\frac{2\kappa\epsilon_0\phi}{qN_A}\right)^{1/2} \tag{2-25}$$

for $N_D \gg N_A$ (*N*+ into lightly doped *P*). ϕ is the total electrostatic potential variation across the junction, i.e., $\phi = \phi_B + V_R$, where V_R is the reverse voltage across the junction.

It should be noted that Equations 2-24 and 2-25 allow for the depletion layer spreading into the lightly doped side of the junction only. Since there is a small but finite spread of the depletion layer into the heavily doped side of the junction, these equations will improve in accuracy with reverse bias.

2.2.10 Junction Capacitance

The junction under zero or reverse bias can be considered to be a parallel plate capacitor; i.e., it is a pair of (semi) conductor regions separated by a dielectric (the depletion region). Thus, with reasonable accuracy, we may invoke the parallel plate capacitor model to calculate the capacitance of a *P-N* junction.

$$C_d = \frac{\kappa\epsilon_0}{X_d} \tag{2-26}$$

where X_d is the depletion region width. Thus, if we assume as before, a single-sided junction, we may use Equations 2-24 or 2-25 to obtain

$$C_d = \sqrt{\frac{q\kappa\epsilon_0 N}{2\phi}} \tag{2-27}$$

where *N* is the dopant concentration on the lightly doped side of the junction. It can be seen from Equation 2-27 that the junction capacitance is a function of the square root of both the doping level and the potential across the junction. It should be noted that with the applied voltage at zero, there will still be a potential of approximately 0.6 V across a junction in silicon, due to the ϕ_B. Evaluating Equation 2-27 for silicon with zero applied voltage ($\phi = 0.6$ V) and $N = 1 \times 10^{15}$ atoms per cc, we obtain a value of 11.9×10^{-9} F/cm^2 or 0.077 pF/mil^2.

2.2.11 Diode Current–Voltage Characteristics

Up to this point, we have essentially considered currents that may flow as the result of applied voltages. Moll[3] and earlier Shockley[4] have considered these currents in detail and have developed an equation commonly known as the *Rectifier Equation.* Equation 2-28 is the form of the rectifier equation for a one-sided P-N^+ junction.

$$J = \frac{qD_n n_p}{L_n} (e^{qV/kT} - 1) \qquad (2\text{-}28)$$

In Equation 2-28, J is the current density across the junction, V is the voltage applied to the junction, and n_p is the concentration of electrons in the P-region (calculated from Equation 2-8). The terms D_n, the electron diffusion constant, and L_n, the diffusion length for electrons, need a little further explanation. D_n is a proportional constant relating the diffusion of electrons under a concentration gradient. It has the dimension of cm^2/s. L_n is defined by

$$L_n = \sqrt{D_n \tau_n} \qquad (2\text{-}29)$$

where τ_n is the minority carrier *lifetime* of the electrons. It should be noted that whenever an electron and a hole encounter one another, they may recombine and both be annihilated. Thus, the lifetime may be considered the mean time during which an electron may exist, after being generated, until it recombines with a hole. Many impurity elements, gold for example, serve as *recombination centers* or places in the lattice where the rate of recombination is enhanced; thus, the minority carrier lifetime of such gold doped material is significantly reduced.

If the junction is forward biased, i.e., the applied voltage (V) is greater than approximately 0.6 V the exponential term in Equation 2-28 becomes large compared to one and the forward current density (J_F) will be

$$J_F = \frac{qD_n n_p}{L_n} e^{qV/kT} \qquad (2\text{-}30)$$

which shows the commonly observed exponentially increasing current with applied voltage. On the other hand, if the junction is reversed biased, V is negative, the -1 term will dominate, and the equation will reduce to

$$J_R \simeq -\frac{qD_n n_p}{L_n} \qquad (2\text{-}31)$$

This approximation is often referred to as the *saturation current* density. Note that it is independent of voltage, as is observed in good *P-N* junctions. Figure 2-9 shows the typical current-voltage characteristics of a *P-N* junction.

2.2.12 Avalanche Breakdown

Equation 2-31 will apply to a *P-N* junction under reverse bias only until the field across the depletion region is so great that it can excite electrons directly from the valence band to the conduction band, or until the carriers are accelerated in the field to such a degree that they are capable of carrier pair generation on impact with a silicon atom, resulting in an avalanche effect.

The first mechanism, commonly known as the zener effect, is observed only in those cases where both sides of the junction are heavily doped, so that relatively little depletion layer spread occurs. Hence, very high fields can occur at quite low voltages. True zener breakdown is observed in silicon only in those junctions which show reverse breakdown in the 3 to 7 V range.

In the more lightly doped junctions, the fields do not build up so quickly, and the multiplication effects in the latter mechanism dominate. The following empirical expression has been generated to fit this observation.

$$M(V) = \frac{1}{1 - \left(\dfrac{V}{V_{BN}}\right)^n} \tag{2-32}$$

where $M(V)$ is the multiplication factor as a function of voltage, V_{BN} is the breakdown voltage, V is the applied voltage, and the exponent n is 2 or 4, depending on whether the avalanching species are electrons or holes,

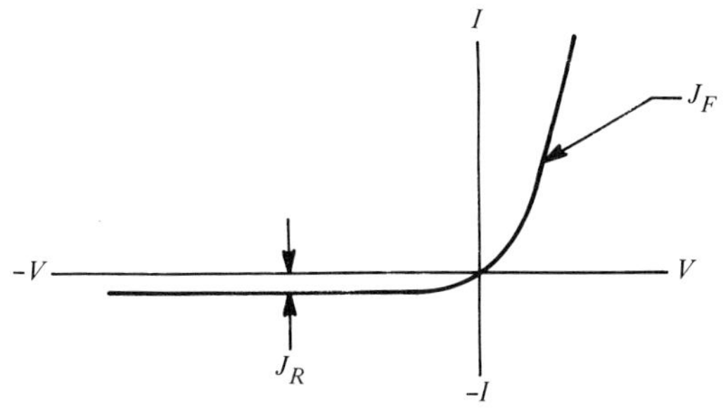

Figure 2–9 Current-Voltage characteristics of *P-N* junction.

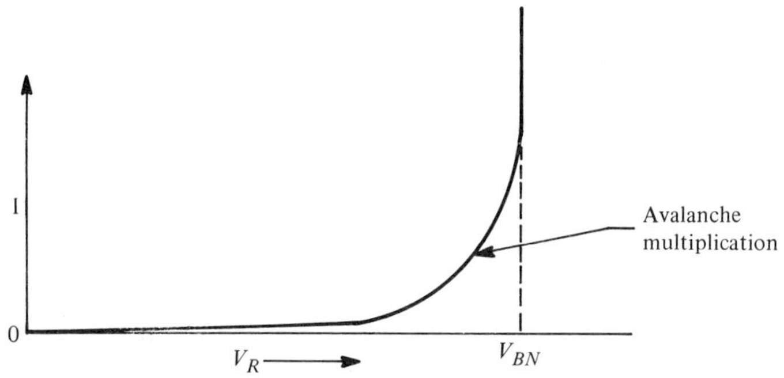

Figure 2–10 Typical avalanche breakdown characteristics of *P-N* junction.

respectively. Figure 2-10 shows the typical avalanche breakdown characteristics of a *P-N* junction.

2.3 ANALYSIS OF THE MOS TRANSISTOR

The concepts necessary for an understanding of the operation of MOS transistors will be developed in this section. Following this the current-voltage characteristics of the MOS transistor will be derived. The starting point for this analysis is conventional semiconductor theory as outlined in the previous section. In the conventional theory, the phenomena of interest occur relatively deep within the bulk of the silicon crystal; therefore, effects due to the boundaries or edges of the semiconductor are not usually considered.

The MOS transistor, in contrast to the bipolar transistor and to the junction field effect transistor, is fabricated at the surface of the silicon crystal and its operation depends on the presence of the semiconductor to insulator boundary. MOS, which stands for metal oxide semiconductor, describes the essential structure of the transistor as it is usually fabricated, although the more general description as a conductor (or semiconductor)-insulator-semiconductor structure would include other forms that are sometimes considered.

Two versions of the transistor may be conceived: one in which current is carried by electrons, called an *N*-channel transistor, and the other in which the current carriers are positive charges (holes), called a *P*-channel transistor. The latter type is by far the more common, and accordingly, the following discussion and analysis are based on what is conventionally called the *P*-channel enhancement mode transistor.

The analysis begins with a discussion of the manner in which the semiconductor surface modifies the classical semiconductor theory and introduces some important concepts relating to the surface, before the MOS transistor equations are derived.

A more detailed discussion of some pertinent semiconductor surface theory is presented in the Appendix, together with an analysis of the relationship between the conductance characteristic of an MOS transistor and the conductance and capacitance measurements performed by the semiconductor surface physicist.

For further study of semiconductor theory and the electrical characteristics associated with semiconductor surfaces the reader is referred to Grove.[5] The solution to Poisson's equation at the semiconductor surface was first presented by Kingston and Neustadter[6] and is covered in comprehensive texts by Many, *et al.*[7] and Frankl.[8] The analysis used here follows the approach of Sah and Pao.[9-11]

2.3.1 The Semiconductor Below the Surface

We start by considering the conditions within a semiconductor lying just inside a plane surface, and with the semiconductor in thermal equilibrium. For a *P*-channel MOS transistor, the semiconductor bulk material is *N*-type silicon and the insulator on the surface is silicon dioxide, SiO_2. Deep in the bulk the mobile charge carriers are electrons; their density is determined by considerations discussed in Section 2.2.5. The electrical condition of the silicon near the surface is described as being in accumulation, depletion, or inversion, according to whether the mobile charge density at (but inside) the surface is greater than, less than, or of opposite type to that in the bulk. Figure 2-11 illustrates these three conditions in terms of the electron energy bands, charge distribution, and electric field.

(a) Accumulation

In accumulation of an *N*-substrate the energy bands bend downward in a narrow region close to the surface. Associated with this is an electric field which increases the concentration of electrons at the surface, giving rise to a net charge Q_A (per unit area) in the narrow accumulation region. Figure 2-11a illustrates this condition. Accumulation of charge at the silicon surface can be conveniently brought about by applying a suitable external electric field oriented to attract electrons toward the surface. The silicon surface is still of the same type as the bulk but of higher conductivity.

(b) Depletion

In the depletion condition illustrated in Figure 2-11b the energy bands bend moderately upward at the surface. The electron concentration is

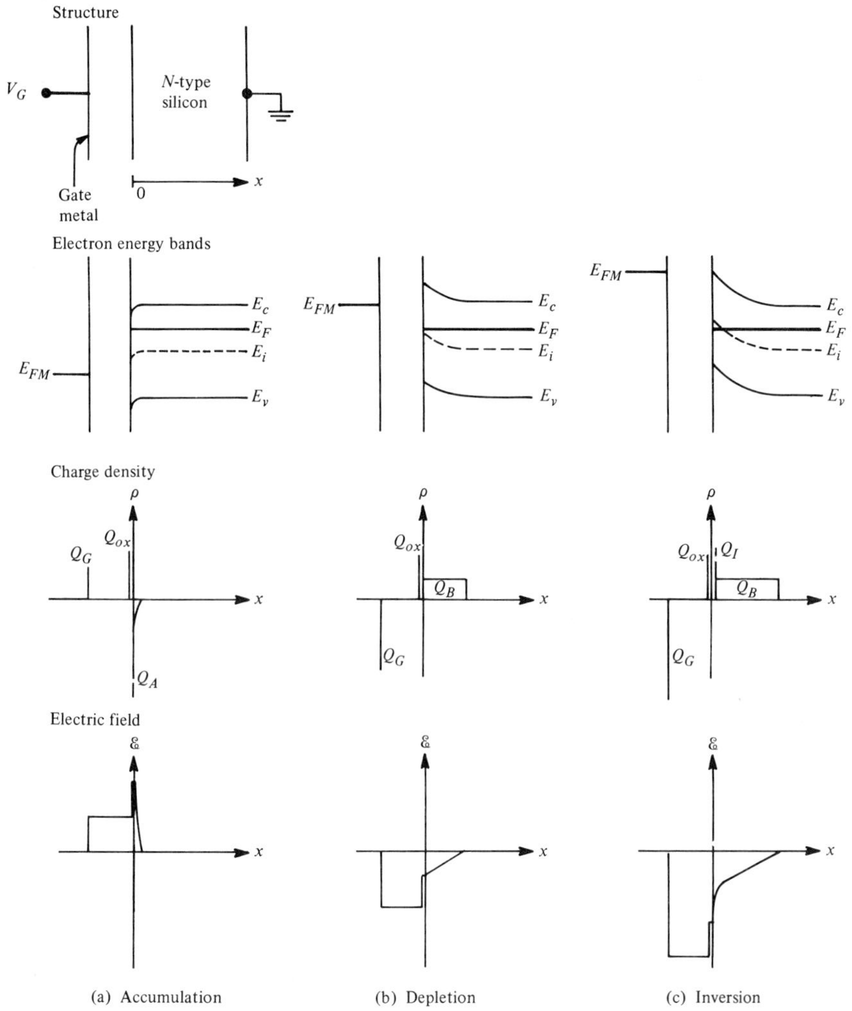

Figure 2–11 Energy, charge, and electric field diagrams for a *P*-channel MOS structure.

consequently less than in the bulk, thus uncovering the fixed positive charge Q_B of the dopant impurity ions.

The characteristics of the depletion region are usually analyzed by using the depletion approximation, in which the depletion region is considered to be completely "depleted" of mobile charges (electrons in this case) and the boundary between this region and the bulk is sharp and abrupt. Under the depletion approximation, Q_B consists of a rectangular charge distribution as shown in Figure 2-11b. In the one-dimensional geometry considered here, the electric field is then a linear function of position, and the electrostatic potential (and electron energy) are parabolic functions of position.

The total charge Q_B per unit surface area contained in a depletion region of thickness X_d is

$$Q_B = qN_D X_d \tag{2-33}$$

The electric field ($\mathcal{E}_s$) at (but inside) the surface of the silicon is obtained from Gauss' law. (The integral of the normal component of the electric displacement vector over a closed surface is equal to the charge enclosed.)

Considering the one-dimensional nature of the present geometry and noting that the electric field is zero in the silicon bulk:

$$\mathcal{E}_s = -\frac{Q_B}{\epsilon_s} \tag{2-34}$$

The negative sign arises from the fact that the outwardly directed vector normal to the surface is in the $-x$ direction.

As in the case of accumulation, an external electric field can be used to place the silicon surface in depletion. In the absence of surface states and work function differences which are discussed in the next section, the application of an electric field given by Equation 2-34 will deplete the surface to a distance X_d given by Equation 2-33.

(c) Inversion

In addition to accumulation and depletion, there is one other possible condition of the silicon surface. This is inversion, in which the surface layer becomes of opposite conductivity type from the substrate. This condition gives rise to the unique characteristics of the MOS transistor.

Starting with a silicon surface in depletion (Figure 2-11b), if one attempts to increase the amount of depletion (by increasing the magnitude of $\mathcal{E}_s$, for example), a point is soon reached at which the intrinsic energy level (E_i) is equal to the electron Fermi level (E_F) at the surface. The surface now has become exactly intrinsic. As the bands are bent further, E_i rises above E_F near the surface, and the density of holes becomes greater than that of electrons (Figure 2-11c). The surface is now said to be inverted. The density of holes and electrons at the surface, or anywhere within the semiconductor for that matter, are determined by the classical Maxwell-Boltzmann statistics (along the lines of Equation 2-11):

$$n = n_i e^{(E_F - E_i)/kT} \tag{2-35a}$$

$$p = n_i e^{(E_i - E_F)/kT} \tag{2-35b}$$

Although the surface is inverted as soon as E_i rises above E_F, the density of holes remains fairly small until E_i has risen considerably above E_F. A significant number of holes do not become available until the exponential term in Equation 2-35b becomes quite large. Consequently, the "strong inversion" approximation is frequently used, in which a strongly inverted surface is considered to be formed only after E_i at the surface is as far above E_F as it was below E_F in the bulk. From Equations 2-35a and 2-35b it is apparent that this is the same as saying that the surface becomes strongly inverted when the hole density at the surface equals the electron density in the bulk, or simply the surface is exactly inverted from the bulk.

As before, the electric field just inside the surface of the silicon is found from Gauss' law

$$\mathcal{E}_s = -\frac{Q_B + Q_I}{\epsilon_s} \tag{2-36}$$

where Q_I is the mobile charge per unit area in the inversion layer and Q_B is the fixed charge of the ionized impurity atoms.

The electrostatic potential (ϕ) in the semiconductor is defined in terms of the potential energy of a charged particle. Since semiconductor energy diagrams are drawn in terms of electron energy (E) and the electron has a charge (^-q), the relation between electrostatic potential and electron energy is $E = {}^-q\phi$. For convenience, we define a zero of potential and in all the following work the Fermi level in the bulk is taken as the potential zero.* Then the potential corresponding to an energy E is

$$\phi = -\frac{E - E_F}{q} \tag{2-37}$$

It is also convenient when referring to a potential variation in the semiconductor to consider the potential corresponding to the intrinsic Fermi level (E_i). Deep within the semiconductor this gives the bulk Fermi potential (ϕ_F) as

$$\phi_F = -\left.\frac{E_i - E_F}{q}\right|_{x > X_d} \tag{2-38}$$

*This convention is by no means universal. Some authors define potential with respect to the intrinsic level in the bulk. The Fermi potential is then the negative of that given by Equation 2-38. In this text, the potential is defined with respect to the Fermi level, since it corresponds to a real measurable voltage that goes to zero when the substrate is grounded.

Thus ϕ_F is positive in an N-type semiconductor and negative in a P-type.

Under the strong inversion approximation, the total bending of the energy bands corresponds to a potential barrier (ϕ_B) of twice the Fermi potential and in the opposite direction

$$\phi_B = -2\phi_F \tag{2-39}$$

This band bending is analogous to the total band displacement in a conventional P-N junction as discussed in Section 2.2.9. The difference is that in the present case the junction is a "field induced junction" formed as a result of the $\mathcal{E}_s$. By our previous definition of ϕ_F, ϕ_B is negative for a P-inversion layer on a N-substrate and positive for an N-layer on a P-substrate.

Once the condition of strong inversion has been achieved the energy bands no longer bend appreciably farther as the inversion layer is made stronger. This results from the fact that the exponential term in Equation 2-35b is now quite large, and small increases in its argument easily give rise to large increases in mobile charge density. Thus, as $\mathcal{E}_s$ is increased, Q_I increases and Q_B and X_d remain substantially constant.

2.3.2 MOS Capacitor

The previous section discussed the electrical nature of the semiconductor surface from a viewpoint inside the semiconductor and showed how the electrical characteristics vary from the bulk to the surface. We now consider some important properties outside the silicon itself.

In fabricating an MOS transistor the surface of the silicon is covered with a thin (usually 1000 to 2000 Å) dielectric layer. In present devices this layer is usually silicon dioxide (SiO_2) which is grown on the surface by a high-temperature oxidation of the silicon, although other materials and different methods of forming the dielectric are sometimes used. This insulating layer is generally referred to as the oxide. On top of this oxide is the "gate" metal, usually aluminum. The entire sandwich forms the MOS capacitor, as illustrated in Figure 2-11.

The electrical characteristics of the oxide are determined by its thickness (t_{ox}) and permittivity (ϵ_{ox}) and by any electrical charges (Q_{ox}) that may be present within the oxide. Electrical charges in the oxide have an important effect on the characteristics of the MOS transistor. A considerable amount of work has been done to investigate the nature of these charges.[12-14] (See also Chapter 12 of Reference 5.) They may arise in several ways: impurity ions incorporated within the oxide, ionized silicon atoms within the oxide, or charges near the silicon to oxide interface due to the termination of the regular silicon crystal lattice. Whatever their origin, extensive data now show that, with proper preparation of the surface, these charges are not mobile and are located close to the surface of the silicon. Accordingly, it is a good approximation to combine all the fixed charges in a sheet of charge, Q_{ox} (coulombs/cm^2), located at the silicon-oxide interface.* This is indicated in the charge diagrams in Figure 2-11. Q_{ox} is thus an effective charge density which modifies the electric field in both the oxide and the silicon. It has an important effect in determining the threshold voltage of the MOS transistor, which is discussed in the next section. Experimentally, it is found that Q_{ox} is independent of substrate type but does depend on the crystallographic orientation of the substrate. For the usual types of thermally grown oxides, Q_{ox} is positive in sign.

If one measures the small signal capacitance between the gate metal and the silicon substrate as a function of the gate to substrate voltage, a curve similar to Figure 2-12 is obtained. The three regions of this curve correspond to the three conditions of the silicon surface illustrated in Figure 2-11. For the present case of an N-type substrate, when the gate voltage is

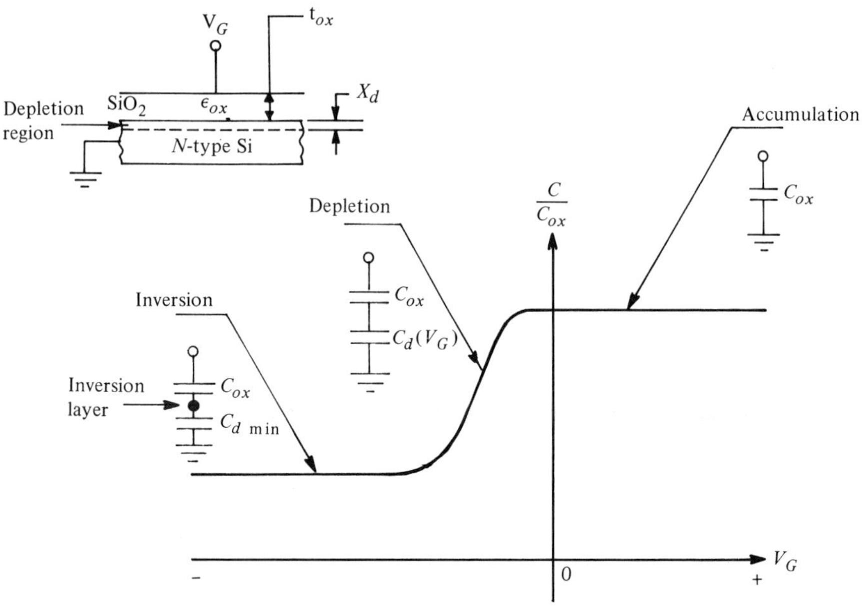

Figure 2–12 Capacitance of MOS structure as a function of gate voltage.

*The charge we call Q_{OX} is often referred to in the literature as surface state charge Q_{SS}, and the term Q_{OX} sometimes refers to only the impurity ions. In order to avoid confusion and sidestep the conflicting definitions of Q_{SS}, we use the term Q_{OX} to represent the combined effect of all the charges associated with the oxide.

positive, the silicon surface is accumulated and acts as a conductor, with the result that the capacitance (per unit area) is just that of a parallel plate capacitor with the oxide as a dielectric: $C = C_{ox} = \epsilon_{ox}/t_{ox}$. When the gate voltage is sufficiently negative to form a depletion region in the silicon, corresponding to Figure 2-11b, the capacitance is that due to the series combination of the oxide and the depletion region capacitance, $1/C = 1/C_{ox} + 1/C_d$. As the depletion region becomes wider its capacitance decreases and the total capacitance decreases. Finally, for a sufficiently negative gate voltage an inversion layer forms as in Figure 2-11c, the depletion layer is at its maximum width, and the capacitance no longer decreases.

The MOS capacitance characteristic, as illustrated in Figure 2-12, has been a particularly useful tool in the study of silicon surfaces. A good analysis and discussion is given in Section 9.2 of Reference 5.

2.3.3 Threshold Voltage

As discussed in the previous sections, the conditions at the silicon surface can be described as either accumulation, depletion, or inversion. The

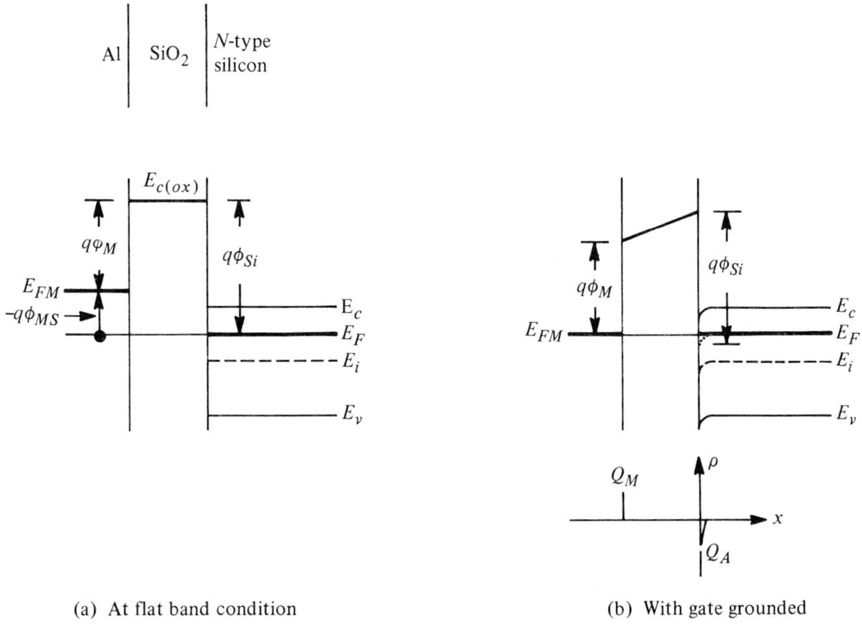

Figure 2–13 Band diagram showing the metal to silicon work function ϕ_{MS} with no oxide charge present.

factors that set the dividing lines between these three conditions will now be considered.

The transition between the accumulation and depletion condition occurs when the energy bands in the silicon continue horizontally up to the surface. This is illustrated in Figure 2-13a, and is called the "flat-band" condition. In this condition there is no electric field in the silicon, and the net charge density in the silicon is zero. Mobile electrons in the *N*-type substrate just balance the positive charges of the donor atoms.

In the absence of any charges Q_{ox} in the oxide, the electric field outside the silicon is zero and any energy bands in the oxide are also horizontal. Note, however, as indicated in Figure 2-13a, that this condition does not occur when the gate voltage is zero. This is a result of the difference between the metal and the silicon electron work functions and gives rise to a metal to silicon potential barrier ϕ_{MS}.

The work function of a material is defined as the energy necessary to remove an electron at the Fermi level from the material in a vacuum. In the case of the metal oxide silicon structure, the appropriate energies to consider are the modified work functions from the Fermi levels in the electrode materials to some oxide conduction band. These are indicated in Figure 2-13a in terms of their equivalent potential barriers, ϕ_M and ϕ_{Si}. The figure shows the conditions at flat band and with no oxide charge present. Thus the gate voltage to achieve flat band is simply the difference of the two potential barriers, which is usually written as

$$V_G = \phi_{MS} = \phi_M - \phi_{Si} \qquad *(2\text{-}40)$$

ϕ_{MS} is often loosely referred to as the metal to silicon work function. It is the gate voltage necessary to counterbalance the work function difference between the gate metal and the silicon substrate.

Barrier heights and ϕ_{MS} for several metals on silicon have been measured by Deal *et al.*[15] For aluminum and *N*-type silicon, ϕ_{MS} is about -0.3 V at typical substrate doping concentrations.

The gate voltage necessary to achieve flat band is further modified by the oxide charge Q_{ox} mentioned earlier. This is illustrated in Figure 2-14. At flat band there is again no electric field in the silicon. To achieve this with a positive Q_{ox}, the gate voltage must be sufficiently negative that an equal negative charge, $-Q_{ox}$, appears on the metal or

*Strictly speaking, ϕ_M and ϕ_{Si} are negative (electron energy in the oxide band is higher than in the metal or silicon) and Equation 2-40 should be written $\phi_{MS} = \phi_{Si} - \phi_M$ which comes directly from Figure 2-13a by noting that $\phi_{MS} + \phi_M = \phi_{Si}$. However common usage takes the barriers ϕ_M and ϕ_{Si} as positive quantities and thus the form of Equation 2-40 gives the correct sign to ϕ_{MS}.

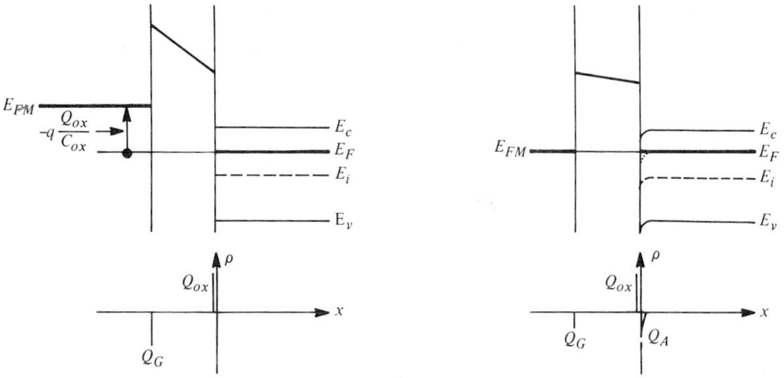

(a) At flat band condition (b) With gate grounded

Figure 2–14 Band diagram showing the effect of fixed charge Q_{ox} associated with the oxide, with no work function difference.

$$V_G = -\frac{Q_{ox}}{C_{ox}} \tag{2-41}$$

With both a work function difference and oxide charge present the gate voltage, V_{FB} to cause the flat band condition is, from the previous two equations,

$$V_{FB} = \phi_{MS} - \frac{Q_{ox}}{C_{ox}} \tag{2-42}$$

When the gate is more negative than V_{FB}, the silicon surface is in depletion and the energy bands begin to bend upward at the surface.

When the intrinsic level (E_i) just touches the Fermi level the surface is intrinsic. With any further increase a P-type inversion layer begins to form as illustrated in Figure 2-11. Strictly speaking, this defines the boundary between the depletion and inversion condition. In practice, however, E_i must be increased appreciably beyond E_F to obtain a significant positive charge density in the inversion layer. Consequently, the "strong inversion" approximation is generally used. As discussed in Section 2.3.1, strong inversion occurs when

$$(E_i - E_F)_{\text{surface}} = (E_F - E_i)_{\text{bulk}}$$

The total potential barrier (ϕ_B) due to band bending at the surface is then

$$\phi_B = -2\phi_F \tag{2-43}$$

The negative sign indicates that the energy bands are bending in a direction opposite to ϕ_F. In the present case, E_i is below E_F in the bulk and the bands bend upward at the surface. ϕ_B is thus a negative number in a *P*-channel device.

The depletion layer which has been formed between the silicon surface and the bulk is analogous to that formed in a one-sided junction, which was discussed in Section 2.2.9. Accordingly, the width of the depletion layer X_d is

$$X_d = \left(\frac{2\epsilon_s[-\phi_B]}{qN_D}\right)^{1/2} \tag{2-44}$$

and it contains a charge Q_B per unit area of

$$Q_B = (2\epsilon_s q N_D[-\phi_B])^{1/2} \tag{2-45}$$

The gate voltage necessary to bring about strong inversion is called the threshold voltage (V_T). This voltage is the flat band voltage (V_{FB}) plus the band bending to reach strong inversion, plus the voltage necessary to generate the electric field that holds the charge Q_B in the depletion region. The depletion region charge Q_B requires a charge $-Q_B$ on the gate metal, which in turn requires a gate voltage $-Q_B/C_{ox}$. Adding these terms, the total gate voltage (V_T) at the onset of strong inversion is

$$V_T = V_{FB} + \phi_B - \frac{Q_B}{C_{ox}} \tag{2-46}$$

and, using Equation 2-42,

$$V_T = \phi_{MS} + \phi_B - \frac{Q_{ox}}{C_{ox}} - \frac{Q_B}{C_{ox}} \tag{2-47}$$

The term Q_B/C_{ox} represents the contribution to the threshold voltage due to the substrate doping and, in the equilibrium case considered so far (no dc voltage on the inversion layer), is given by Equation 2-45. In the more general case considered in the next section a voltage V may reverse bias the channel to substrate junction. Q_B then becomes

$$Q_B = (2\epsilon_s q N_D[-V - \phi_B])^{1/2} \tag{2-48}$$

where V is the voltage (at the quasi-Fermi level) of the *P*-inversion layer side of the junction with respect to the *N*-side. In the present case V and ϕ_B are negative quantities, which makes the bracketed term positive. Q_B is positive for an *N*-substrate.*

*Polarities of these and other quantities for both *P*- and *N*-channel transistors, as well as the important equations are summarized in Table 2-1 given subsequently.

To illustrate the relative importance of the terms in Equation 2-47 and to aid in keeping the proper polarities, V_T will be calculated for a typical transistor on <111> N-type silicon with $N_D = 1 \times 10^{15}$ atoms/cm^3 and $Q_{ox} = +4 \times 10^{11}$ charges/cm^2. The oxide thickness is 1000 Å, $C_{ox} = \epsilon_{ox}/t_{ox} = 3.46 \times 10^{-8}$ F/cm^2, and the gate metal is aluminum. The arrangement of the terms corresponds to Equation 2-47.

$$V_T = (-0.3)V + (-0.6)V - \frac{(+6.40 \times 10^{-8})\ \text{coulomb/cm}^2}{(3.46 \times 10^{-8})\ \text{F/cm}^2} - \frac{(+1.41 \times 10^{-8})\ \text{coulomb/cm}^2}{(3.46 \times 10^{-8})\ \text{F/cm}^2}$$

$$= -0.3V - 0.6V - 1.85V - 0.41V = -3.16V$$

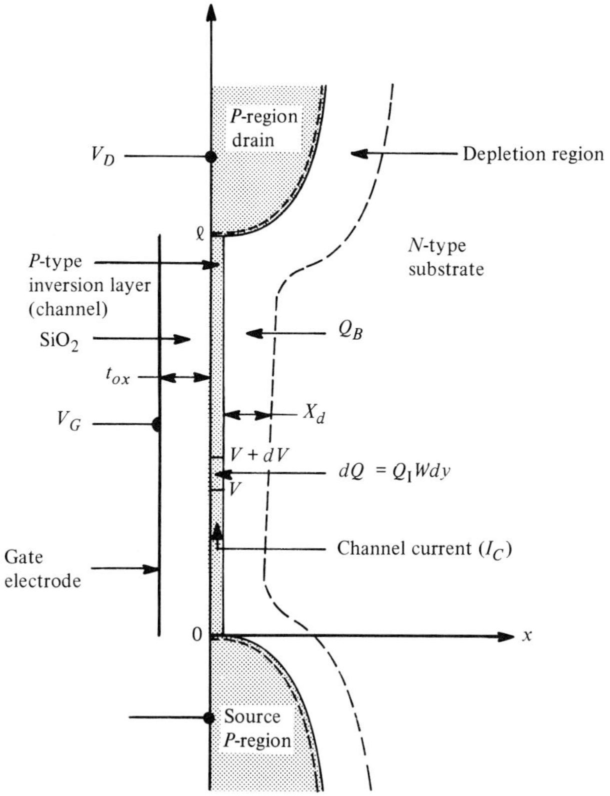

Figure 2–15 MOS transistor structure.

2.3.4 MOS Transistor

Figure 2-15 illustrates the structure of an MOS transistor, together with the coordinate system and some of the important parameters used in the subsequent analysis. This structure is that of the previous discussion with the addition of the *P*-region source and drain. These regions are of the same conductivity type as the inversion layer and provide the terminals for lateral (parallel to the surface) current flow in the inversion layer. This path for current flow is commonly called the transistor channel. Current flow in the channel occurs by conduction of mobile inversion layer charge (holes in this case) under the influence of an electric field along the channel. The inversion layer is quite thin, typically on the order of 100 Å, and so the current flow in an MOS transistor consists of a thin sheet of current parallel to the silicon surface.

The following derivation is a one-dimensional analysis, in which we consider the variation of inversion layer charge and channel voltage in the y direction only. The component of the electric field in the y direction $\mathcal{E}_y$ that propels the mobile charge is, to a first approximation, independent of the component of the field through the oxide in the x direction ($\mathcal{E}_x$) that holds the charge in the inversion layer. This is the "gradual channel" approximation. With it, one can write the mobile charge in the channel as a function of only the gate and channel voltages. The channel current is then determined by the way in which this charge moves under the influence of the lateral electric field ($\mathcal{E}_y$) due to the source and drain voltages.

Consider a small section of the channel region. When the gate voltage is equal to the threshold voltage, a channel just begins to form. Further increase of the gate voltage by an amount ΔV beyond threshold results in a charge per unit area $\Delta Q = C_{ox}\,\Delta V$ on the gate electrode and a charge $-C_{ox}\,\Delta V$ on the opposite side of the oxide capacitor. This is just the mobile charge in the channel. In the general case with a voltage V on the channel (where V corresponds to the quasi-Fermi level for holes in the inversion layer), the voltage across the oxide capacitor is $V_G - V$, and the charge Q_I in the channel inversion layer is

$$Q_I = -C_{ox}(V_G - V_T - V) \qquad \text{(2-49)}$$

This mobile charge gives rise to a conductive layer along the surface of the silicon. The sheet conductivity (σ_S) (mho/sq) of this layer is determined by the amount of charge and the average mobility (μ_p) of this charge.*

*See Section 2.3.13 for a discusstion of sheet resistivity ρ_s. $\sigma_s = 1/\rho_s$. It should also be remembered that the mobility here is that in the inversion layer, and is significantly less than (approximately one-half) the bulk mobility discussed in Section 2.2.7.

$$\sigma_s = \mu_p Q_I = \mu_p C_{ox}(V_G - V_T - V) \tag{2-50}$$

The channel current flows in the y direction and, by Ohm's law, the current density is $J_s = \sigma_s \mathcal{E}_y$, in which the current density is expressed as a sheet current density (A/cm) corresponding to the use of a sheet conductivity. $\mathcal{E}_y$, by the gradual channel approximation, is a function only of the voltage change in the y direction, i.e., $\mathcal{E}_y = -dV/dy$. The current density (J_s) is simply the total channel current (I_C) divided by the width (W) of the conducting channel. By combining these definitions,

$$\frac{I_C}{W} = J_s = \sigma_s \mathcal{E}_y = -\sigma_s \frac{dV}{dy}$$

or

$$I_C dy = W \sigma_s dV \tag{2-51}$$

and using σ_s from Equation 2-50

$$I_C dy = W \mu_p C_{ox}(V_G - V_T - V) dV \tag{2-52}$$

The threshold voltage term (V_T) is given by Equations 2-47 and 2-48. When these are inserted in Equation 2-52 and rearranged slightly

$$I_C dy = W \mu_p C_{ox} \left(V_G - \phi_{MS} - \phi_B + \frac{Q_{ox}}{C_{ox}} - V + \frac{(2\epsilon_s q N_D[-V - \phi_B])^{1/2}}{C_{ox}} \right) dV \tag{2-53}$$

Equation 2-53 can be integrated directly. We take the general case where both the source and drain are at arbitrary voltages V_S and V_D. Integrating from source to drain, the limits of integration are $y = 0$, $V = V_S$ at the source and $y = \ell$, $V = V_D$ at the drain.

$$I_C \int_0^\ell dy = W \mu_p C_{ox} \left\{ \left(V_G - \phi_{MS} - \phi_B + \frac{Q_{ox}}{C_{ox}} \right) \int_{V_S}^{V_D} dV - \int_{V_S}^{V_D} V dV + \frac{(2\epsilon_s q N_D)^{1/2}}{C_{ox}} \int_{V_S}^{V_D} (-V - \phi_B)^{1/2} dV \right\} \tag{2-54}$$

$$I_C = \mu_p C_{ox} \frac{W}{\ell} \left\{ \left(V_G - \phi_{MS} - \phi_B + \frac{Q_{ox}}{C_{ox}} \right) (V_D - V_S) - \frac{1}{2} (V_D{}^2 - V_S{}^2) - \frac{2(2\epsilon_s q N_D)^{1/2}}{3C_{ox}} ([-V_D - \phi_B]^{3/2} - [-V_S - \phi_B]^{3/2}) \right\} \tag{2-55}$$

V_D and V_S are measured with respect to the substrate and I_C flows in the $+y$ direction from source to drain.

Within the assumptions used in the derivation, Equation 2-55 is the substantially complete and general MOS transistor equation. Although most of these assumptions were discussed at the time they were introduced, for convenience, the assumptions are listed below.

a. depletion approximation
b. one-dimensional geometry; depletion region spreading around the source, drain, and sides of the channel neglected
c. strong inversion approximation
d. Q_{ox} constant; particularly Q_{ox} assumed independent of the amount of band bending
e. effective surface mobility (μ_p) constant
f. gradual channel approximation
g. channel current composed only of drift current; diffusion currents neglected

An important additional restriction on Equation 2-55 must also be noted. The derivation was based on the condition, implicit in Equation 2-49, that a well-formed channel is present everywhere between the source and drain. This is not always the case; the situation when current flow occurs without a well-formed channel is termed saturation, as discussed in the next section. In the present case, the transistor is in the nonsaturation condition, sometimes called either the linear or the triode region.

We note that the threshold voltage discussed in the previous section and defined by Equation 2-47 does not appear intact in Equation 2-55. As Equation 2-47 shows, part of the threshold voltage is a fixed quantity (depending only on fabrication conditions) and is not a function of device operating conditions. This fixed part of the threshold voltage appears explicitly in the channel current equation in the form ($V_G - V'_T$), where $V'_T = \phi_{MS} + \phi_B - Q_{ox}/C_{ox}$. The remaining part of the threshold voltage is due to the fixed charge Q_B in the depletion region. This charge is a function of the voltage along the channel, which depends on the drain and source voltages. As a result, this portion of the threshold voltage becomes incorporated in the last term of the channel current equation.

It should also be noted that Equation 2-55 includes the "body effect" (see Section 2.4.6). This again occurs because the bulk charge (Q_B) which gives rise to the body effect has been included in the derivation.

We now consider several variations of the MOS transistor channel current equation. If the source is electrically connected to the substrate, $V_S = 0$ and Equation 2-55 becomes

$$I_{CSO} = \mu_p C_{ox} \frac{W}{\ell} \left\{ \left(V_G - \phi_{MS} - \phi_B + \frac{Q_{ox}}{C_{ox}} - \frac{1}{2} V_D \right) V_D - \frac{2(2\epsilon_s q N_D)^{1/2}}{3C_{ox}} \left([-V_D - \phi_B]^{3/2} - [-\phi_B]^{3/2} \right) \right\} \tag{2-56}$$

which is the form in which the MOS transistor equation is usually derived.[5,10,16] Conversely, with a voltage V_S on the source, and the drain connected to the substrate, $V_D = 0$ and Equation 2-55 again reduces to the same form with reversed polarity.

$$I_{CDO} = -\mu_p C_{ox} \frac{W}{\ell} \left\{ \left(V_G - \phi_{MS} - \phi_B + \frac{Q_{ox}}{C_{ox}} - \frac{1}{2} V_S \right) V_S - \frac{2(2\epsilon_s q N_D)^{1/2}}{3C_{ox}} \left([-V_S - \phi_B]^{3/2} - [-\phi_B]^{3/2} \right) \right\} \quad (2\text{-}57)$$

These two equations illustrate the complete symmetry of the MOS transistor. This symmetry is, of course, also apparent in Equation 2-55 and is to be expected from physical considerations.

Adding Equations 2-56 and 2-57,

$$I_C = I_{CSO} + I_{CDO} \quad (2\text{-}58)$$

Thus, the current in the general case with a voltage on both the drain and source can be decomposed into the current through two simpler transistors, one with the source grounded and the other with the drain grounded. This is illustrated in Figure 2-16.

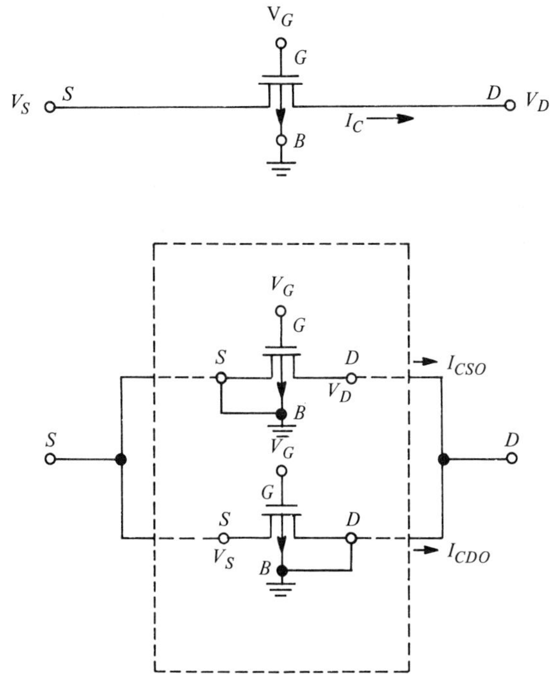

Figure 2–16 Decomposition into two transistors, one with source grounded and the other with drain grounded.

A major simplification occurs in the current equation if the charge Q_B in the depletion region can be neglected, or at least considered constant. To see this, we take the case where the source is grounded ($V_S = 0$) and assume that Q_B is constant all along the channel and has the value at the source, which is given by Equation 2-45. This is the charge Q_{BO} in a junction with no applied voltage

$$Q_{BO} = (2\epsilon_s q N_D[-\phi_B])^{1/2} \tag{2-59}$$

Equation 2-54 then becomes

$$I_C \int_0^\ell dy = W\mu_p C_{ox} \left\{ \left(V_G - \phi_{MS} - \phi_B + \frac{Q_{ox}}{C_{ox}} + \frac{Q_{BO}}{C_{ox}} \right) \int_0^{V_D} dV - \int_0^{V_D} V dV \right\} \tag{2-60}$$

which integrates to

$$I_C = \mu_p C_{ox} \frac{W}{\ell} \left\{ \left(V_G - \phi_{MS} - \phi_B + \frac{Q_{ox}}{C_{ox}} + \frac{Q_{BO}}{C_{ox}} \right) V_D - \frac{1}{2} V_D^2 \right\} \tag{2-61}$$

or

$$I_C = \mu_p C_{ox} \frac{W}{\ell} \left\{ (V_G - V_T)V_D - \frac{1}{2} V_D^2 \right\} \tag{2-62}$$

where V_T is the threshold voltage as derived in Equation 2-47. Equation 2-62 was derived for the case where the source is at ground potential.

A similar result is also obtained from the general transistor Equation 2-55 if both V_O and V_S are small compared to ϕ_B. Making use of the approximation that $(1 + x)^n = 1 + nx$ for small x, the bracketed terms in Equation 2-55 are written in the form

$$(-\phi_B)^{3/2} \left[\frac{V}{\phi_B} + 1 \right]^{3/2} \approx (-\phi_B)^{3/2} \left[\frac{3V}{2\phi_B} + 1 \right]$$

whereupon the last term inside the braces in Equation 2-55 reduces to

$$\frac{(2\epsilon_s q N_D[-\phi_B])^{1/2}}{C_{ox}} (V_D - V_S) = \frac{Q_{BO}}{C_{ox}} (V_D - V_S)$$

Equation 2-55 then becomes

$$I_C = \mu_p C_{ox} \frac{W}{\ell} \left\{ (V_G - V_T)(V_D - V_S) - \frac{1}{2} (V_D^2 - V_S^2) \right\} \tag{2-63}$$

where V_T is again as defined as above. Equation 2-63 shows that the simple model with constant Q_B is exact for small drain and source voltages.

Equation 2-62 is commonly called the Sah Equation[9] after its first application to the MOS transistor. Because of its reduced complexity, this is the form of the channel current equation that is usually used in most design work. In spite of the rather gross approximation of constant Q_B, Equation 2-62 gives a reasonably good fit for many cases encountered in digital circuit design, particularly if electrical measurements on completed transistors are used to determine the transistor parameters at operating conditions close to those in the final application.

Figure 2-17 compares the complete expression (Equations 2-56 or 2-55 with $V_S = 0$) with the simplified Equation 2-62 for a set of typical parameters.

2.3.5 Saturation

The results of the preceding section have been based on the condition that a well-formed channel existed everywhere under the gate. Under appropriate conditions, it is possible for drain current to flow even though

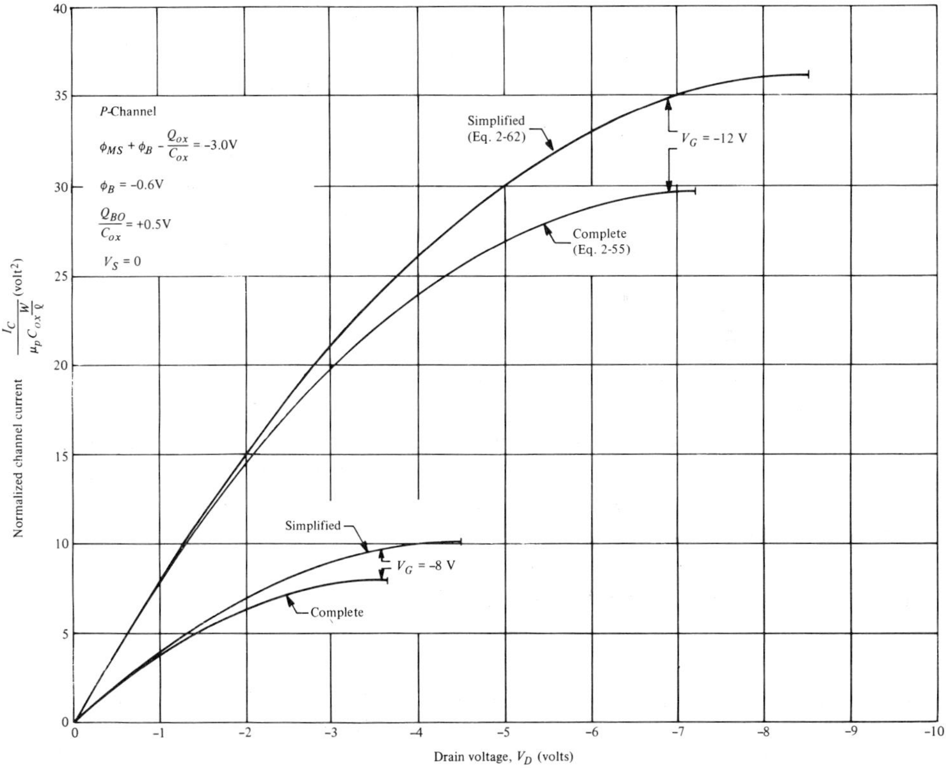

Figure 2–17 Comparison of complete MOS transistor equation with the simplified model for some typical parameters.

an inversion layer may not be formed in a small region under the gate. This situation is called channel current "saturation" or channel "pinch-off" and represents operating conditions substantially different from the non-saturated condition of the previous analysis.

Consider a P-channel transistor with the source grounded and a fixed voltage $V_G < V_T$ on the gate so that a channel is formed. The drain voltage (V_D) is increased (in the negative direction) from zero. Initially, a linear resistive channel exists between the source and drain, and the drain current (I_C) increases linearly with drain voltage. This was illustrated near the origin in Figure 2-17. As the drain voltage becomes more negative, the channel voltage (V) in the region of the channel near the drain also becomes more negative; consequently, less of the gate electric field is available to hold charges in the channel. The conductivity in that region of the channel is thus less, which causes the characteristic curve to flatten.

Eventually a drain voltage will be reached at which the electric field in the oxide near the drain end of the channel will be insufficient to hold any charge in the inversion layer; all of the charge in the silicon will now be contained in the depletion region. At this point, the analysis of the preceding section is no longer valid. The channel becomes pinched off, and the transistor is now in saturation. The drain voltage at which pinch-off occurs is called the drain pinch-off voltage (V_{DP}). Any further increase (negatively) in the drain voltage now results in the additional voltage appearing across the depleted (pinched-off) region adjacent to the drain. The voltage on the channel side of this pinch-off region remains at V_{DP}.

The current through the transistor is determined by that portion of the channel between the source and the pinch-off point. Thus, assuming that the length of the channel between the source and the pinch-off point does not change, the drain current remains constant. Current flow in the well-formed part of the channel occurs by drift, wherein the carriers (holes) drift under the influence of the lateral electric field along the channel. At the pinch-off point, the carriers are injected into the depletion region that surrounds the drain and are then quickly swept by the large electric field over to the drain.

At the edge of saturation, and adjacent to the drain, the charge Q_I in the inversion layer given by the Equation 2-49 is zero,

$$Q_I = -C_{ox}(V_G - V_T - V) = 0 \tag{2-64}$$

At this point, V_{DP} is just equal to the voltage (V) in the channel adjacent to the drain, i.e., $V = V_{DP}$. Thus, Equation 2-64 gives the drain voltage at the edge of saturation as

$$V_{DP} = V_G - V_T \tag{2-65}$$

This expression is complete, providing that one remembers that V_T as given by Equation 2-47 is also a function of the channel voltage, which in the present case is equal to the drain voltage at the pinch-off point.

Taking first the simplified theory developed in the previous section, in which V_T is assumed to be constant, the drain voltage at saturation from Equation 2-65 is substituted into the channel current Equation 2-62 to give the current in saturation as

$$I_{CP} = \frac{1}{2} \mu_p C_{ox} \frac{W}{\ell} (V_G - V_T)^2 \tag{2-66}$$

Note that this is independent of V_D; thus the drain characteristic curves are horizontal lines. This is shown in Figure 2-18, together with the separation into the saturated and unsaturated regions.

Also indicated in Figure 2-18 is a third region which may be called "cutoff." In this condition, the gate voltage is smaller than the V_T and so the channel current is always zero. The cutoff condition may be considered as an extension of saturation, in which the entire channel is pinched-off rather than just a single point at the drain, as in saturation. In some cases, with voltages on both the source and drain, cutoff can

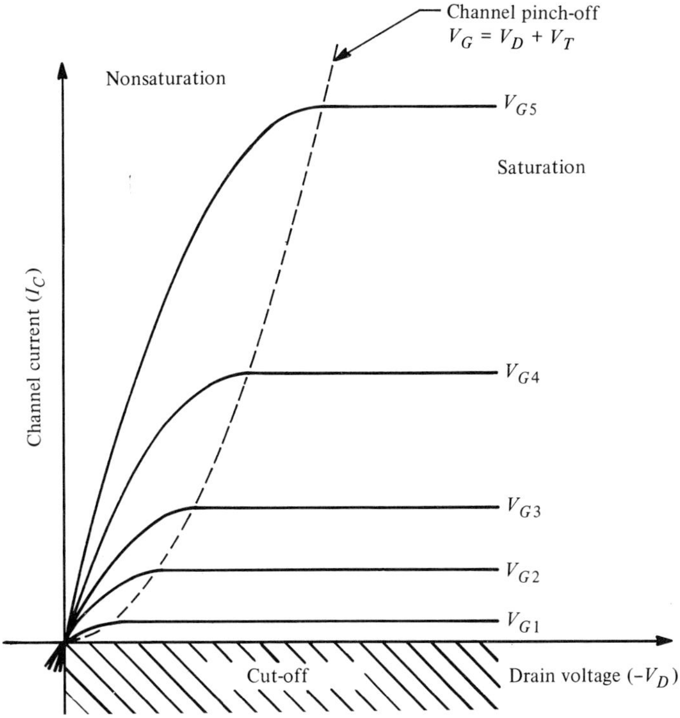

Figure 2–18 Drain characteristic curves illustrating the saturated and non-saturated regions of operation of an MOS transistor.

occur because the channel becomes pinched-off at the source end as well as the drain end.

The three regions of operation may be easily identified with the gate voltage necessary to place the transistor in that condition:

Cutoff	$\lvert V_G \rvert < \lvert V_T \rvert$ Gate voltage is insufficient to form any channel.
Saturation	$\lvert V_T \rvert \leq \lvert V_G \rvert \leq \lvert V_D + V_T \rvert$ Gate voltage is sufficient to form a channel at the source end but not at the drain end.
Nonsaturation	$\lvert V_D + V_T \rvert < \lvert V_G \rvert$ Gate voltage is large enough to form a channel everywhere.

The preceding discussion considered saturation in terms of the simplified theory. For the case of the complete theory in which the channel current is given by Equation 2-55, the dependence of the depletion layer charge (Q_B) on the channel voltage must be explicitly included. Accordingly, the substitution of Equation 2-48 into Equation 2-47 and then that result substituted into Equation 2-64 with $V = V_{DP}$ at pinch-off, gives

$$V_G - \phi_{MS} - \phi_B + \frac{Q_{ox}}{C_{ox}} + \frac{(2\epsilon_S q N_D[-V_{DP} - \phi_B])^{1/2}}{C_{ox}} - V_{DP} = 0 \tag{2-67}$$

which relates the drain and gate voltages at the edge of saturation.

Solving for V_{DP},

$$V_{DP} = V_G - \phi_{MS} - \phi_B + \frac{Q_{ox}}{C_{ox}} + \frac{\epsilon_S q N_D}{C_{ox}^2}\left\{\left[1 - \frac{2C_{ox}^2}{\epsilon_S q N_D}\left(V_G - \phi_{MS} + \frac{Q_{ox}}{C_{ox}}\right)\right]^{1/2} - 1\right\} \tag{2-68}$$

This value of channel voltage at the pinch-off point in Equation 2-55 will give the channel current in saturation. As in the case of the simplified theory, the transistor characteristics in saturation will be horizontal lines. However, pinch-off occurs at smaller values of drain voltage than in the simplified model. This is illustrated in Figure 2-17, in which the channel current curves have been terminated just at the pinch-off point as calculated from Equations 2-65 and 2-68 for the simplified and the complete transistor model.

2.3.6 Summary

Table 2-1 summarizes the important results from the previous discussions. The equations apply to both *P*-channel and *N*-channel transistors, enhancement and depletion mode, by taking the appropriate signs as indicated. All voltages are taken with their true sign and measured with respect to the substrate.

2.4 DEVICE PARAMETERS AND CHARACTERISTICS

2.4.1 Simple dc Model of the MOS Transistor

In the preceding sections, the basic MOS transistor equations were developed by considering the physical operation of the device from the standpoint of conventional semiconductor theory. In this section, we will study the characteristics of *P*-channel MOS transistors and related structures from a circuit design standpoint, considering the device parameters which are important in the design of an MOS integrated circuit.

The starting point for this analysis is the basic Sah model[9] of the MOS transistor derived in the previous section. Equations 2-62 and 2-66 give the channel current in the nonsaturated and saturated regions, respectively. These are rewritten below for a *P*-channel transistor in a format commonly used for circuit design. With the source as the voltage reference node and positive current flowing into a device terminal, the drain to source current (I_{DS}) is the negative of the channel current (I_C). Also, using $C_{ox} = \epsilon_{ox}/t_{ox}$, the transistor current equations become

Nonsaturation Region $\quad |V_{GS} - V_T| > |V_{DS}|$

$$I_{DS} = -\left(\frac{\mu_p \epsilon_{ox}}{2t_{ox}}\right)\left(\frac{W}{\ell}\right)[2(V_{GS} - V_T)\, V_{DS} - (V_{DS})^2] \qquad (2\text{-}69)$$

Saturation Region $\quad |V_{GS} - V_T| \leqq |V_{DS}|$

$$I_{DS} = -\left(\frac{\mu_p \epsilon_{ox}}{2t_{ox}}\right)\left(\frac{W}{\ell}\right)(V_{GS} - V_T)^2 \qquad (2\text{-}70)$$

where μ_p = average surface mobility of holes in the channel
t_{ox} = thickness of the oxide over the channel
ϵ_{ox} = permittivity of the oxide
ℓ = length of the channel in the direction of current flow
W = width of the channel
V_{GS} = gate to source voltage
V_{DS} = drain to source voltage
V_T = threshold voltage

To simplify these equations, it is convenient to define two new parameters, k' and k.*

*The terminology k and k' arose with the first commercial development of MOS integrated circuits. Although these terms may perhaps not be ideal, they are in sufficiently common use that it seems reasonable to retain them here.

Table 2–1 Summary of Important MOS Equations for Both *N*- and *P*-Channel Transistors

	P-Channel	*N-Channel*
Substrate doping type	N	P
Terminal polarities (V_G, V_D, V_S) in normal operation	−	+
V_T Threshold voltage for enhancement mode transistors	−	+
Q_B Bulk charge in depletion region	+	−
Q_{ox} Effective oxide and interface charge	Almost always +	Almost always +
ϕ_F Potential in bulk of intrinsic level with respect to Fermi level	+	−
ϕ_B Band bending at beginning of inversion ($\approx -2\phi_F$)	−	+
ϕ_{MS} Metal (Aluminum) to silicon work function	−	−
μ Effective surface mobility	$\mu_p = +$	$\mu_n = -$
N Substrate doping concentration	N_D	N_A

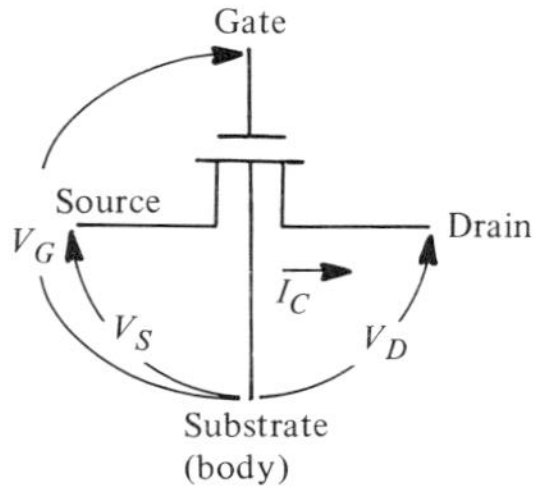

Threshold voltage $\quad Q_B = \pm(\mp 2\,\epsilon_s qN[V + \phi_B])^{1/2}$ $\quad V$ = channel to substrate voltage ($V < 0$ for *P*-channel, $V > 0$ for *N*-channel)

Bulk depletion region charge* $\quad V_T = \phi_{MS} + \phi_B - \dfrac{Q_{ox}}{C_{ox}} - \dfrac{Q_B}{C_{ox}}$

Nonsaturated, channel current

Complete expression*

$$I_C = \mu C_{ox}\frac{W}{\ell}\left\{\left(V_G - \phi_{MS} - \phi_B + \frac{Q_{ox}}{C_{ox}}\right)(V_D - V_S) - \frac{1}{2}(V_D^2 - V_S^2) - \frac{2(2\,\epsilon_s qN)^{1/2}}{3\,C_{ox}}\left([\mp V_D \mp \phi_B]^{3/2} - [\mp V_S \mp \phi_B]^{3/2}\right)\right\}$$

Simplified expression, Source at ground

$$I_C = \mu C_{ox}\frac{W}{\ell}\left\{(V_G - V_T)V_D - \frac{1}{2}V_D^2\right\}$$

Saturation, drain pinch-off voltage

Complete expression*

$$V_{DP} = V_G - \phi_{MS} - \phi_B + \frac{Q_{ox}}{C_{ox}} \pm \frac{\epsilon_s qN}{C_{ox}^2}\left\{\left[1 \mp \frac{2\,C_{ox}^2}{\epsilon_s qN}\left(V_G - \phi_{MS} + \frac{Q_{ox}}{C_{ox}}\right)\right]^{\frac{1}{2}} - 1\right\}$$

Simplified expression

$$V_{DP} = V_G - V_T$$

Saturation, channel current

$$I_{CP} = \frac{1}{2}\mu C_{ox}\frac{W}{\ell}(V_G - V_T)^2$$

*Note: Use upper sign for *P*-channel transistors, lower sign for *N*-channel.

$$k' = \frac{\mu_p \epsilon_{ox}}{2t_{ox}} \tag{2-71}$$

$$k = k' \frac{W}{\ell} \tag{2-72}$$

The drain current expressions then become

Nonsaturation Region $\quad |V_{GS} - V_T| > |V_{DS}|$

$$I_{DS} = -k[2(V_{GS} - V_T)\, V_{DS} - (V_{DS})^2] \tag{2-73}$$

and

Saturation Region $\quad |V_{GS} - V_T| \leq |V_{DS}|$

$$I_{DS} = -k(V_{GS} - V_T)^2 \tag{2-74}$$

The parameter k' is not a design variable, since it is fixed by the process. The parameter k, however, is a design variable that is dependent upon device geometry. The values of the parameters V_T and k' will depend on the manufacturing process. Two versions of the basic P-channel MOS process are in common use, the high (or standard) threshold process and the low threshold version. Typical values of the process parameters are:

	High Threshold	*Low Threshold*
V_T(volts)	−4.0	−2.0
$k'(A/\text{volt}^2)$	2×10^{-6}	2×10^{-6}

By using these values in Equations 2-73 and 2-74, the transistor drain characteristics can be calculated. The fact that the characteristics of the device depend upon the ratio of W/ℓ is of major importance in the design of MOS circuits.

The dimension ℓ in these equations is the electrical effective channel length, which is the distance between the source and drain P-regions. Because the P-region diffusion proceeds laterally as well as vertically, this distance is less than the masking dimension (L) which is the dimension designed into the photomask. These are illustrated in Figure 2-19. The amount of lateral diffusion is usually taken as being equal to the junction depth X_j, giving

$$\ell = L - 2X_j \tag{2-75}$$

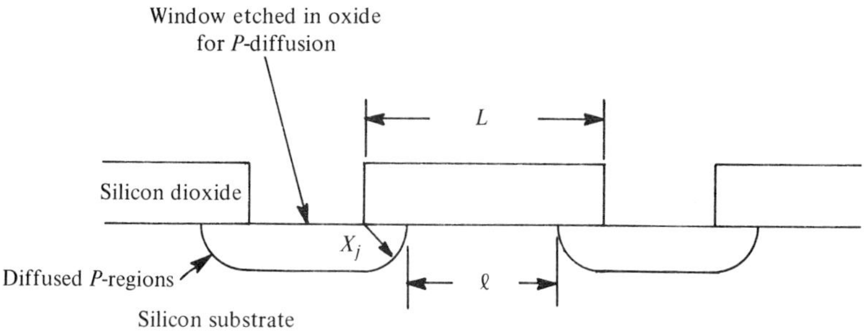

Figure 2–19 MOS transistor after *P*-region diffusion step, illustrating channel length dimensions.

2.4.2 MOS Transistor Drain Characteristic Curves

Figures 2-20 and 2-21 illustrate some conventional *P*-channel MOS transistor drain characteristics for several geometries and for both a standard threshold (Figure 2-20) and a low threshold voltage (Figure 2-21) process. In these figures, the substrate is connected to the source.

The transistor with a $W = 20$ mils and $\ell = 3.8$ mils, (Figures 2-20a and 2-21a), is a test structure which is sufficiently large that second-order effects are small. The depletion layer widening is negligible and the characteristics in the saturated region are almost perfectly horizontal. By way of contrast, note the appreciable slope of the saturated region curves for the transistors with $\ell = 0.16$ mil (Figures 2-20b and 2-21b). This is caused by the decrease in channel length as more negative drain voltages widen the depletion layer around the drain, and is very similar to base-width modulation (Early effect) in a bipolar transistor.

Figures 2-20b and 2-21b are curves for transistors with a $W = 2.0$ mils and $\ell = 0.16$ mil which are typical geometries of transistors used as inverters in many complex circuits. Figures 2-20c and 2-21c show characteristics of transistors with a $W = 0.4$ mil and $\ell = 1.8$ mils which are typical values for a fairly low current device often used as a pull-up or load transistor.

Figure 2-22 is the same transistor as in Figure 2-20b ($W = 2.0$ mils and $\ell = 0.16$ mil) with the drain voltage increased to show the breakdown characteristics.

Figure 2-23 shows how the drain characteristics of a typical inverter transistor change with temperature. This transistor is the same one as in Figure 2-21b ($W = 2.0$ mils and $\ell = 0.16$ mil). These temperature variations can be described in terms of the temperature variation of V_T and k'.

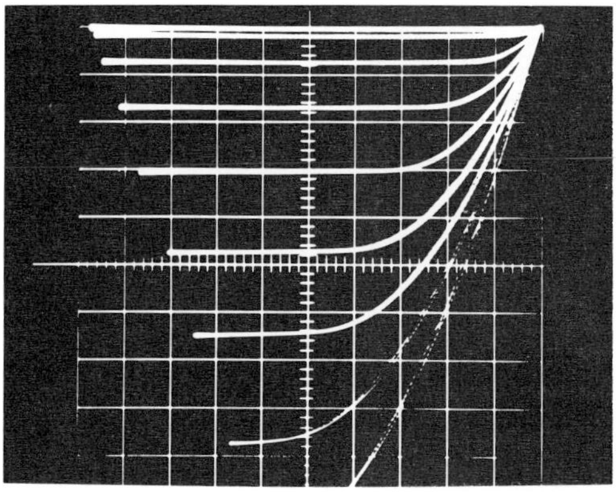

Large test transistor

W = 20 mils, ℓ = 3.8 mils

$\frac{W}{\ell} = 5.3$

Vert. I_{DS} = 200 μA/div

Horiz. V_{DS} = 2V/div

Step V_{GS} = 2V/step

Last step = -20 V

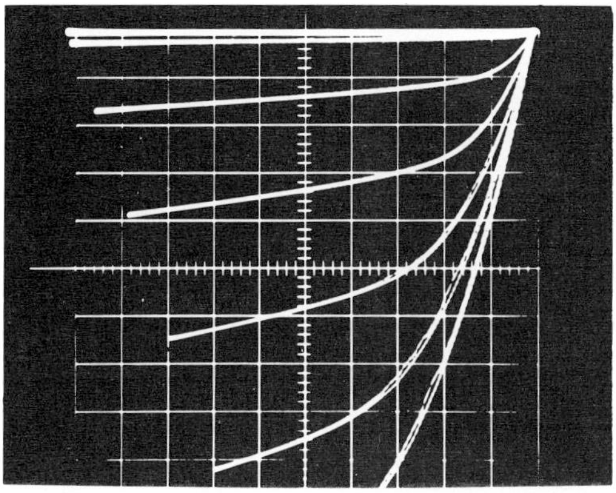

Typical inverter transistor

W = 2.0 mils, ℓ = 0.16 mil

$\frac{W}{\ell} = 12.5$

Vert. I_{DS} = 200 μA/div

Horiz. V_{DS} = 2V/div

Step V_{GS} = 2V/step

Last step = -16 V

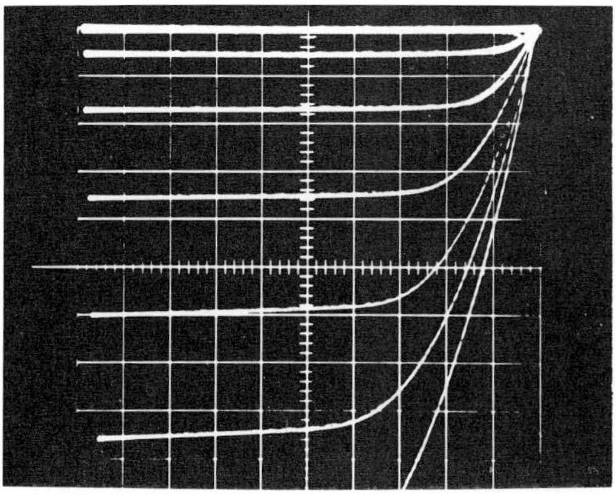

Typical pull-up transistor

W = 0.4 mil, ℓ = 1.8 mils

$\frac{W}{\ell} = 0.22$

Vert. I_{DS} = 5 μA/div

Horiz. V_{DS} = 2V/div

Step V_{GS} = 2V/step

Last step = -18 V

Figure 2–20 Typical drain characteristics of *P*-MOS transistors—standard threshold voltage process.

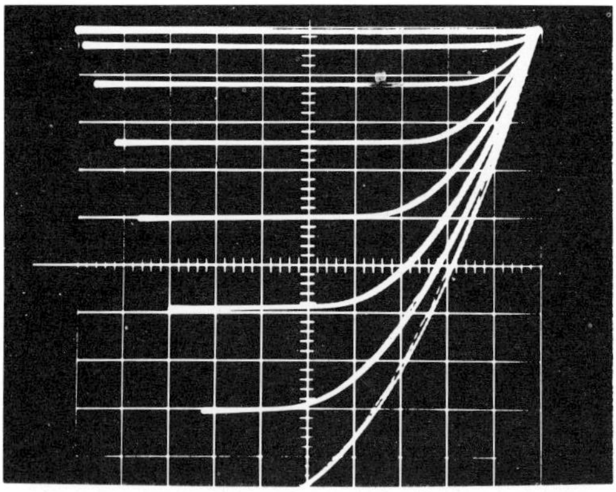

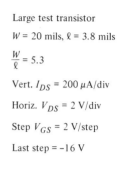

Large test transistor

W = 20 mils, ℓ = 3.8 mils

$\frac{W}{\ell} = 5.3$

Vert. I_{DS} = 200 μA/div

Horiz. V_{DS} = 2 V/div

Step V_{GS} = 2 V/step

Last step = −16 V

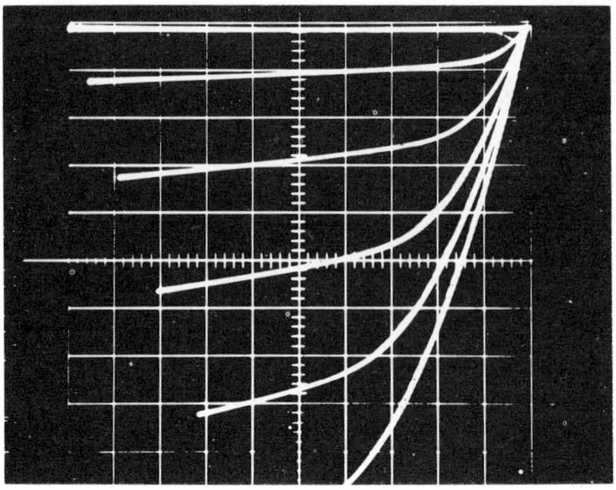

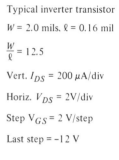

Typical inverter transistor

W = 2.0 mils, ℓ = 0.16 mil

$\frac{W}{\ell} = 12.5$

Vert. I_{DS} = 200 μA/div

Horiz. V_{DS} = 2V/div

Step V_{GS} = 2 V/step

Last step = −12 V

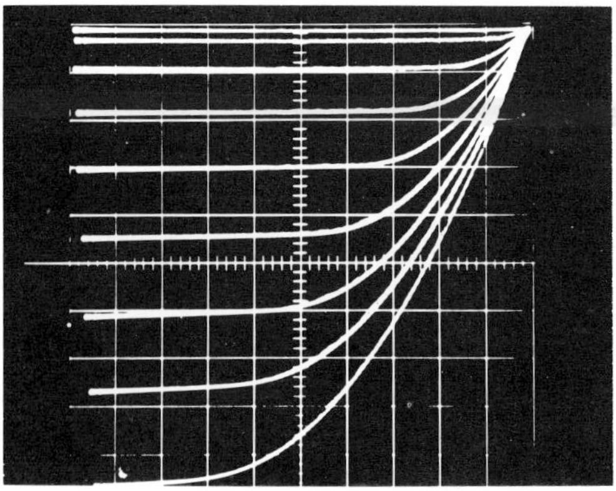

Typical pull-up transistor

W = 0.4 mil, ℓ = 1.8 mils

$\frac{W}{\ell} = 0.22$

Vert. I_{DS} = 10 μA/div

Horiz. V_{DS} = 2 V/div

Step V_{GS} = 2 V/step

Last step = −18 V

Figure 2–21 Typical drain characteristics of *P*-MOS transistors—Low threshold voltage process.

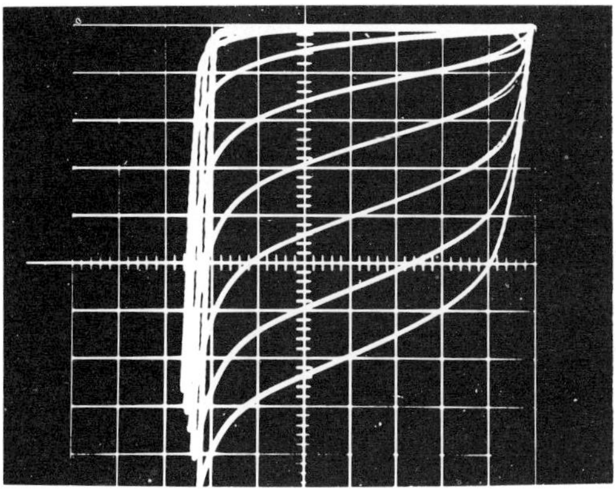

Vert. I_{DS} = 500 μA/div

Horiz. V_{DS} = 10 V/div

Step V_{GS} = 2 V/step

Last step = −18 V

Figure 2–22 Typical drain characteristics to breakdown for inverter transistor in Figure 2–20b.

2.4.3 Conduction Factor (*k*′)

From the theory previously developed, the dc model for the MOS transistor divides the operating characteristics into the nonsaturation and the saturation regions. The nonsaturation region is also called the linear or the triode region and represents those combinations of voltages and currents for which a continuous resistive channel exists between the source and drain. The saturation region is also called the pinch-off region, inasmuch as the channel conductivity is zero near the drain.

By combining Equation 2-72 into Equations 2-73 and 2-74, the drain current in each region can be written as

Nonsaturation Region $|V_{GS} - V_T| > |V_{DS}|$

$$I_{DS} = -k' \frac{W}{\ell} [2(V_{GS} - V_T)\, V_{DS} - (V_{DS})^2] \qquad (2\text{-}76)$$

Saturation Region $|V_{GS} - V_T| \leqq |V_{DS}|$

$$I_{DS} = -k' \frac{W}{\ell} (V_{GS} - V_T)^2 \qquad (2\text{-}77)$$

In these equations, it is important to remember that the true threshold voltage is the voltage which just begins to form a channel. By conven-

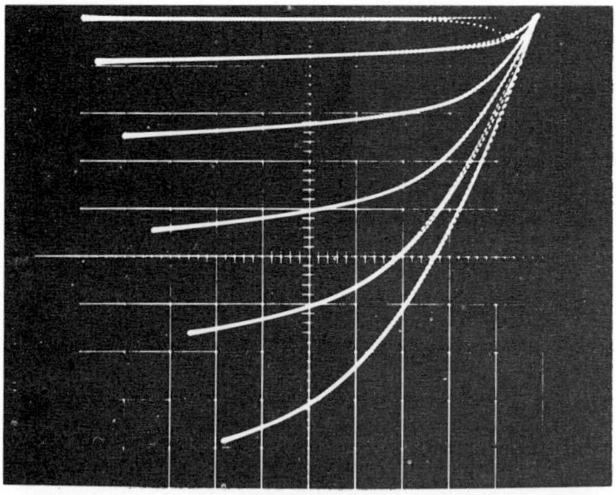

Inverter transistor

$W = 2.0$ mils, $\ell = 0.16$ mil

Low threshold voltage process

Vert. $I_{DS} = 200\ \mu A/div$

Horiz. $V_{DS} = 2\ V/div$

Step $V_{GS} = 2\ V/step$

Last step = -12 V

Same scale for all curves.

$T = +125°C$

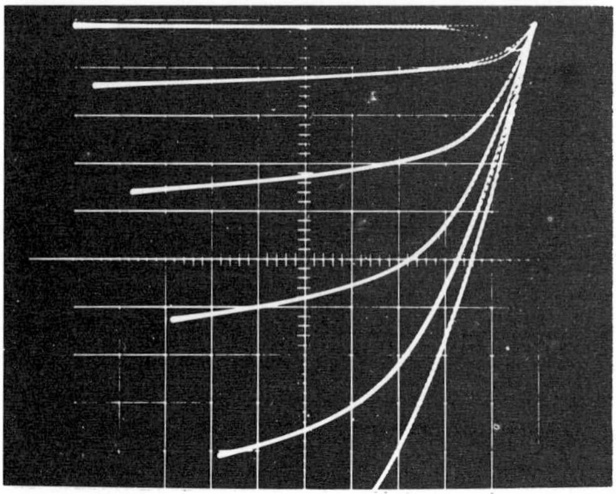

$T = +25°C$

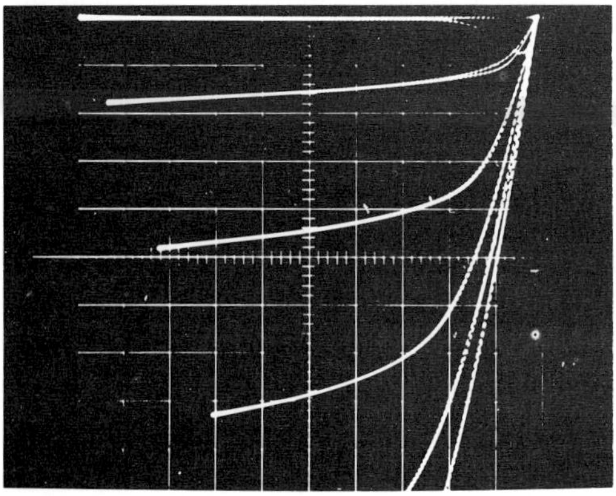

$T = -55°C$

Figure 2–23 Typical drain characteristics at various temperatures for inverter transistor in Figure 2–21b.

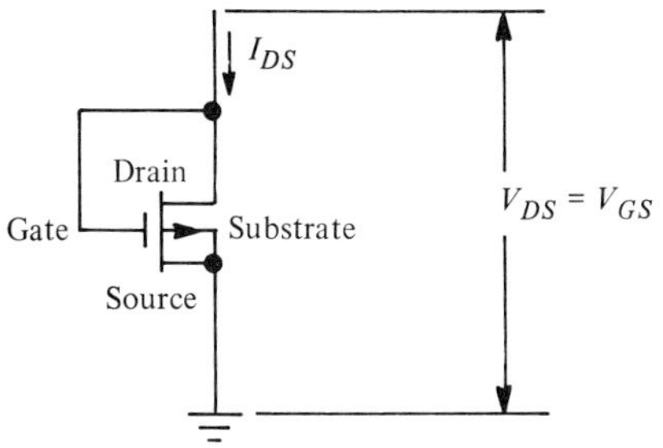

Figure 2–24 Two-terminal connection for saturated operation.

tional usage on most data sheets V_T is the voltage which gives some finite drain current (usually 1 μA). The importance of this distinction is that V_T is a true process parameter and is independent of the geometry of a transistor, whereas V_T at 1 μA depends somewhat on the geometry. For a transistor with $W = 2.0$ mils and $\ell = 0.16$ mil, the true V_T is about 0.2 V less than V_T at 1 μA. The difference increases as W/ℓ decreases. The correct parameter to use in the equations is the true threshold voltage (V_T), although the quantity V_T at 1 μA as given on data sheets is frequently used.

The simplest arrangement for measuring an MOS transistor is to connect the gate with the drain, which makes $V_{GS} = V_{DS}$ and places the transistor in the saturation region (see Figure 2-24). This is a connection commonly used in production measurements on test transistors. When the square root is taken of both sides of Equation 2-77, a plot of $\sqrt{|I_{DS}|}$ gives a straight line with a slope of $(k'(W/\ell))^{1/2}$ which crosses the horizontal axis at V_T.

$$-\sqrt{|I_{DS}|} = \sqrt{k' \frac{W}{\ell}}\,(V_{GS} - V_T) = \sqrt{k}(V_{GS} - V_T) \qquad \text{(2-78)}$$

Such curves for a typical device from a standard and a low threshold process are shown in Figure 2-25. These curves are from a transistor of $W = 2.0$ mils and $\ell = 0.16$ mil, which are typical dimensions for test transistors in production circuits, as well as for inverter transistors. The quantity k can be determined by rewriting Equation 2-78 for points P_1 and P_2 (see Figure 2-26). Leaving off the absolute value signs and taking all quantities as magnitudes

$$\sqrt{I_{DS1}} = \sqrt{k}(V_{GS1} - V_T) \qquad \text{(2-78a)}$$

and

$$\sqrt{I_{DS2}} = \sqrt{k}(V_{GS2} - V_T) \qquad \text{(2-78b)}$$

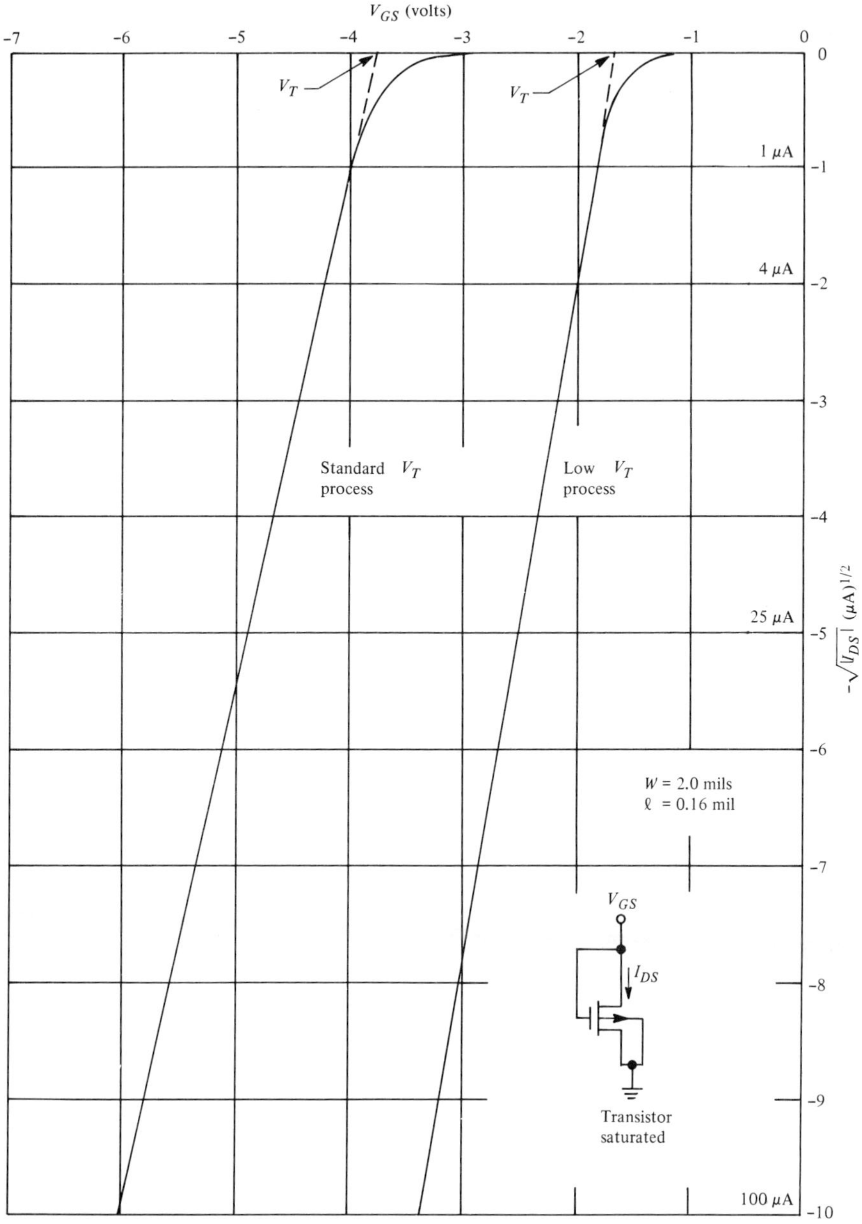

Figure 2–25 Typical curves of $\sqrt{I_{DS}}$ vs. V_{GS} for saturated operation.

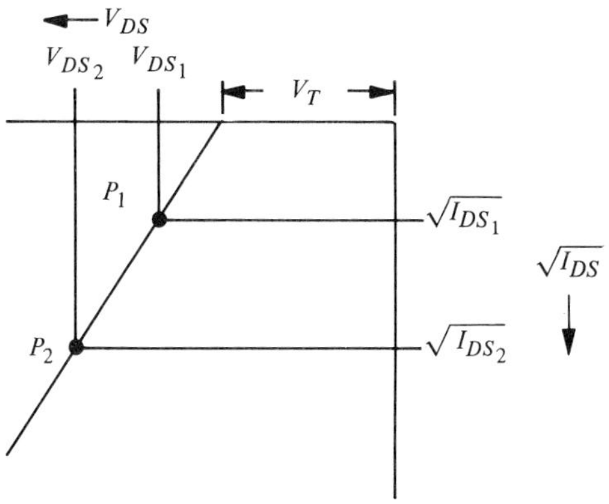

Figure 2–26 Determination of *k*.

Subtracting Equation 2-78a from Equation 2-78b and solving for k gives

$$k = \left(\frac{\sqrt{I_{DS2}} - \sqrt{I_{DS1}}}{V_{GS2} - V_{GS1}}\right)^2 \tag{2-79}$$

Thus the gain parameter k may be calculated from measurements on a saturated transistor at two values of drain current. The conduction factor k' is then obtained from the geometry of the device by Equation 2-72.

Also k may be obtained directly from the slope of the $\sqrt{I_{DS}}$ vs. V_{GS} curves of Figure 2-25 by differentiating Equation 2-78,

$$\frac{d(-\sqrt{I_{DS}})}{dV_{GS}} = \sqrt{k} \tag{2-80}$$

This shows that the parameter k is just the square of the slope of Figure 2-25.

Equation 2-78 is useful for studying device characteristics in the saturation region. In the nonsaturation region, curves of channel conductance versus gate voltage are useful. We define the large signal channel conductance as $G = I_{DS}/V_{DS}$, and from Equation 2-76:

$$G = \frac{I_{DS}}{V_{DS}} = -k' \frac{W}{\ell} [2(V_{GS} - V_T) - V_{DS}]$$

$$= -2k' \frac{W}{\ell} \left[V_{GS} - V_T - \frac{1}{2} V_{DS}\right]$$

As V_{DS} approaches zero, the conductivity (G_0) is that at the origin of the drain characteristic curves:

$$G_0 = \left.\frac{V_{DS}}{I_{DS}}\right|_{V_{DS}\to 0} = -2k'\frac{W}{\ell}(V_{GS} - V_T) \tag{2-81}$$

Typical curves of G_0 vs. V_{GS} are shown in Figure 2-27. These curves were obtained in a circuit which holds $V_{DS} = 50$ mV and uses I_{DS} as a measure of G_0. In this type of measurement, the allowable value of V_{DS} is set by considering the complete device model discussed in Section 2.3.4. A value of $|V_{DS}| << |\phi_B|$ is used to avoid appreciable errors that may arise from the 3/2 power terms in Equation 2-55. 50 mV is commonly used.

Differentiating Equation 2-81,

$$\frac{dG_0}{dV_{GS}} = -2k'\frac{W}{\ell} = -2k \tag{2-82}$$

shows that the conduction factors k' and k can also be obtained from curves like Figure 2-27. Thus k is simply one-half the slope of the G_0 versus V_{GS} curve.

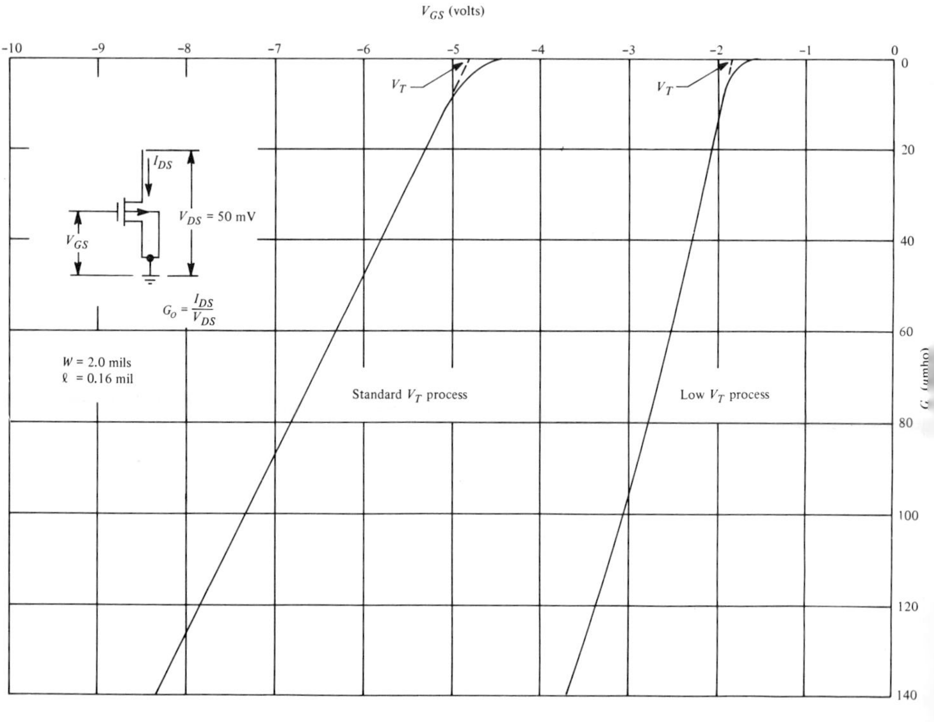

Figure 2–27 Typical curves of channel conductance (G_o) for unsaturated operation.

When G_0 vs. V_{GS} curves are plotted for large gate voltages, curvature away from a straight line is sometimes appreciable, since mobility (and thus k') decreases with a large gate electric field. When this happens, the slope is usually measured in the lower voltage region where the curves are fairly linear.

Figure 2-28 shows how some G_0 vs. V_{GS} curves change with temperature. Note that the true threshold voltage (V_T) decreases with increasing temperature and that the channel conduction factor (which is proportional to the slope of the curves) also decreases. The decreasing conductance corresponds to an increasing channel resistance; thus, there is one operating point where the two temperature variations cancel out. It is difficult to take advantage of this in the design of digital circuits, but designers should note that when the threshold voltage is measured at some finite current, as is usually done, this temperature cancelling effect will make the apparent threshold voltage variation with temperature appear to be less than it actually is.

From the slopes of curves such as those in Figure 2-28, the temperature variation of k' is obtained. The results for several geometries in both a standard voltage and a low voltage process are shown in Figure 2-29

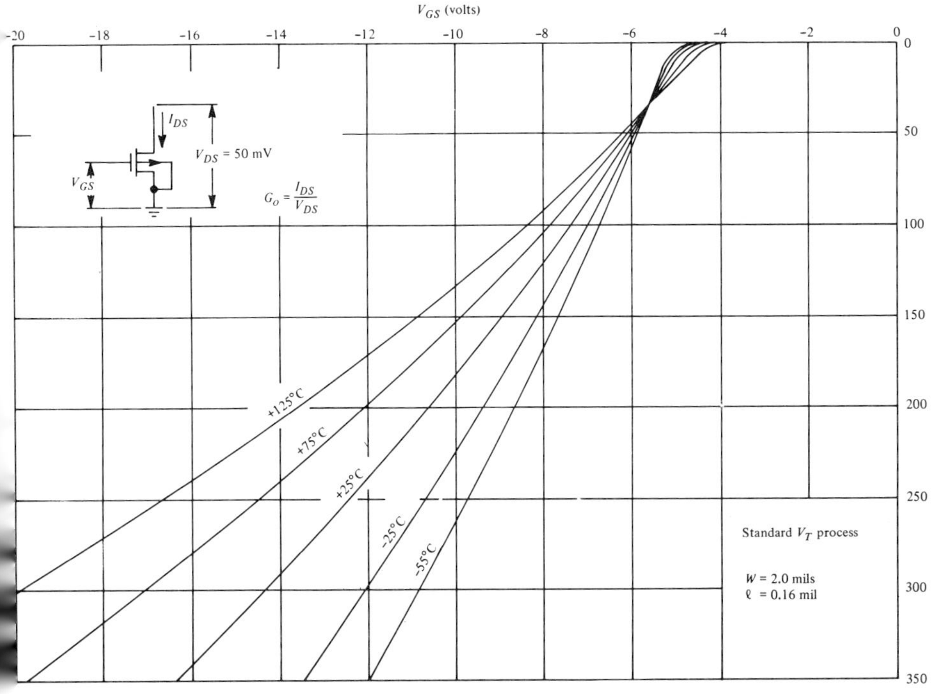

Figure 2–28 Typical channel conductance curves at various temperatures.

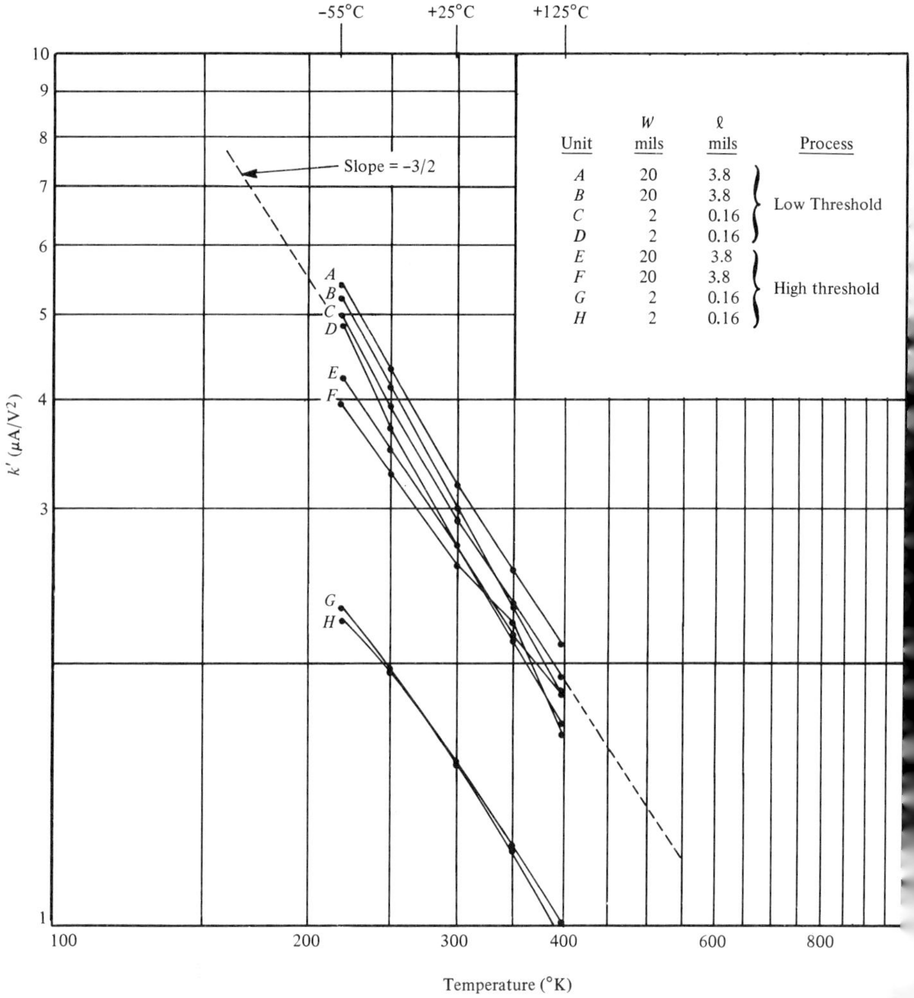

Figure 2–29 *k′* variation with temperature.

which plots k' as a function of absolute temperature on logarithmic scales.

Also indicated in Figure 2-29 is a straight line with a slope of $-3/2$, from which it is apparent that the experimental data have the same slope. Thus k' varies as the inverse 3/2 power of the absolute temperature. From the definition of k' (Equation 2-71), the average hole mobility (μ_p) also has the same temperature variation. This is in agreement with the theory for mobility due to lattice scattering which predicts a lattice mobility (μ_L) that varies as $T^{-3/2}$ (See Section 13-4 of Ref. 17).

Figure 2-29 shows that this theoretical temperature variation is an excellent fit and that the temperature variation of k' is given by

$$\frac{k'}{k_0'} = \left(\frac{T}{T_0}\right)^{-3/2} \tag{2-83}$$

where k_0' = value of k' at room temperature,
T = absolute temperature (°K),
T_0 = room temperature (298°K).

Equation 2-83 is plotted in Figure 2-30 to scales that are somewhat more convenient for design purposes.

2.4.4 Threshold Voltage (V_T)

The same G_0 vs. V_{GS} curves (Figure 2-28) used to obtain the variation of k' with temperature can be used to find how the threshold voltage (V_T) varies with temperature. To do this, the G_0 vs. V_{GS} curves are extrapolated linearly back to the V_{GS} axis ($G_0 = 0$) to obtain V_T. The results are plotted

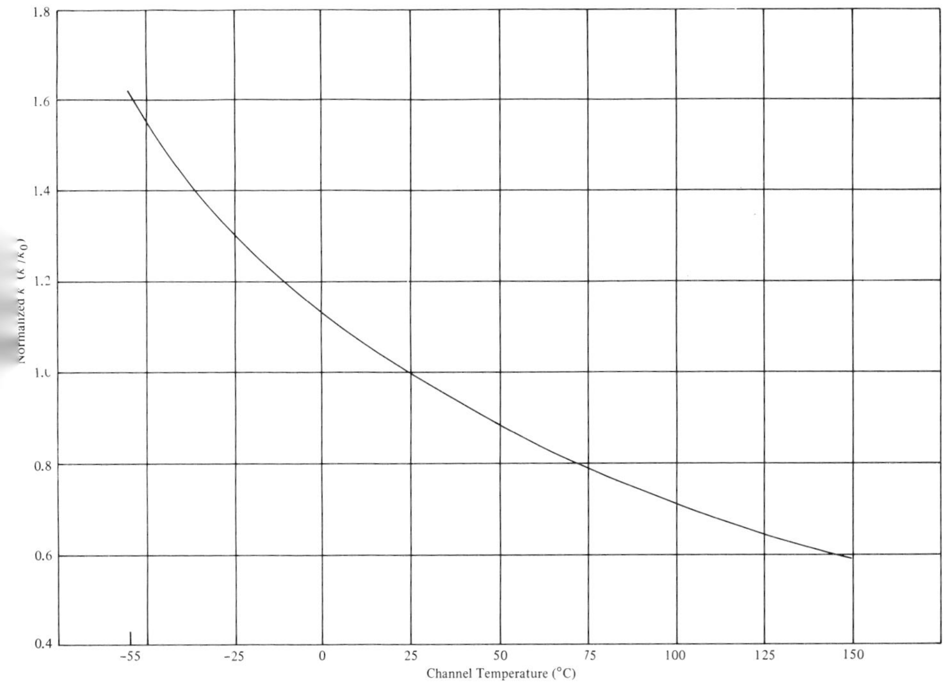

Figure 2–30 Normalized conduction factor (k') as a function of temperature.

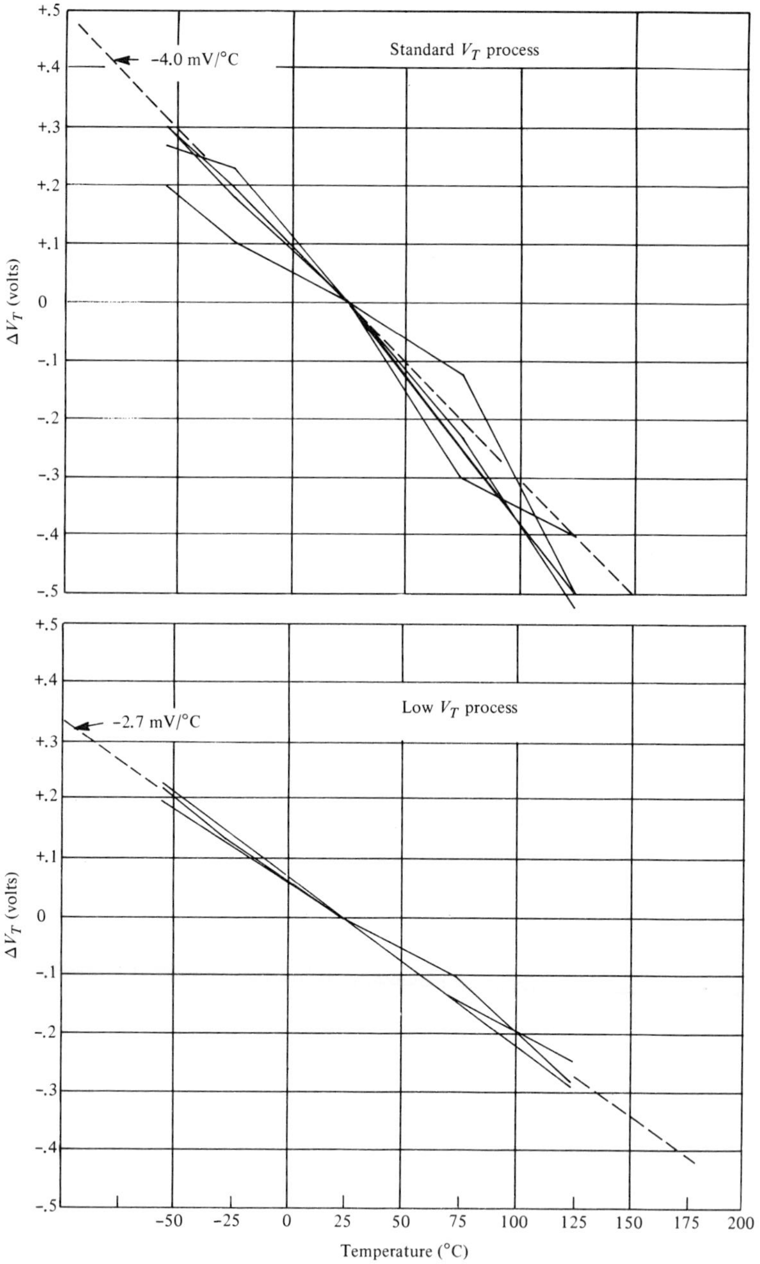

Figure 2–31 Typical variation of threshold voltage (V_T) with temperature.

in Figure 2-31 for the same eight transistors of various geometries and processes that were used in Figure 2-29. Also indicated in Figure 2-31 are the average temperature coefficients for these two processes. These values are typical of many MOS processes, but for exacting designs, values of the threshold voltage temperature coefficients should be obtained from the manufacturer.

As mentioned in Section 2.4.3, the temperature variation of V_T measured at some finite current will appear to be somewhat less than the value shown in Figure 2-31 because of the cancelling effect of the k' temperature variation.

There is one other characteristic of MOS transistors which may, in some designs, be important in connection with the threshold voltage. This is the fact that the transistor turn-on at threshold is not completely sharp, but is rather a smooth transition between the nonconducting and conducting states. Figures 2-25 and 2-27 show this transition clearly.

In the derivation of the MOS transistor equations, the strong inversion approximation (see Section 2.3.3) is used, in which a conducting channel is considered to be formed when the silicon surface is as strongly inverted, *P*-type in the present case, as the bulk material was *N*-type. No channel is assumed to exist at any smaller gate voltage. This is valid after turn-on, but is not strictly correct just below threshold. Channel formation is actually a gradual process.

For most ratio type digital circuits, this gradual channel formation is not important since the currents involved are quite small. However, for ratioless circuits or when a voltage must be stored for a long period of time on a drain node, these small currents may become important. In such cases, Figure 2-32 may be used to estimate the magnitude of the current in a "partially formed" channel. These curves are for the same two transistors as in Figure 2-25 and are typical for the very low current region.

2.4.5 Small Signal MOS Transistor Parameters

The small signal transistor parameters are defined in terms of partial derivatives of the terminal voltages and currents. Since MOS transistor characteristic curves somewhat resemble those of pentode vacuum tubes, the same terminology has been retained for conventional usage. The important parameters are defined below. These definitions are valid in all regions of operation; however, they are mainly useful in the saturated region, and most of the discussion will be restricted to that region.

The mutual transconductance (g_m) describes a change of I_{DS} for a change of V_{GS} with constant V_{DS}.

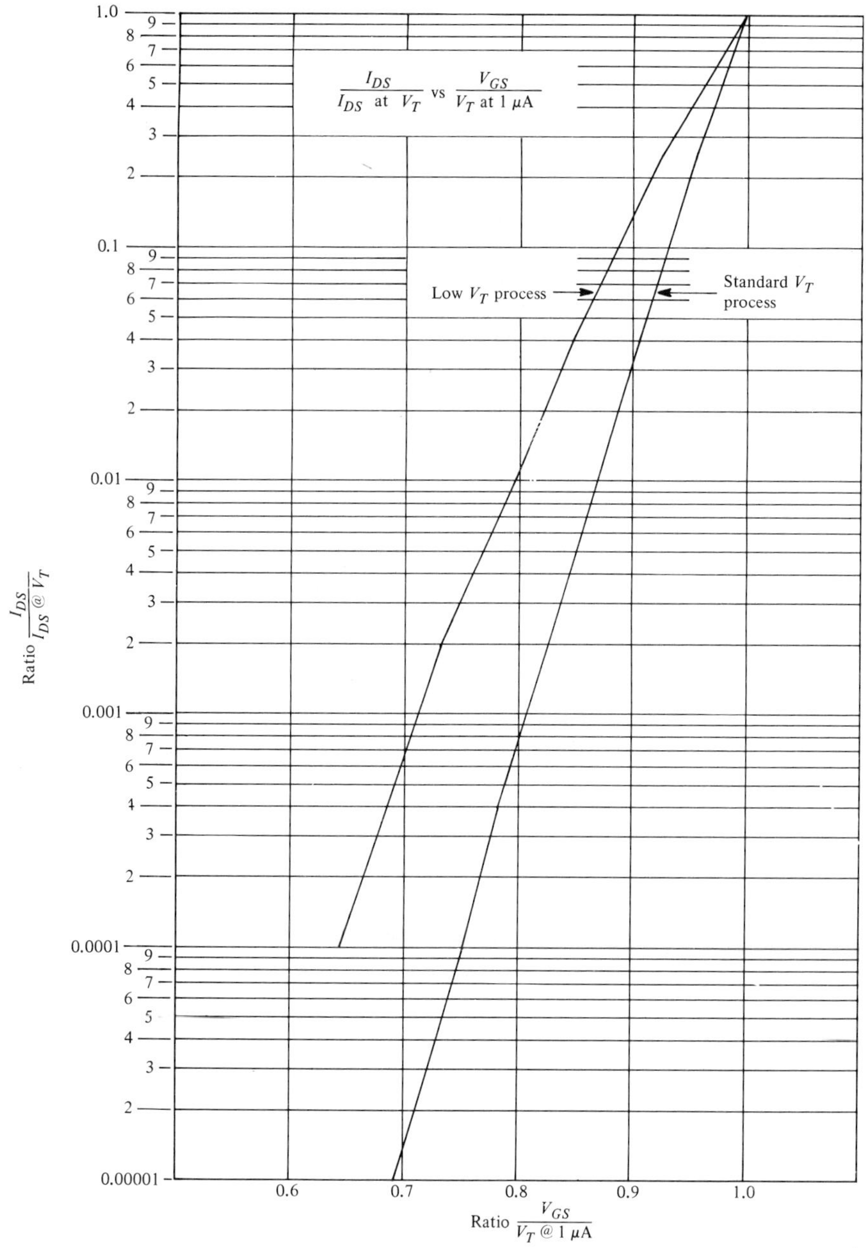

Figure 2–32 Current in a partially formed transistor channel.

$$g_m = \frac{\partial I_{DS}}{\partial V_{GS}} \tag{2-84}$$

g_m can be determined graphically from the drain characteristics in the saturation region as shown in Figure 2-33. Note that in a real transistor g_m is dependent on the choice of both the V_{DS} and I_{DS} levels.

The g_m in the region illustrated in Figure 2-33 can be calculated as follows:

$$g_m = \frac{I_{DS2} - I_{DS1}}{V_{GS2} - V_{GS1}} = \frac{670 - 480}{9 - 8} = 190\ \mu\text{mhos}$$

Applying the definition of g_m to the drain current equation in saturation (Equation 2-74) gives

$$g_m = \frac{\partial I_{DS}}{\partial V_{GS}} = \frac{\partial}{\partial V_{GS}}\,[-k(V_{GS} - V_T)^2] \tag{2-85}$$

$$g_m = -2k(V_{GS} - V_T) \tag{2-86}$$

Combining this with Equation 2-74 gives, finally,

$$g_m = 2k\sqrt{\frac{-I_{DS}}{k}} = 2\sqrt{-kI_{DS}} \tag{2-87}$$

The negative sign results from the fact that I_{DS} is negative for a P-channel transistor.

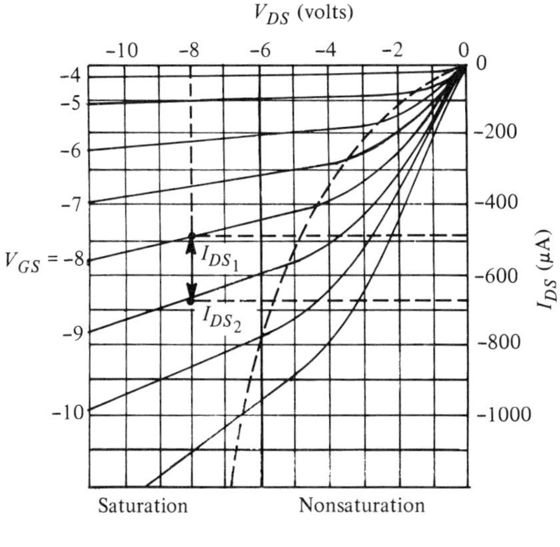

Figure 2–33 Graphical determination of g_m.

The dynamic drain resistance (r_d) relates a change of V_{DS} to a change of I_{DS} for a constant value of V_{GS}.

$$r_d = \frac{\partial V_{DS}}{\partial I_{DS}} \tag{2-88}$$

r_d can be determined graphically from the drain characteristics in the saturation region as shown in Figure 2-34.

For this example:

$$r_d = \frac{V_{DS2} - V_{DS1}}{I_{DS2} - I_{DS1}} = \frac{9 - 7}{520 - 475} = \frac{2V}{45\ \mu\text{A}} = 44.5\ \text{k}\Omega$$

For typical devices, r_d can range from 10 to 500 kΩ. Infinite values for r_d would be expected from the theoretical expression since the drain current is independent of drain voltage in the saturation region.

The mechanism that gives rise to the finite drain resistance is generally channel length modulation as discussed by Crawford[18] (Section 2-2). As the drain voltage is made more negative, the depletion layer around the drain widens (i.e., extends farther toward the source), resulting in a shorter channel length (ℓ). From the definition of the transistor conduction factor $k = k'\ W/\ell$, we see that a decrease in ℓ will increase k, giving rise to an increased current.

When MOS transistors are fabricated on high resistivity substrates, an additional mechanism may contribute to the finite drain resistance. This

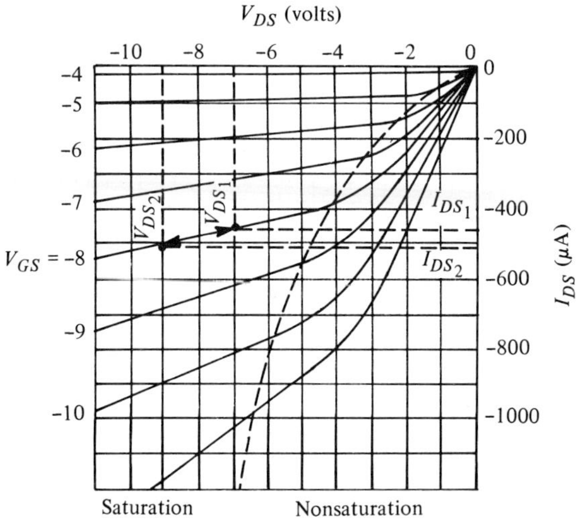

Figure 2–34 Graphical determination of dynamic drain resistance (r_d).

is an electrostatic feedback from the drain through the substrate to the back side of the channel, and is analyzed in Wallmark and Johnson[19] (Section 5-2).

One other small signal parameter that may be defined for the saturated transistor is the amplification factor μ which relates a change of V_{DS} to a change of V_{GS} for a constant value of I_{DS}.

$$\mu = \frac{\partial V_{DS}}{\partial V_{GS}} \tag{2-89}$$

This may be determined graphically in a manner similar to the above examples, or from the relation

$$\mu = g_m r_d \tag{2-90}$$

A parameter of considerable interest in switching applications is the dynamic on-resistance (r_{on}) of the device when it is turned on and operating in the nonsaturation region.

$$r_{\text{on}} = \frac{\partial V_{DS}}{\partial I_{DS}} \qquad \text{nonsaturation} \tag{2-91}$$

This parameter can be evaluated from Equation 2-73 for the nonsaturation region,

$$r_{\text{on}} = \frac{1}{\dfrac{\partial I_{DS}}{\partial V_{DS}}} = \frac{1}{\dfrac{\partial}{V_{DS}}\{-k[2(V_{GS} - V_T)\, V_{DS} - (V_{DS})^2]\}} \tag{2-92}$$

$$r_{\text{on}} = -\frac{1}{2k(V_{GS} - V_T - V_{DS})} \tag{2-93}$$

This expression is valid anywhere in the nonsaturation region, the negative sign again reflecting the fact that in a *P*-channel transistor the terminal voltages and currents are negative. Since the drain characteristic curves are linear near the origin, a commonly specified quantity is the channel on resistance (R_0) at the origin ($V_{DS} = 0$).

$$R_0 = -\frac{1}{2k(V_{GS} - V_T)} \tag{2-94}$$

By substituting Equation 2-86 into Equation 2-94,

$$R_0 = \frac{1}{g_m} \tag{2-95}$$

Equation 2-95 indicates that for a constant value of V_{GS}, R_0 in the nonsaturation region is the reciprocal of g_m in the saturation region.

2.4.6 Body Effect

The commonly used term "body effect" refers to the changes in the transistor characteristics when a bias voltage is applied between substrate (body) and source. This effect is of considerable importance in many circuit applications of MOS devices. In all the previous discussions, the substrate is connected to the source and the combination is considered as ground potential. When the source and substrate are at different potentials (substrate reverse biased with respect to the source), the depletion region between the channel and substrate widens and contains more charge; consequently, the gate electric field necessary to form a channel is increased. Thus the apparent threshold voltage of the transistor is increased.

Figure 2-35 shows how a typical G_0 vs. V_{GS} curve is affected by the substrate or body to source voltage. The translation of the curves with

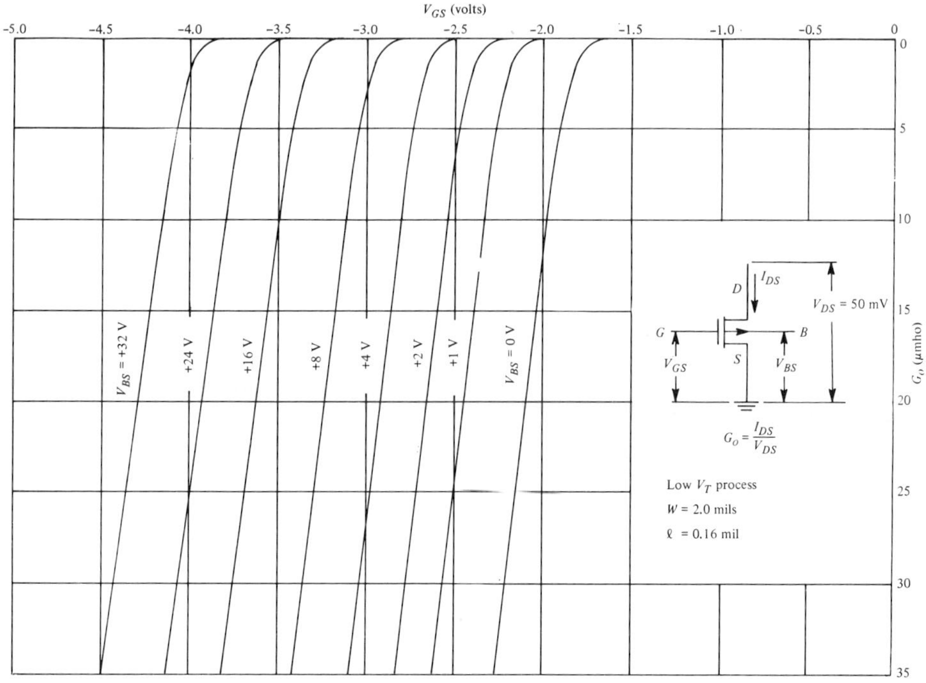

Figure 2–35 Effect of substrate bias on a typical curve of channel conductivity vs. gate voltage.

no change in slope indicates that k' does not change; only the apparent V_T changes. As can be seen from Figure 2-35, increasingly positive body to source voltages significantly change the turn-on voltage.

The results of measurements on several geometries are shown in Figures 2-36a and 2-36b for a standard and a low voltage process. These curves include data obtained from G_0 vs. V_{GS} curves, as in Figure 2-35, and also measurements of conventional V_T at 1 μA. ΔV_T is the change in threshold voltage due to the substrate bias. The apparent or effective threshold voltage then is

$$V_T(\text{effective}) = V_T(V_{BS} = 0) + \Delta V_T \tag{2-96}$$

One common engineering design approximation used to describe the body effect is given by

$$\Delta V_T = -\frac{1}{2}\sqrt{V_{BS}} \tag{2-97}$$

This is also plotted in Figures 2-36a and 2-36b, showing that Equation 2-97 gives a fair average value for the body effect in these devices, but that there is considerable variation with geometry. The main utility of

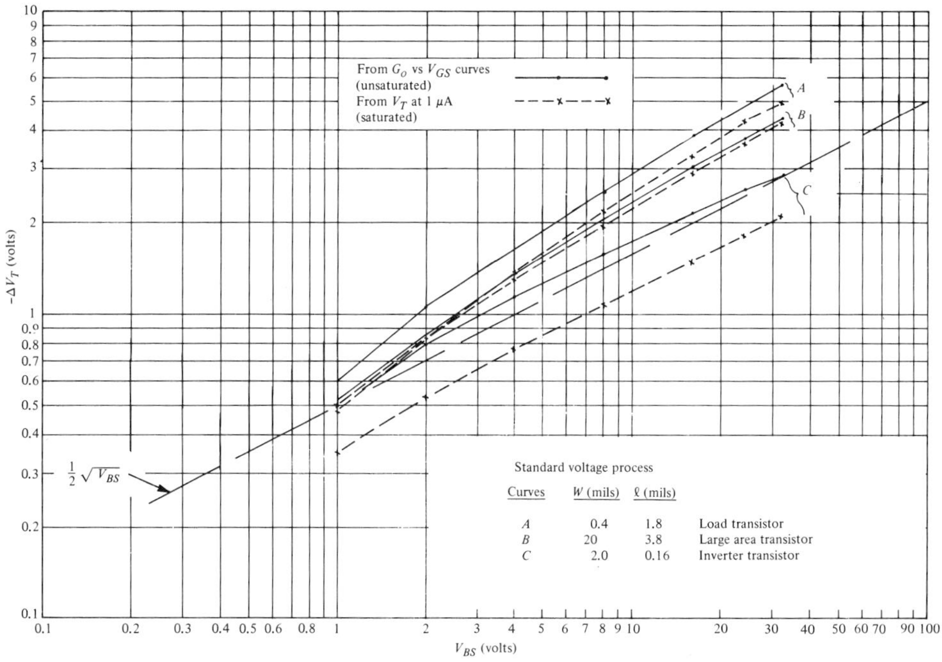

Figure 2–36a Typical body effect curves, standard voltage process.

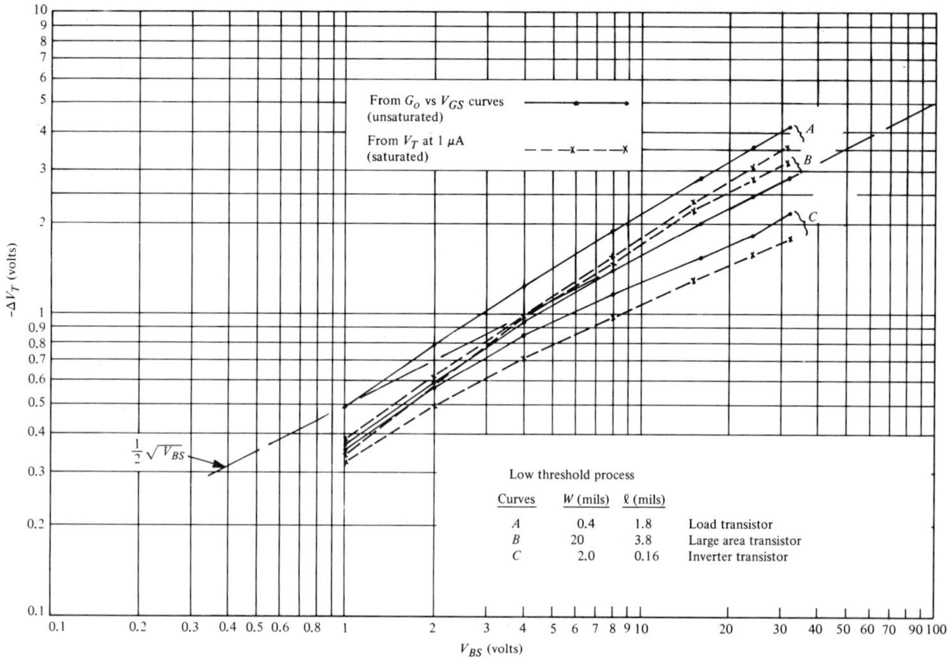

Figure 2–36b Typical body effect curves, low voltage process.

Equation 2-97 is its relative simplicity in many engineering design situations.

Some insight into the nature of the body effect can be obtained from the following analysis. When an inversion layer is formed in an MOS structure there is a *P-N* junction between the surface inversion layer (transistor channel) and the electrically neutral bulk material. Since this junction is formed by the action of the electric field at the surface of the silicon it is referred to as a "field induced" junction in contrast to the more conventional "metallurgical" junction around the source and drain. This field induced junction has all the properties of conventional *P-N* junctions, the transistor channel being the *P*-side and the substrate the *N*-side, in our present *P*-channel transistor.

As in a conventional junction, a depletion region forms between the *P*- and the *N*-regions, in which the electric field sweeps away all mobile electric charge, thus uncovering the fixed charge of the ionized donor atoms in the substrate. Because the surface inversion layer is much thinner than the depletion layer in the substrate, the junction is a one-sided step junction. Using the formula (see Sections 2.2.9 and 2.3.3) for the depletion region width (X_d) in a one-sided step junction,

$$X_d = \left(\frac{2\epsilon_s \phi_T}{qN_D}\right)^{1/2}$$

the total charge Q_B (per unit surface area) in the depletion region is

$$Q_B = qN_D X_d = (2\epsilon q_s N_D \phi_T)^{1/2} \tag{2-98}$$

in which q is the magnitude of the electronic charge, ϵ_s is the permittivity of silicon, N_D is the density of ionized donor atoms, and ϕ_T is the total potential drop across the junction. The total junction potential (ϕ_T) is the sum of the applied reverse bias (V_{BS}) and the equilibrium junction potential (ϕ_B) which is typically about 0.7 V in silicon. In these equations ϕ_B is taken as positive.

$$\phi_T = V_{BS} + \phi_B \tag{2-99}$$

The change in gate voltage necessary to support a change ΔQ_B in the charge under the gate oxide capacitor is

$$\Delta V_T = -\frac{\Delta Q_B}{C_{ox}} = -\frac{1}{C_{ox}}\left[Q_B(\text{at } V_{BS}) - Q_B(\text{at } V_{BS} = 0)\right] \tag{2-100}$$

where $C_{ox} = \epsilon_{ox}/t_{ox}$ is the capacitance per unit area of the gate oxide.

Combining Equation 2-99 with Equation 2-98 into Equation 2-100 gives, finally,

$$\Delta V_T = -\frac{(2\epsilon q_s N_D)^{1/2}}{C_{ox}}\left[(V_{BS} + \phi_B)^{1/2} - (\phi_B)^{1/2}\right] \tag{2-101}$$

Equation 2-101 fits the experiment curves of Figures 2-36a and 2-36b quite well for the large area transistors ($W = 20$ mils, $\ell = 3.8$ mils). For this case, the preceding one-dimensional analysis applies. For the smaller transistors, the lateral dimensions are no longer much larger than the width of the depletion region and the geometry becomes a two- (or three-) dimensional field problem.

2.4.7 MOS Capacitance

The MOS capacitance occurs between the metal gate electrode and the transistor channel, which is located in the silicon surface. It involves the electric field in the gate oxide, and, since this is the same field that forms the conducting transistor channel, the MOS capacitance is closely related to the intrinsic operation of the transistor. Accordingly, all the capacitances associated with an actual transistor are conveniently divided into two classes, intrinsic and parasitic. This section will consider the intrinsic MOS capacitance. Parasitic capacitances are discussed in Section 2.4.9.

The intrinsic capacitances, inherent in the basic physical operation of the device, are associated with the charges stored on the gate electrode and in the channel itself. These capacitances are illustrated in Figures 2-37a and 2-37b. Theoretical equations for these capacitances can be derived from the basic physical model of the device. Sah[9] gives this derivation and the resultant equations. These capacitance values vary with voltage because the charge distribution on the gate and in the channel changes as the voltages change. The results of this analysis are summarized in Table 2-2. The capacitance symbols are shown with a "prime" sign to emphasize that they are the intrinsic part of the total capacitance per unit area appearing between the terminals.

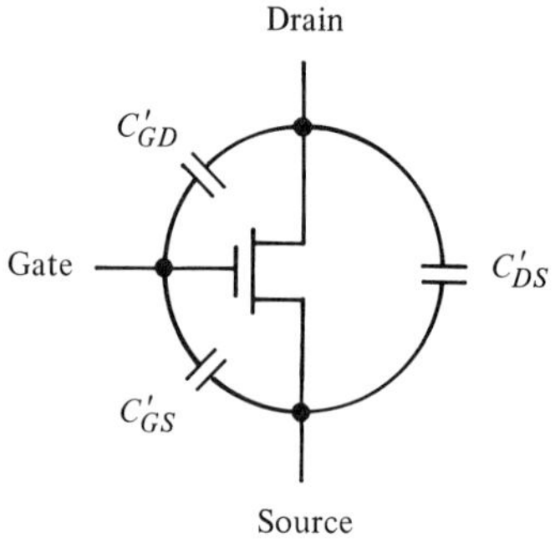

Figure 2–37a Schematic representation of intrinsic capacitances.

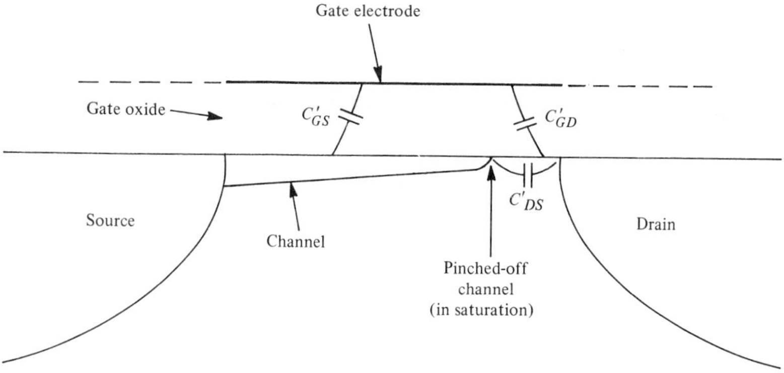

Figure 2–37b Location of the intrinsic capacitances.

Table 2–2 Summary of Capacitance Values

Intrinsic capacitance	*Range of values in the nonsaturation region*	*Value in saturation region*
C'_{GD}	0 to $\frac{1}{2} C_{ox}$	0
C'_{GS}	$\frac{1}{2} C_{ox}$ to $\frac{2}{3} C_{ox}$	$\frac{2}{3} C_{ox}$
C'_{DS}	0	0

C_{ox} is the capacitance per unit area of the gate oxide parallel plate capacitor and is expressed as

$$C_{ox} = \frac{\epsilon_{ox}}{t_{ox}} = \frac{\kappa_{ox}\epsilon_0}{t_{ox}} \tag{2-102}$$

where ϵ_{ox} = permittivity of the oxide
κ_{ox} = dielectric constant of the oxide (= 4 for SiO_2)
ϵ_0 = permittivity of free space = 8.85×10^{-6} pF/μ
t_{ox} = oxide thickness.

In nonsaturation operation, the intrinsic gate to drain capacitance (C'_{GL}) and gate to source capacitance (C'_{GS}) will vary anywhere within the ranges given above depending on the bias conditions involved. In design work, it is customary to take the maximum value in the range as applying over the whole nonsaturation region. The errors involved are quite small, since the capacitance to the intrinsic channel has a fairly large parasitic overlap capacitance in parallel with it.

For inverter transistors with a minimum gate ℓ (typically around 0.2 mil) it is often reasonable to make the additional simplification that the channel region area is small compared to the overlap region area, and that the total gate oxide capacitance (C_{ox}) is simply overlap capacitance, half of which is to the source and the other half to the drain. The overlap capacitance and other parasitic capacitances are discussed in Section 2.4.9.

2.4.8 Capacitance vs. Voltage (C–V) Curves

When there is no (or very little) current in a transistor channel, the entire channel inversion layer is at the same potential and considerable simplification results in the capacitance theory. The capacitance between a metal gate electrode and a silicon surface has been extensively covered in the literature. A good discussion appears in Grove[5] (Chapter 9.2).

Figure 2-38 shows curves of the small signal gate capacitance (C_G) vs. gate voltage (V_{GS}) for a typical standard threshold voltage process. The curve labeled isolated capacitor has no *P*-region in contact with, or connected to, the inversion layer under the gate. The transistor curve has a source and drain, both grounded, adjacent to the inversion layer. The general shape of these curves may be understood by considering the gate capacitance to be a series connection of C_{ox} and possibly a depletion layer capacitance (C_d) (see Figure 2-39). The formation of the accumulation, depletion, and inversion layers was discussed in Section 2.3.1.

At the right side (V_{GS} positive) of the curves (Figure 2-38), the silicon surface is strongly accumulated, and the capacitance is that due to the oxide, $C_G = C_{ox}$. As V_{GS} becomes more negative, a depletion layer begins to form under the gate region, starting as a very thin (high capacitance) layer and gradually increasing in thickness, thus decreasing in capacitance. Finally when V_{GS} approaches the threshold voltage, an inversion layer begins to form and the depletion layer no longer grows. In the case of the isolated capacitor, C_G now levels off at its minimum capacitance given by the series connection of the oxide and the depletion layer capacitance:

$$\frac{1}{C_G} = \frac{1}{C_{ox}} + \frac{1}{C_d} \qquad \text{(2-103)}$$

In the case of the MOS transistor, the source (or drain) *P*-region adjacent to the inversion layer provides an electrical connection to the inversion layer and the capacitance returns to that of the pure oxide ($C_G = C_{ox}$).

When an MOS capacitor is designed into a circuit, it is important to note whether it is an isolated capacitor or if there is an adjacent *P*-region which will give it a characteristic like the transistor. The usual design procedure with a transistor type structure is simply to take $C_G = C_{ox}$ for all gate voltages and neglect the dip in the capacitance characteristic. This is reasonable for digital circuits since the end points of signal transitions are usually well outside the dip in the transistor *C–V* curve. With isolated capacitors the nonlinearity of the capacitor will be important; consequently, this type of structure is not generally recommended for circuit design.

The thin oxide over *P*-region capacitor illustrated in Figure 2-39c is not an intrinsic capacitance, but is included here for comparison with the aforementioned MOS structures. This capacitance is not voltage dependent, and would thus seem ideal for design purposes. The disadvantage of

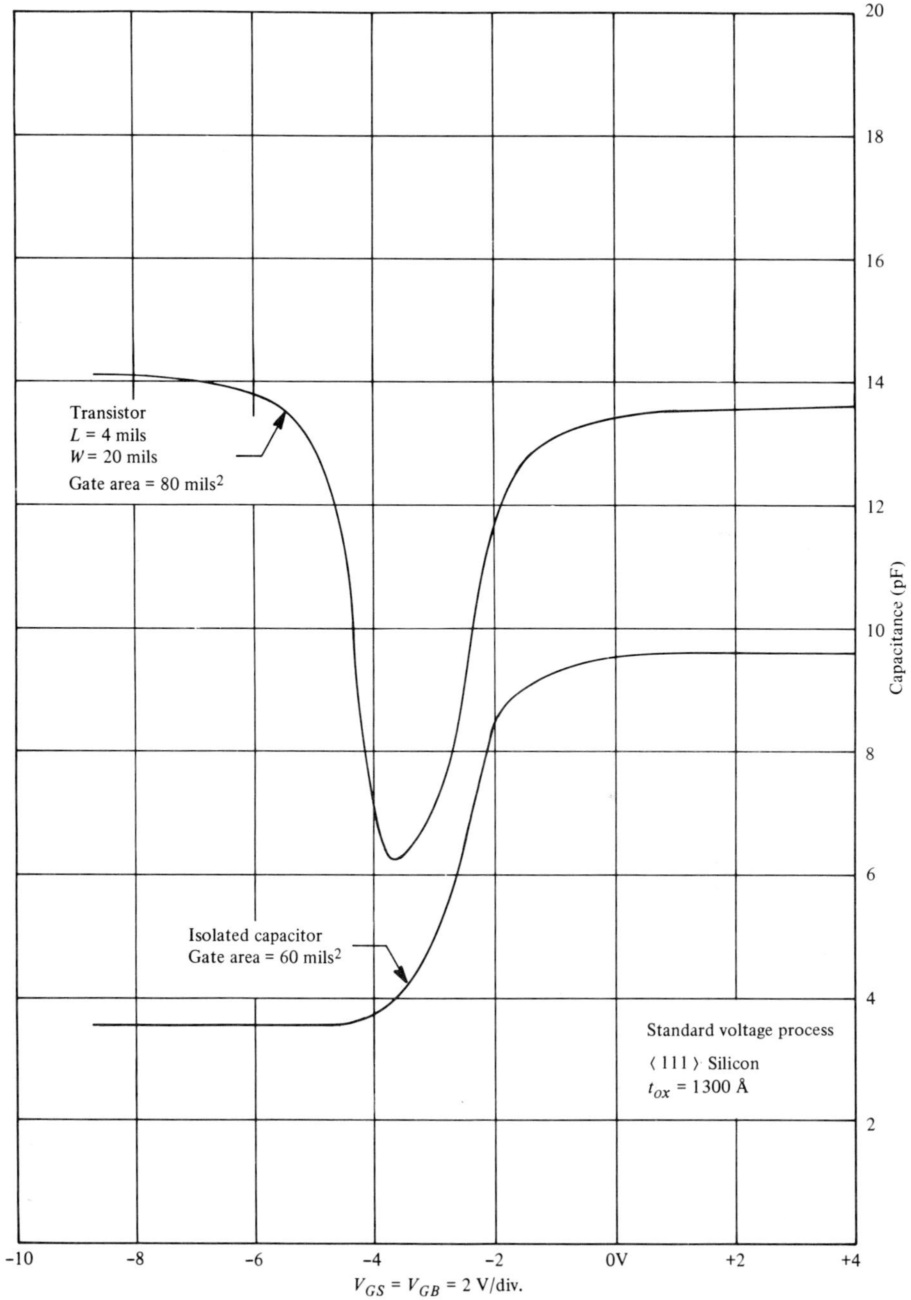

Figure 2–38 Typical variation of intrinsic gate capacitance with gate voltage.

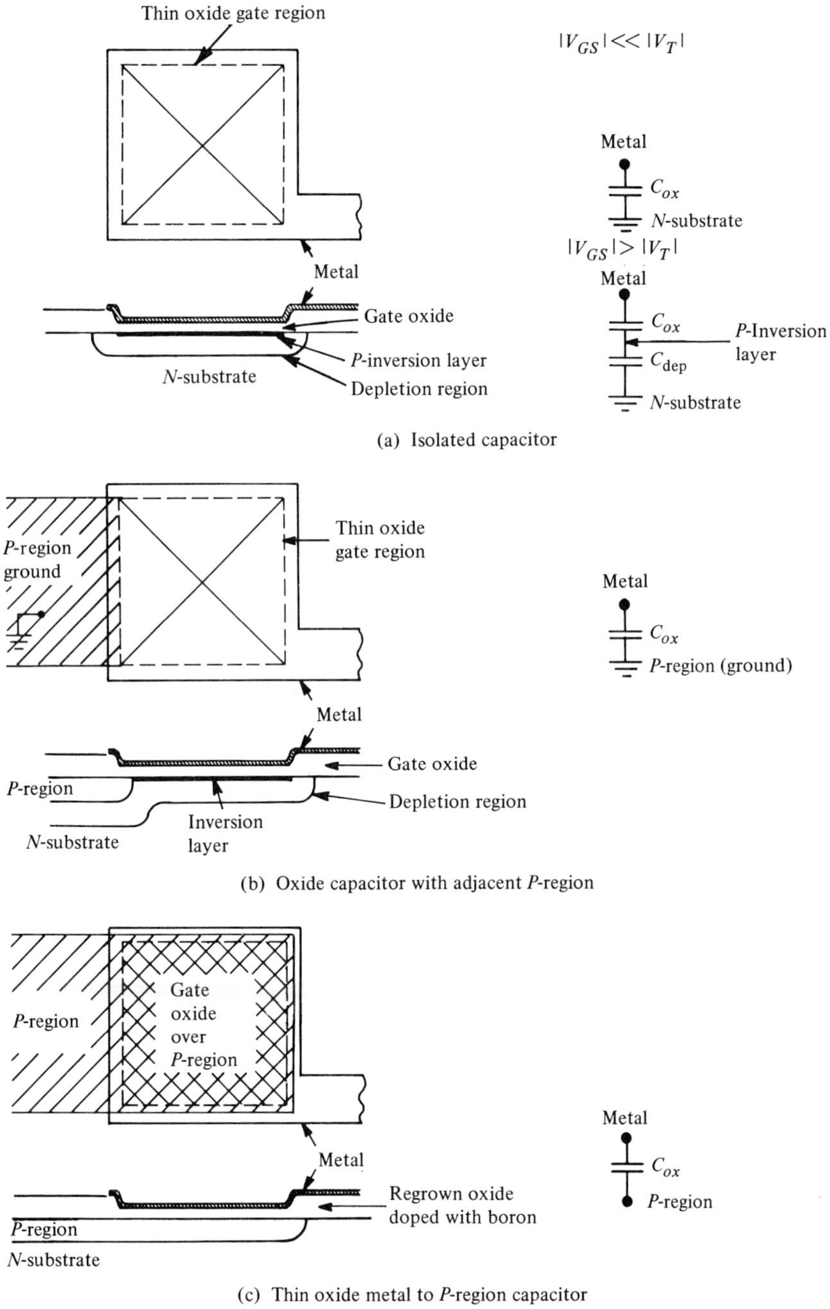

Figure 2–39 Forms of thin oxide capacitors.

the thin oxide over *P*-region capacitance is that its oxide is grown from the heavily doped *P*-region. Consequently, this oxide often contains more defects than does pure gate oxide. Thin oxide over *P*-capacitors frequently have a higher incidence of shorts and low breakdowns than normal gate oxide capacitors.

In spite of this problem, thin oxide over *P*-capacitors are sometimes used, for example, when capacitance is required between two nodes, neither of which is at ground potential. In general, however, when one electrode of the capacitor is at ground potential the arrangement of Figure 2-39b is preferred.

2.4.9 Parasitic Capacitance

In addition to the intrinsic MOS capacitance, several other capacitances are important in determining the performance of a circuit. These are listed below and illustrated in Figure 2-40 in terms of a layout and its corresponding circuit structure. Figure 2-41 shows a schematic of the same structure with all capacitors included.

In the list below, approximate values of capacitance are indicated. These values are typical of many present MOS processes which use thick oxide over the field and over *P*-regions. In actual design calculations, exact values for the particular MOS process should be obtained from the manufacturer.

C_{MOS} Metal-thin oxide-silicon capacitance. Approximately 0.2 pF/mil^2. This is the capacitance that actually forms the transistor channel and is intrinsic to device operation. (See Section 2.4.7.)

C_{PN} Junction capacitance, *P*-region to substrate. Approximately 0.1 pF/mil^2. Capacitance is a function of junction voltage. (See Section 2.4.10.) This is frequently one of the most significant of the stray or parasitic capacitances.

C_M Metal over field oxide capacitance. Similar to C_{MOS} but, because field oxide is approximately 10 times the thickness of gate oxide, C_M is approximately 0.02 pF/mil^2. C_M can become significant in chips with complex interconnections and large metal areas. C_M is not significantly voltage dependent for voltages less than the field inversion threshold voltage V_{TF} discussed in Section 2.4.14.

C_{MP} Metal-thick oxide-*P*-region capacitance. Approximately 0.02 to 0.03 pF/mil^2. Usually slightly larger than C_M since the thick oxide over *P*-region is generally somewhat thinner than field oxide. This is the capacitance that frequently contributes noise coupling from one signal to another in crossovers.

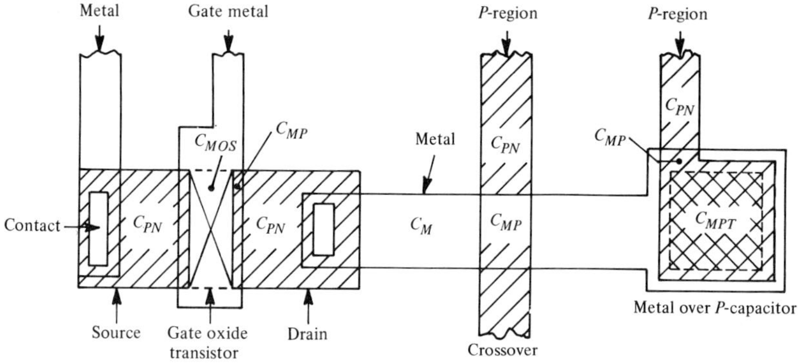

(a) As Drawn on a Composite

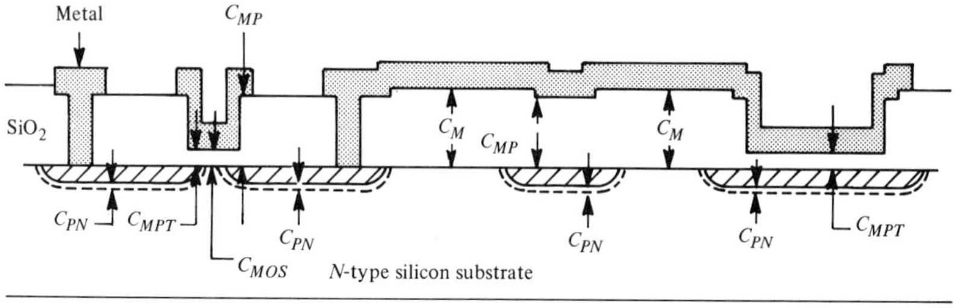

(b) Idealized cross section

Figure 2–40 Location of capacitance.

C_{MPT} Metal-thin oxide-P-region capacitance. Approximately 0.2 pF/mil^2. Same as the zero bia C_{MOS}, since oxide thicknesses are similar, but is not voltage dependent.

In a transistor, C_{MPT} is sometimes called the overlap capacitance, and occurs where the thin gate oxide overlaps the source or drain P-regions. This overlap arises in two ways. The first, due to lateral diffusion of the P-regions under the gate oxide, is always present. The second occurs when the gate mask is intentionally overlapped onto the source or drain to allow a greater mask misalignment tolerance in manufacturing. C_{MPT} is a very important capacitance in many transistors, giving rise to significant noise spikes in coupling transistors and to "Miller effect" feedback in inverter transistors.

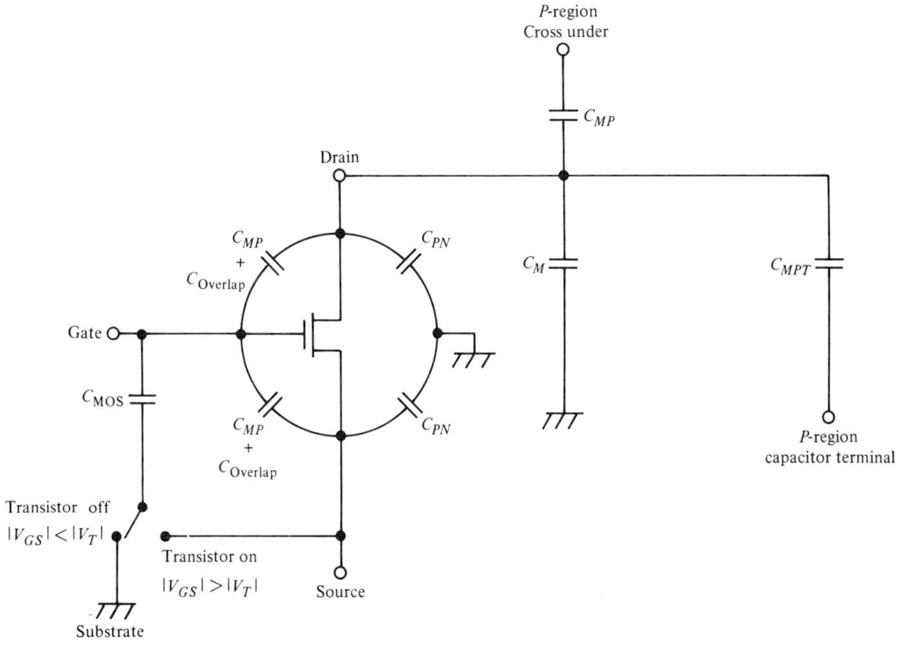

Figure 2–41 Schematic corresponding to Figure 2–40 showing all capacitances.

2.4.10 Junction Capacitance

Figure 2-42 illustrates the cross section of a typical *P*-region as it might appear in an interconnect crossunder or as the source or drain of a tran-

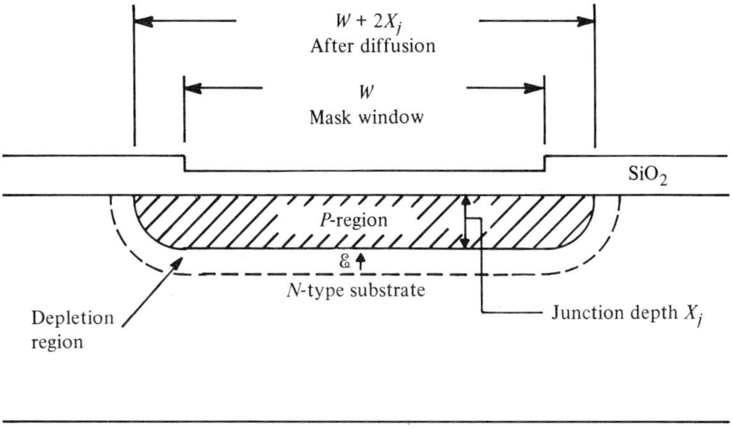

Figure 2–42 Cross section illustrating junction capacitance.

sistor. In normal operation of MOS circuits, the *P*-region is reverse biased (negative with respect to the N-substrate), and is then electrically isolated from the substrate by a depletion region in which all mobile charged carriers have been removed by the electric field across the depletion region. The junction capacitance appears across this depletion region and is frequently sufficiently large that it is one of the major limitations on the performance of MOS circuits. The subject of junction formation and calculation of depletion layer width and capacitances are discussed in Sections 2.2.9 and 2.2.10.

Note in Figure 2-42 that lateral diffusion causes the actual *P*-region width to be larger than the width as drawn on the composite. For small geometries this increase can be significant. $W + 2X_j$ is commonly used in calculating the junction area; this is sufficiently accurate for most design problems, although a more exact calculation would include the curved area at the sides of the *P*-region.

A depletion region also appears between a transistor channel (field induced *P*-region) and the substrate. This additional junction capacitance may add to that appearing on a source or drain node, and should be included in the total capacitance calculation. For example, when a coupling or a multiplexer transistor is turned on, the total junction capacitance loading the signal is that due to the areas of the source and the drain (including lateral diffusion on the appropriate sides) plus the gate area.

The thickness of a junction depletion layer depends on the voltage across the junction; consequently, the junction capacitance will be a function of the junction voltage. This variation is expressed as

$$\frac{C_j}{C_{j0}} = \left(1 - \frac{V}{\phi_B}\right)^n \tag{2-104}$$

in which C_j = junction capacitance at voltage V
$C_{j0} = C_{PN}$ = capacitance at zero bias
V = junction voltage (V is negative for reverse bias)
ϕ_B = built-in junction potential (taken as positive here)
n = exponent depending on type of junction:
$n = -1/2$ for an abrupt step junction
$n = -1/3$ for a linearly graded junction

Figure 2-43 shows a plot of the normalized capacitance (C_j/C_{j0}) vs. junction voltage (V) for three different units from both a standard and a low voltage process. The solid curve represents the average of the units.

Values for ϕ_B and n in Equation 2-104 can be found in the following manner. Equation 2-104 is rewritten as

$$\log \frac{C_j}{C_{j0}} = n \log \left(1 - \frac{\phi_B}{V}\right) \tag{2-105}$$

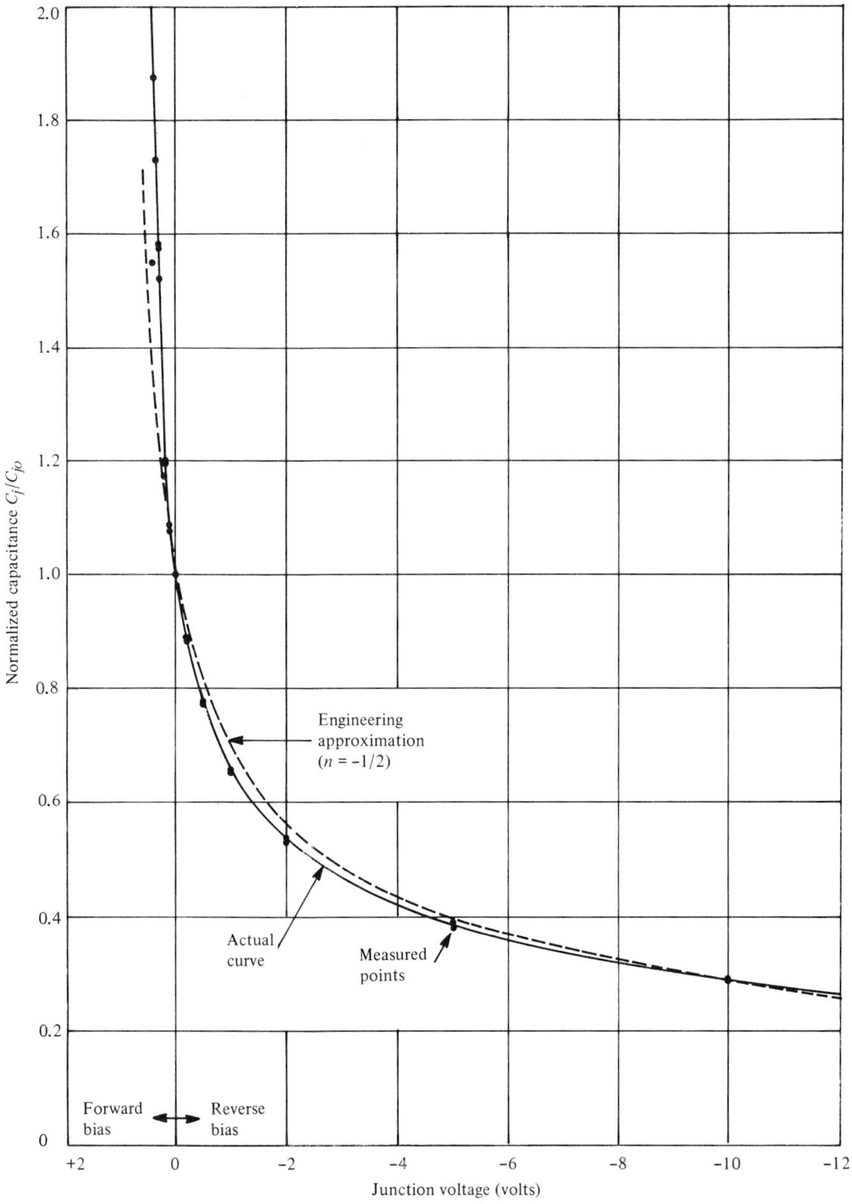

Figure 2–43 Variation of junction capacitance with voltage for typical diffused junctions.

We now assume a value of ϕ_B and plot the measured values of C_j/C_{j0} vs. $(1 - V/\phi_B)$ on log-log scales. If the estimate of ϕ_B is correct the resultant curve will be a straight line with a slope of n. Figure 2-44 shows this for one of the devices from the previous figure and for several values of ϕ_B. From this we see that a value of $\phi_B = 0.6$ V gives the best straight-line fit over the range most useful in circuit designs. The curvature at the extreme left of Figure 2-44 is a result of the junction becoming strongly forward-biased ($V \approx +0.4$ V) and so the capacitance measurements are of questionable accuracy. In the range of voltages from +0.3 to −20 V the $\phi_B = 0.6$ V curve forms an excellent straight line with a slope corresponding to $n = -0.440$. We note that this value of n indicates a junction profile somewhere between a step and a linearly graded junction, which is expected for a diffused junction.

The values of n and ϕ_B determined by the above procedure are not very convenient for engineering calculations. As a result, the approximation that the junction capacitance varies inversely as the square root of voltage is often used. The dashed curve in Figure 2-43 shows the approximation with $n = -\frac{1}{2}$ and $\phi_B = 0.92$ V, which is perhaps the most useful fit for the voltages at which MOS circuits usually operate. This engineering

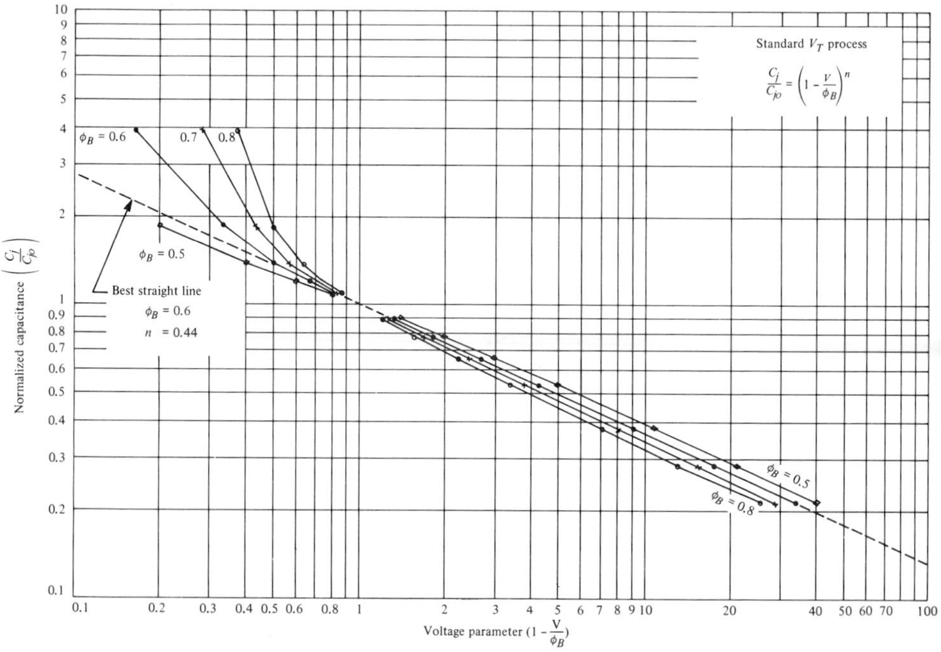

Figure 2–44 Calculation of junction capacitance variation.

approximation was selected to fit the exact expression at 0 and at −10 V bias. The following table summarizes the results for junction capacitance voltage variation, and compares the approximation with the exact values for the typical junctions of Figure 2-43.

Equation	$\frac{C_j}{C_{j0}} = \left(1 - \frac{V}{\phi_B}\right)^n$	
Accurate values	$\phi_B = 0.6$ V,	$n = -0.04$
Common approximation	$\phi_B = 0.92$ V,	$n = -1/2$.

2.4.11 Junction Leakage and Breakdown

In this section we consider some characteristics that arise from the fact that non-ideal junctions are involved in the fabrication of MOS transistors. Leakage and breakdown characteristics are sometimes of considerable importance in LSI chip design. The analysis of these characteristics is complicated by the silicon surface, which makes the problem two-dimensional, and the presence of the gate electrode, which modifies the electric field distribution in the junction region. A good analysis of some of these problems is found in Grove,[5] (Chapter 10). For the present, we consider breakdown and leakage from a circuit design standpoint.

In some MOS LSI circuits, leakage currents can be effectively neglected. This may be the case, for example, in conventional static logic circuits that operate at current levels in the range of 10 to 100 μA, and may have junction leakages on the order of 10 to 100 nA. In other designs this is not the case, and leakage currents become a major performance limitation. This occurs frequently in ratioless circuits and in circuits that make use of charge storage on a gate node for temporary memory. In a conventional dynamic MOS shift register, for example (see discussion of dynamic logic in Section 5.3 of Chapter 5), charge is lost from the storage nodes due to leakage currents in the coupling transistor drain junction. The rate at which this charge is lost will determine the maximum memory time, which then sets the minimum clock frequency for reliable operation of the register.

Leakage current in silicon *P-N* junctions varies greatly with temperature. A common design formula is that the leakage current doubles for each 10°C rise in junction temperature. This is a reasonable average rule for most present MOS processes, although particular processes may vary slightly; values between 8° and 12°C are sometimes quoted. The proper temperature to use in calculating the leakage current is that at the silicon die surface. This may be considerably higher than the ambient temperature due to internal power dissipation in the device.

In addition to its dependence on temperature, the junction leakage current in MOS transistors is also a function of junction and gate voltages and *P*-region area and shape. The usual procedure in most design problems is to characterize the leakage current in terms of junction (*P*-region) area, neglecting variation with other parameters. If this is done for geometries typical of actual circuits, and at a voltage near the upper range at which the circuits will operate, the resultant leakage data will be fairly representative. The leakage is thus specified as a current per unit area of *P*-region. Then in a design problem the total leakage current is calculated from the *P*-region area and the temperature.

Many MOS processes in common use have typical junction leakage currents of a few (in the range of 1 to 3) pA/mil^2 at room temperature. However, it is not generally advisable to design around these low values, because there is considerable variation from unit to unit. A factor of 10 variation (both up and down) from the typical is perhaps not uncommon for a good production process. Thus, for reliable circuits a comfortable safety factor is desirable, and worst-case design values typically fall in the range of 10 to 100 pA/mil^2 at room temperature and with a bias of 15 V.

If the drain voltage of an MOS transistor is increased beyond the voltage at which the leakage current is measured, the current will increase gradually until finally a condition is obtained where further small increases in voltage produce large increases in current. This is the breakdown characteristic.

In MOS transistors breakdown can occur by two possible mechanisms, avalanche breakdown and drain to source punch-through. Avalanche breakdown occurs between the *P*-region and the substrate, and results when the electric field in the junction depletion region is large enough to accelerate charge carriers sufficiently to ionize neutral silicon atoms. The new charge carriers thus generated are themselves accelerated; the whole process rapidly builds up until the current is limited by the external circuit.

In punch-through breakdown the reverse bias drain voltage widens the depletion region around the drain. This depletion region grows into the relatively lightly doped substrate in all directions, including toward the source. When this growing depletion region sufficiently touches (actually merges with) the depletion region around the source, charge carriers (holes) can be injected directly from the source into the drain depletion region where they are then swept by the electric field over to the drain. Since there is a plentiful supply of carriers in the source, the current will increase until it is limited by the external circuit. Note that punch-through occurs directly between the source and the drain regions. The current does not flow to the substrate.

In MOS transistors designed with minimum channel lengths (ℓ), either type of breakdown—avalanche or punch-through—may predominate, depending on the particular parameters. Transistors with longer ℓ, such as load devices, will always show avalanche breakdown. Neither type of breakdown is harmful as long as the device does not become overheated. Breakdown may also occur in the presence of a normal conducting channel, as the transistor characteristics in Figure 2-22 illustrate.

Figure 2-45 shows some typical breakdown characteristics that apply to MOS integrated circuits. The curve labeled "normal junction avalanche breakdown" applies to a normal *P*-region with thick oxide over its entire boundary and is the usual junction breakdown characteristic for a planar diffused junction.

In an MOS transistor structure the presence of the thin gate oxide and the gate metal considerably modifies both the normal avalanche and the punch-through breakdown. The curve labeled "initial gate enhanced avalanche breakdown" illustrates the effect of the gate in enhancing or modulating the normal breakdown, giving rise to the name "gate modulated breakdown" for this mechanism. Gate modulated breakdown also has the same sharp knee of a good quality planar junction and usually exhibits an additional characteristic referred to as "creep." If the breakdown current that passes through the device is increased beyond a few microamperes, the characteristic slowly shifts or creeps toward higher voltages. This is a nonreversible process, and simply results in a new breakdown characteristic displaced to the right. Typical characteristics may sometimes be shifted

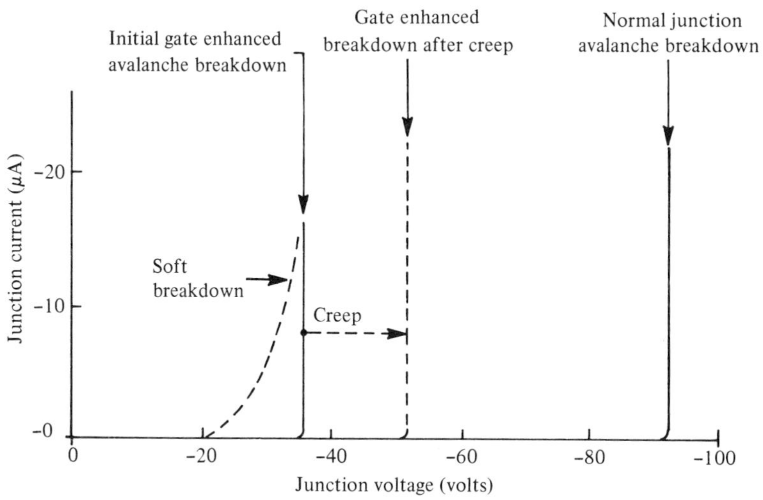

Figure 2–45 Breakdown characteristics.

20 to 30 V, after which the current to cause further creep becomes excessive.

Two possible mechanisms are sometimes proposed to explain this effect: (1) localized heating or microplasmas in the junction region under the gate and (2) localized ionic drift in the oxide due to very large electric fields in the oxide over the junction. Creep is of relatively little concern to the designer, since it improves the breakdown characteristics. The only exception to this occurs when the gate modulated breakdown mechanism is used for static charge protection, which is discussed in the next section.

Also indicated in Figure 2-45 is a "soft breakdown" characteristic, which is occasionally observed and is considered to be due to defects within the depletion region. In complex MOS arrays the field inversion phenomenon (see Section 2.4.14) sometimes gives rise to a characteristic which may be confused with a soft junction breakdown.

2.4.12 Gate Protection

The gate electrode of an MOS transistor is completely isolated from the remainder of the transistor by the SiO_2 layer which forms the dielectric in the gate to substrate capacitor. If too high a voltage is applied to the gate, the dielectric breakdown field is exceeded and the silicon dioxide beneath the gate will rupture. This type of breakdown results in permanent destruction of the device and causes a short between the gate metal and either the substrate or a *P*-region.

The breakdown field in carefully prepared oxides is around 10^7 V/cm. Many MOS processes that are in common use have gate oxide thicknesses in the range of 1000 to 1400 Å, which leads to typical breakdown voltages in the range of 100 to 140 V. To provide a reasonable safety margin, most manufacturers specify that the maximum voltage applied betweeen the gate electrode and any other device terminal be limited to from 60 to 80 V.

The small size and the fact that the gate of an MOS transistor is an almost ideal capacitor makes the devices sensitive to damage by stray electrostatic charges. Such charges can be generated by normal handling and are accentuated by nylon clothing and rubber-soled footwear.

Two approaches can be taken to minimize the possibility of accidentally destroying these devices. One method is to take special precautions to avoid exposure of the gate to high voltages or electrostatic charges. The other is to incorporate a protective device or circuit into the input (gate) lead. The latter approach, which is discussed in detail later, is recommended wherever the resultant reduction in gate input impedance is of little concern.

The first approach requires strict handling procedures, thorough grounding of equipment, and careful selection of clothing and material that may come in contact with the MOS devices. Typical measures are:

a. Check all equipment that comes in contact with MOSFET devices for excessive voltages; use soldering irons with grounded tips; and adjust or modify power supplies to eliminate excessive voltage transients that might appear when the supplies are turned on or off.
b. Verify that work benches and the cases of all equipment such as power supplies, oscilloscopes, meters, etc., are grounded.
c. Train personnel to ground themselves for a moment by touching the metal frame of a table or other ground source before handling MOS devices. When an MOS device is to be plugged into a socket, hold the device case with one hand and the circuit chassis or ground with the other.
d. While handling MOS devices, protect them by wrapping their leads with a conductive material (such as a short strip of aluminum foil or a metal washer) to connect the leads together.
e. Personnel handling MOS devices should not wear untreated nylon or other synthetic smocks. Cotton or synthetics that have had a conductive treatment are acceptable.

In addition to these measures, various types of antistatic sprays or rinses have been tried. Although most manufacturers use some of the above handling procedures in various degrees, they are clearly a nuisance for the user and do not always give complete protection. Consequently the incorporation of protective devices into the circuit itself is highly recommended.

Protective Devices

Several types of protective devices are available to prevent damage to the gate oxide from static charges and accidentally applied voltages. These are generally used on all MOS array inputs and are often indicated on data sheets as zener diodes, although they are not true zener diodes. Protective devices may also be used on array outputs, although the output transistor often serves as its own protective device.

All protective devices require adding some *P*-region to the protected line. Thus they will appreciably lower the input resistance. For an unprotected MOS transistor gate the input resistance is very high, on the order of 10^{14} to 10^{16} Ω. When some *P*-region is added, this input resistance may decrease to the order of 10^{10} Ω. This is of little concern in digital circuits. For electrometer applications where very low input leakage currents are required, there are MOS transistors available without protective devices.

The most common protective device uses the gate modulated junction breakdown mechanism (discussed in the previous section). One of the early versions of this type of protective device is shown in Figure 2-46a and is

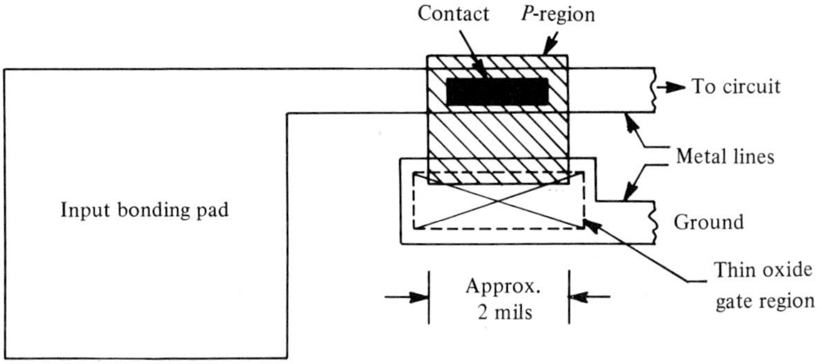

(a) Early version

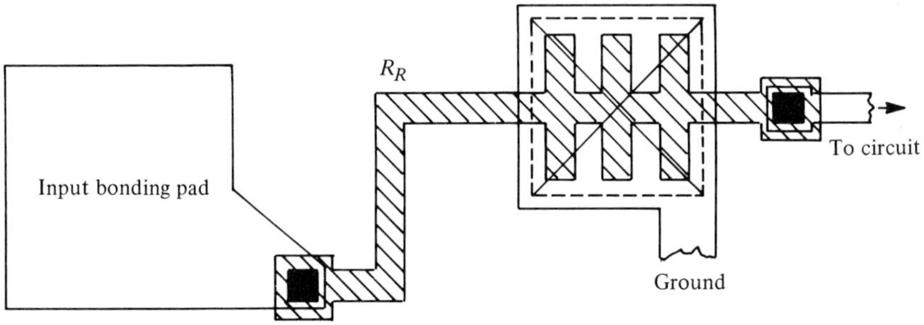

(b) Improved version

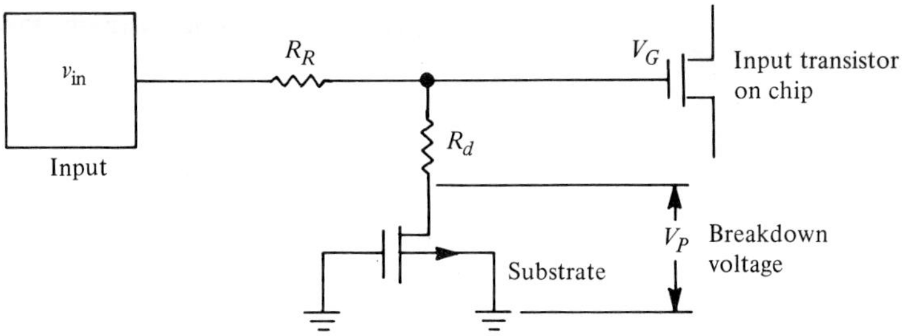

(c) Equivalent circuit

Figure 2–46 Protective devices based on gate modulated junction breakdown.

still sometimes used. This device is small and easy to incorporate on a complex array.

An improved version of the same type of protective device is shown in Figure 2-46b. This device incorporates a series resistance, typically on the order of 1 to 3 kΩ, to limit peak currents during breakdown. This resistance is simply a diffused *P*-region which is explained in the following section. In addition to the series resistance, this device has a considerably longer region in which the gate modulated breakdown can occur. This increases the current handling capability of the device. A double-sided E configuration device is shown here, although various shapes may be used. The main objective is to obtain a long protective breakdown boundary, which typically is in the range of 6 to 10 mils.

Figure 2-46c shows an equivalent circuit for a protected input. R_R is the resistance of the series *P*-region, and R_d is the dynamic resistance of the gate modulated breakdown mechanism. R_d is a function of the current through the device. At small currents, on the order of a few microamperes, R_d may be on the order of 10 kΩ, decreasing to about 1 kΩ at current levels above 10 mA. From the equivalent circuit

$$V_G = V_P + (v_{\text{in}} - V_P)\frac{R_d}{R_R + R_d} = \frac{v_{in}}{1 + \dfrac{R_R}{R_d}} + \frac{V_P}{1 + \dfrac{R_d}{R_R}} \tag{2-106}$$

where V_P is the protective device breakdown voltage, and V_G the voltage appearing on the gate which is to be protected. By rearranging this equation

$$v_{in} = V_G\left(1 + \frac{R_R}{R_d}\right) - V_P\left(\frac{R_R}{R_d}\right) \tag{2-107}$$

which gives the allowable input voltage (v_{in}) for a specified maximum gate voltage V_G. Typical values for some of the parameters might be $V_P = -50$ V, $R_d = 1$ kΩ, $R_R = 2$ kΩ, $V_G = -80$ V, giving $v_{in} = -140$ V.

In some circuits an upper limit on R_R may be set by speed considerations; the combination of R_R and the total capacitance on the protected node forms a low-pass filter. For chip input signals this *RC* time constant is typically on the order of 10 ns which is usually negligible, but for signals with a large capacitance (clock lines, for example) the speed limitation may be severe. In these cases the protective device is frequently designed with no series resistor. The protective device may also be designed around some breakdown mechanism other than the gate modulated breakdown devices discussed above.

Two other devices that are gaining favor are the source to drain punch-through device and the field inversion transistor (see Section 2.4.14). These are illustrated in Figure 2-47 and may be analyzed in a manner similar to

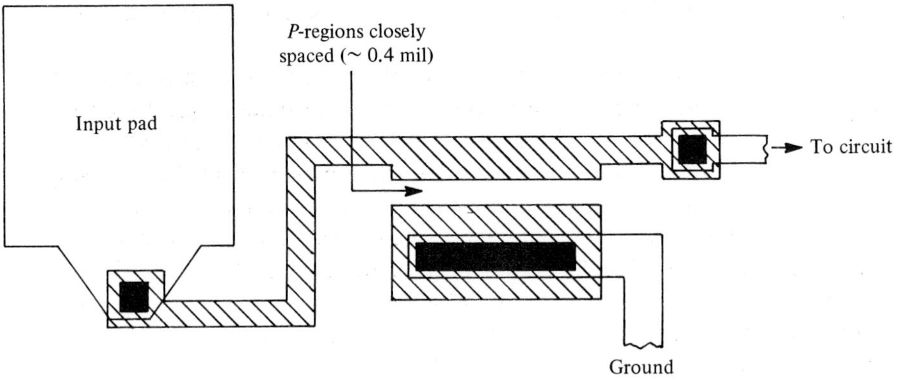

(a) P to P punch-through device

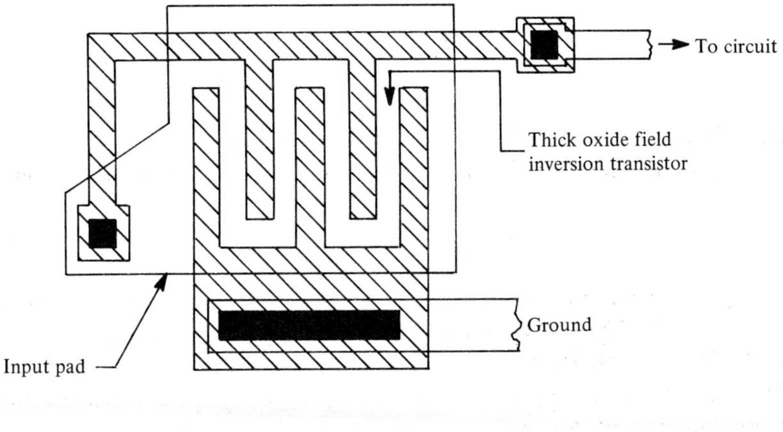

(b) Field inversion device

Figure 2–47 Other types of gate protective devices.

that used previously. The source to drain punch-through characteristics are fairly similar to the gate modulated breakdown characteristics, and the preceding analysis applies directly. The field inversion device has a square law characteristic the same as a normal MOS transistor but with considerably higher threshold voltage and lower k' because the "gate" oxide is so much thicker. One attractive feature of the field inversion device is that, being an MOS transistor, it may be designed to give the desired characteristics by choosing an appropriate W and ℓ; however, the low k' usually results in a fairly large device. Figure 2-47 shows such a device designed underneath an input bonding pad.

Both the punch-through and the field inversion protective devices have one characteristic that makes them less convenient to use than the gate modulated device. Current flow in these devices is from a drain to a source region rather than from a single P-region to the substrate. Thus they require a solid metal ground connection capable of carrying the full breakdown current. This can sometimes cause considerable inconvenience in layout. Note that the grounded gate in the modulated junction breakdown devices does not conduct current, and so P-region crossunders are frequently used for a ground connection to these gates.

When positive voltages are applied to any of the devices discussed here, the characteristic is that of a forward-biased silicon diode since all of these devices are formed by a diffused P-region in an N-substrate.

2.4.13 Diffused Resistors

This section covers the characteristics of diffused resistors as they are available in conventional MOS processing. A good discussion of diffused regions in general may be found in Warner and Fordemwalt[20] (Section 3-3).

a. Sheet Resistance

For a cubic or rectangular bar of length (L), width (W), and height or depth (X_j), the resistance (R) between the two ends of the bar is

$$R = \rho \frac{L}{A} = \rho \frac{L}{X_j W} \tag{2-108}$$

where ρ = average resistivity (Ωcm),
A = cross-sectional area = $X_j W$.

For a diffused resistor of a fixed junction depth (X_j), the value of the resistor can only be changed by varying L or W, the topological geometry of the resistor. Therefore, in the design of the shape of diffused resistors, a more convenient quantity than ρ can be used by combining ρ and X_j and defining a new variable ρ_s from Equation 2-108 as follows:

$$R = \frac{\rho}{X_j} \frac{L}{W} = \rho_s \frac{L}{W} \tag{2-109}$$

where $\rho_s = \dfrac{\rho}{X_j}$ = sheet resistance (ohms/square = $\Omega/\square$).

We note that the units of sheet resistance are in ohms, and that the term "per square" is usually included as a mental key only and is not part of the dimension of ρ_s. The concept of sheet resistance simply means that any square, of whatever size, will have a resistance of ρ_s between opposite edges. These "squares of resistance" can be combined in series and/or parallel and the net resistance calculated from the corresponding combination of discrete resistors, provided that corner and spreading effects are negligible.

The sheet resistance is fixed by the nature of the diffusion process used, and varies considerably between manufacturers. Values from 50 to 150 Ω/□ are quoted, with perhaps the vicinity of 100 Ω/□ being most common. P-region sheet resistance also varies with temperature. Figure 2-48 shows some typical curves. The temperature coefficient of resistance depends somewhat on the resistance itself; therefore, for design purposes, exact values should be obtained for the particular process to be used.

b. Lateral P-Region Diffusion

A typical low value, diffused resistor is shown in Figure 2-49. L is the

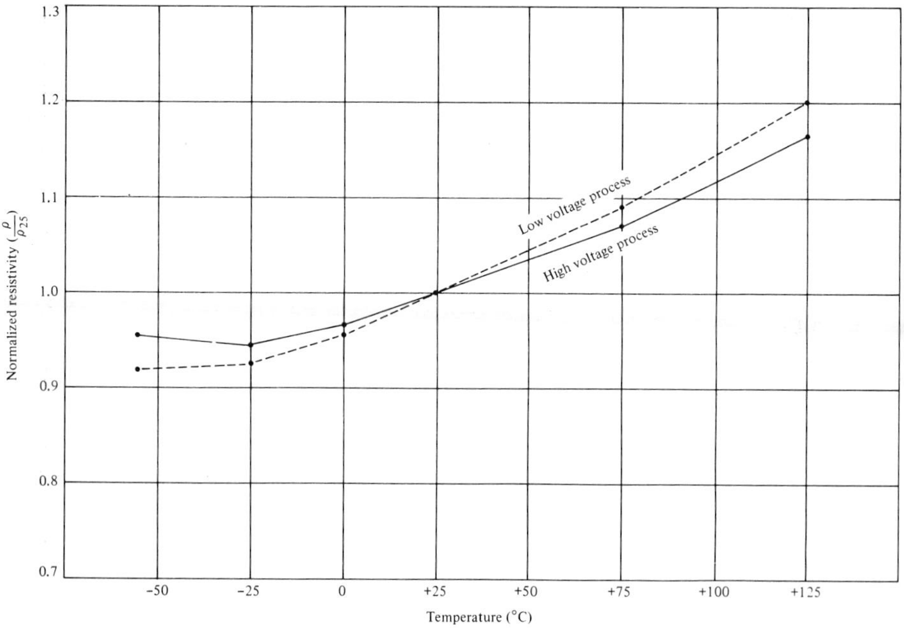

Figure 2–48 Typical variation of *P*-region resistivity with temperature.

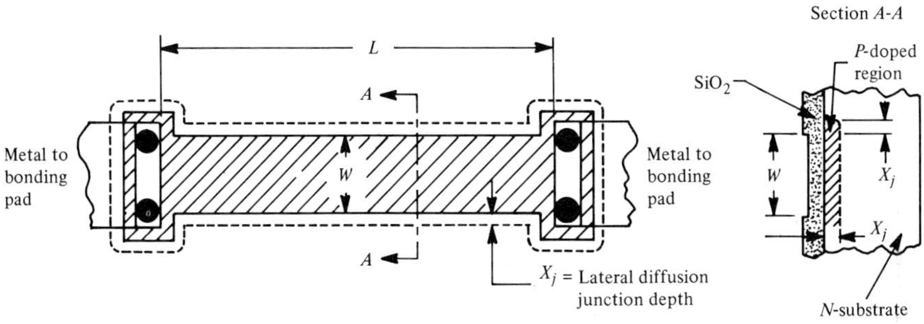

Figure 2–49 Diffused resistor.

length of the resistor as shown on the composite diagram as well as the actual resistor length. W is the width of the resistor as drawn on the composite and the width of the area that is exposed through the etched oxide to the dopant during predeposition. Therefore, after diffusion, the actual width of the resistor will be $W + 2X_j$. This results in

$$R = \frac{\rho_s L}{W + 2X_j} \tag{2-110}$$

which is in contrast to Equation 2-109. However, we note that the lateral diffusion does not result in a uniformly doped layer all the way out to the side of the P-region. Consequently, the final answer lies somewhere between Equations 2-109 and 2-110. Accordingly,

$$R = \frac{\rho_s L}{W + \alpha X_j} \tag{2-111}$$

where $\alpha(0 \leq \alpha \leq 2)$ is a geometrical factor to account for the graded nature of the resistor boundary. The limiting cases when $\alpha = 0$ and 2 correspond to Equations 2-109 and 2-110.

By making measurements on resistors with differing widths (W) and constant L, it is possible to check Equation 2-111. Figure 2-50 shows the results of such measurements. In addition to these data, the figure also indicates Equation 2-111 for the two limiting cases of $\alpha = 0$ and 2 and for the case of $\alpha = 0.3$ This corresponds to the measurements to within an accuracy sufficient for calculating the values of resistors typical in MOS complex circuits.

We note finally that, in a resistor with lateral diffusion on both sides, the maximum value of α is 2.0, which would correspond to the full lateral diffusion contribution. Thus, an actual value of $\alpha = 0.3$ means that effectively $0.3/2.0 = 0.15 = 15$ percent of the lateral diffusion contributes to the current conduction in the resistor. For resistors wider than about 0.4

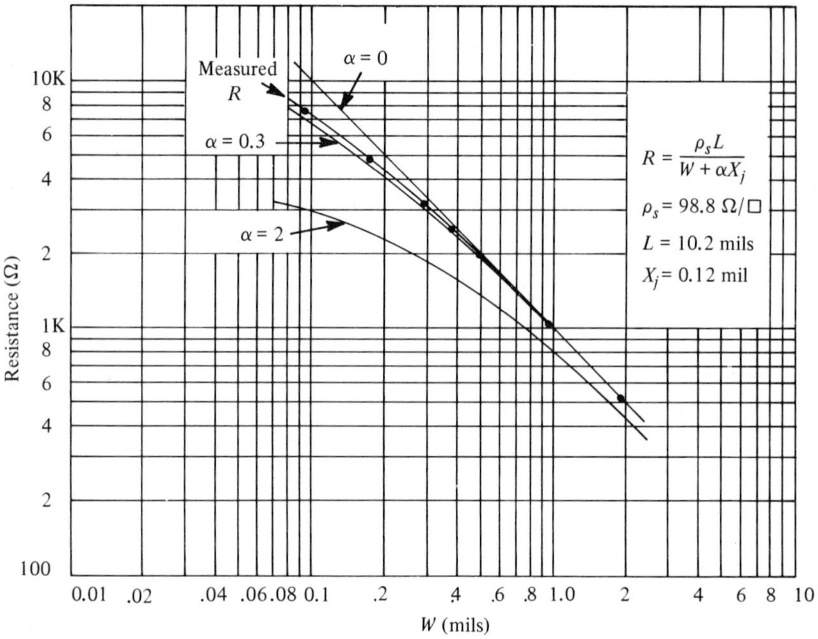

Figure 2–50 Effect of lateral diffusion on a *P*-region resistor.

mil, the lateral diffusion can be neglected and Equation 2-109 used. In narrower resistors the lateral diffusion has a significant effect and Equation 2-111 should be used.

2.4.14 Parasitic Transistors

In addition to the characteristics and parameters of the MOS device previously discussed, there are several other factors that must be considered in designing MOS LSI circuits. Undesired parasitic transistors may be created by poor layout practices or may even be impossible to avoid in the layout of some arrays. Both parasitic MOS and bipolar transistors are possible, and a consideration of their characteristics is essential for trouble-free designs.

a. Field Inversion

In a normal MOS transistor, the conducting channel is formed by the action of an electric field through the thin oxide over the gate region. The thick-

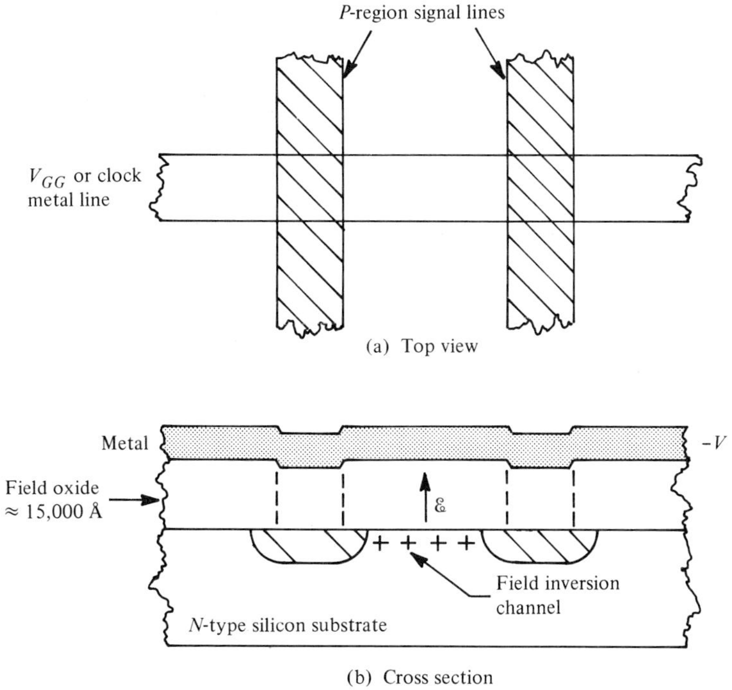

Figure 2–51 Field inversion channel formation.

ness of the oxide determines the amount of electric field required to form a channel. If the oxide is very thick, it is still possible to form a channel, provided the gate voltage is sufficiently large. Thus it may be possible to form a transistor channel in some designs where none is intended. This is illustrated in Figure 2-51 and is called a field inversion transistor. Field refers to the thick oxide field that surrounds the normal transistors.

Figure 2-52 shows some curves where the square root of the channel current of the field transistor is plotted vs. gate voltage, as was done in Section 2.4.4 for a normal transistor. These curves were taken on a device with a channel width of 2 mils and electrical channel length (ℓ) of 0.36 mil. From the slope of these curves one can calculate an effective conduction factor (k') and the true threshold voltage (V_{TF}) for field inversion in the same manner as for the normal transistors discussed previously.

One should note the curvature near the bottom of Figure 2-52, which indicates that appreciable field inversion channel current starts to flow at voltages somewhat less than the field inversion threshold voltage as usually defined on data sheets. In designs where these small currents are important,

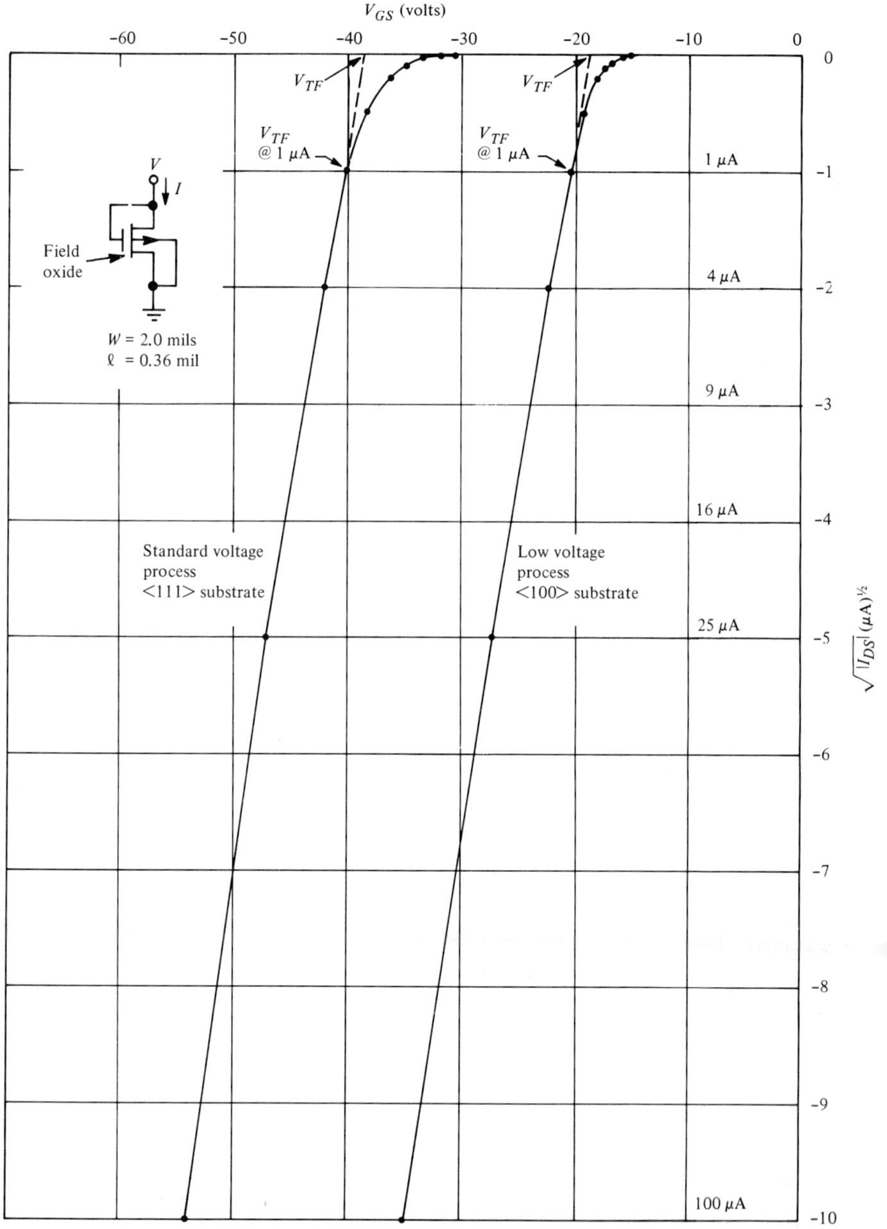

Figure 2–52 Typical field inversion transistor characteristics

one should remember that V_{TF} as usually listed is the voltage to cause 1 μA of current to flow.

Figure 2-53 shows the nature of the gradual increase in field inversion currents at voltages below V_{TF}. For any field inversion transistor, this curve gives the ratio of channel current at a voltage less than V_{TF} at 1 μA to the current that would flow at that V_{TF}. The following design example shows how this curve is used.

Figure 2-54 illustrates a case in which the field inversion transistor has an important effect on the circuit operation. A sample calculation will show how the magnitude of this effect may be estimated, given the following typical parameters:

$$V_{\phi} = -28 \text{ V}$$
$$V_{DD} = -15 \text{ V}$$
$$V_{TF} = -30 \text{ V} \qquad \text{at } 1\ \mu\text{A}$$

Field inversion channel dimensions are

$$W = 0.3 \text{ mil}$$
$$\ell = 0.76 \text{ mil}$$

The leakage current path is treated as a current generator. This is a conservative procedure, since for the given values of V_{DD} and V_{ϕ}, the field inversion transistor will be unsaturated and, consequently, will have a somewhat smaller current than calculated from Figure 2-53. From this figure at a voltage ratio of $28/30 = 0.93$ we can see that the field inversion (FInv) channel will conduct 12 percent of the current at full V_{TF}. Since this test device has a $W = 2.0$ mils and an $\ell = 0.36$ mil, the ratio of the geometry in the problem to the geometry in the test device is

$$\frac{I_{\text{Field Inversion}}}{I_{\text{Test Device}}} = \frac{\left(\frac{W}{\ell}\right)\text{FInv}}{\left(\frac{W}{\ell}\right)\text{Test}} = \frac{\frac{0.3}{0.76}}{\frac{2.0}{0.36}} = 0.071$$

Combining this with the above 12 percent from Figure 2-53 and the fact that V_{TF} is specified at 1 μA, the field inversion leakage current in this example is

$$I_{\text{FInv}} = 0.071 \times 0.12 \times 1\ \mu\text{A} = 8.5 \text{ nA}.$$

This field inversion leakage current will discharge the memory capacitor (C_M) shown in Figure 2-54, thus limiting the maximum time that information can be stored. Again considering the field inversion channel as a current generator, the voltage on C_M will decay by

$$\Delta V = \frac{I_{\text{FInv}} \times T\phi_2}{C_M} \text{ (volts)} \qquad (2\text{-}112)$$

where $T\phi_2$ is the width of the ϕ_2 clock.

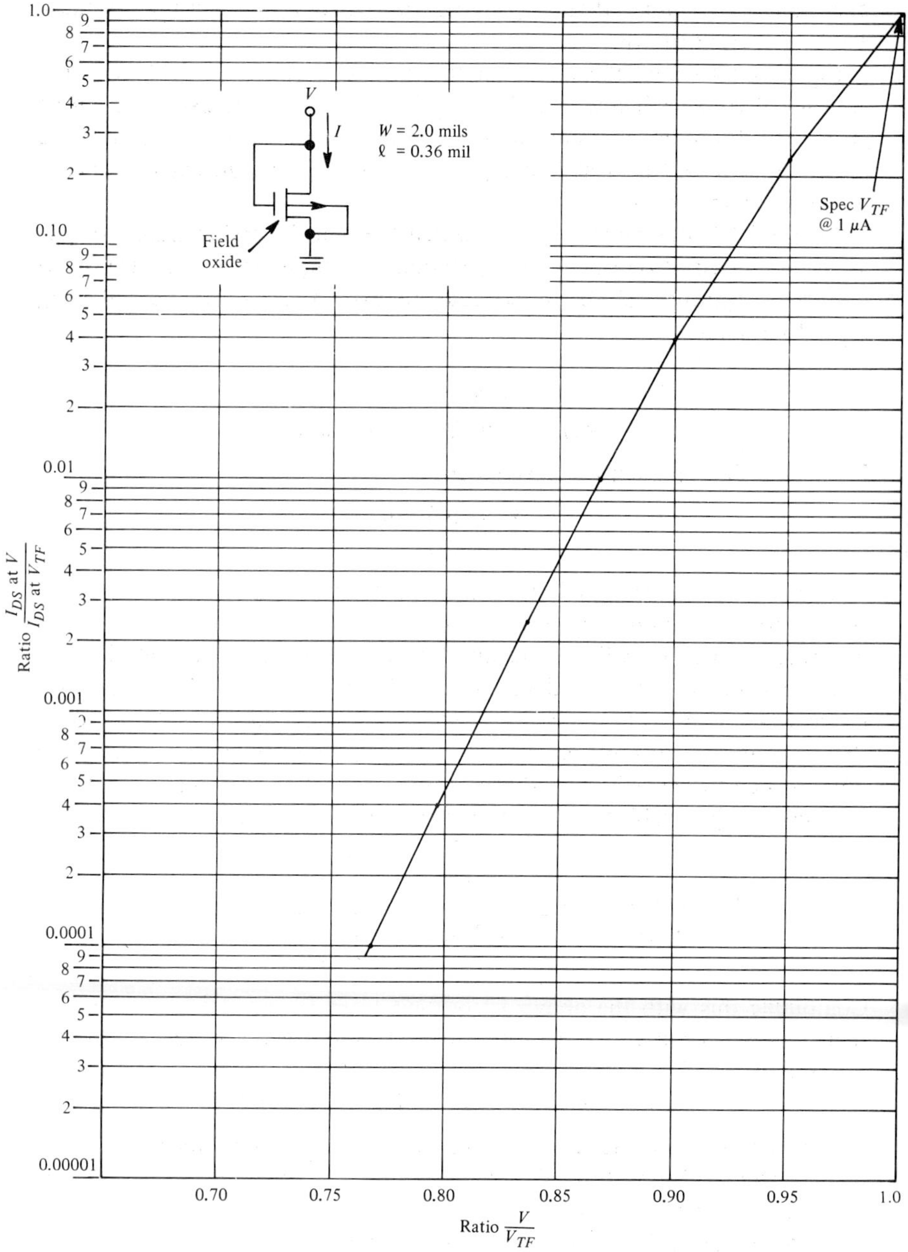

Figure 2–53 Typical gradual turn-on characteristics of field inversion channels.

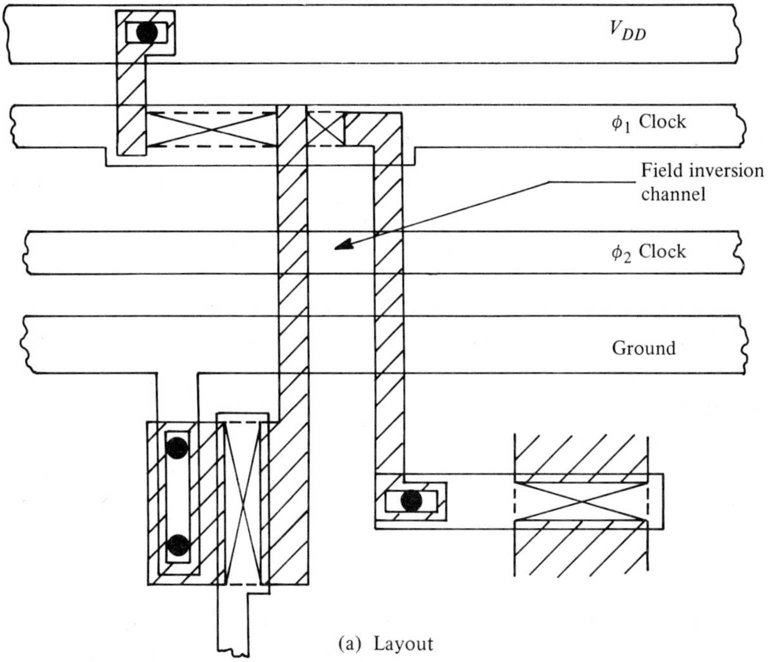

(a) Layout

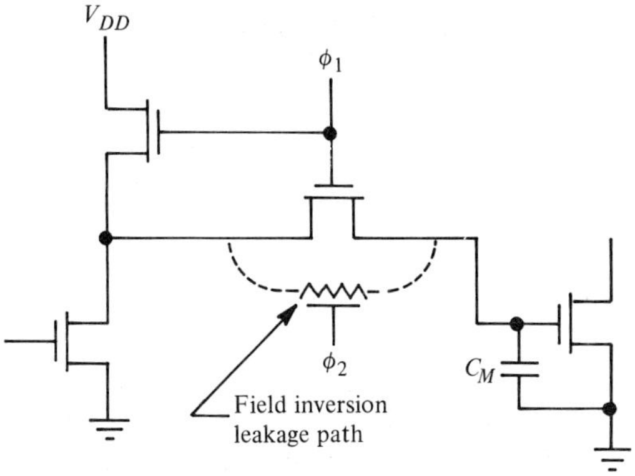

(b) Equivalent schematic

Figure 2–54 Example of field inversion problem.

Note that in this example the field inversion channel only exists during the ϕ_2 clock time. At other times, ϕ_2 is at ground potential and no channel exists.

b. Lateral PNP Transistor Action

The basic MOS *P*-channel transistor can, under certain conditions, operate as a bipolar *PNP* transistor. The transistor so formed is a lateral *PNP* transistor, wherein the source and drain act as the emitter and collector of the *PNP* transistor, and the substrate acts as the base. This lateral *PNP* transistor has a relatively low β, but it does exist and can operate when a *P-N* junction is forward-biased.

Figure 2-55 illustrates a typical circuit where *PNP* transistor action can arise. The *PNP* transistor effect can also occur when fast rise time inputs or clocks have overshoot or ringing that goes far enough above ground to forward-bias a junction.

Figure 2-56 shows some data illustrating how the *PNP* transistor α varies with channel length (L). The base transport factor α is defined as in a conventional transistor by:

$$\alpha = \frac{I_C - I_{CO}}{I_E} \approx \frac{I_C}{I_E} \tag{2-113}$$

since leakage current (I_{CO}) is negligible for most silicon transistors.

An estimate can be made of the effect of the *PNP* transistor in Figure 2-55 as follows. At the trailing edge of the clock, node A attempts to go above ground by an amount $V_\phi - V_{DD}$. Neglecting the forward-biased diode drop of 0.7 V, this gives a charge $Q_A = C_A\ (V_\phi - V_{DD})$, which is injected from node A into the *PNP* transistor base (substrate). Of this charge, a fraction α is collected by the collector (node B), thus reducing the charge stored on C_B by an amount

$$\Delta Q_B = \alpha Q_A = \alpha C_A\ (V_\phi - V_{DD})$$

and giving a voltage loss

$$\Delta V = \frac{\Delta Q_B}{C_B} = \alpha \frac{C_A}{C_B}\ (V_\phi - V_{DD}) \tag{2-114}$$

Depending on the parameters, the amount of charge ΔQ_B and resultant voltage ΔV lost from memory node B may be enough to cause the following device to turn off sufficiently so that its output is adversely affected. In conventional ratio type circuits, this is rarely a problem, since C_A is usually just parasitic overlap capacitance and is quite small. In addition, stray capacitance to ground (not shown in Figure 2-55) on node A serves to limit the forward-bias transient. However, in some designs, C_A is intention-

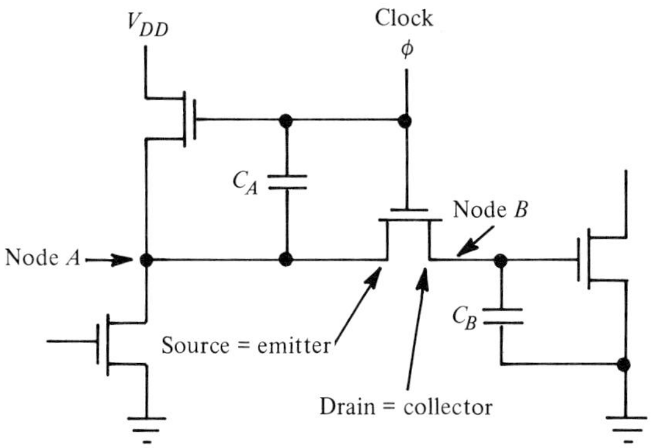

(a) Circuit diagram

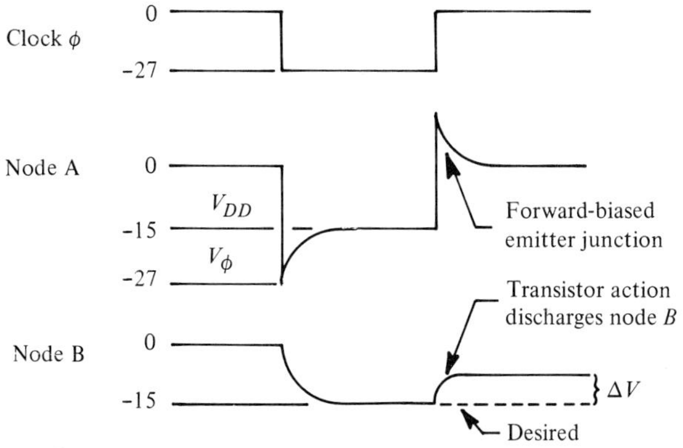

(b) Waveforms

Figure 2–55 *PNP* transistor action.

ally made large; and in these cases, the *PNP* transistor action can be significant.

2.4.15 MOS Parameter Test Methods

In the process of manufacturing complex MOS arrays, it is desirable to have some way of conveniently measuring the more important electrical parameters. Since these measurements are difficult or impossible to make

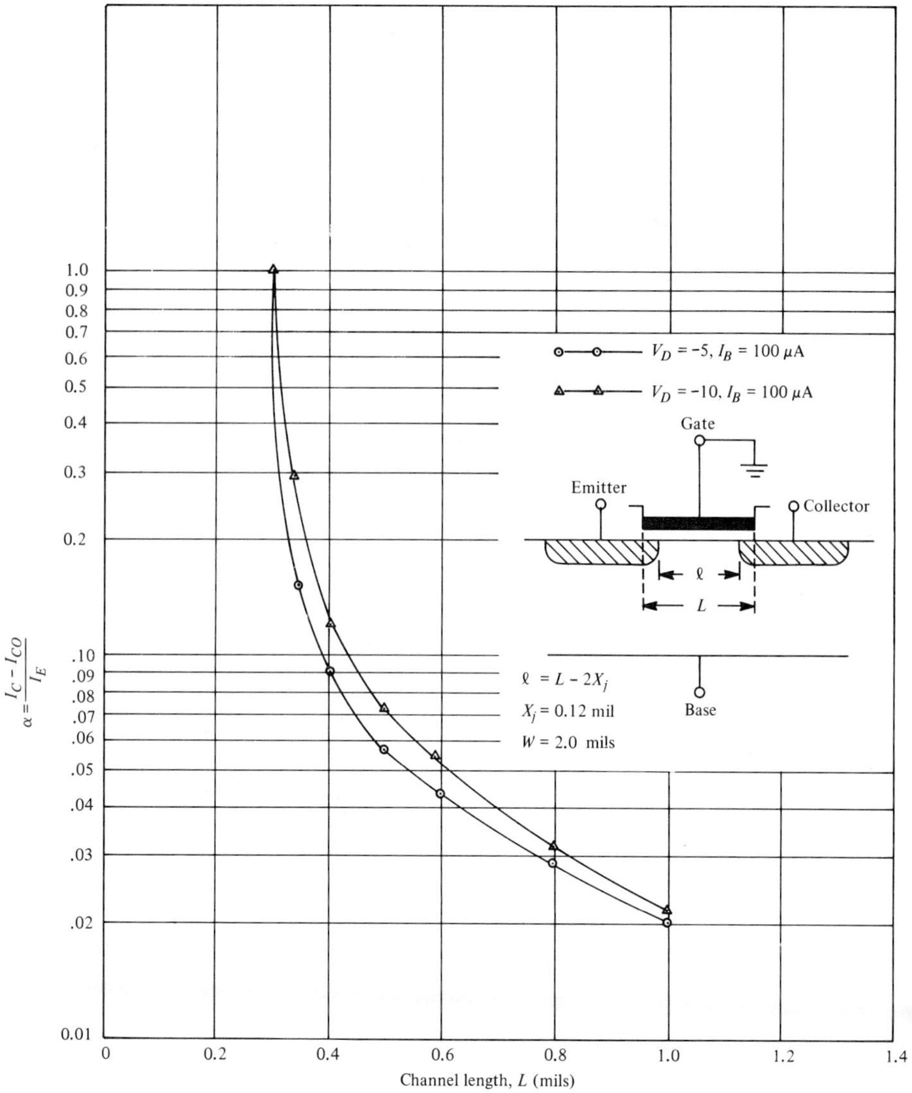

Figure 2–56 Lateral *PNP* transistor α as a function of *L*.

on completely interconnected circuits, it is common to incorporate some type of test structure or test transistor into the wafer. Measurements on this test structure can be used to help evaluate the suitability of the wafer and to determine parameter values and distributions. These types of measurements are referred to as "wafer mapping."

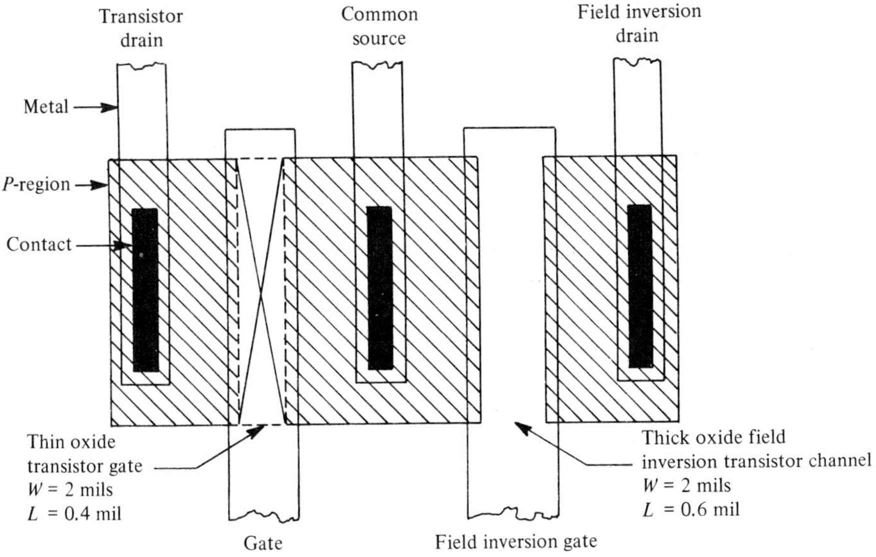

Figure 2–57 Typical test device.

Two general applications of test devices are common. In one of these, a relatively simple structure is incorporated in each individual circuit die. Figure 2-57 illustrates such a device. The test structure has metal pads which are used for electrical probing while the circuit is still in the wafer form. These pads are not bonded when the circuit is packaged, so the test device is not available for testing in a final packaged device. However, the capability of probing the test device after the package has been opened is useful in the analysis of defective units. This type of test device is kept as small and simple as possible since it appears on each die.

The other type of test structure is much larger (usually about the same size as a circuit die) and is placed at selected positions on a wafer instead of the normal circuit. This is done by replacing the normal circuit mask by a test pattern mask during certain steps in the photoreduction process. After wafer mapping, the test dice are discarded when the circuit dice are packaged. This type of test device allows a great variety of different structures. In addition to the basic mapping transistors, there are sometimes very large transistors, different types of capacitors, line resolution patterns, and even simple complete circuits, such as several stages of shift register. Thus quite complete data about a wafer may be obtained from these structures; but after they are separated from the circuit dice, diagnostic trouble shooting and failure analysis are more difficult.

The following list outlines those parameters that are most often measured during wafer mapping.

Threshold Voltage (V_T at 1 μA)

This measurement is invariably made with the transistor saturated and $V_{GS} = V_{DS}$, (Figure 2-58). The voltage is measured at a definite current (usually 1 μA) and is thus close to, but different from, the true threshold voltage extrapolated to zero current. The symbol V_{GST} and a current level of 10 μA are sometimes used.

Channel Conductivity

Two methods are common. The first (Figure 2-59) uses the saturated connection as above and fixes the value of the term ($V_{GS} - V_T$) in the saturated drain current equation (Equation 2-74).

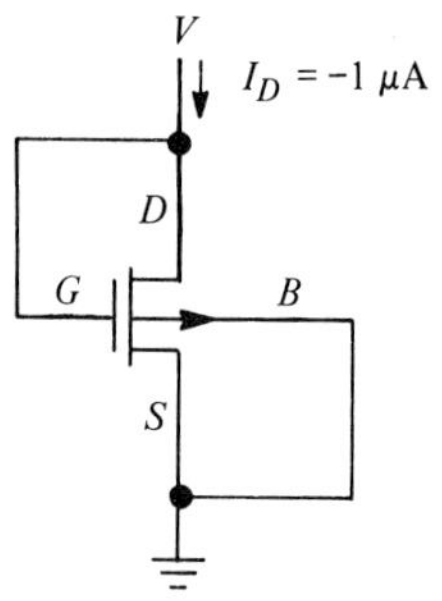

Figure 2-58 V_T diagram.

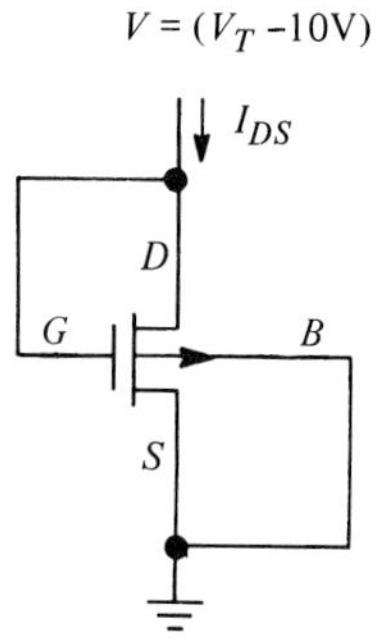

Figure 2-59 First *k'* diagram.

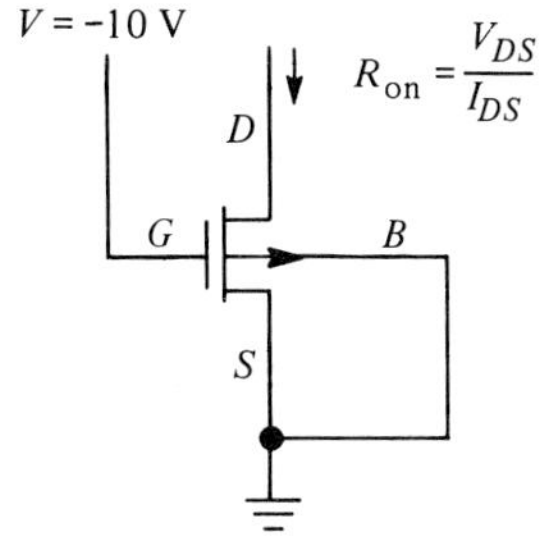

Figure 2–60 Second k' diagram.

$$I_{DS} = -k(V_{GS} - V_T)^2$$

A measurement of I_{DS} is thus proportional to the transistor conduction factor (k). The intrinsic channel conduction factor (k') is then found from Equation 2-72.

$$k' = k\,\frac{\ell}{W} = -\frac{\ell}{W}\cdot\frac{I_{DS}}{(V_{GS} - V_T)^2}$$

This method has the practical advantage that the measured quantity I_{DS} is easily converted to the design parameter k'. Since this is a saturated measurement, the values will be slightly conservative (lower) compared to those in an unsaturated measurement. The disadvantage is that it is a two-step measurement that requires a prior measurement of V_T.

The second method (Figure 2-60) uses a nonsaturated condition and measures the drain resistance with a fixed value of gate bias. From Equation 2-94 the condition factor is

$$k = -\frac{1}{2R_{\text{on}}(V_{GS} - V_T)}$$

This type of measurement is very easy to instrument with a digital meter and gives values that are accurate near the origin of the transistor characteristics.

Transistor Drain Breakdown (BV_{DSS})

This is the usual breakdown measurement (Figure 2-61) and gives the lower of either the punch-through or the gate modulated junction breakdown as discussed in Section 2.4.11.

Gate Modulated Junction Breakdown (BV_{PNG})

This measurement is sometimes used in addition to BV_{DSS} (Figure 2-62). Drain to source punch-through may or may not occur; but no current exists as long as the source lead is open.

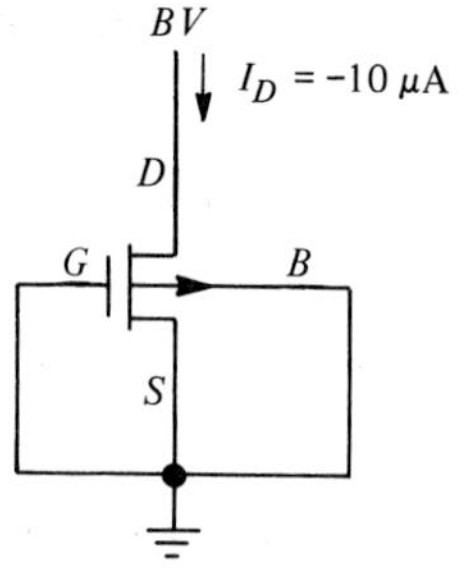

Figure 2–61 BV_{DSS} diagram.

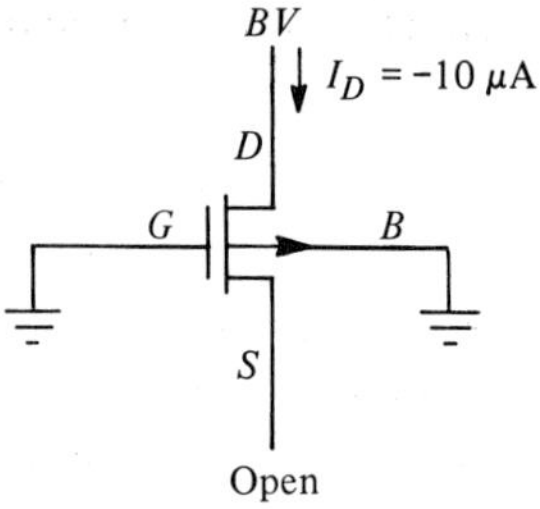

Figure 2–62 BV_{PNG} diagram.

Field Inversion Threshold (V_{TF})

This measurement is the same as V_T on a conventional transistor. The source to drain spacing in this test structure (Figure 2-63) must be larger than the normal thin oxide transistor in order to avoid source to drain punch-through.

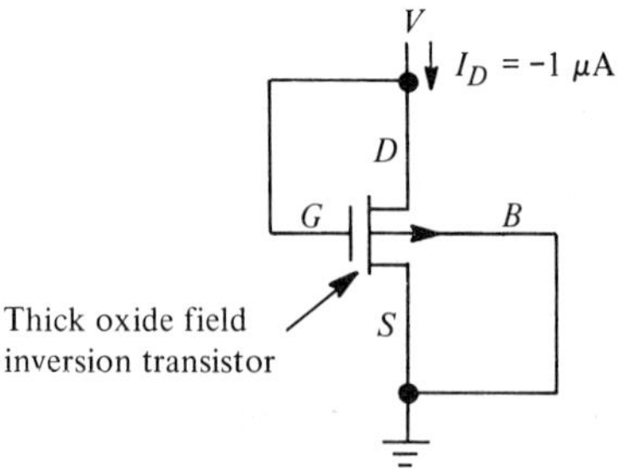

Figure 2–63 V_{TF} diagram.

Junction Breakdown (BV_{PN})

This measurement can be made on the field inversion structure by leaving the source lead open (Figure 2-64). Since BV_{PN} is usually so much greater than the normal transistor breakdowns BV_{DSS} and BV_{PNG}, it is of little interest for design purposes but is useful for monitoring the manufacturing process.

P-Region Resistivity (ρ_p)

A simple diffused resistor is sometimes included in a test pattern for measuring resistivity. This is calculated from the dimensions of the diffused resistor and the measured resistance by Equation 2-109 or 2-111. Two variations may be considered. A long narrow resistor most closely approximates diffused resistors used as load elements in a circuit, and measurements on these will give production spread information directly applicable to resistor design. If the resistor is closer to a square shape and has large dimensions compared to the usual masking dimensional tolerances, it serves as an accurate monitor of the intrinsic process sheet resistivity.

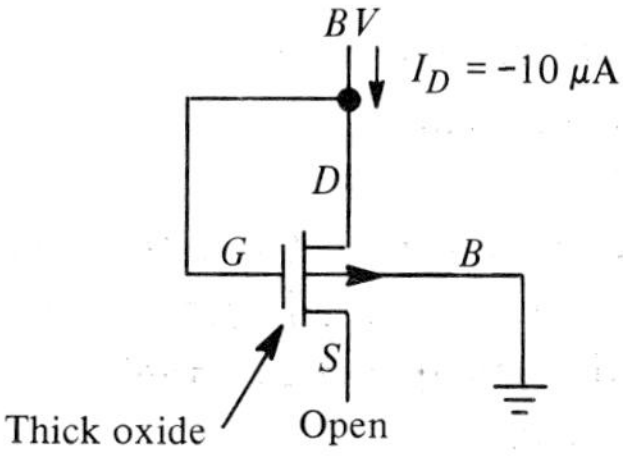

Figure 2–64 BV_{PN} diagram.

NOMENCLATURE

A	Area
B	Body (substrate)
C	Capacitance, usually per unit area
C_d, C_j	Junction depletion layer capacitance
C_{MOS}	Capacitance of the MOS capacitor
C_{ox}	Capacitance of the gate oxide
D	Drain
D_n, D_p	Diffusion constant for electrons, holes
E	Electron energy
E_c	Electron energy of the lower edge of the conduction band
E_F	Electron energy of the Fermi level
E_g	Energy gap of the forbidden band
E_i	Electron energy of the Fermi level in intrinsic material
E_v	Electron energy of the upper edge of the valence band
$\mathcal{E}$	Electric field
G	Gate; channel conductance
G_0	Channel conductance at zero drain voltage
I	Current
I_C	Channel current
I_{CP}	Channel current at pinch-off
I_{DS}	Drain to source current
J	Current density
J_F	Current density in a forward biased junction
J_R	Current density in a reversed biased junction
J_S	Sheet current density in the channel
$\mathfrak{k}$	Boltzmann's constant
k	Transistor conduction factor
k'	Intrinsic channel conduction factor
ℓ	Length of a conducting channel
L	Channel length before diffusion
L_n, L_p	Electron, hole diffusion length
$M(V)$	Avalanche multiplication factor as a function of voltage
n	Free electron concentration
n_i	Free electron concentration in intrinsic silicon
n_p	Free electron concentration in acceptor doped silicon
N_A	Density of acceptor atoms
N_c	Density of energy levels in the conduction band
N_D	Density of donor atoms
N_v	Density of energy levels in the valence band

p	Free hole concentration
p_n	Free hole concentration in donor doped silicon
q	Magnitude of the electronic charge
Q	Charge; charge per unit surface area
Q_A	Charge in accumulation layer
Q_B	Charge in depletion layer under the channel
Q_{BO}	Depletion layer charge in thermal equilibrium
Q_I	Mobile charge in the inversion layer
Q_{ox}	Fixed charges near the Si—SiO_2 interface due to the presence of the oxide over the silicon
R	Resistance
S	Source
t_{ox}	Thickness of the oxide
T	Absolute temperature
V	Voltage; voltage of an incremental section of the transistor channel
V_{BN}	Avalanche breakdown voltage
V_{BS}	Body to source voltage
V_D	Drain to substrate voltage
V_{DP}	Drain voltage at pinch-off
V_{DS}	Drain to source voltage
V_{FB}	Flat band voltage
V_G	Gate to substrate voltage
V_P	Protective device breakdown voltage
V_S	Source to substrate voltage
V_T	Threshold voltage
V_{TF}	Field (thick oxide) threshold voltage
W	Width of a conducting channel
x	Distance into the silicon substrate
X_d	Depletion layer thickness
X_{dn}	Depletion layer thickness on the n side
X_{dp}	Depletion layer thickness of the p side
X_j	Distance of the metallurgical junction below the surface
y	Distance along the surface in the direction of current flow
α	Base transport factor; resistor side diffusion parameter
ϵ_0	Permittivity of free space
ϵ_{ox}	Permittivity of the oxide
ϵ_s	Permittivity of silicon
κ	Dielectric constant
μ_n, μ_p	Mobility of electrons, holes; also average mobility in the inversion layer
ρ	Charge density; resistivity

ρ_s	Sheet resistivity
σ	Conductivity
σ_s	Sheet conductivity
τ_n	Minority carrier life time of an electron
ϕ	Electrostatic potential
ϕ_B	Equilibrium junction potential; band bending at the surface
ϕ_F	Potential of the intrinsic level with respect to the Fermi level in the bulk
ϕ_{Fn}	The Fermi potential in an *n* region
ϕ_{Fp}	The Fermi potential in a *p* region
ϕ_{MS}	Metal to silicon work function

REFERENCES

1. R. A. Evans, et al., "Physical/Electrical Properties of Silicon," *ASD-TDR-63-316,* **V,** Research Triangle Institute, Durham, N.C., 1964.
2. J. C. Irvin, "Resistivity of Bulk Silicon and of Diffused Layers in Silicon," *Bell Sys. Tech. J.,* **41,** 387–410 (March 1962).
3. J. L. Moll, "The Evaluation of The Theory for The Voltage Current Equations of *P-N* Junctions," *Proc. IRE,* **46,** 1076–1082 (June 1958).
4. W. Shockley, "The Theory of *P-N* Junctions in Semiconductors and *P-N* Junction Transistors," *BSTJ,* **28,** 435–489 (July 1949).
5. A. S. Grove, *Physics and Technology of Semiconductor Devices,* John Wiley & Sons, New York, 1967.
6. R. H. Kingston and S. F. Neustadter, "Calculations of the Space Charge, Electric Field, and Free Carrier Concentration at the Surface of a Semiconductor," *J. Appl. Phys.,* **26,** 718–720 (June 1955).
7. A. Many, Y. Goldstein, and N. B. Grover, *Semiconductor Surfaces,* American Elsevier, New York, 1965.
8. D. R. Frankl, *Electrical Properties of Semiconductor Surfaces,* Pergamon Press, London, 1967.
9. C. T. Sah, "Characteristics of the Metal-Oxide-Semiconductor Transistor," *IEEE Trans. Electron Devices,* **ED-11,** 324–345 (July 1964).
10. C. T. Sah and H. C. Pao, "The Effects of Fixed Bulk Charge on the Characteristics of Metal-Oxide-Semiconductor Transistors," *IEEE Trans. Electron Devices,* **ED-13,** 393–409 (April 1966).
11. H. C. Pao and C. T. Sah, "Effects of Diffusion Current on Characteristics of Metal-Oxide (Insulator)-Semiconductor Transistors," in *Solid State Electronics,* **9,** 927–937, Pergamon Press, London (1966).
12. E. Kooi, *The Surface Properties of Oxidized Silicon,* Springer-Verlag, New York, 1967.
13. A. G. Revesz and K. H. Zaininger, "The Si—SiO_2 Solid—Solid Interface System," *RCA Rev.,* **29,** no. 1 (March 1968).

14. P. V. Gray, "The Silicon—Silicon Dioxide System," *Proc. IEEE,* **57,** no. 9, 1543–1551, (September 1969).

15. B. E. Deal, E. H. Snow, and C. A. Mead, "Barrier Energies in Metal-Silicon Dioxide-Silicon Structures," in *J. Phy. Chem. Solids,* **27,** 1873–1879, Pergamon Press London (1966).

16. H. K. J. Ihantola, "Design Theory of a Surface Field-Effect Transistor," *Tech. Report No. 1661-1,* Stanford Electronics Labs, Palo Alto, California, August 1961.

17. A. J. Dekker, *Solid State Physics,* Prentice-Hall, New York, 1962.

18. R. F. Crawford, *MOS FET in Circuit* Design, McGraw-Hill, New York, 1967.

19. J. T. Wallmark and H. Johnson, *Field Effect Transistors,* Prentice-Hall, New York, 1966.

20. R. M. Warner, Jr. and J. N. Fordemwalt, eds., *Integrated Circuits, Design Principles and Fabrications,* McGraw-Hill, New York, 1965.

GENERAL REFERENCE

R. S. C. Cobbold, *Theory and Application of Field Effect Transistors,* John Wiley & Sons, New York, 1970.

3 MOS Processing

3.1 INTRODUCTION

In this chapter, a comparison is made of the processes currently used to manufacture MOS integrated circuits. The conventional *P*-channel MOS, by virtue of its widespread manufacture, is used as a common reference in this comparison. A typical *P*-channel process is therefore discussed in some detail, preceded by a brief review of basic techniques common to the manufacture of any silicon semiconductor device. Electrical characteristics and fabrication methods that affect the design and manufacture of circuits are then examined for the various MOS processes.

3.2 BASIC SILICON PROCESSING TECHNIQUES

Silicon has become the principal semiconductor material primarily because a dense, uniform oxide which is stable over a wide range of temperatures can be grown on its surface. This property makes it possible to use the oxide as a barrier to selectively mask against the diffusion of dopants. A description of processing techniques that employ this characteristic follows.

3.2.1 Silicon Material, Diffusion, Oxidation

Silicon semiconductor devices are fabricated on thin, single-crystal substrates called "wafers." These wafers are sliced from a cyclindrical ingot

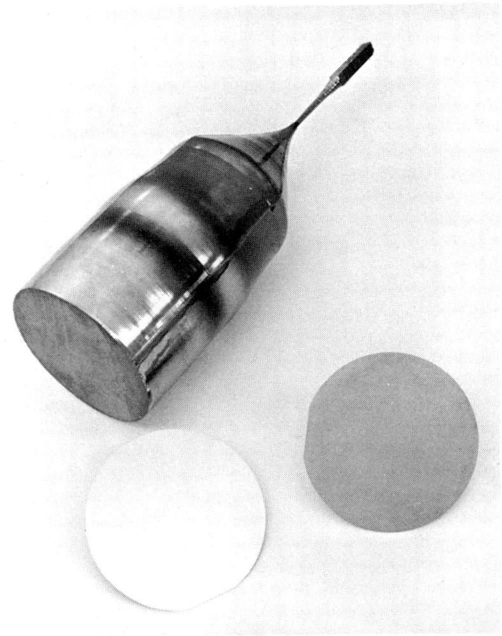

Figure 3–1 Silicon wafers.

of silicon (Figure 3-1) and lapped and polished before processing.[1] Depending on crystal size, wafers vary in diameter from less than one inch to several inches. Electrical characteristics of the wafer substrates are determined by the doping type and concentration, which are controlled during the crystal growing process.

Electrical junctions are formed in the substrate by introducing the opposite type of dopant to a concentration that exceeds the bulk doping by means of solid state diffusion. Junctions can be obtained in selected regions by masking those areas of the silicon surface that are not to be doped with a suitable thick layer of silicon dioxide.

The oxide layer is grown by exposing the wafer to an oxidizing ambient of oxygen or water vapor, or both, at an elevated temperature (typically in the range 900° to 1200°C). This is done in open-ended cylindrical quartz tubes that are resistance-heated in special "diffusion furnaces" (Figure 3-2). These furnaces provide precise control of temperature (deviations are less than ± 1°C) over a considerable length of the tube called the "flat zone." Oxidation conditions are maintained so that an excess of oxidizing reactants exist in the tube atmosphere. This is done

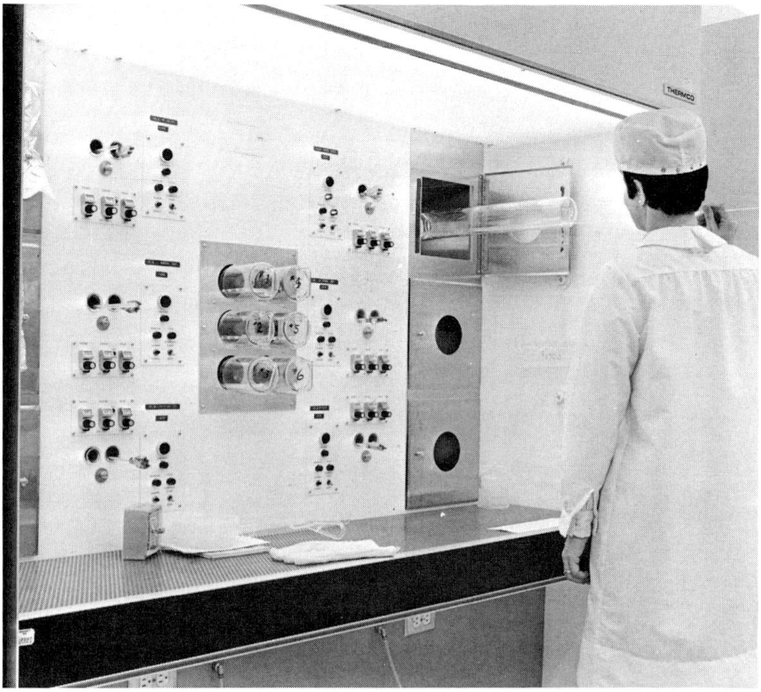

Figure 3–2 Diffusion furnaces.

by introducing the reactants at one end of the furnace tube and allowing free exhaustion at the other end. Under this condition, temperature remains the controlling factor in oxidation, making it possible to grow a uniform oxide layer on many wafers within the furnace flat zone. Once the oxide is suitably patterned by a technique to be discussed shortly, the wafer is doped by diffusion techniques.

The diffusion of dopants is usually accomplished by a two-step process.[2] The first, called "predeposition," is used to introduce dopant impurities to only a shallow depth in the silicon. This is done in a manner similar to oxidation. The furnace atmosphere is now comprised of a gaseous compound of the dopant atom, which coats the wafer surface. The compound in turn reacts with the silicon to form a dopant-rich surface layer, which serves as a source of dopant for further diffusion in the second step. This step, called "drive-in diffusion," diffuses the dopant impurities deeper into the silicon to the desired concentration profile. Diffusion is generally done in an inert gas atmosphere, but an oxidation step usually follows to form a protective layer over the diffused regions.

3.2.2 Masking and Etching

In order to use silicon dioxide as a selective diffusion barrier, it is necessary to define the oxide coverage on the silicon material. This is accomplished by the so-called photolithographic[3] or masking process. The process patterns the oxide in a sequence that employs three key materials: a photographic "mask"; a photosensitive material called "photoresist"; and a silicon dioxide etching acid. In short, the pattern on the mask is imaged on the photoresist film which, in turn, protects specific areas of the oxidized wafer from attack, when it is exposed to an oxide etchant. These three materials and their use are described in the following paragraphs.

Masks are photographically produced plates that have opaque device size patterns repeated on them (Figure 3-3). Typically they are obtained by the following procedure. First, a "ruby" master, as described in Sec-

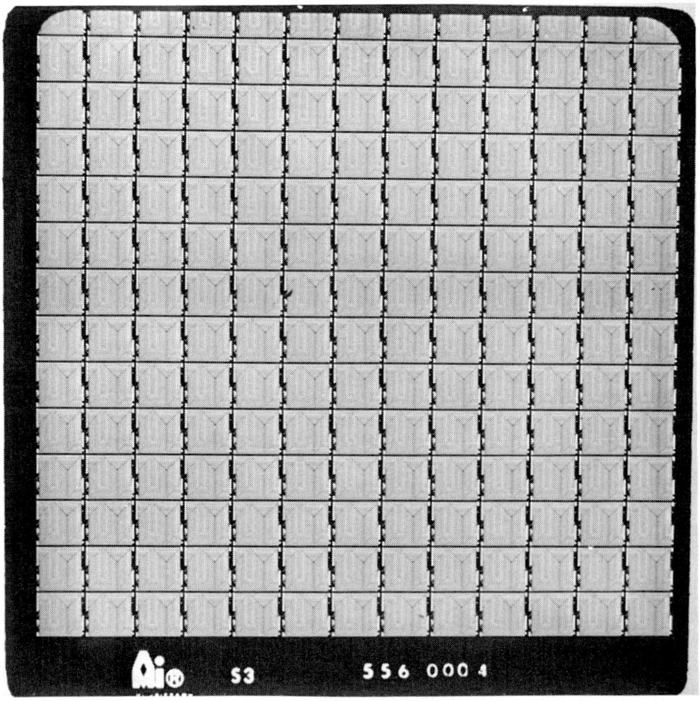

Figure 3–3 Integrated circuit mask.

tion 8.4 of Chapter 8, is produced for each circuit level or mask required. The pattern on each ruby is then reduced to final device size by a two-step photographic reduction. Since the ruby is either transparent or opaque to a suitable monochromatic light, the initial reduction is performed by transmitted illumination, recording the image on a high resolution, high

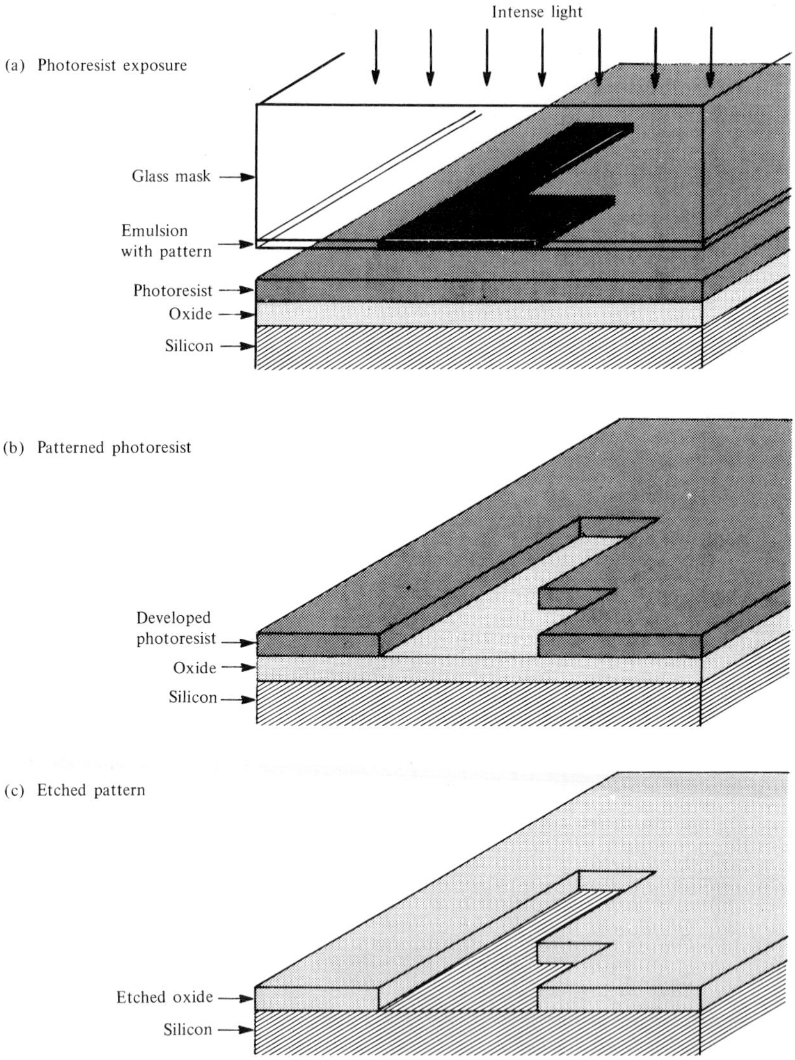

Figure 3–4 Masking process.

contrast photographic plate. The first reduction image is reduced a second time, while simultaneously repeating the pattern in matrix fashion, a process called "step and repeat." The positioning of repeated patterns must be done to exacting tolerances, since the alignment of successive device levels depends on exact pattern positioning across the entire mask. The second step reduction pattern is then contact-printed to form the final masks or "working plates." The working plates can be comprised of any of several materials, the most common being emulsion on glass. More durable opaque thin films, such as chrome metal, are also extensively used. The masks can be produced to have opaque patterns on a clear background or vice versa, depending on the material to be removed in subsequent etching and the type of photoresist used.

Photoresist, the second material, is an etchant resistant polymer that is photosensitive when applied in thin layers. This is usually done centrifugally by placing a drop of liquid resist on the wafer and spinning it to form a uniform film. The resist can then be patterned as follows. If the resist is of the "negative" type, it will be soluble in certain liquid developers unless polymerized by exposure to intense light. By means of contact printing with the appropriate mask (Figure 3-4a), the polymerization will not occur under the opaque mask areas. Conversely, if the resist is of the "positive" type it will be soluble in developers only in areas exposed to light. The photoresist and suitable mask can thus be used to obtain a device-pattern, etch-resistant film which is used, in turn, for selective etching of the silicon dioxide film (Figure 3-4b).

The material used to etch the oxide is a solution of hydrofluoric acid (HF). Usually, it is used in a buffered form containing ammonium fluoride to ensure more constant etching characteristics. When the photoresist-patterned wafer is exposed to this solution, the oxide will be attacked in areas not covered by resist. For diffusion masking, the oxide is entirely removed from the regions to be doped down to the bare silicon substrate, which is not affected by the etchant (Figure 3-4c).

Oxide etching may be required for extra diffusions, contact openings, or thinning oxide regions. These remasking steps must be performed so that subsequent layers are accurately located with respect to preceding ones. This registration is ensured by a technique called "mask alignment." With a mask aligner, (Figure 3-5) which consists of an optical microscope and electromechanical servos, it is possible to align precisely the new pattern visible on the mask to the preceding pattern on the wafer. This is done by viewing through the transparent portions of the mask. The photoresist-coated wafer is then exposed and the development and etch procedure is repeated. Alignment of the masks is often facilitated by the incorporation of "alignment marks." By superposing these marks,

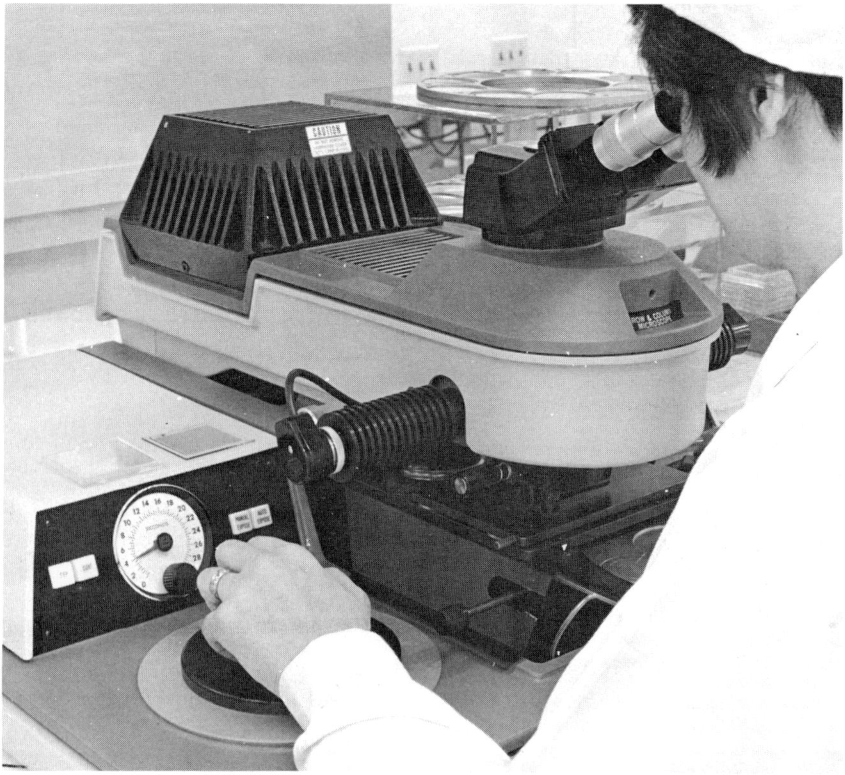

Figure 3–5 Mask aligner.

device geometries can be aligned without the need for familiarity with device topology.

Masking techniques are also employed when the diffused regions are to be connected to form circuits. The interconnection medium used is a metal thin film or combination of films. After etching contact openings through the oxide down to the diffused regions, the wafer is coated with a thin metal film, usually by evaporation. The metal to be deposited is heated in a vacuum system until it vaporizes, uniformly coating the wafer surface. Aluminum is the metal used most often. The metal film is then defined into the desired pattern by the same photoresist methods as described previously, but using a suitable metal etch.

3.2.3 Assembly

The separation of individual circuits from the wafer substrate is most commonly done by a procedure called "scribing and breaking." This is

Figure 3–6 Wafer scribing.

accomplished by taking advantage of the natural tendency of the silicon wafer to cleave along certain crystal planes. During the ingot sawing operation, the wafer was given an asymmetry, or flat, to index the orientation of the crystal planes. Subsequent masking operations take this orientation into account. When the completed wafer is oriented and scored with a diamond scribe (Figure 3-6), inducing localized crystallographic stress, it is broken along the stress lines into separate fragments called "dice" (Figure 3-7). Newer methods of dicing such as laser scribing and "sawing" are also gaining acceptance.

The next step is dice encapsulation. Besides being a protective enclosure for the device, a package must also provide a means for connecting the contact points on the die to the external leads (Figure 3-8). The dice are mounted on package substrates, usually by means of thermal solder type attachment. This "die attach" step, in addition to mechanically

Figure 3–7 Diced wafer.

securing the die in the package, provides an electrically conductive connection to the silicon substrate. This also allows the heat generated in the circuit to be dissipated by conduction into the package.

Connection to the contact points or "pads" on the die is then accomplished by wire bonding. In this "lead bonding" operation, one end of a fine metal wire (typically 1 mil in diameter) is attached to the die pads and the other end is attached to the package bonding "lands." The package is fabricated in such a manner that the lands in turn are connected to the external package leads. Lead bonding is repeated for each of the pads to be connected (Figure 3-9). Other more automated methods by which all connections are made simultaneously have been developed; but these require special wafer preparation or highly specialized packages.

The last assembly operation is to finalize the encapsulation of the die by isolating the sensitive circuit from its environment. By attaching a lid

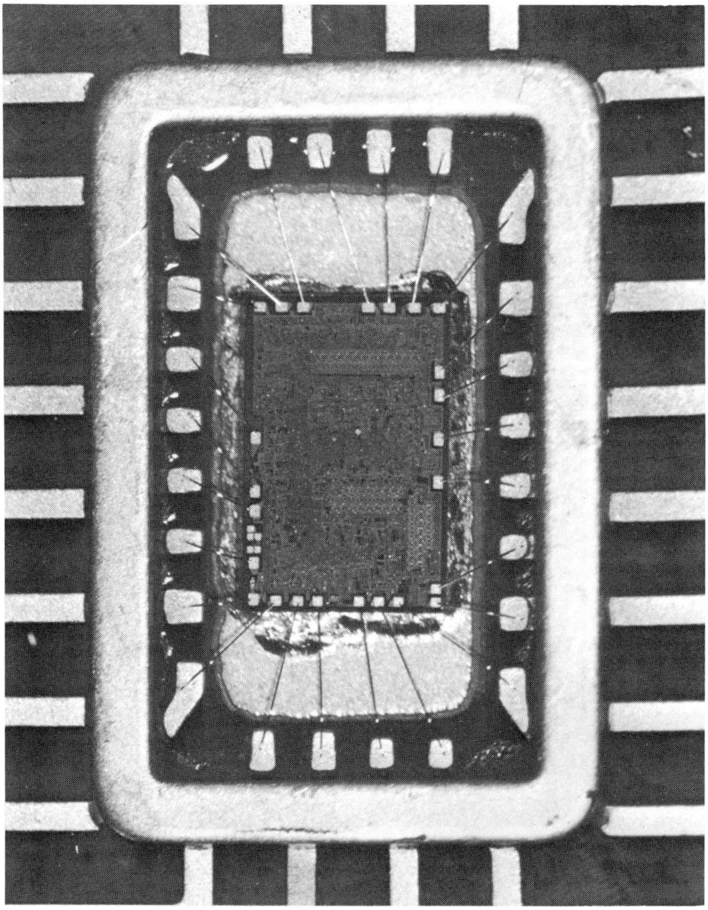

Figure 3–8 Packaged die.

over the package substrate while it is in an inert gas ambient, the die is hermetically sealed. Direct encapsulation of the die in plastic materials is also possible for certain kinds of circuits, but this often requires special wafer preparation.

3.3 TYPICAL *P*-CHANNEL MOS PROCESS

In the evolution of processes used to manufacture large-scale integrated circuits, *P*-channel MOS has become the industry standard. The dominance of *P*-channel over *N*-channel is based upon the ease with which the sur-

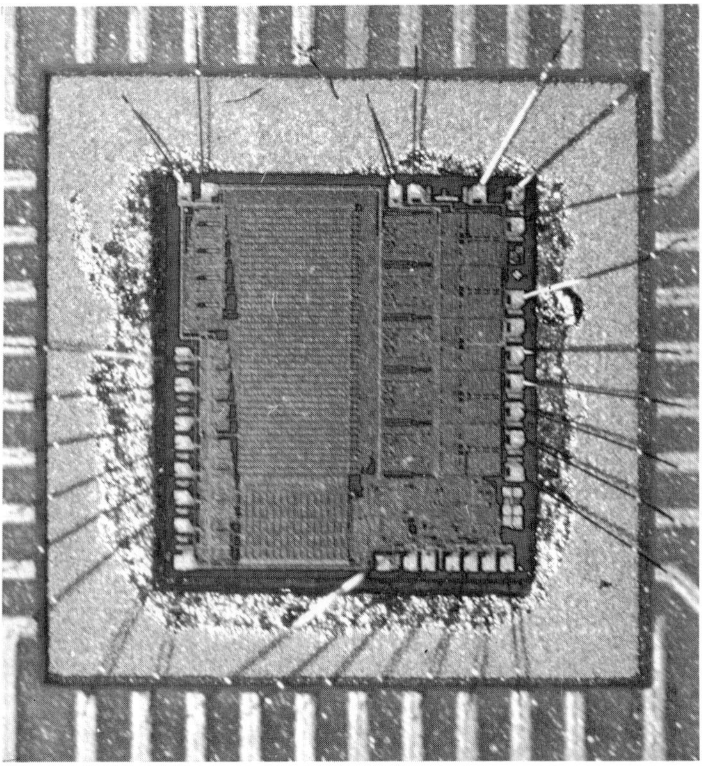

Figure 3–9 After lead bonding.

face properties of oxidized silicon can be controlled in the *P*-channel process. Typical manufacturing variations in the surface state density, which caused variations in the threshold voltage of *P*-channel transistors (see Section 2.3), were often sufficient to change *N*-channel transistors from enhancement to depletion mode. Thus, although *N*-channel MOSFETs offered theoretically superior electrical performance (see Section 3.4.2), the reproducible enhancement mode characteristics of *P*-channel became the basis of digital integrated circuit development.

Since its inception, several variations of the basic *P*-channel process technology have been developed. In order to understand better the characteristics of other *P*-channel processes, and the fabrication of MOSFETs in general, the fabrication sequence of a typical "thick oxide" *P*-channel MOS process (see Figure 3-10) will be described in the following section. The term "thick oxide" refers not to the field oxide, but to the final oxide

A Wafer fabrication

1. Starting material:<111>silicon (Figure 3-1)
2. Initial oxidation (Figure 3-11)
3. 1st (*P*-region) mask (Figures 3-12 and 3-13)
4. Boron predeposition (Figure 3-15)
5. Diffusion-oxidation (Figure 3-17)
6. 2nd (gate) mask (Figures 3-18 and 3-19)
7. Gate oxidation (Figure 3-20)
8. 3rd (contact) mask (Figure 3-21)
9. Metalization (Figure 3-23)
10. 4th (metal) mask (Figure 3–24)
11. Passivation deposition
12. 5th (pad) mask

B MOS Assembly

1. Wafer evaluation
2. Die sort
3. Scribing and breaking (Figure 3-6)
4. Optical inspection
5. Die attach operation (Figure 3-29)
6. Lead bonding operation (Figure 3-30)
7. Optical inspection
8. Sealing (Figure 3-31)
9. Hermeticity testing
10. Environmental testing
11. Final electrical testing

Figure 3–10 *P*-channel process sequence.

($\sim$ 1 μ) over the diffused *P*-regions. In the discussion of wafer fabrication, diagrams of a single MOS device at appropriate stages will be used as illustration. The advantages of "thick oxide" MOS as compared to earlier "thin oxide" MOS will be discussed in a later section.

3.3.1 Wafer Fabrication

The starting material for *P*-channel processing is high-quality, polished *N*-type wafers in a narrow resistivity range. The selected resistivity is a compromise between minimizing depletion region spreading, necessary for compact geometries, and maintaining sufficiently high junction breakdown voltages. As reviewed in the Section 2.4.4, lower than expected breakdown voltages are obtained due to gate modulation of the avalanche process, for a given substrate resistivity. As a result, the typical starting material might be in the range 3 to 6 Ωcm, with <111> crystal orientation. This silicon crystal orientation was first used to commercially fabricate *P*-channel MOSFETs, although <100> orientation is now also used (see Section 3.4.3).

After the wafer is thoroughly cleaned, a relatively thin oxide is grown

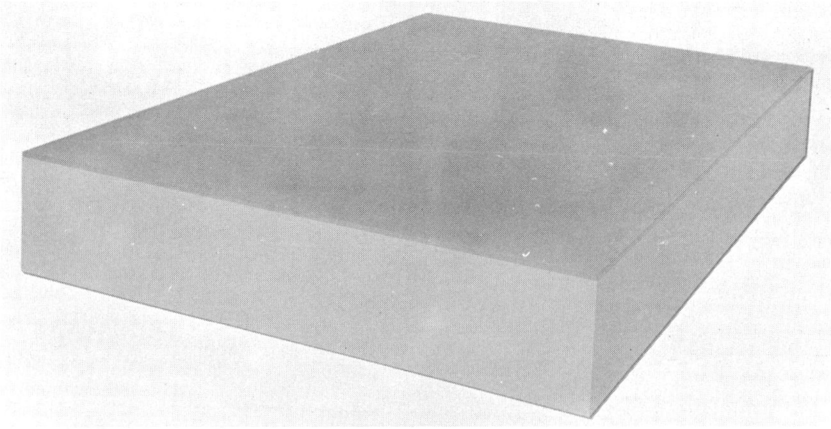

Figure 3–11 Oxidized silicon.

over its entire surface to serve as a mask against subsequent diffusion (Figure 3-11). Boron, which is used as the *P*-type dopant, is effectively masked by oxide layers on the order of 1000 Å. However, the initial oxide is often made substantially thicker in order to avoid growing extra field oxide later in the process.

After initial oxidation, the wafer proceeds to the first masking step, the *P*-diffusion mask. This mask serves to define the source and drain regions of all the MOS transistors on a die and the "crossunders" or *P*-region submetal interconnects. Since this masking and etching operation determines the source to drain spacing, or transistor length (L), it must be tightly controlled.

The wafer is first spin-coated with liquid photoresist and then baked to drive off the resist solvents. It is then ready for exposure of the first mask (Figure 3-12). Since the wafer has no previous pattern on it, this step requires no mask alignment, per se. The overall mask pattern, however, must be oriented to the flat of the wafer as necessary for post-fabrication wafer dicing operations. If a photoresist of a "negative" type is used, the areas under the opaque regions of the mask are unchanged when the wafer is exposed to an intense light source. Subsequent developing of the photoresist by means of dipping in or spraying with a suitable solvent removes the resist from these unexposed areas. The polymerized resist which was under the clear areas of the mask then remains in the desired pattern (Figure 3-13).

After a second bake to drive off residual developer and to complete polymerization of the photoresist, the wafer is ready for etching. The oxide etchant consists of a solution of hydrofluoric acid (HF) and

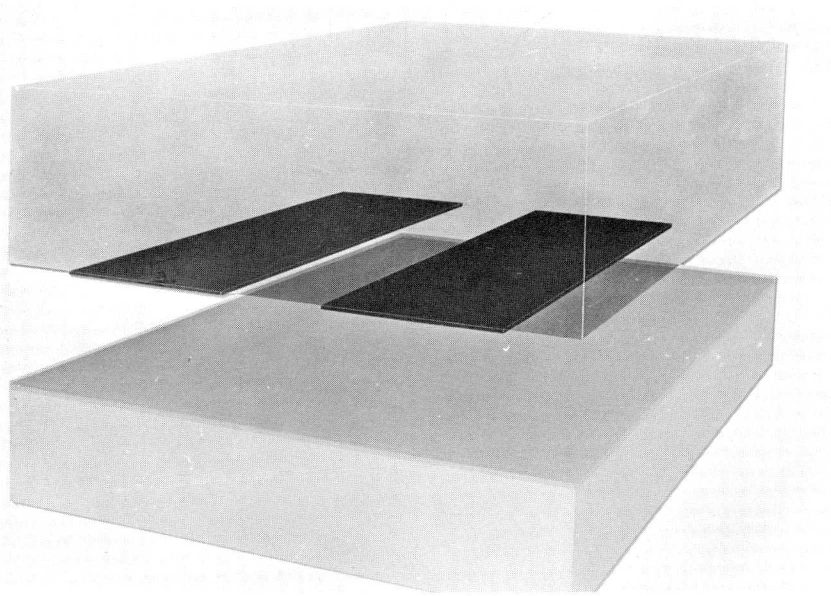

Figure 3–12 Exposure of 1st (*P*-region) mask.

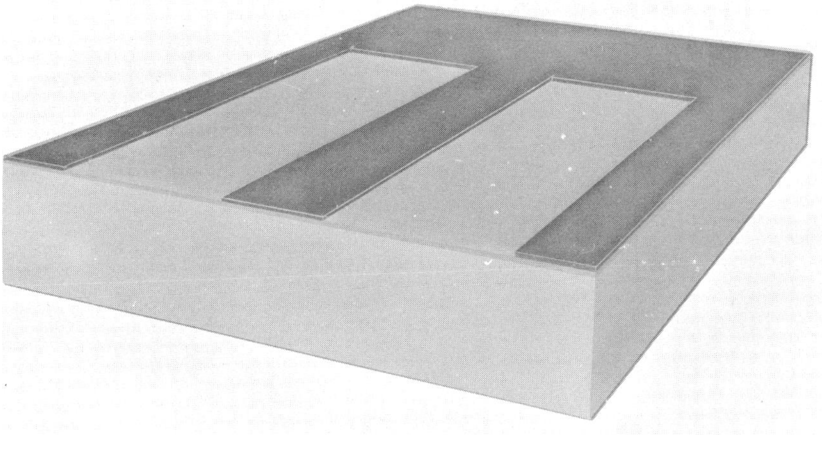

Figure 3–13 Photoresist developed (1st mask).

ammonium fluoride (NH_4F). The wafer is etched in this solution until all the oxide is removed from those areas that are free of photoresist. The photoresist is subsequently removed (Figure 3-14) in a suitable strip-

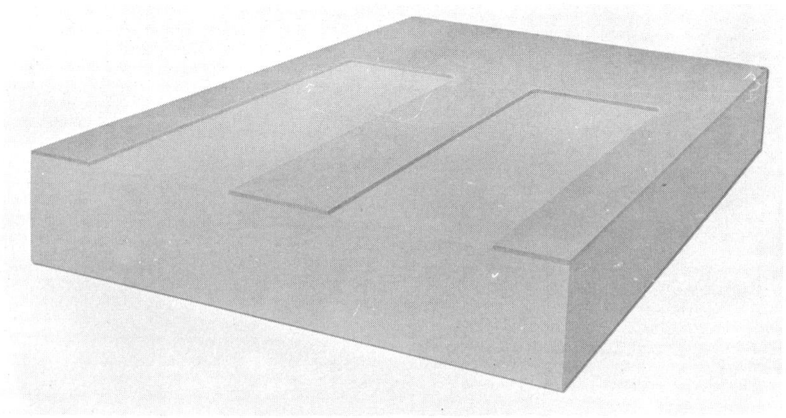

Figure 3–14 1st mask etched and photoresist removed.

ping solution, often a mixture of inorganic acids. Other commercially manufactured mixtures are also available. After a rinsing and drying procedure, the wafer is ready for boron predeposition.

Predeposition is performed in a diffusion furnace typically in the temperature range of 1000° to 1100°C. The surface concentration of the boron layer is a function of deposition temperature. Gaseous compounds of boron, such as diborane (B_2H_6) and boron trichloride (BCl_3), or liquid compounds, such as boron tribromide (BBr_3), are commonly used sources of boron dopant. Gaseous dopants may be directly introduced into the furnace diluted by an inert gas. Liquid dopants are used by bubbling an inert gas carrier through the liquid. Oxygen and/or water vapor is also introduced with BBr_3 and BCl_3 dopants to prevent halogen pitting of the silicon surface. Pitting occurs because the halogen compounds normally formed with silicon are volatile at elevated temperatures. When an oxidizing atmosphere during doping is maintained instead, boric oxide is thought to be formed by the following reaction (where X represents a halogen)

$$4BX_3 + 3O_2 \rightarrow 2B_2O_3 + 6X_2$$

Boric oxide in turn reacts with silicon to give free boron by the next reaction

$$2B_2O_3 + 3Si \rightarrow 3SiO_2 + 4B$$

Boric oxide and silicon dioxide together form a glassy layer which remains

Figure 3–15 After predeposition.

on the wafer after predeposition. This is removed by dipping the wafer in hydrofluoric acid solution which has no effect on the boron that has already been diffused into the silicon. The wafer is now ready for diffusion-oxidation as shown in Figure 3-15.

The diffusion-oxidation serves not only to diffuse the dopant impurities further into the silicon, but also to grow a protective layer of oxide over the sensitive *P*-doped regions. At this point there are two different concepts for obtaining the final thick oxide over the diffused regions. One approach, the thermal oxide method, is to continue the diffusion oxidation until the desired amount of oxide is grown. The alternate approach, the deposited oxide method, is to discontinue the diffusion-oxidation, having grown little or no oxide layer over the *P*-regions, and then deposit the additional oxide required.

These oxide depositions, or "chemical vapor depositions" as they are called, may also be done in either of two basic ways. One is a high-temperature method. The high-temperature deposited oxides (typically 800° to 1100°C) are done in an epitaxial type reactor (Figure 3-16). The wafers are located on an inductively heated graphite susceptor within a quartz reaction tube. A silicon compound, such as silicon tetrachloride ($SiCl_4$) or silane (SiH_4), that is diluted in an inert gas carrier reacts with an oxidant, such as carbon dioxide, to produce the silicon dioxide deposition on the hot substrates. Careful control of reactant composition, temperature uniformity, and gas flow are essential to deposition uniformity.

The low-temperature (typically 300° to 500°C) oxides are deposited

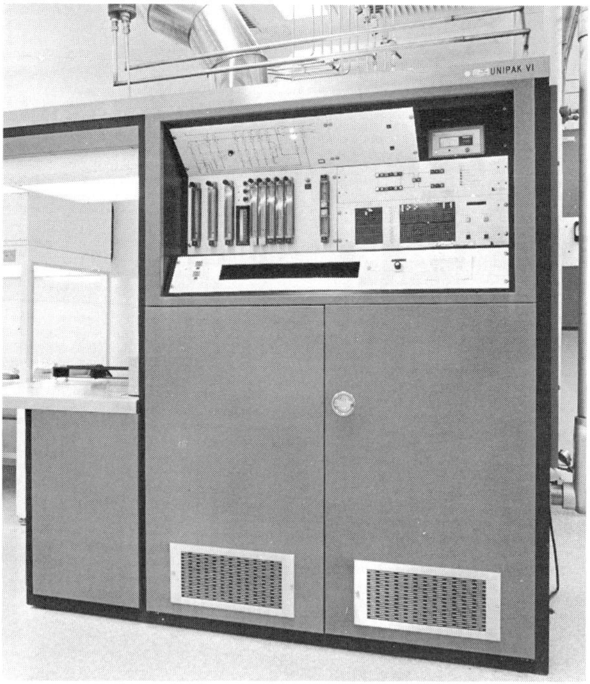

Figure 3–16 Epitaxial reactor.

in an analogous fashion, using a resistance heated reactor, but based on the decomposition of silane in oxygen. The characteristics of these two oxides may differ, but either serves to provide a thick *P*-region oxide and thickened field oxide.

If the alternate deposited oxide method is used, it is generally necessary to stabilize the oxide by performing a "gettering" step. It has been shown that a phosphorus doped glass layer will attract and trap mobile ionic contaminants (principally alkali ions).[4] Without this immobilization of ions, the characteristics of the MOS transistor could change as a function of temperature and/or applied electric field. Instead, use of thermally grown "clean" oxides is preferred since a phosphorus stabilization step is not required. With either the thermal or deposited oxide concept, final *P*-region oxide and field oxide thicknesses are typically at least 1 and 1.5 μ respectively (Figure 3-17).

At this point, the wafer is ready for the second or gate mask. This mask serves to thin or remove the oxide from the gate and *P*-region contact areas. Etching a thick oxide down to the thin gate oxide required (typically in the range of 1000 to 2000 Å) is usually impractical. The

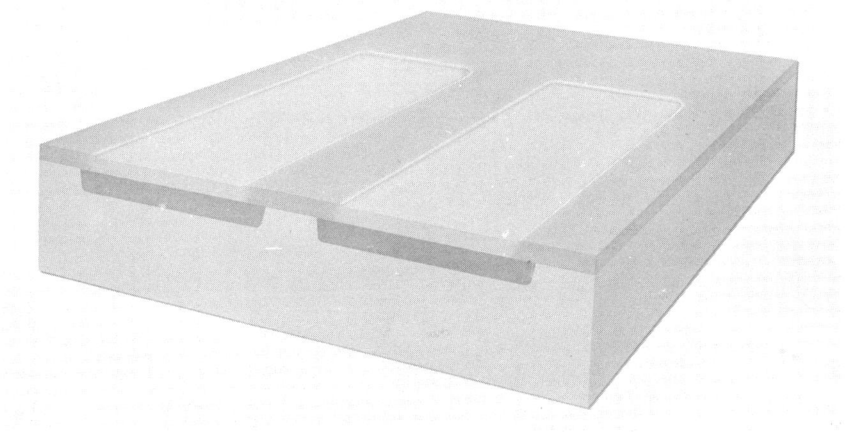

Figure 3–17 After *P*-region diffusion-oxidation.

oxide is thus generally completely removed and subsequently regrown to the final gate oxide thickness.

The masking and etching operation is performed in a manner similar to the previous masking step. However, alignment of the second mask to the first pattern is now of concern (Figure 3-18). Since the thin gate

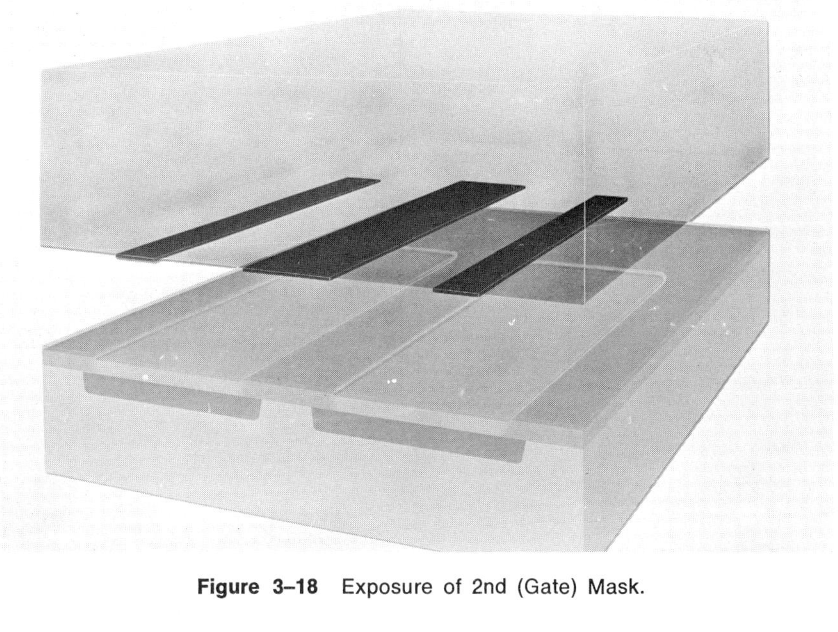

Figure 3–18 Exposure of 2nd (Gate) Mask.

oxide region must extend to both source and drain for an MOS transistor to function, the alignment of the gate mask must be very precise. The tolerance for misalignment depends on intentional gate overlap, as set forth in the "design rules," and the amount of lateral diffusion of the source and drain. These parameters vary significantly between manufacturers. Since gate overlap is a source of parasitic capacitance (see Section 2.4.9), designed overlap is usually kept to a minimum. Thus the lateral diffusion dimension, typically 2 to 3μ, is the only misalignment tolerance.

Since the oxide to be removed is very thick, the duration of the gate etch operation is long. The photoresist process and etch conditions must thus be carefully controlled to prevent "undercutting." This term refers to excess beveling of the etched oxide edge, which can pose a severe problem to the control of the gate dimensions. Since the silicon surface is bared by this etch, cleanliness of all procedures at this point is also critical. After completion of the etch and photoresist removal, the wafer appears as in Figure 3-19.

The gate oxidation step immediately follows the gate masking step. Oxidation of this thin layer is usually done in a dry oxygen atmosphere. This oxidation step serves not only to obtain accurately the final gate oxide thickness, but also to determine the surface state density (Q_{ox}) of the oxide-silicon interface (see Section 3.4.4). Since the control of this charge density is a dominant factor in the value and variation of the final MOS threshold voltage, it is essential that this operation be reproducible. The

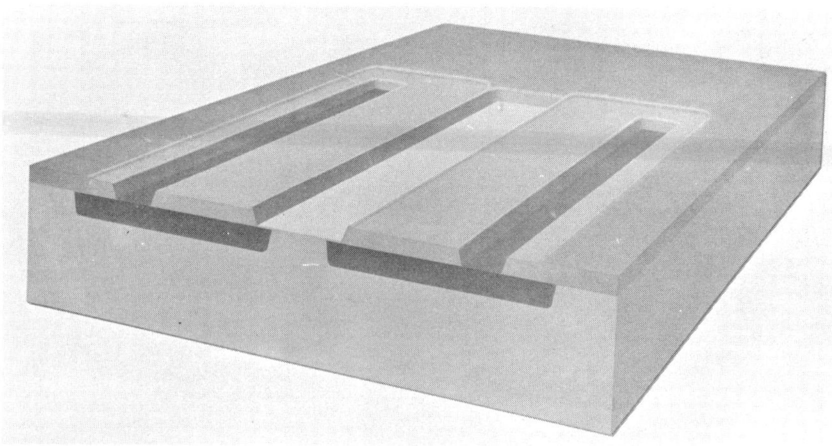

Figure 3–19 2nd mask etched and photoresist removed.

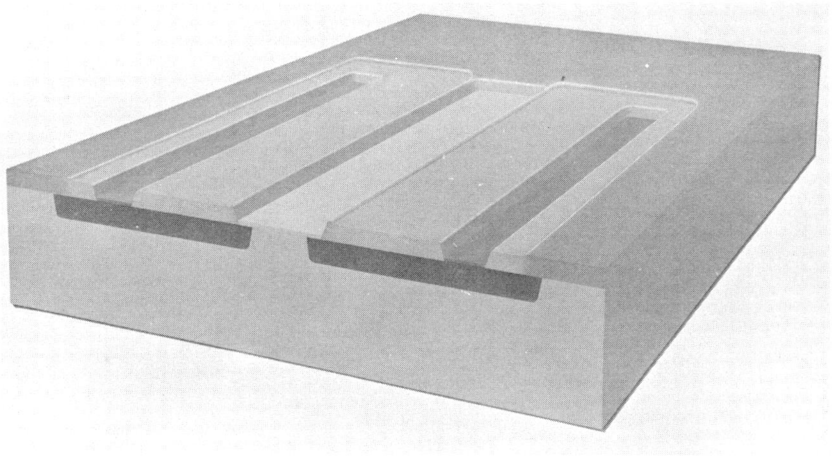

Figure 3–20 After gate oxidation.

gate oxidation step also simultaneously diffuses the junctions to their final depth (Figure 3-20).

Upon completion, the wafer proceeds to the third or oxide cutout masking step. Since a thin layer of oxide was grown in the *P*-region contact areas at the same time as the gate oxide was grown, it must now be removed. This masking and etching step is not of a critical nature, except for the obvious fact that the oxide in the "cutouts" must be entirely removed. Any residual oxide will pose contact problems at the interconnection step. Also, any voids in the resist coverage could result in pinholes being etched in the gate oxide, appearing later as shorts. Application of the third mask, completes all operations that control the oxide profiles (Figure 3-21).

Interconnect metalization is applied to the wafer by vacuum evaporation. Aluminum is the most common metalization used. By means of an electron beam evaporator (Figure 3-22), ultrapure films on the order of 1 μ thick are deposited. The wafer at this point is shown in Figure 3-23. Due to the deep oxide steps associated with the *P*-region contact cutouts, metalization step continuity is a prime consideration. Various evaporator fixtures have been designed to provide varying angles of incidence between wafer substrates and evaporator source during the metalization cycle to assure even distribution of the deposited metal.

A fourth masking step is used to delineate the interconnects and bonding pads (Figure 3-24). Precise edge definition and freedom from defects is necessary in this photoresist step; otherwise shorts from "metal bridg-

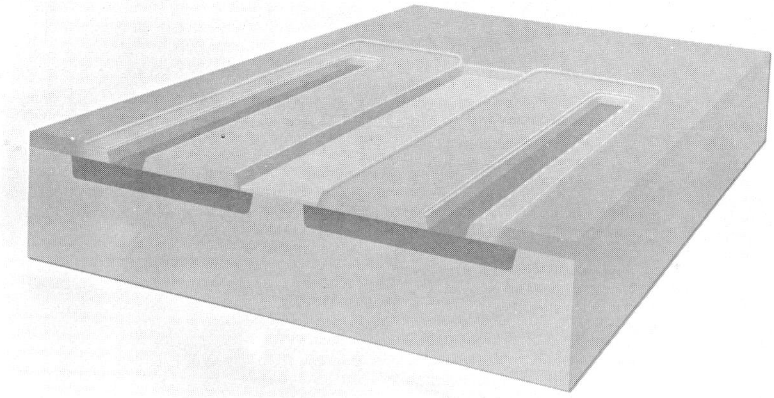

Figure 3–21 3rd (contact) mask etched and photoresist removed.

Figure 3–22 Electron beam evaporator.

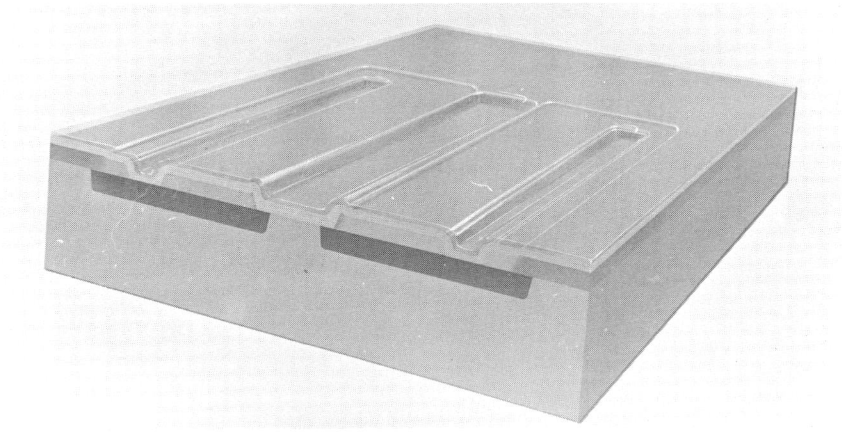

Figure 3–23 After metallization.

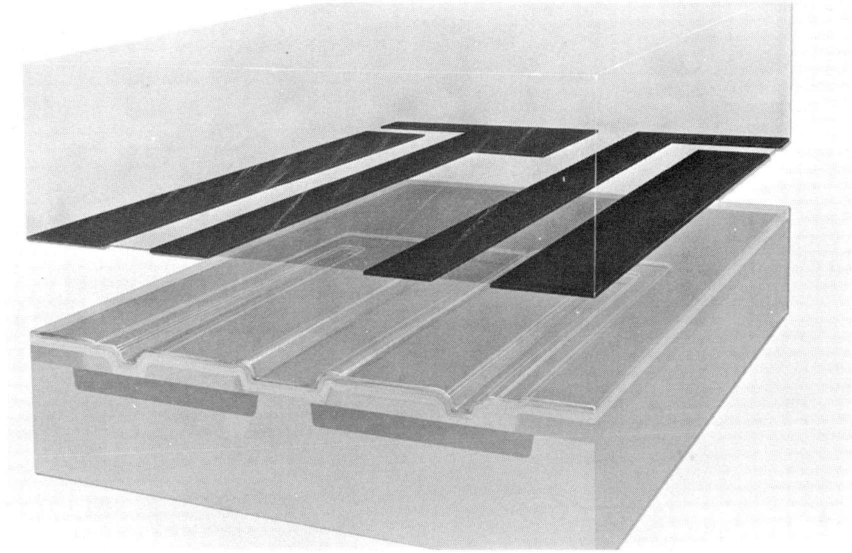

Figure 3–24 Exposure of 4th (metal) mask.

ing" or opens from "metal voids" may result. The accuracy and resolution of interconnect masking also, to a great extent, influence the packing density of an MOS array.

Overlap tolerance of the metal mask is also designed such that all thin gate oxide areas will be completely covered. This is necessary to prevent the tendency of uncovered thin oxide areas to cause an MOSFET to remain in on-condition after gate bias is removed.

After metal mask, the wafer is ready for an optional oxide overcoat for protection against scratches. This passivation layer is applied by the same method as that used for the low temperature deposited oxide previously described. Since the deposition temperature is well below the melting point of aluminum, the passivation is entirely compatible with the metalization pattern. A fifth and final masking operation is required to create pad "windows" for removing this oxide from the bonding pad regions. The completed MOSFET is shown in Figure 3-25 and is electrically complete, ready for testing and assembly.

3.3.2 Assembly and Test

Post-fabrication assembly and test operations for MOS ICs are similar to those employed for other semiconductor devices, with the exception of two very important requirements: (1) protection from damage by static electric charge; and (2) the necessity for highly specialized functional test equipment. The former necessitates specialized care, such as grounding all handling and assembly equipment and shorting the package leads together except during testing. Even with designed-in gate protective devices, these precautions reduce the incidence of voltage transient rup-

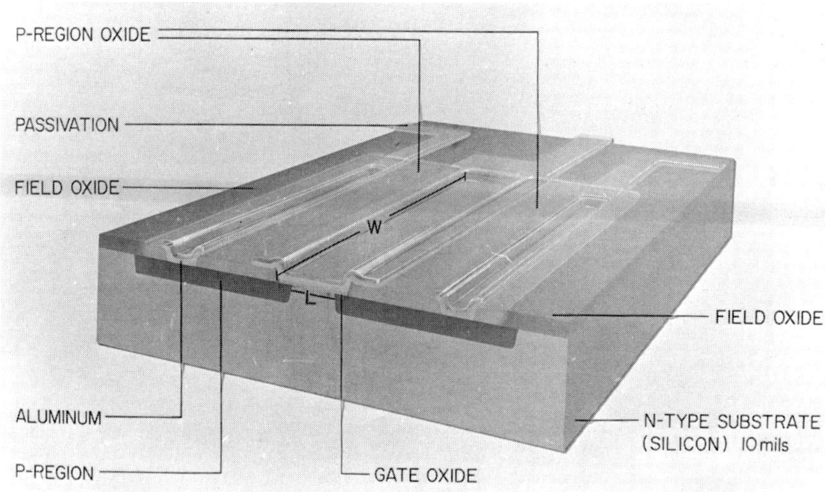

Figure 3–25a Completed MOSFET.

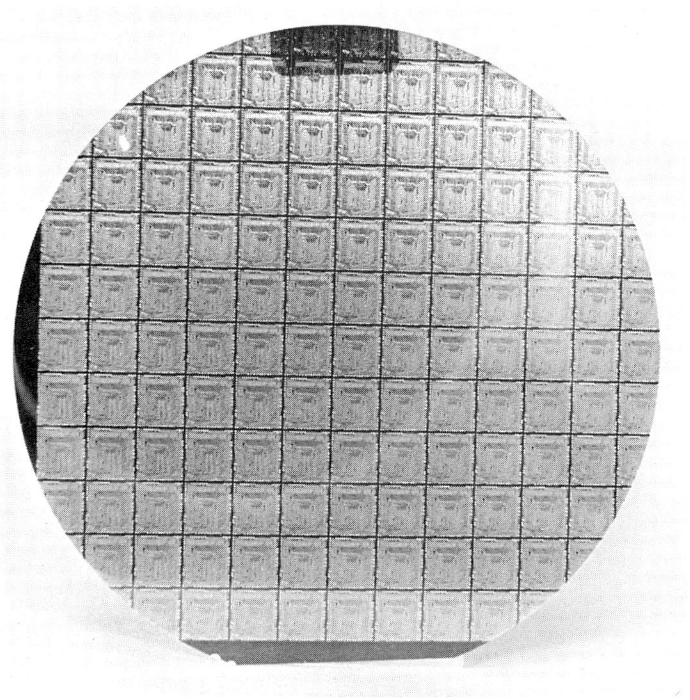

Figure 3–25b Completed wafer.

tures of the thin oxide in MOSFET gates. The latter requirement, test equipment, has resulted in the development of computer programmable automatic functional testers to cope with the task of testing many different types of highly complex arrays. A description of various LSI testers is contained in Chapter 6, Section 6.5.4.

To monitor the electrical characteristics of the wafer fabrication process, a parameter test is performed prior to any assembly screening operations. In this wafer evaluation test, discrete MOS parameters such as threshold voltage, conduction factor, parasitic field threshold, and breakdown voltage of a special test transistor are checked. This transistor may be located either within the perimeter of the complex array, and thus repeated on all dice, or occupy a die location to itself and be interspersed in a few places on the wafer. Either method provides a parameter profile across the wafer. If all parameters are within the range of the design limits, the wafer undergoes a wafer functional screening or "die sort."

During this test, each circuit or die on the wafer is completely tested functionally. The nonfunctioning units are marked, usually with an ink dot, and eliminated after the wafer is scribed and separated into individual dice.

An optical inspection (Figure 3-26) of the functionally good dice follows the scribing. This is done at high magnification to sort out devices that are not cosmetically perfect. These defective dice would otherwise be potential reliability failures or high performance operation failures. The typical visual defects fall in categories of photomasking defects, handling damage, and contamination defects. Photomasking defects such as metal or *P*-region interconnects that are abnormally narrowed or widened along their length are screened, as well as similar defects in the transistor gate metal and source-drain regions. Also screened are handling damages in the form of scratches to the active regions, or cracks or chips in the die from the dicing operation, and contamination defects such as foreign particles and chemical residues. The optically good dice that pass

Figure 3–26 Optical inspection of dice.

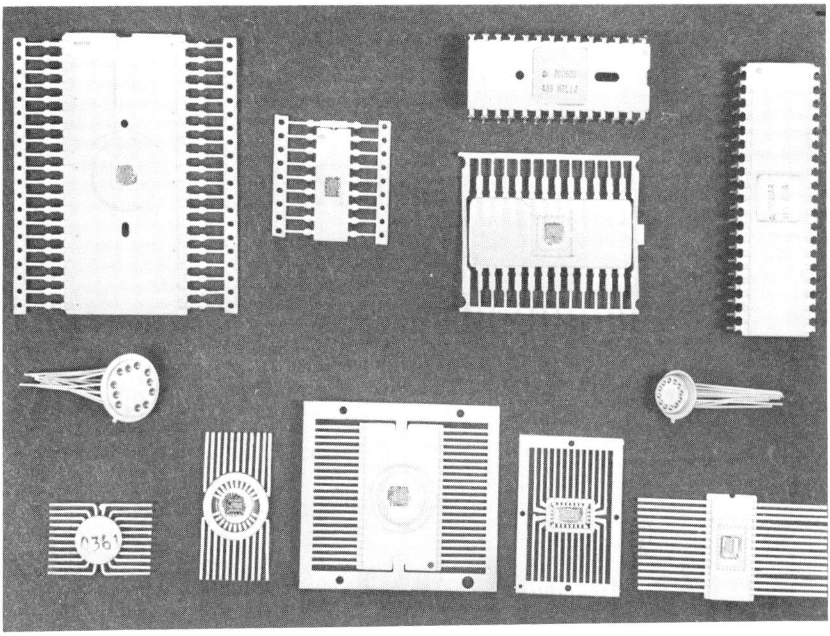

Figure 3–27 MOS packages.

these inspections then proceed to the die attach operation, the first of the packaging steps.

Packages commonly used for MOS encapsulation are of three basic types, metal "TO" cans, "flatpacks," and ceramic and plastic dual in-line packages or "DIPs." Photographs of each are shown in Figure 3-27. An innovation in flat packages, the "edgemount," is shown in Figure 3-28. It features easy plug-in installation and comparatively low initial cost. The dice are mounted in packages by means of a gold alloy solder pre-form. The package substrate and solder pre-form are heated to the eutectic temperature of the alloy. The silicon die is then oriented to the package bonding lands and fused in place (Figure 3-29). A rigid mounting for the die and a substrate connection is thereby effected.

The lead-bonding operation follows. There are two common methods for connecting the package pins to the device pads: "ultrasonic" and "thermocompression" bonding. Ultrasonic bonding is a frictional fusing process. By scrubbing a fine aluminum bonding wire onto the surface of the aluminum bonding pad at high vibrational frequency, a fused attachment results. By repeating the procedure on the appropriate package bonding land, the necessary connection is completed.

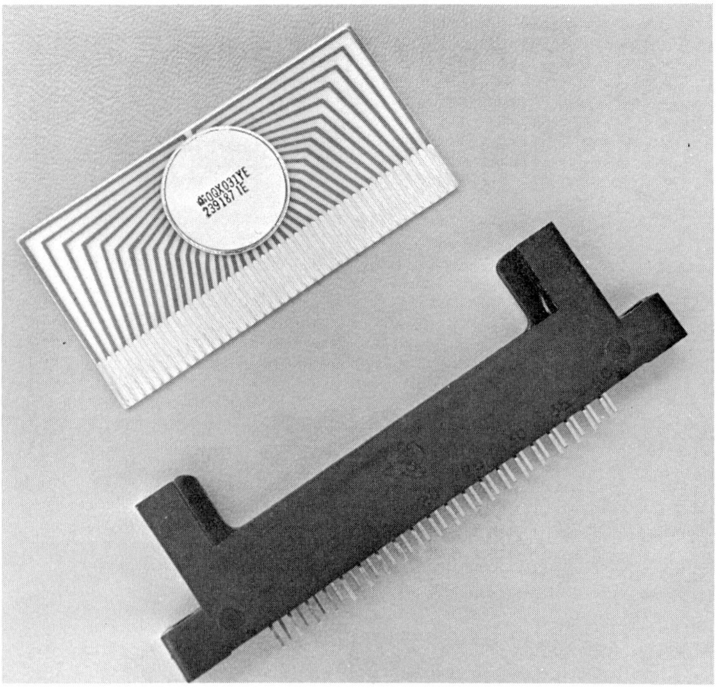

Figure 3–28 Edgemount package and connector.

Thermocompression bonding is an alloy-fusing process. First, the temperature of the packaged die is raised by placing it on a hot stage. Second, by heating the end of the gold bonding wire and pressing it into contact with the surface of the bonding pad, the gold and aluminum fuse together, attaching the wire to the pad. Lead-bonding machines have been developed, based on either bonding technique, to perform semiautomatically the lead attachment operation (Figure 3-30)

After all the package to die connections have been made, a final optical inspection is made to assure that the pad to package bonding is correct and the die is optically perfect. The packages are then hermetically sealed. Flatpacks and DIPs are sealed by soldering on a lid by passing the package through a sealing furnace (Figure 3-31) with solder pre-form and lid clipped in place. TO cans are sealed by welding on the lid with specially designed equipment in an inert gas atmosphere. After sealing, packages go through a series of hermeticity, shock, vibrational, and other environmental tests.

Final electrical testing of both dc parameters, such as leakage and

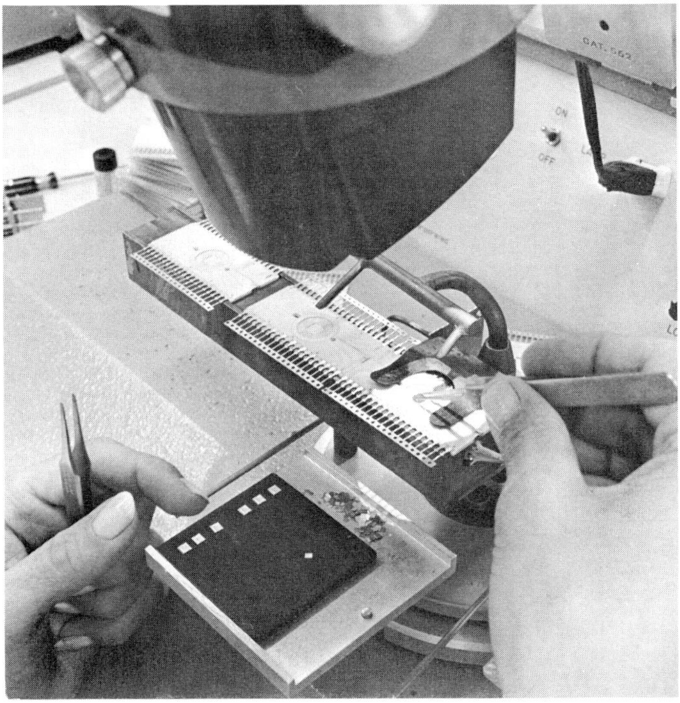

Figure 3–29 Die attach operation.

breakdown voltage, and complete functional circuit operation is then performed. Because the device is in packaged form, more stringent electrical testing is possible than during the die sort operation. Devices that pass final electrical testing are ready for shipment, except for reliability testing as specified by the customer.

3.4 DESCRIPTION AND COMPARISON OF MOS PROCESSES

Of the basic MOS types, *P*-channel, as previously mentioned, was the first to be viable in the manufacture of integrated circuits. Because of this, *P*-channel has been the basis for the development of many MOS process variations. The relationship of MOS processes is shown in Figure 3-32. In this section, the other basic type of MOSFET, *N*-channel, and its combination with *P*-channel, to form complementary MOS (CMOS), will be reviewed first. The remainder of this section will be devoted to *P*-channel processes that were developed as a result of new or improved technologies.

Figure 3–30 Lead bonder.

3.4.1 *N*-Channel

Initial development work based on the MOS concept was directed at producing *N*-channel rather than *P*-channel transistors. Figure 3-33 shows an *N*-channel MOSFET. The fabrication sequence for the *N*-channel process is essentially identical to that for *P*-channel. The reason for initially selecting *N*-channel was the theoretical advantage of employing electron conduction vs. hole conduction in silicon.

As shown in Chapter 2, the conductivity factor (k') depends on the mobility of the majority carriers in the field induced channel, assuming the oxide thickness (t_{ox}) and permittivity (ϵ_{ox}) are not changed.

$$k' = \frac{\mu\epsilon_{ox}}{2t_{ox}} \qquad (2\text{-}71)$$

Although the effective mobility of the carriers in the thin inversion region (i.e., surface mobility) is only approximately one-half the value of the respective bulk mobility, mobility for electrons is on the order of three

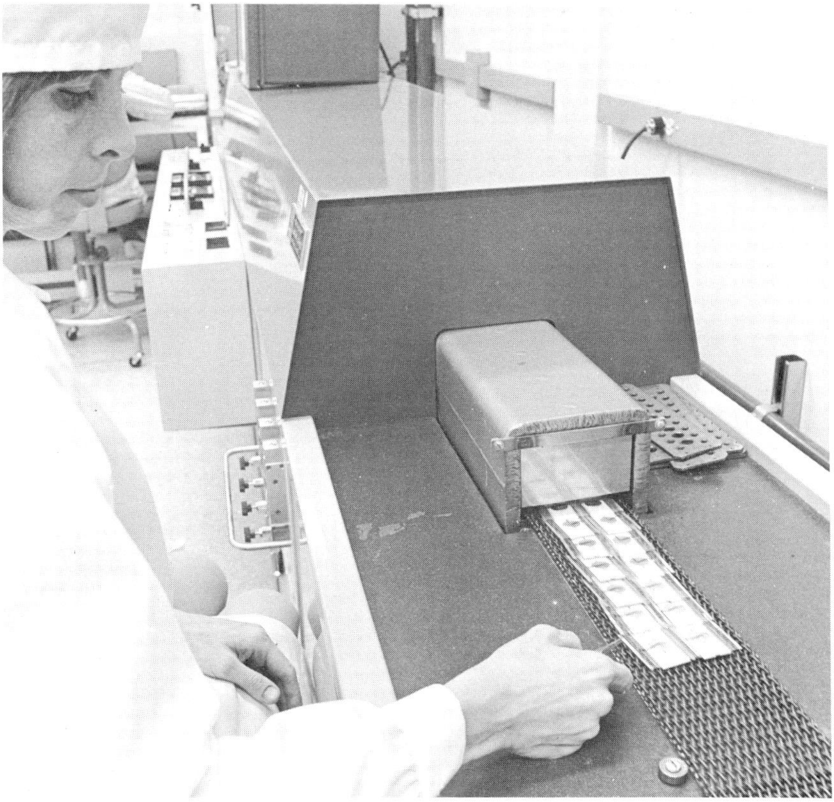

Figure 3–31 Sealing furnace.

times the respective hole mobility. Because of this mobility ratio, k', and hence the performance of an N-channel MOSFET is, theoretically, significantly better than the comparable P-channel MOSFET. The mobility advantage of N-channel is difficult to achieve, however, as the following discussion will show.

As mentioned in the opening remarks of Section 3.3, control of the threshold voltage of the N-channel MOSFET is a greater problem than with P-channel. The charge terms, Q_B and Q_{ox} in the V_T equation repeated below (see Chapter 2, Section 2.2.3), are both positive and hence additive for P-channel MOS.

$$V_T = \phi_{MS} + \phi_B - \frac{Q_{ox}}{C_{ox}} - \frac{Q_B}{C_{ox}} \quad (2\text{-}47)$$

Enhancement mode characteristics, i.e., $V_T < 0$, are thus ensured.

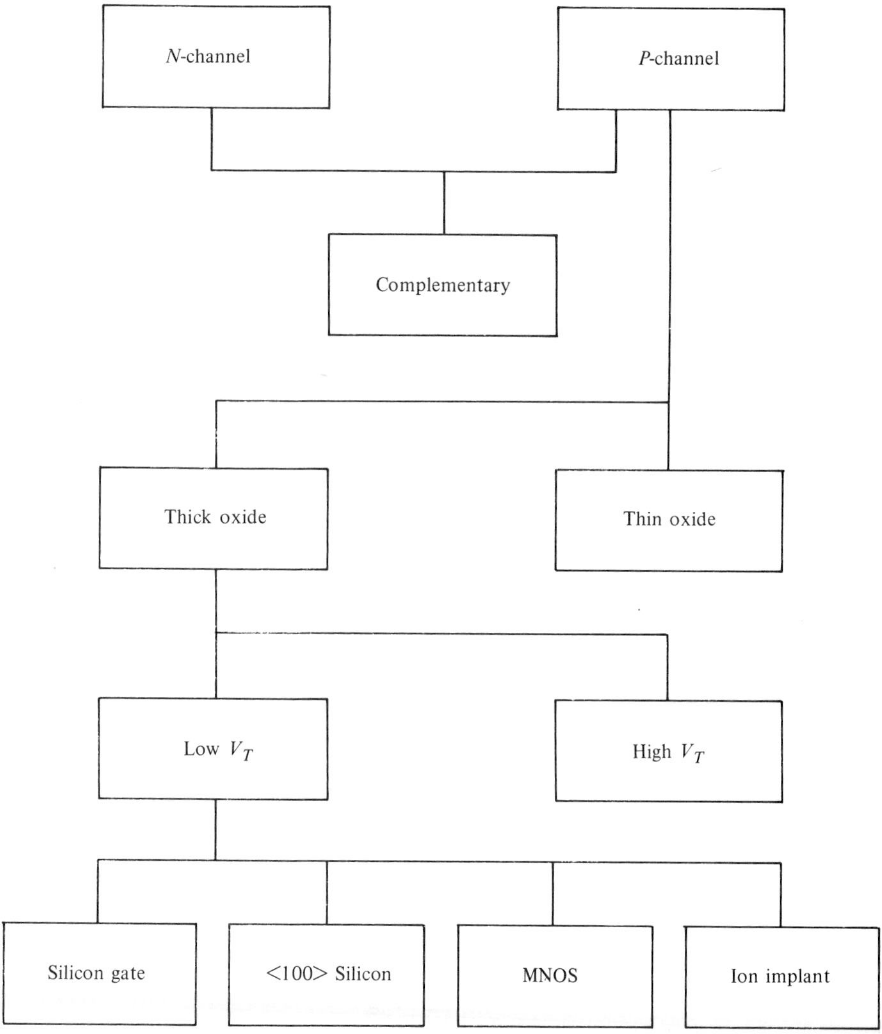

Figure 3–32 Relationship of MOS processes.

For N-channel, however, the bulk doping charge term, Q_B, is negative, corresponding to a P-type substrate. In order to obtain enhancement mode characteristics, i.e., $V_T > 0$, the contribution of Q_B must be such that

$$\phi_{MS} + \phi_B - \frac{Q_{ox}}{C_{ox}} - \frac{Q_B}{C_{ox}} > 0 \tag{3-1}$$

or

$$-Q_B > Q_{ox} - (\phi_{MS} + \phi_B)C_{ox} \tag{3-2}$$

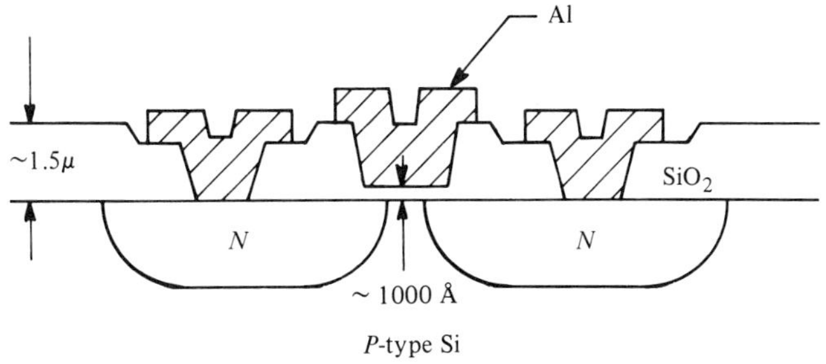

Figure 3–33 *N*-channel MOSFET.

The following example illustrates the relative magnitude of these contributing terms.

Using <100> orientation silicon for lowest surface state density, Q_{ox}/q is on the order of $+1.0 \times 10^{11}$ charges/cm² (i.e., $Q_{ox} = 1.6 \times 10^{-8}$ coulomb/cm²). For typical substrate doping levels, the sum of ϕ_{MS} and ϕ_B (both slowly varying as a function of dopant concentration) is on the order of −0.3 V. Using $C_{ox} = 3.5 \times 10^{-8}$ F/cm² (corresponding to $t_{ox} \approx 1000$ Å), the inequality (Equation 3-2) becomes

$$-Q_B > 1.6 \times 10^{-8} \text{ coulomb/cm}^2 - (-0.3 \text{ V})(3.5 \times 10^{-8} \text{ F/cm}^2)$$

or

$$-Q_B > 2.65 \times 10^{-8} \text{ coulomb/cm}^2$$

For the inequality to be satisfied, the substrate doping (N_A) must be greater than 3.3×10^{15} cm^{-3}, corresponding to about 4 Ωcm *P*-type.

For a reasonable enhancement V_T value, say 1 V, Q_B becomes $\approx 6.2 \times 10^{-8}$ coulomb/cm² corresponding to $N_A \approx 1.5 \times 10^{16}$ cm^{-3} (i.e., about 1 Ωcm). In practice, however, the substrate doping level cannot be chosen this high; if it is, the transistor breakdown (gate modulated breakdown BV_{NPG}) becomes unacceptably low (≈10 V). The result is that for acceptable device breakdown voltages, with the typical Q_{ox} value used in the example, the substrate doping level (and hence Q_B) cannot be chosen to provide a V_T that is more than a marginal enhancement mode device. But one parameter, the gate capacity (C_{ox}), can be varied to improve V_T. An increase in the gate thickness (t_{ox}) can increase V_T to a more suitable value while maintaining reasonable breakdown voltages.

With $N_A = 7 \times 10^{15}$ cm^{-3} (i.e., $Q_B = 4 \times 10^{-8}$ coulomb/cm²) for the bulk doping in the previous example (providing a $BV_{NPG} \approx 25$ V), a V_T of about +0.4 V would be achieved with the 1000 Å gate oxide (i.e., $C_{ox} = 3.5 \times 10^{-8}$ F/cm²). In addition by using a 2000 Å gate oxide instead (i.e., $C_{ox} = 1.75 \times 10^{-8}$ F/cm^2) the same value of Q_B provides a more acceptable value of V_T, about 1 V.

However, if V_T is improved by increasing the gate oxide thickness, the conductivity factor (k') is decreased. Thus it is difficult to realize the full performance benefit of electron mobility on k′ while maintaining acceptable values of V_T.

Another consideration which can be both a circuit design limitation and an asset for N-channel is the "body effect." From Section 2.4.6, the increase in V_T (ΔV_T) with source to substrate bias (V_{BS}) is given in Equation 2-101 repeated below.

$$\Delta V_T = \frac{(2q\epsilon_s N_A)^{1/2}}{C_{ox}} [(|V_{BS} + \phi_B|)^{1/2} - (|\phi_B|)^{1/2}] \qquad (2\text{-}101)$$

The coefficient for multiplying the voltage terms is proportional to the square root of the substrate acceptor doping level (N_A).

As the previous V_T calculation showed, a substrate doping level of about 7×10^{15} cm^{-3} was required. Compared to a doping level of about 1×10^{15} cm^{-3} for P-channel, the relative coefficient for ΔV_T ($(2q\epsilon_s N_A)^{1/2})/C_{ox}$) is 2.6 times greater for N-channel than for P-channel.

In addition, if a thicker gate oxide is used (i.e., lower gate capacitance), to provide a V_T of 1 V as in the previous example, the coefficient for the body effect is further increased. For a gate oxide thickness of 2000 Å, the body effect coefficient increases by an additional factor of two. Thus, the rate of increase in effective V_T due to the body effect can be five times as great as for P-channel. This increase in V_T can cause severe circuit design problems.

The body effect phenomenon, however, can also be used advantageously to control V_T. By means of a fixed source to substrate bias (V_{BS}), the threshold voltage of a marginally enhancement mode transistor, say V_T a few tenths of a volt, can be increased to a more practical value. Using the body effect in this manner reduces the undesirable increase in V_T caused by further increase of V_{BS} typically encountered in circuit applications.

Based on the parameters from the previous sample V_T calculation, effective V_T (i.e., $V_T + \Delta V_T$) is shown in Figure 3-34 as a function of V_{BS}. Also shown is the body effect increase of V_T for a typical high threshold P-channel MOSFET. Curve A is the effective V_T characteristic of the normally 0.4 V V_T N-channel MOSFET. Curve A' is curve A displaced horizontally by 1 V V_{BS}. That is, it represents the body effect characteristic of the same device with the addition of a fixed 1 V body bias. This value of V_{BS} provides a 1 V initial V_T, the same as device B.

Comparing curve A' with curve B, it is apparent that the magnitude of

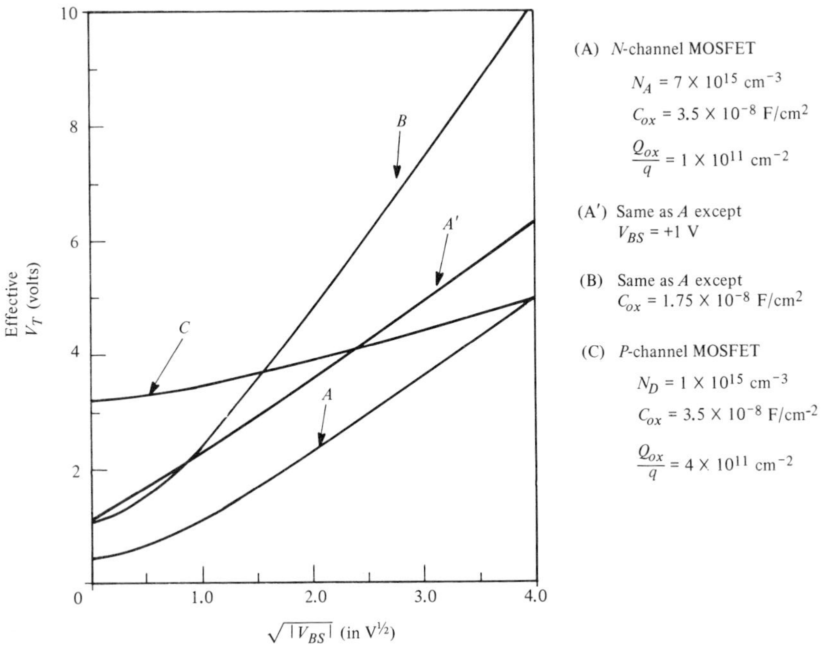

Figure 3–34 Effective V_T vs. body to source bias.

the body effect is greatly reduced by using the fixed body bias approach to V_T control rather than changing C_{ox}. Notice also that a typical high threshold *P*-channel transistor (curve *C*) has an effective V_T comparable to an *N*-channel transistor (A′) when V_{BS} is about 16 V.

Thus, it is apparent that the magnitude of the body effect remains a substantial problem with *N*-channel, at least with the parameters chosen for this example. Also, to achieve a more practical V_T and reduce the body effect by operating with a fixed body bias an extra supply voltage is required, thereby making *N*-channel less desirable than *P*-channel in systems design.

In addition to the problems based on theoretical considerations, certain practical fabrication problems also make *N*-channel processing difficult. The first of these is the tendency for boron dopant in the *P*-type substrate to redistribute during thermal oxidation. Decrease of the surface concentration of the substrate may promote the tendency for low parasitic turn-on (V_{TF}), depending on the arrangement and temperature of thermal oxidations. Also, difficulty in making good aluminum contacts to the *N*-regions and spurious lateral transistor effects present problems both in the processing and design with *N*-channel. Of course, the basic requirement of an

N-channel process is the necessity of achieving a low, reproducible surface state density.

3.4.2 Complementary

Complementary MOS (CMOS) evolved as a result of circuit design innovation using the combination of opposite polarity MOSFETs (P- and N-channel). In terms of application of conventional processing technology, CMOS is the most sophisticated of the MOS processes. The basic transistor structure and the basic fabrication sequence are shown in Figure 3-35. Because opposite polarity transistors are used to perform circuit functions, complementary offers increased speed and low standby power dissipation.[5] Balanced against these advantages is the complexity of the process.

The most important step in fabricating CMOS is the "P-well" diffusion for the N-channel portions of the circuit. These deep, low doping concentration P-diffusions require extreme precision, since dopant concentration regulates both the V_T and breakdown voltage of the N-channel MOSFETs.

Control of the surface concentration of this diffusion step (and therefore Q_B) and the surface density (Q_{ox}) of the gate oxides also determine the relative threshold voltages of the P- and N-channel transistors. As seen in the N-channel analysis, an increase in Q_{ox} would lower the V_T of N-channel transistors while raising the V_T of P-channel transistors. Therefore, it is difficult to match the magnitude of the V_T of these two transistor types. This makes CMOS the most sensitive of MOS processes to control.

CMOS does have an advantage over single polarity MOSFETs (especially N-channel) in one regard, body effect. By using opposite polarity transistors to perform certain circuit functions (e.g., pull-up or load transistors which are subject to the body effect when a single polarity MOS is used), the body effect is avoided altogether.

In terms of processing yield, the necessity of extra masking and diffusion steps for CMOS places it at a disadvantage compared to P-channel MOS. CMOS is also less efficient in terms of chip area. The P-wells, because of the large lateral extent of the diffusion, consume a great deal of chip area similar to the isolation diffusions required in bipolar transistor technology.

Thus the performance advantages of CMOS are offset, to an extent, by the sacrifices in process complexity and topological density.

3.4.3 Thick Oxide vs. Thin Oxide *P*-Channel

The first P-channel MOS processes used to manufacture integrated circuits were of the "thin oxide" type. That is, the thickness of the final oxide

1. *P*-well mask
2. 1st boron deposition and diffusion-oxidation
3. *P*-channel source and drain mask
4. 2nd boron deposition and diffusion-oxidation
5. *N*-channel source and drain mask
6. Phosphorus deposition and diffusion-oxidation
7. Gate mask
8. Gate oxidation and contact mask
9. Metalization and metal mask

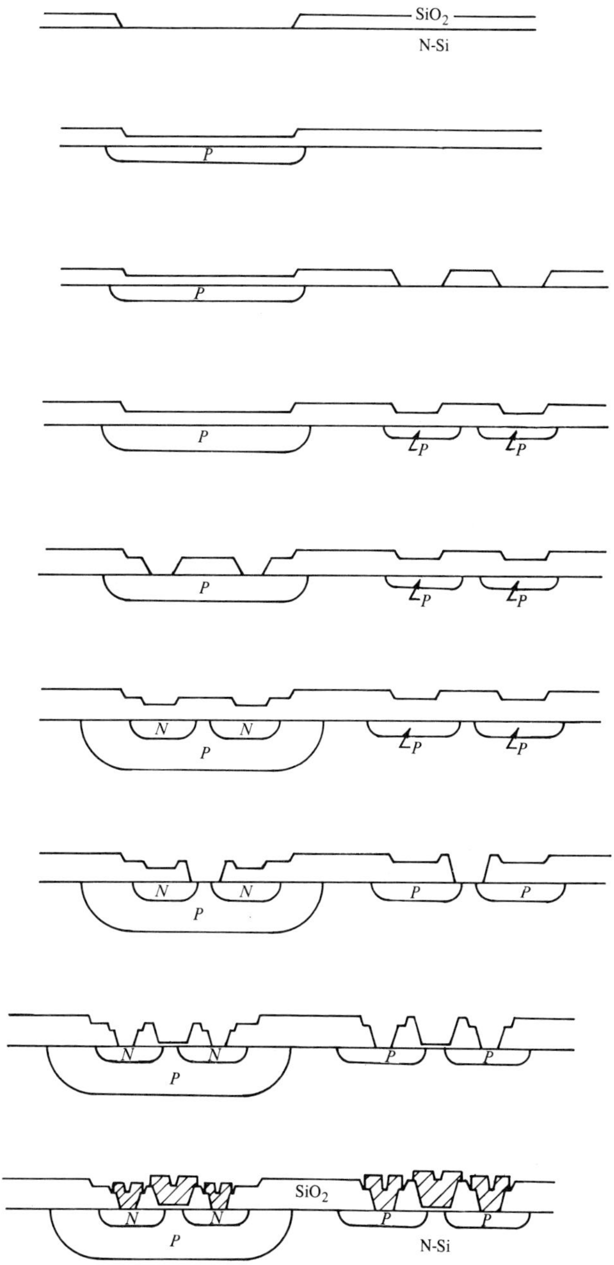

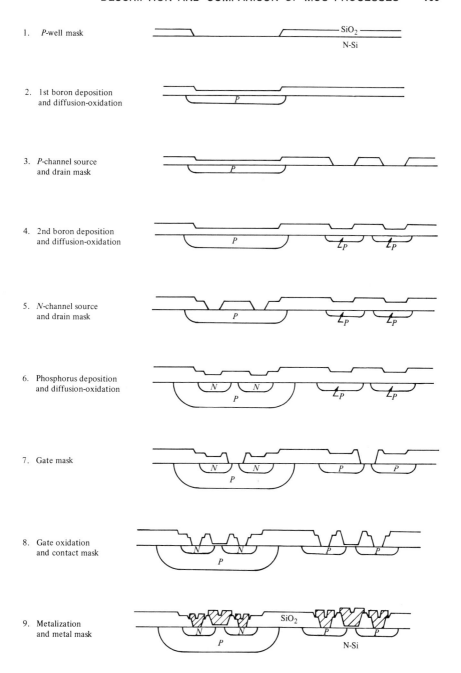

Figure 3–35 Complementary MOS fabrication sequence.

over the diffused *P*-regions was comparable to that of the gate oxide. A cross-sectional view of the fabrication sequence is shown in Figure 3-36. The fabrication sequence is less complex than the one employed for thick oxide MOS discussed in Section 3.3 However, the thin *P*-region oxide adversely affected the manufacturing yield and imposed electrical design limitations.

The manufacturing yield losses were attributed to metal to *P*-region shorts and "briding." Metal to *P*-region shorts occurred as a result of the greater incidence of pinholes on the thin oxide layer caused by the growing of oxide over doped *P*-regions. Because of these pinholes, the oxide was easily ruptured by low voltages that occurred during normal testing or as a result of accumulation of static charge.

The problem of bridging was brought about by the large difference in oxide thickness between the field and *P*-region areas which limited the

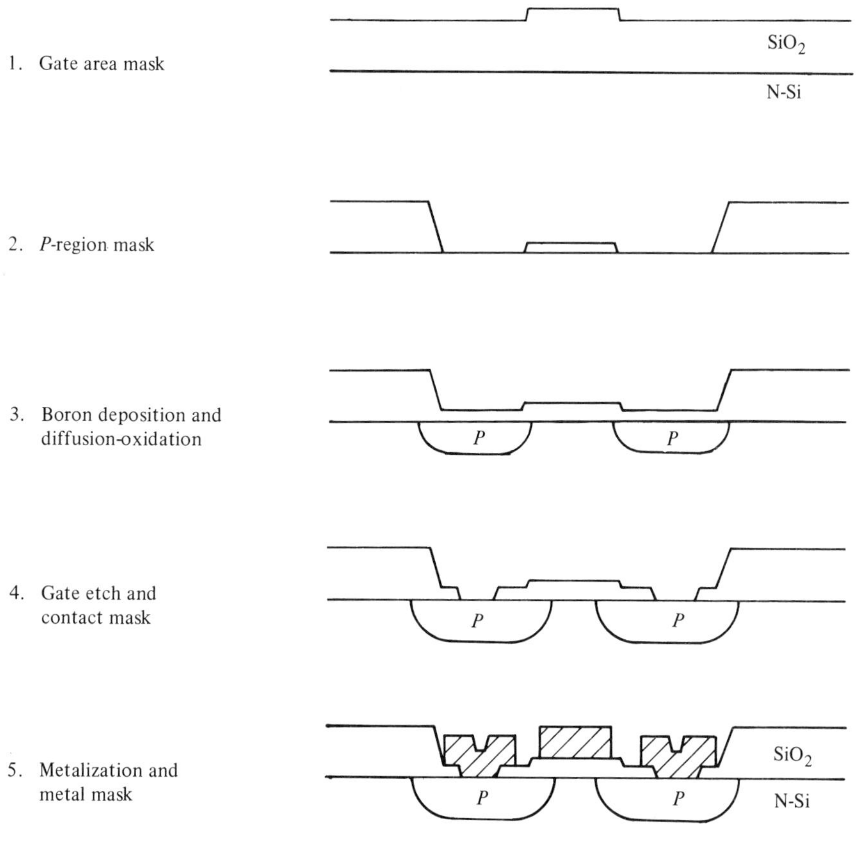

Figure 3–36 Thin oxide *P*-channel fabrication sequence.

degree of contact between the mask and wafer during the photomasking step. The associated loss of resolution resulted in incomplete removal of aluminum between adjacent interconnects in the *P*-region areas during etching of the metal mask. Because of this, there was a greater tendency toward bridging, which caused shorts between adjacent metal interconnects, thereby lowering the manufacturing yield significantly.

The principal circuit design limitation was high metal to *P*-region parasitic capacitance. Since the *P*-region oxide was approximately as thick as the gate oxide, the associated capacitance was equally high, thereby significantly degrading circuit performance.

Because of the disadvantages of the thin oxide process, it was highly desirable to thicken the *P*-region oxide. The first thick oxide devices were produced in 1967. At that time, chemical vapor deposition of high quality oxides was not practical in a manufacturing environment. Therefore, these processes were based on *P*-region oxides that were thermally grown. Subsequent thick oxide processes were adaptations of the basic thin oxide processes using deposited oxides.

In addition to solving the process and design problems mentioned, the thick oxide process yielded another advantage. Because of the thickened *P*-region oxide, gate overlap capacitance was reduced, thereby improving device performance. Improvement in breakdown voltage characteristics also resulted in reduced leakage. Because of these improvements, the thick oxide *P*-channel process has become the dominant manufacturing process of the MOS integrated circuit industry.

3.4.4 High and Low Threshold Processes

After the successful implementation of thick oxide P-channel, the next significant evolution of MOS technology was the development of low threshold voltage, bipolar compatible, *P*-channel processes. The first processes developed, which produced low ($\approx$2 V) V_T were based on $<100>$ rather than $<111>$ orientation silicon. Subsequent low V_T processes have been developed using other variations of the basic thick oxide *P*-channel process. To understand the effect of changing substrate crystal orientation and other structural changes of the MOS system, let us review the basic parameters of the threshold equation (Equation 2-47).

$$V_T = \phi_{MS} + \phi_B - \frac{Q_{ox}}{C_{ox}} - \frac{Q_B}{C_{ox}}$$

From Section 2.3.3,

ϕ_{MS} is the metal to silicon work function.

ϕ_B is the potential that corresponds to the band bending at inversion.

Q_B is the charge contribution of substrate depletion region.

C_{ox} is the gate oxide capacitance.
C_I is used in place of C_{ox} for the more general case where the dielectric is other than silicon oxide.

ϕ_{MS} is a function of the gate electrode material, most commonly aluminum, and, to a slight degree, a function of the substrate doping level (see Equation 2-40). ϕ_B is also a slowly varying function of the doping level (N_D) (see Equations 2-38 and 2-39). In fact, it is often convenient to combine ϕ_{MS} and ϕ_B, since the sum is nearly constant for a given gate electrode system.

Q_{ox} is most strongly a function of the substrate crystal orientation. It is also a function of oxidation and annealing conditions.[6-8]

Q_B is a function of the substrate doping level (see Equation 2-45). The range of substrate concentrations is strongly limited by practical considerations of junction breakdown voltage and depletion region spreading. Therefore, Q_B is not a strong controlling factor of V_T for conventional *P*-channel technology (ion implantation is a special case to be mentioned later). Lastly, C_I is a function of the thickness(es) and the dielectric constant(s) that comprise the gate insulator.

The following tabulation summarizes the different low threshold *P*-channel technologies based on the corresponding V_T controlling parameters.

Technology	*V_T Controlling parameter*
<100> Silicon	Q_{ox}
MNOS	C_I
Silicon gate	ϕ_{MS}
Ion implantation	Q_B

The first two variations of *P*-channel technology will be discussed in this section. The other two low V_T processes, silicon gate and ion implantation, are discussed in separate sections inasmuch as these new technologies involve more than the achievement of low threshold voltages.

Low V_T *P*-channel was first developed by changing the substrate from <111> to <100> crystal orientation silicon. Although a typical value for the oxide charge density Q_{ox}/q for <111> silicon was 4 x 10^{11} charges/cm^2, <100> silicon provides a lower value, typically 1.5 x 10^{11} charges/cm^2. When Q_{ox} is replaced by (1.6 x 10^{-19} coulomb) (1.5 x 10^{11} cm^{-2}) = 2.4 x 10^{-8} coulomb/cm^2, in the threshold voltage calculation example of Section 2.3.3, a V_T of 2.0 V is obtained. At the same time that a low V_T is achieved, however, a lower parasitic field oxide threshold

(V_{TF}) is obtained. With a 1.5 μ field oxide (i.e., $C_{ox} = 0.23 \times 10^{-8}$ F/cm^2) the value for V_{TF} is 17.4 V. Because the minimum value can be significantly below this and thus uncomfortably low for circuit design, two approaches were taken toward improving V_{TF}.

The first approach was to increase the field oxide thickness. By increasing t_{ox} to a practical upper limit of nearly 2 μ, the typical V_{TF} increased to a more reasonable value of 22 V. The second approach to increasing V_{TF} was to incorporate "field doping." This technique involves an extra masking and diffusion-oxidation step to increase the bulk doping selectivity in regions between the individual transistors. This technique can be one of two forms. With one form, a mask is designed for doping all regions except those of the gate and *P*-regions (true field doping). The doping is then done prior to other masking steps, using a slow diffusing *N*-type impurity, such as antimony (see Figure 3-37). The concentration of this field doping must be tightly controlled to be only slightly greater than the bulk doping, so as not to make the junction breakdowns undesirably low, and yet provide sufficient increase of V_{TF}. The other field doping method, the use of "channel-stops" (see Figure 3-38) does not require such precise

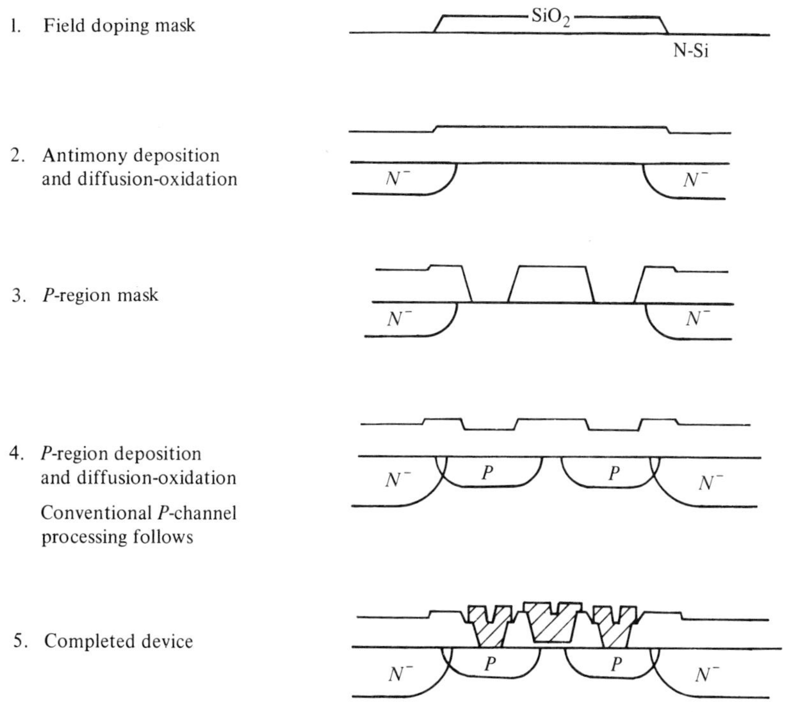

Figure 3–37 Field doping technique.

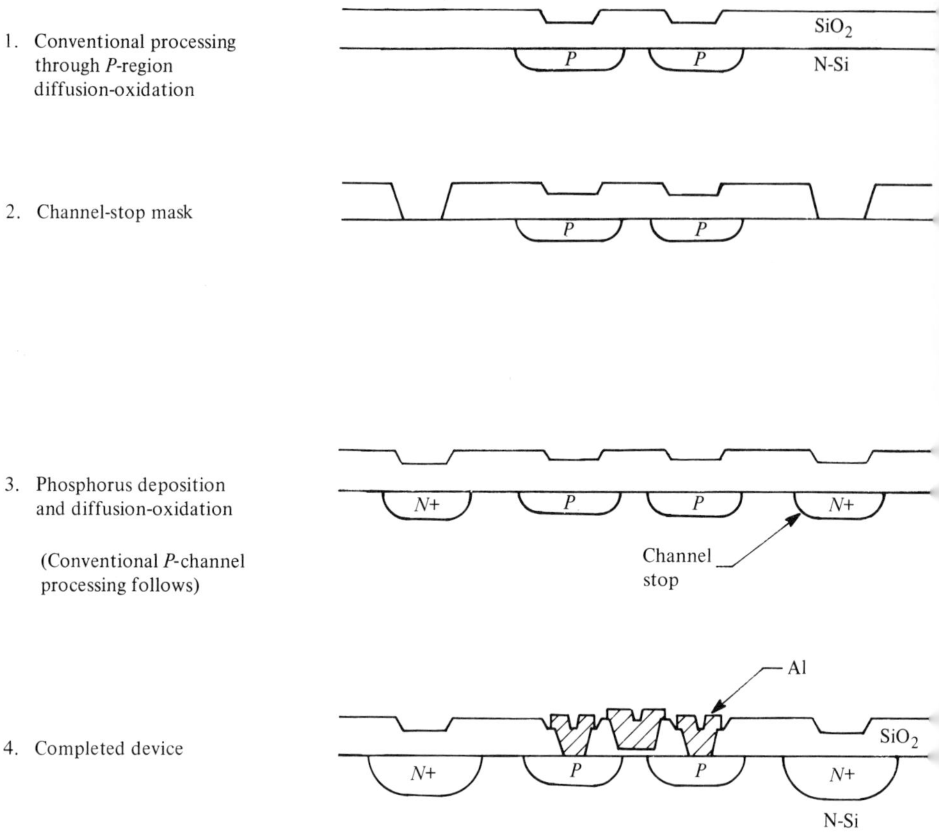

Figure 3–38 Channel-stop technique.

control of the enhanced (N^+) diffusions. In this case, smaller selective areas (often bars) are doped (usually using phosphorus) to a higher concentration, but the separation between these areas and the P-regions must be wider, since low voltage breakdowns occur when P-region depletion region spreading is restricted. The increased separation, of course, increases the size of the circuit.

A trade-off in topological design was thus made for the advantage of easier processing with the channel-stop technique. Field doping, on the other hand, does not require extra chip area; but either technique requires an extra masking and diffusion step, compared to the ultra-thick field oxide method for achieving reasonable V_{TF}.

A second process type was subsequently developed to achieve low voltage P-channel without the V_{TF} problems of $<100>$ silicon. This is the MNOS

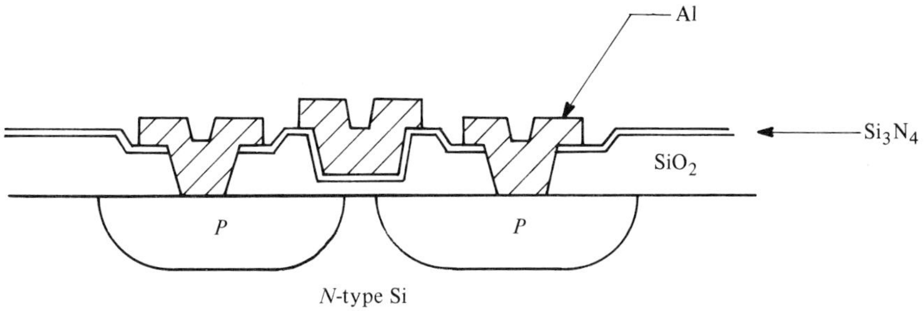

Figure 3–39 MNOS structure.

(Metal Nitride Oxide Silicon) technology. As the name implies, a significant change is made to the basic gate structure of the MOS transistor (see Figure 3-39). A layer of silicon nitride dielectric (Si_3N_4) is added to conventional silicon dioxide to create a "sandwich" of dielectric with a capacitance per unit area significantly higher than that of oxide alone.

As seen in the threshold equation, an increase in C_{ox} (now C_I) will suitably lower V_T. Since conventional gate oxides were already as thin as practical, from the standpoint of pinhole and susceptibility to static charge rupture, the only other way to increase the gate capacitance was to use an insulator with higher dielectric constant. One of the materials investigated was silicon nitride, which has a dielectric constant of about 7[9]. Chemical vapor deposition techniques were developed, using the reaction between silane and ammonia, to produce uniform, high quality, nitride thin films.[10] Direct replacement of oxide with nitride as the gate insulator did not prove feasible; however, threshold shifts occurred as a function of applied bias at room temperature due to carrier trapping at the silicon-silicon nitride interface.[11]

It was found that these threshold hysteresis effects could be avoided by growing a thin layer (a few hundred angstroms) of oxide on the gate silicon before the nitride is deposited.[11] The resulting dielectric "sandwich" has a gate capacitance given by:

$$C_I = \frac{1}{\dfrac{1}{C_{ox}} + \dfrac{1}{C_N}} = \frac{K_{ox}K_N\epsilon_0}{K_N t_{ox} + K_{ox} t_N} \tag{3-3}$$

However, when a gate dielectric comprised of a thin layer of oxide (e.g., 400 Å) was used with a substantially thicker layer (e.g., 800 Å) of Si_3N_4, V_T shifts occurred under bias as a function of time. This phenomenon is due to charge accumulation at the oxide-silicon nitride interface and occurs as a result of the difference between the electronic conductivities of the two materials.[12] To avoid this effect, the oxide thickness of the MNOS structure was increased to be substantially greater than that of the nitride (e.g., 600 Å oxide, 400 Å nitride). Because of this restriction on the relative thicknesses of the dielectrics, full advantage cannot be taken of the higher dielectric constant of silicon nitride.

Together with processing improvements to reduce Q_{ox}, the MNOS structure has been successfully applied to produce low V_T (~ 2 V) while maintaining the higher V_{TF} of $<111>$ material (~ 25 V), compared to the conventional $<100>$ low voltage *P*-channel process. Because of the higher gate oxide capacitance of the dielectric sandwich, the conductivity factor (k') is also higher (~ 25 percent). In addition, silicon nitride included in the gate dielectric offers some degree of protection from alkali ion drift instability. With current "clean oxide" technology, this is normally not a serious problem, but the use of silicon nitride might be an advantage in this respect when the devices are encapsulated in plastic. However, these improvements are made at the expense of the extra processing required to deposit reproducibly and etch contact openings through the thin nitride layer. Furthermore, variations in nitride deposition thickness can seriously affect both the range of V_T produced and the long-term reliability of MNOS devices because of electronic conductivity instability.

3.4.5 Ion Implantation

As mentioned in the preceding section, it is not possible to achieve low threshold voltage P-channel MOSFETs by altering only the substrate doping level. The development of a new technology, ion implantation of semiconductor dopants, has made available threshold voltage control in a way never previously possible. For a discussion of ion implantation technology, see Mayer et al.[13]

With ion implantation techniques, it is possible to enhance or compensate the substrate doping in selected transistor areas. This is accomplished by masking those regions that are not to be implanted with a suitably thick medium (e.g., metal, field oxide, or photoresist) to stop the penetration of the ion beam. A typical apparatus for ion implantation is shown in Figure 3-40.

Ion implantation technology can readily produce low V_T on normally high threshold, $<111>$ material.[14] The ion implantation step is performed just before the contact masking step (Figure 3-41). Boron ions are implanted through the gate oxide to compensate the doping of the transistor channel regions. Because of the thick oxide in other regions, the boron ions do not penetrate as far as the silicon substrate. Subsequent processing steps remain the same as for regular *P*-channel.

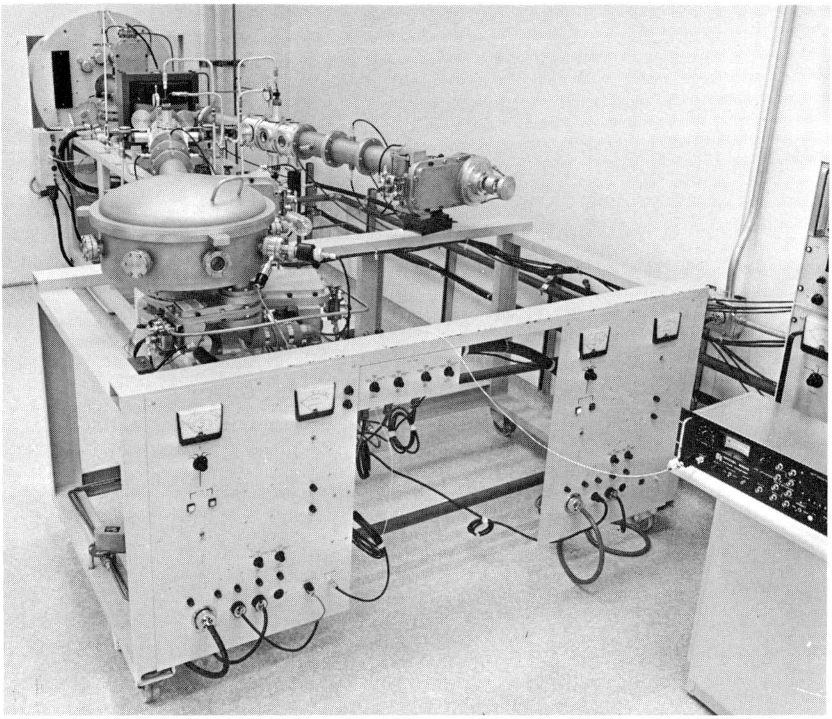

Figure 3–40 Ion implanter.

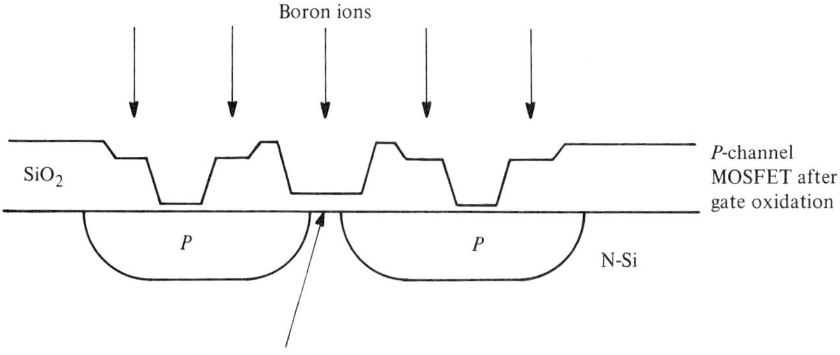

Figure 3–41 V_T modification by ion implantation.

The ion dose required to achieve low V_T on <111> silicon can be calculated directly from the V_T equation. If a 2 V threshold is desired, again using the example of the Section 2.3.3, we have

$$-2\text{ V} = -0.9\text{ V} - \frac{(Q_B + 6.4 \times 10^{-8}\text{ coulomb/cm}^2)}{3.5 \times 10^{-8}\text{ F/cm}^2}$$

or $Q_B = -2.5 \times 10^{-8}$ coulomb/cm^2. That is, the net charge in the channel depletion region must be negative, corresponding to a net *P*-type concentration. Since Q_B for the 10^{15} cm^{-3} substrate is 1.4×10^{-8} coulomb/cm^2, the net charge density that must be implanted is then 3.9×10^{-8} coulomb/cm^2. This corresponds to a boron doping density (boron is singly ionized) of about 2.5×10^{11} ions/cm^2. Because not all implanted ions become electrically active in the silicon lattice, the actual implanted dose would be somewhat higher.

At first thought, a net *P*-type doping in the channel region seems to imply that there should be a residual channel. However, the net carrier density is not just dependent on the dopant density, but is a function of the surface potential produced by the work function and surface state charge (see Equation 2-35a). The net result is that when there is no applied bias on the gate, the channel region is still induced to be *N*-type.

The application of ion implantation to produce threshold modification of basic *P*-channel MOS is a very significant one. Ion implantation requires a minimum of extra fabrication steps, only the implantation step itself plus a short anneal to allow boron atoms to move into substitutional locations in the crystal lattice. Depending on the implanted dose of boron ions, the threshold voltage can be modified at will, while still retaining the high V_{TF} of a $<111>$ *P*-channel process. In fact, one of the most exciting applications of ion implantation is the capability to selectively produce depletion mode MOSFETs for special circuit applications. This simply requires an extra masking step followed by implantation with an appropriately higher boron dose.

Contrasted to earlier semiconductor applications of ion implantation, MOS threshold voltage modification does not produce significant damage to the crystal lattice. The required doses (10^{11}–10^{12} ions/cm^2) and the beam energies ($\sim$50 keV) are below those that produce undesired effects such as increased leakage currents.

The leakage side effect has been a problem when ion implantation is used in the fabrication of self-aligned gates. This technique employs the gate metal to protect the gate oxide while extensions to the normal diffused *P*-regions are implanted.[15] The process sequence is outlined in Figure 3-42. The dose required to provide the necessary values of sheet resistance to the *P*-region extensions is quite high, ($>10^{13}$ ions/cm^2) and causes significant damage to the crystal lattice. Because the aluminum gate electrode must be in place prior to the implantation step, subsequent high temperature annealing to remove crystal damage cannot be performed. As a result, while MOSFETs fabricated by this self-aligned gate technique promise a substantial increase in performance because of elimination of gate overlap

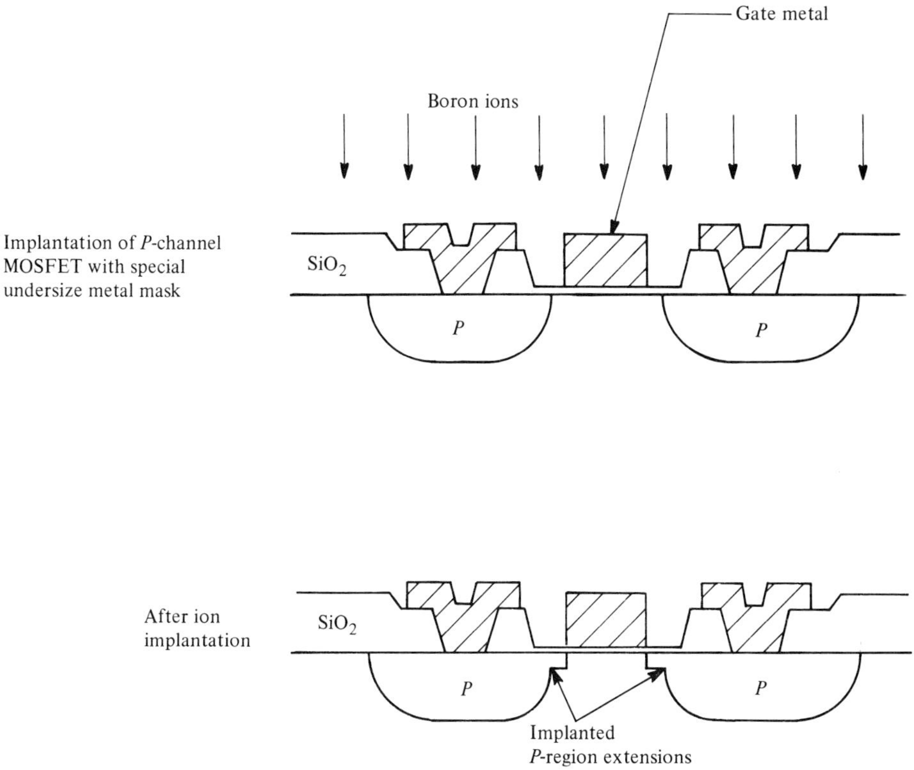

Figure 3–42 Self-aligned gates by ion implantation.

capacitance, the leakage current side effects has limited its success in integrated circuit applications to date.

A solution to the leakage problem has been proposed by applying the ion implanted self-aligned gate technique to the next technology to be discussed, the silicon gate MOS. By using a polycrystalline gate instead of aluminum, post-implantation high temperature annealing is possible.

3.4.6 Silicon Gate

Another parameter that controls V_T is the work function (ϕ_{MS}) between the gate metal and the substrate silicon. ϕ_{MS} depends on the metal of the gate electrode and the doping level of the substrate silicon as shown by the band diagram (with aluminum gate electrode, Figure 3-43). For aluminum and silicon $\phi_{MS} \simeq 0.6\text{ V} - \phi_{FS}$,[16] where ϕ_{FS} is the Fermi potential

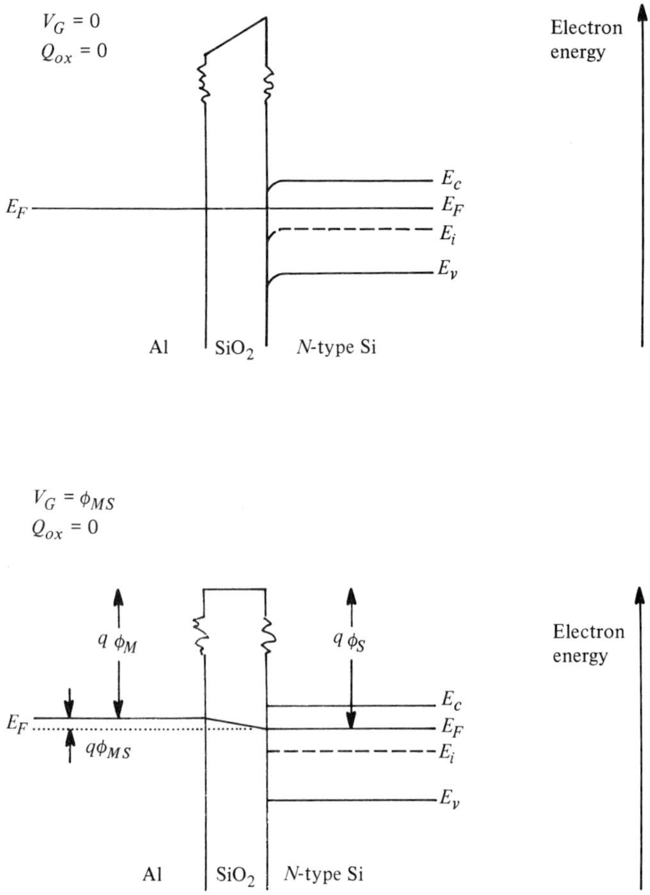

Figure 3–43 Band diagram for the Al-SiO_2-Si system.

of the substrate silicon. For a typical substrate doping of 10^{15} cm^{-3}, the result is $\phi_{MS} = -0.3$ V.

When the aluminum electrode is replaced by doped polycrystalline silicon (in this instance *P*-type), the work function between the gate electrode and substrate silicon is altered.[17] Assuming a band analysis is valid for the polycrystalline silicon (Figure 3-44), the effective work function ($\phi_{S'S}$) of the silicon-oxide-silicon system is given by $\phi_{S'S} = \phi_{FG} - \phi_{FS}$, where ϕ_{FG} is the Fermi potential in the gate silicon.

By doping the gate electrode *P*-type during the normal source-drain deposition-diffusion to a concentration of about 10^{19} cm^{-3}, the effective work function becomes $\phi_{S'S} = +0.5\text{ V} - (-0.3\text{ V}) = +0.8$ V. When

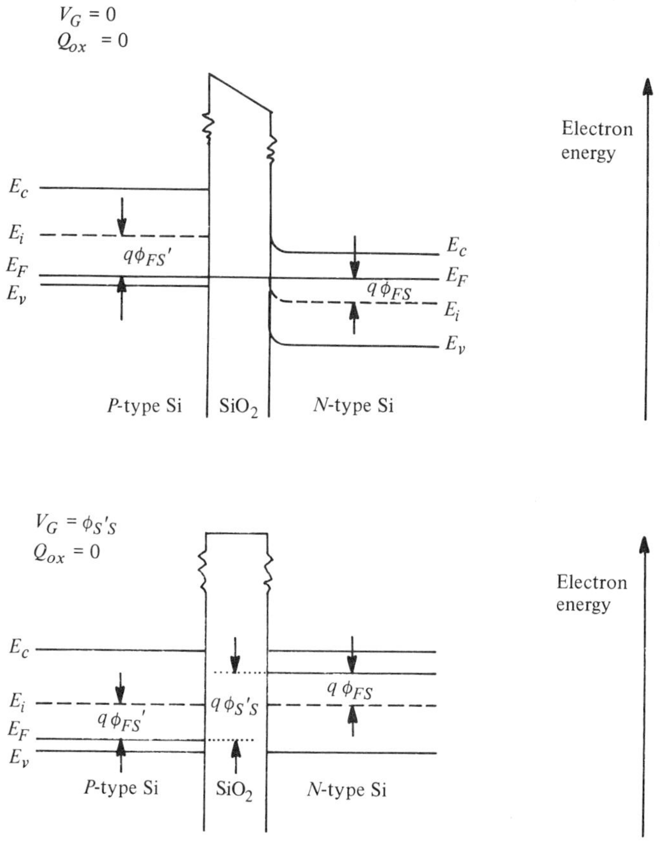

Figure 3–44 Band diagram for the Si-SiO_2-Si system.

this is compared to ϕ_{MS} for the equivalent aluminum gate structure, it can be seen that a change of +1.1 V is obtained. With all other parameters of a *P*-channel transistor the same and the conventional aluminum gate replaced with a P-doped, polycrystalline silicon gate, a V_T reduction of 1.1 V would result. For the *P*-channel example of Section 2.3.3, this would result in a V_T of 2 V. Silicon gate MOS thus represents another way of achieving low V_T, *P*-channel MOS.

In addition to providing a reduction in threshold voltage, the use of polycrystalline silicon for the gate electrode provides two other very distinct process advantages. These both stem from the fact that poly-silicon, in contrast to aluminum, can withstand high temperature processing steps, which enables a complete reorganization of the MOS fabrication sequence (Figure 3-45).

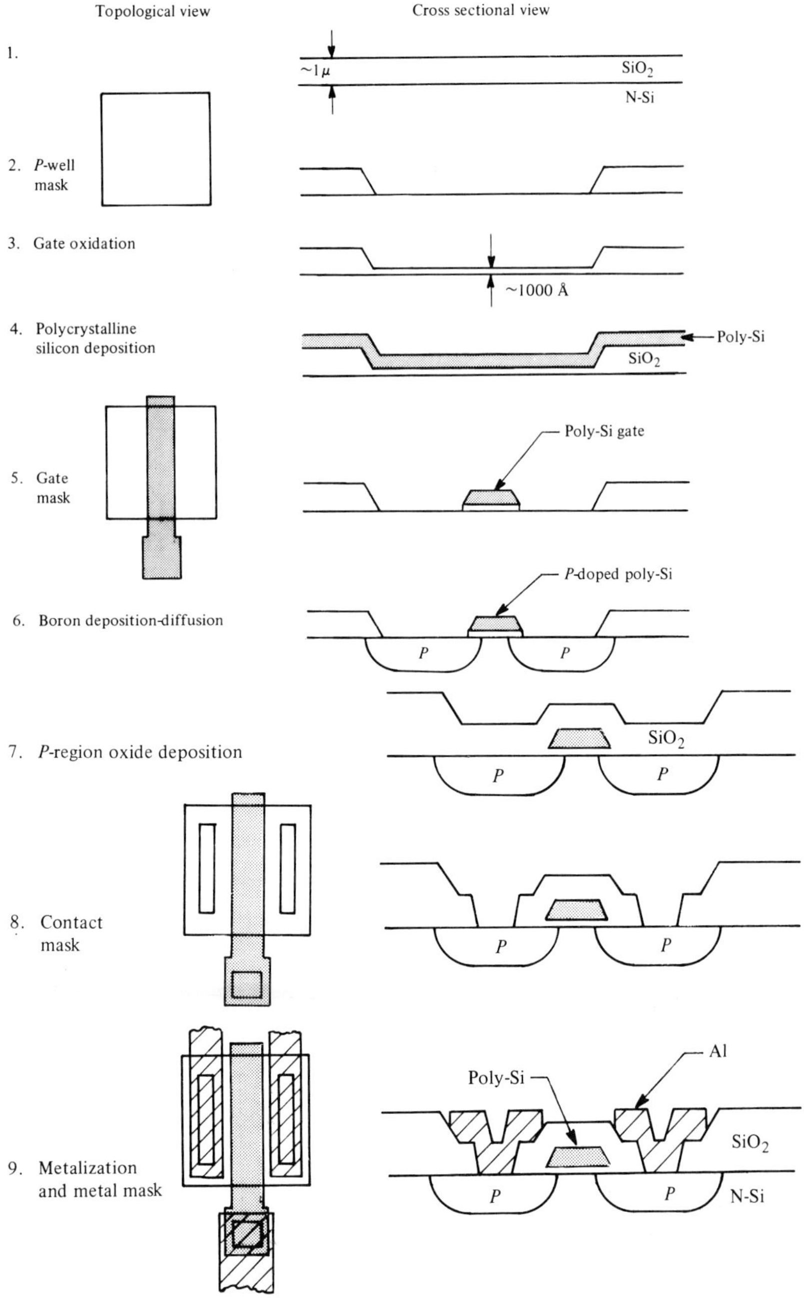

Figure 3–45 Silicon gate MOS fabrication sequence.

The initial oxidation step of the silicon gate process serves not only as a diffusion mask for doping of the P-regions, but it also forms the final "field oxide" thickness for possible poly-silicon interconnects. Consequently, it is advantageous to make this initial oxide as thick as practical to maintain the V_{TF} of parasitic poly-silicon field oxide transistors as high as possible. The practical upper limit of the initial oxide thickness is set by masking resolution achievable with the inherently "out of contact" *P*-well.

The deposition of poly-silicon after the gate oxidation step is done in an epitaxial-type reactor (Figure 3-16). Polycrystalline silicon is readily deposited on top of oxide under the same conditions that yield an epitaxial silicon layer deposited directly on a single crystal substrate.[18]

All subsequent processing steps are performed in a manner similar to conventional *P*-channel MOS. The total number of masking steps required is the same in both processes. Because of the silicon gate processing sequence, the *P*-diffusions are self-aligned to the gate structure. The overlap of the gate over source and drain regions is then only the amount of the lateral diffusion. Since much of the high-temperature processing occurs prior to the *P*-region deposition step, the diffusion depth can be minimal ($\sim 1\ \mu$). Therefore silicon gate processing as compared to conventional P-channel offers better circuit performance due to a substantial decrease in the parasitic capacitances from gate to drain and source.

Because the poly-silicon gate is doped during the *P*-region deposition diffusion, a moderately thick poly-silicon lay ($\sim 0.5\ \mu$) can have a sheet resistance sufficiently low ($<100\ \Omega/\square$) to be a useful additional circuit interconnect medium. This is particularly true since the poly-silicon structure is passivated during the *P*-region oxide deposition step, enabling aluminum interconnection to be freely routed over poly-silicon, even transistor gates. The capacitance between metal and poly-silicon is as low as the metal to *P*-region capacitance since the oxide is the same thickness.

The poly-silicon does not form a complete additional interconnect layer, however. Poly-silicon cannot be crossed over *P*-regions without forming a transistor gate, due to the fabrication sequence of the process. Even with this restriction, the additional "half-layer" of interconnect afforded by silicon gate technology provides a substantial topological advantage, compared to conventional *P*-channel, for the design of large-scale integrated circuits.

Silicon gate MOS does have one slight disadvantage, compared to aluminum gate *P*-channel, in that it does not inherently provide the option of designing high value metal to *P*-region capacitors. By appropriately sequencing the gate and contact masks in a conventional *P*-channel process, it is possible to achieve the fabrication of *P*-regions with thin gate oxide over

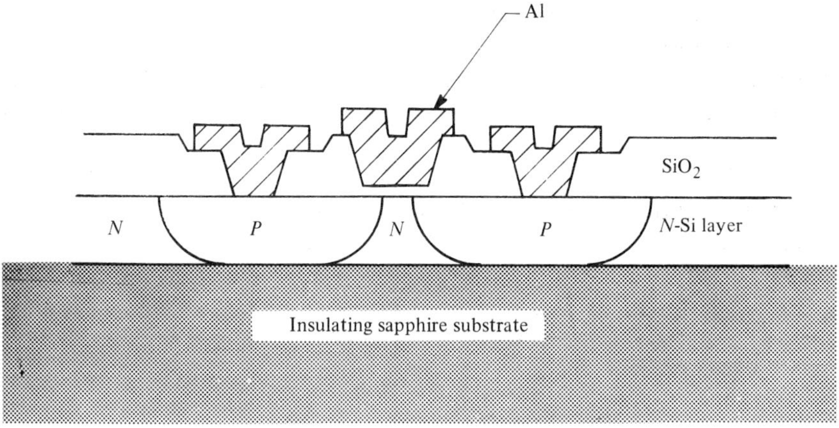

Figure 3–46 Silicon-on-sapphire (SOS) MOS structure.

them. These useful capacitors are not available for circuit design with silicon gate technology, unless processing steps are added, thereby restricting some circuit types (e.g., "ratioless" shift registers). Also, compared to conventional aluminum gate *P*-channel, silicon gate does require some processing steps of greater complexity, particularly the poly-silicon deposition step.

In view of the performance and topological design advantages mentioned, it is apparent that silicon gate technology is a most promising evolution of *P*-channel MOS technology.

3.4.7 Other Technologies

The new MOS-like technology of charge coupled devices[19] looks promising for incorporation with conventional MOS, particularly for shift register type applications. Functional densities may be substantially improved as a result.

Performance improvements are also likely. One of these may be the emergence of silicon-on-sapphire (SOS) MOS.[20] This technology promises greatly reduced parasitic capacitance. Figure 3-46 shows a cross section of an SOS transistor. Of course, the major technical problem to be overcome with SOS technology is the economical manufacture of high-quality substrates.

REFERENCES

1. W. R. Runyan, *Silicon Semiconductor Technology,* McGraw-Hill, New York, 1965.
2. A. S. Grove, *Physics and Technology of Semiconductor Devices,* John Wiley & Sons, New York, 1967.

3. R. M. Warner and J. N. Fordemwalt, eds., *Integrated Circuits, Design Principles and Fabrication,* McGraw-Hill, New York, 1965.

4. D. R. Kerr et al., "Stabilization of SiO_2 Passivation Layers with P_2O_5" *IBM J. Res. Dev.,* **8,** 376 (September 1964).

5. M. H. White and J. R. Cricchi, "Complementary MOS Transistors," *Solid-State Electron.,* **9,** 991–1008 (October 1966).

6. B. E. Deal et al., "Characteristics of the Surface-State Charge (Q_{ss}) of Thermally Oxidized Silicon," *J. Electrochem. Soc.,* **114,** 266 (March 1967).

7. A. G. Revesz and K. H. Zaininger, "The Influence of Oxidation Rate and Heat Treatment on the Si Surface Potential in the Si-SiO_2 System," *IEEE Trans, Electron Devices,* **ED-13,** 246 (February 1966).

8. P. Balk, P. J. Burkhardt, and L. V. Gregor, "Orientation Dependence of Built-In Surface Charge on Thermally Oxidized Silicon," *Proc. IEEE,* **53,** 2133 (1965).

9. G. A. Brown, W. C. Robinette, Jr., and H. G. Calson, "Electrical Characteristics of Silicon Nitride Films Prepared by Silane-Ammonia Reaction," *J. Electrochem. Soc.,* **115,** 948 (September 1968).

10. V. Y. Doo, D. R. Nichols, and G. A. Silvey, "Preparation and Properties of Pyrolytic Silicon Nitride," *J. Electrochem. Soc.,* **113,** 1279 (December 1966).

11. B. E. Deal, P. J. Fleming, and P. L. Castro, "Electrical Properties of Vapor-Deposited Silicon Nitride and Silicon Oxide Films on Silicon," *J. Electrochem. Soc.,* **115,** 300 (March 1968).

12. D. Frohman-Benchkowsky and M. Lenzlinger, "Charge Transport and Storage in Metal-Nitrate-Oxide-Silicon (MNOS) Structures," *J. Appl. Phys.,* **10,** 3307 (1969).

13. J. Mayer, L. Eriksson, and J. Davies, *Ion Implantation in Semiconductors Silicon and Germanium,* Academic Press, New York, 1970.

14. J. Macdougall, K. Manchester, and R. Palmer, "Ion implanation Offers a Bagful of Benefits for MOS," *Electronics,* **43,** 86–90 (June 22, 1970).

15. R. W. Bower et al., "MOS Field Effect Transistors Formed by Gate Masked Ion Implantation," *IEEE Trans. Electron Devices,* **ED-15,** no. 10, 757 (October 1968).

16. B. E. Deal, and E. H. Snow, "Barrier Energies in Metal-Silicon Dioxide-Silicon Structures," *J. Phys. Chem. Solids,* **27,** 1873 (1966).

17. F. Faggin and T. Klein, "Silicon Gate Technology," *Solid-State Electronics,* **13,** 1125 (1970).

18. B. M. Berry, et al., "Integrated Silicon Device Technology," *ASD-TDR-63-316,* **IX.** Epitaxy, Research Triangle Institute, Durham, N.C., 1965.

19. W. S. Boyle and G. E. Smith, "Charge Coupled Semiconductor Devices," *Bell Sys. Tech. J.,* 587 (April 1970).

20. E. C. Ross and C. W. Mueller, "Extremely Low Capacitance Silicon Film MOS Transistors," *IEEE Trans. Electron. Devices,* **ED-13,** 379 (March 1966).

4 MOS Circuit Design Theory

4.1 INTRODUCTION

This chapter presents the fundamentals of using the MOSFET as a circuit element. Device characteristics are examined and, based on these characteristics, techniques are developed for dc and transient analyses.

The MOS device is mainly used as a digital circuit element. The simple digital inverter circuit is analyzed in detail in this chapter. Both graphical and analytical approaches to inverter design are presented. First, the inverter dc transfer characteristics are derived and discussed; then the transient model is derived and practical methods of transient analysis are developed. Design examples are used throughout the chapter to illustrate the principal concepts.

In addition to the simple digital inverter, the MOS device is frequetly used in various other circuit configurations. Important considerations that relate to several of these applications are discussed. Included are output circuits as well as a generalized ratioless circuit.

Another application of the MOS device is in analog circuits, including amplifiers and multiplexers. Although these aspects of MOS usage are not discussed in this chapter, they should not be completely overlooked since MOS device characteristics, such as high input impedance and bilateral current flow, can be used advantageously in analog circuits.

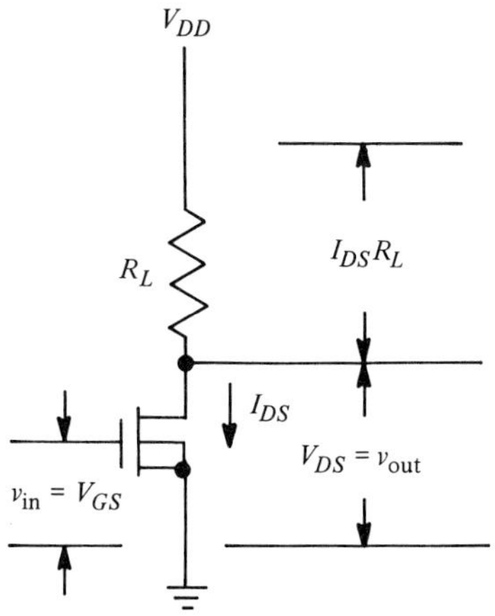

Figure 4–1 Basic inverter with a passive linear load resistor.

4.2 BASIC DIGITAL INVERTER

The basic inverter circuit with a passive linear load resistor is shown in Figure 4-1.

The v_{out} vs. v_{in} transfer characteristics for this inverter can be determined graphically from the load line and the drain characteristics of the device. As an example, consider the typical curves shown in Figure 4-2 for a device with a V_T of 4.0 V. Suppose that $V_{DD} = 12$ V and $R_L = 40$ kΩ. The first step in the analysis is to draw the load line on the characteristics as shown on the graph. The load line equation is $V_{DS} = V_{DD} - I_{DS} R_L$. The v_{out} vs. v_{in} curve can now be constructed point by point from the load line, shown in Figure 4-3.

The transfer characteristics for this MOS device and any other load resistor can be obtained in a similar manner. Examples are shown in Figure 4-4, where load lines are drawn for load resistances of 20, 40, and 120 kΩ. The resulting transfer characteristics are plotted in Figure 4-5.

Generally, it is obvious from these curves that the larger the load resistance the more steeply the transfer characteristics fall and the lower the final output voltage. Furthermore, from the examples shown, it is clear that a load resistance of less than 20 kΩ would be completely unsatisfactory in an inverter circuit because of the poor transfer characteristics. A reasonable load resistance for this input device would be about 40 kΩ.

4.2.1 Diffused Load Resistor

At first one might be tempted to employ an ordinary diffused load resistor similar to that used in bipolar integrated circuits to obtain the 20 kΩ load

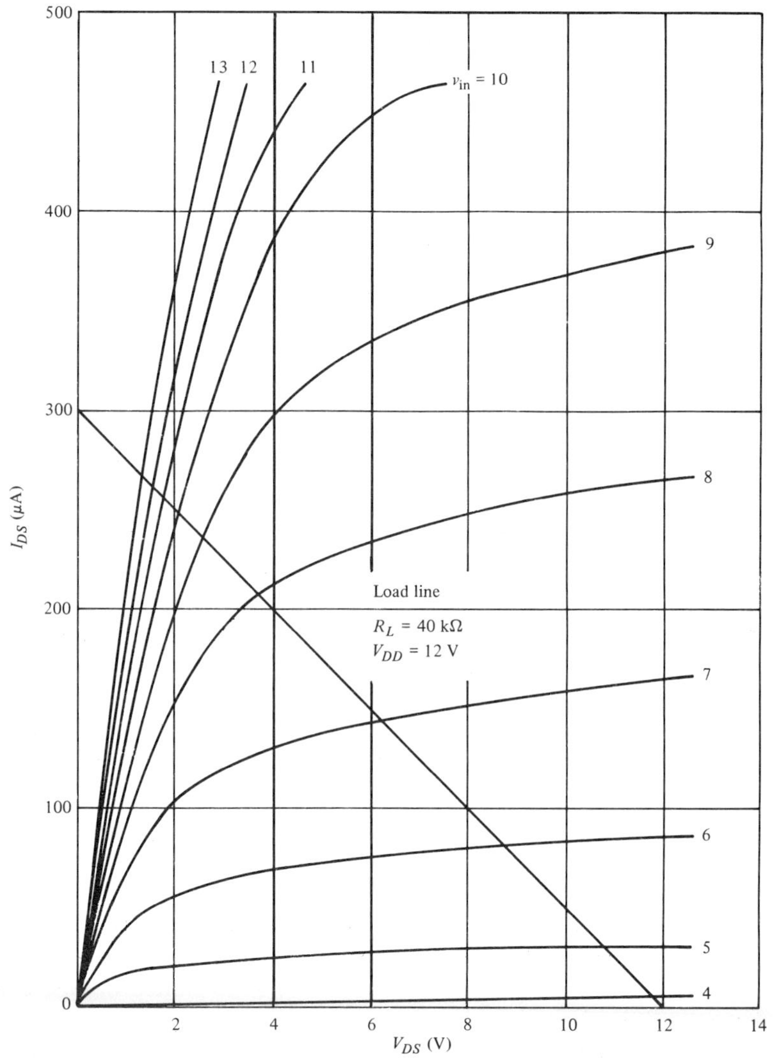

Figure 4–2 Drain characteristics of MOS inverter with a passive linear resistor load line.

needed for the example. A typical value of sheet resistance for a standard conventional diffused resistor is approximately 100 Ω/□. Thus to obtain a 20 kΩ integrated resistor it would be necessary to have 200 squares in series. The minimum width of such a resistor is limited by the manufac-

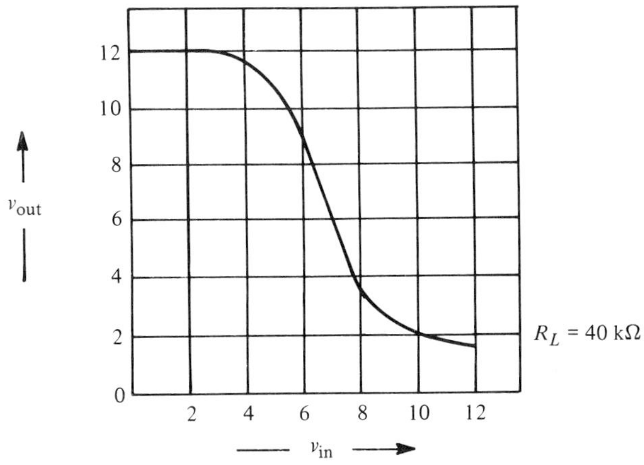

Figure 4–3 Typical transfer characteristics of MOS inverter with a passive linear load resistor.

turing process to approximately 0.3 mil. As a result, a 20 kΩ resistor might be typically 0.3 mil wide and 60 mils long; thus, it would occupy an area approximately 18 mils². In an integrated circuit this is a very large element. By comparison, the area of the MOS input device itself would be typically about 1 mil², or approximately 1⁄18 the size of the diffused load resistor.

Another limiting factor is that the distributed capacitance between the resistor junction and substrate substantially lowers the frequency response of an amplifier with a diffused load resistor. This is not a significant problem when an MOS load resistor is used.

4.2.2 MOS Load Resistor

The overall size of the inverter circuit and its load can be greatly reduced by using an MOS device as an active load device, as shown in Figure 4-6. The gate of the MOS load may be connected to V_{DD} or to a separate supply, V_{GG}. The characteristics of the MOS load resistor can be determined from the graphical drain characteristics. As an example, consider an MOS load resistor with the drain characteristics shown in Figure 4-7 and a V_T of 3.5 V. First, suppose that $V_{GG} = V_{DD}$ (the gate and drain are connected to a common supply). For this case, it is clear that $V_{GS} = V_{DS}$ for the load device. The locus of points where $V_{GS} = V_{DS}$ is the operating V-I characteristic of the load, which is shown in Figure 4-7. Note that the

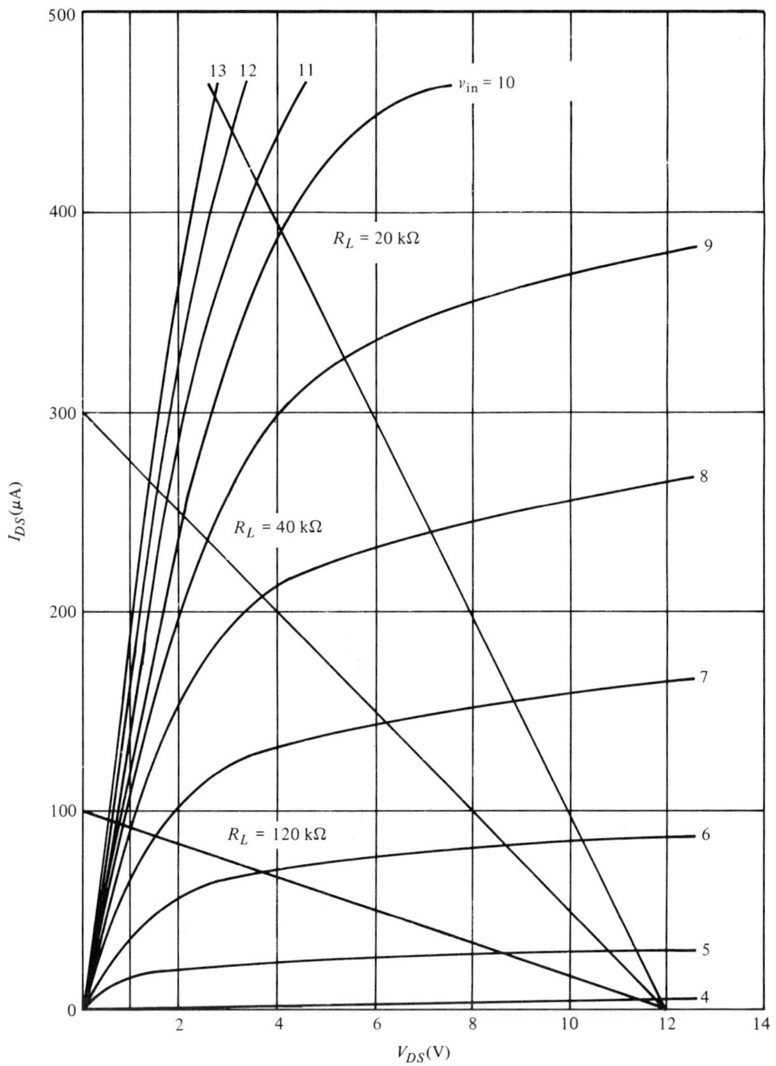

Figure 4–4 Drain characteristics of MOS inverter with various load resistors.

device used as the load is not identical to the MOS input device in that it has a much higher impedance. This aspect will be discussed later.

The performance of the inverter with an MOS load can also be analyzed graphically. The first step is to draw the load line of the MOS load on the input device characteristics as shown in Figure 4-8 for a V_{DD} of 12 V. To do this, the load line equation $v_{out} = V_{DD} - V_{Load}$ is used. Since the

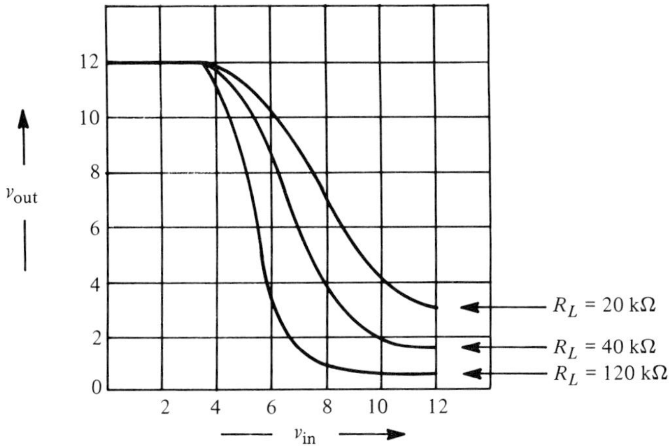

Figure 4–5 Transfer characateristics of MOS inverter with various passive linear load resistors.

load and input device drain currents must be equal, values of V_{Load} at various drain currents are obtained from the load characteristics in Figure 4-7 and corresponding values of v_{out} are calculated from the load line equation. Table 4-1 gives values of v_{out} and V_{Load} as a function of drain current for the V-I characteristic shown in Figure 4-7.

Table 4–1 v_{out} **and** V_{Load} **as the Function of Drain Current**

I_{DS} (μA)	V_{Load}	$v_{\text{out}} = V_{DD} - V_{\text{Load}}$
0	3.5	8.5
5	6.5	5.5
10	7.7	4.3
15	8.6	3.4
20	9.3	2.7
25	9.9	2.1
30	10.5	1.5
35	11.0	1.0
40	11.6	0.4
44	12.0	0.0

The transfer charactersistics shown in Figure 4-9 can now be obtained from the load line and the input device characteristics shown in Figure 4-8. The resultant transfer characteristic corresponds quite closely to that of a 120 kΩ linear diffused resistor (see Figure 4-6) but the MOS load is much

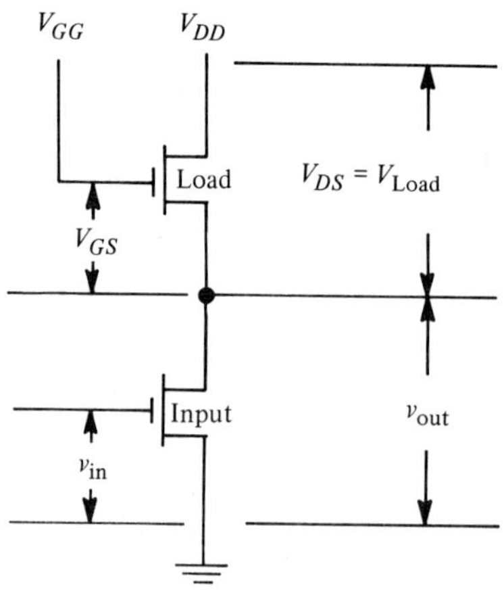

Figure 4–6 Inverter circuit with an MOS load device.

smaller. The total area required for both the MOS input devices and its MOS load is approximately 2 mils^2.

In the example just discussed, the on-resistance of the load device is about 20 times that of the input device. If the load device and input device had the same resistance, the transfer characteristic would be of little practical value. As a result, for a usuable transfer characteristic, it is necessary to adjust the topology of the inverter pair so that the resistance of the load is large compared to that of the input device.

Note in Figure 4-9 that the maximum output voltage is a gate threshold voltage drop less than the drain supply voltage of the MOS load device. It is interesting to examine what determines the maximum output voltage for the inverter with an MOS load device. Since the gate voltage of the input device is less than the threshold voltage, the only drain current is that due to leakage. The load device essentially turns on; but the current is limited by the leakage of the input device to a very low value, typically a few nanoamperes.

In Figure 4-8 the V-I characteristics of the load device intersects the drain leakage current curve at approximately 0 μA because of the scale used, at which point there is a voltage drop equal to V_T across the load device.

In the preceding example, the gate of the load device was connected to the drain where $V_{GG} = V_{DD}$. The characteristics for any other value of V_{GG} can be determined graphically in a similar manner. Examples are shown in Figure 4-10, where V_{GG} is greaterd than V_{DD} and $V_T = 4$ V.

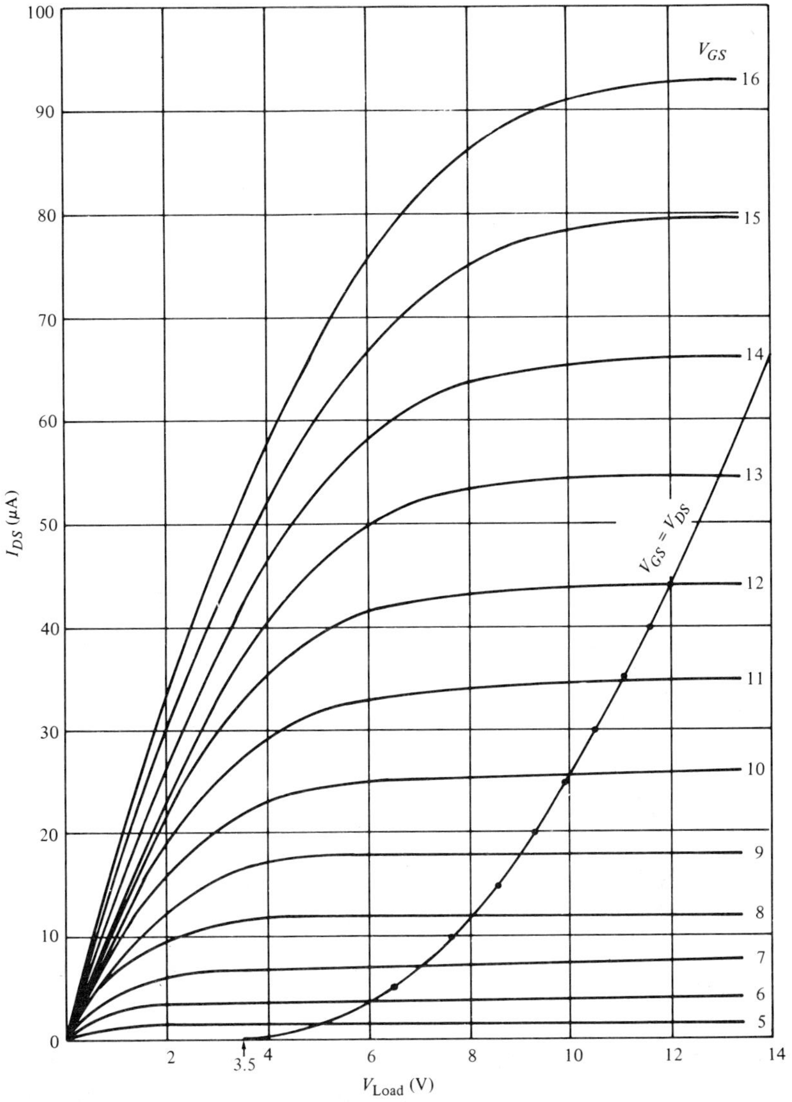

Figure 4–7 V-I characteristics of MOS load device with common gate and drain.

One interesting point to note from the load device characteristics is their relative linearity as a function of $|V_{GG} - V_{DD}|$. As the value of $|V_{GG} - V_{DD}|$ increases, the load device characteristics become more linear. The significance of this point will be discussed later in this chapter.

Transfer curves are shown in Figure 4-11 for various bias conditions

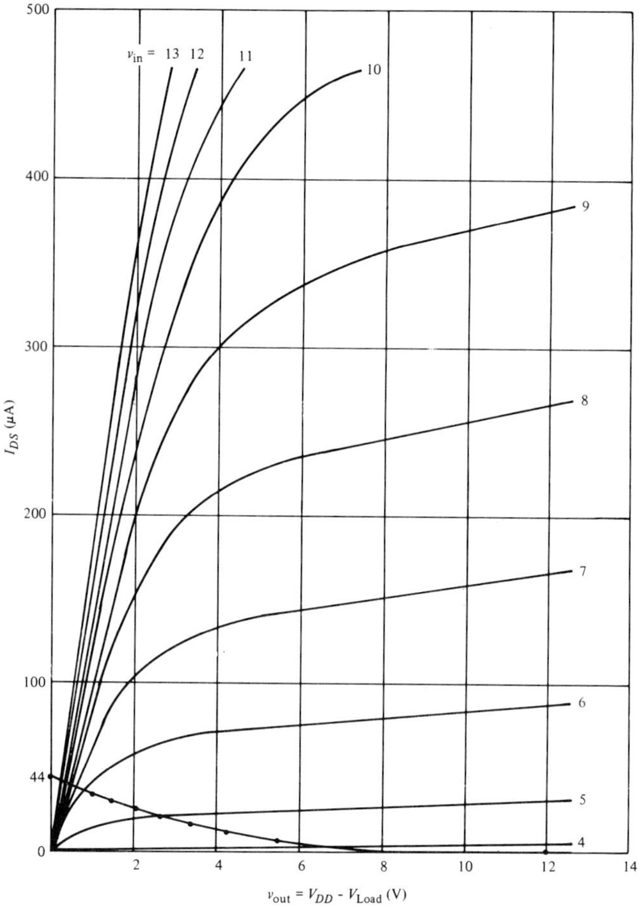

Figure 4-8 Drain characteristics of MOS input device with load line of MOS load device where $V_{GG} = V_{DD}$.

on the MOS load device. The curves can be obtained by plotting the V-I curves of the load device as load lines on the characteristics of the input device.

The maximum output level of the inverter is determined by V_{GG} and is equal to V_T less than V_{GG} until it reaches V_{DD}. Since the off-level cannot be greater than V_{DD}, it remains at this level for all $|V_{GG} - V_T| > |V_{DD}|$. In other words, the voltage drop across the load device becomes zero for this bias condition.

When the biasing is such that $|V_{GG} - V_T| \leqq |V_{DD}|$ the load device operates in the saturation region, as shown in the V-I characteristics in

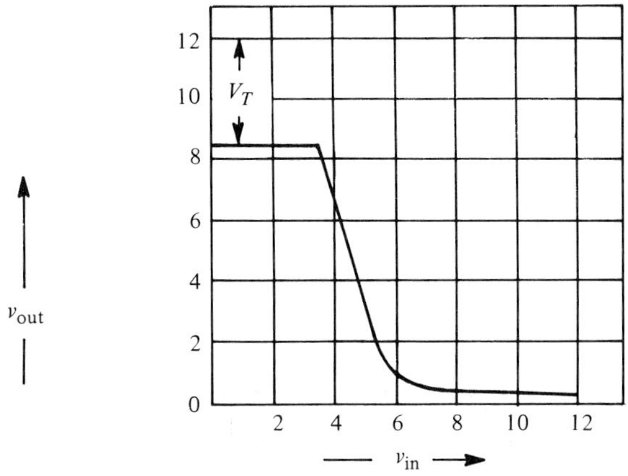

Figure 4–9 Transfer characteristics of MOS inverter circuit with MOS load where ($V_{GG} = V_{DD}$).

Figure 4-12. Frequently in saturation operation, the gate and drain of the load device are connected and $V_{GG} = V_{DD}$. Another example is found in the push-pull circuit. When the biasing is such that $|V_{GG} - V_T| > |V_{DD}|$, the load device operates in the nonsaturation region, as the V-I characteristics show. The load will be in nonsaturation operation if V_{GG} is more than a threshold greater than V_{DD}.

When the load device is biased in the nonsaturation region:

1. The maximum output voltage is equal to V_{DD}. This results in power dissipation advantages since, for given low and high levels, a lower supply voltage may be used.
2. The V-I curve becomes more linear. This provides faster transient response characteristics.

4.2.3 Analytical Voltage Transfer Characteristics

In this section, the analytical expressions for the voltage transfer characteristics of an MOS inverter device and its associated load will be developed. Figure 4-13 shows a generalized MOS inverter and its voltage transfer characteristics. Normalized curves will also be plotted for the various inverter parameters. These will be used later as design curves. The analysis will be based on the theoretical equations derived from the elementary physical model.

Three different types of loads will be analyzed: (1) linear resistance load; (2) saturated MOS load; and (3) nonsaturated MOS load. These

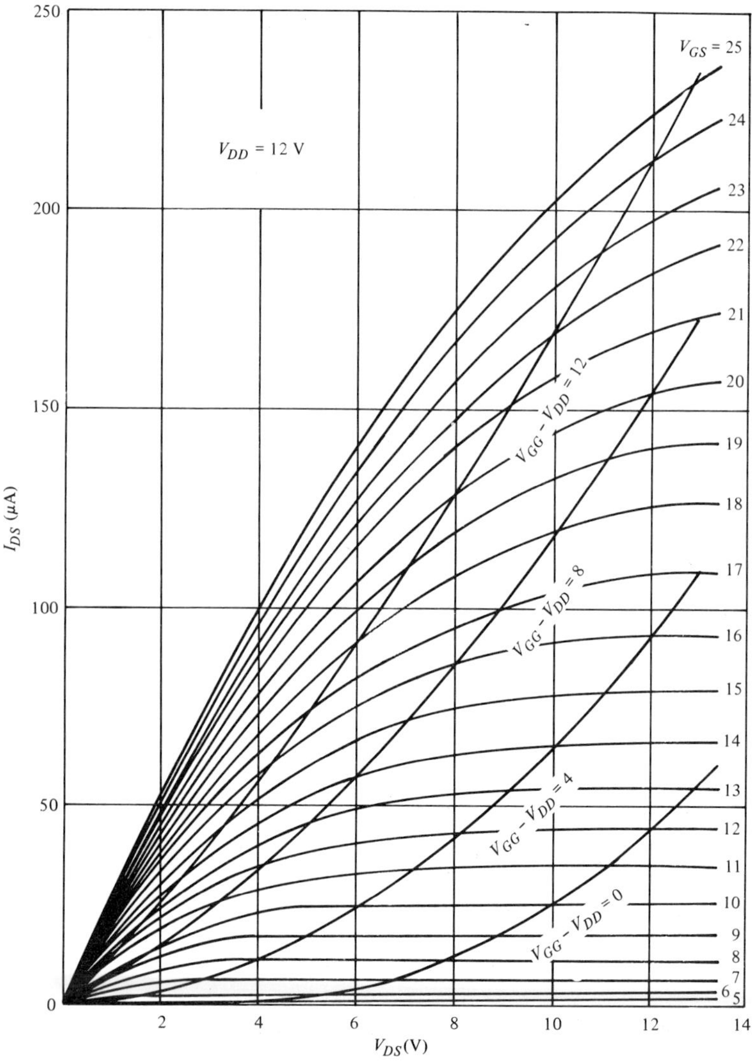

Figure 4–10 Drain characteristics of MOS load device with bias curves for various values of V_{GG} greater than V_{DD}.

are shown in Figure 4-14. After each of these types of loads is analyzed independently, it will be shown that a general equation can be written for analysis of all cases. This general expression is particularly useful in simplifying the transient analysis.

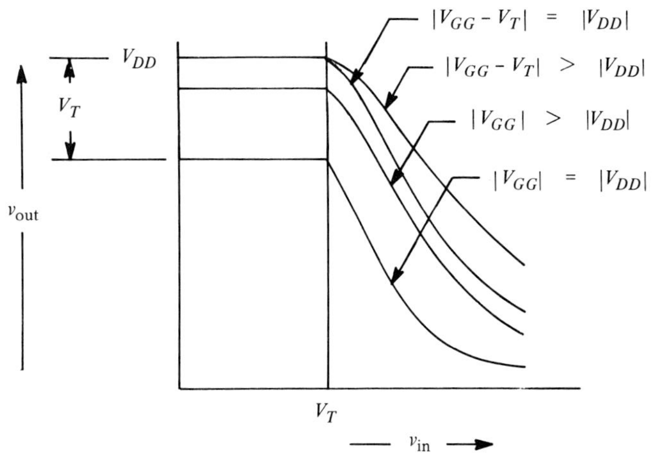

Figure 4–11 Transfer characteristics of MOS inverter circuit with an MOS load.

To determine the theoretical voltage transfer characteristics for each of the three types of loads:

1. Write the equation for the current in the input device. Since the device will be in a saturated condition when the input voltage is small

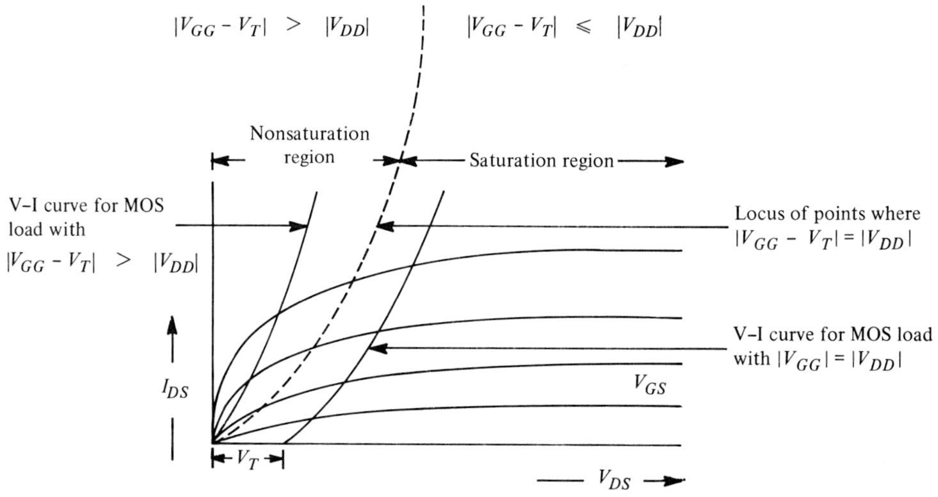

Figure 4–12 Drain characteristics of MOS load device with superimposed V–I curves.

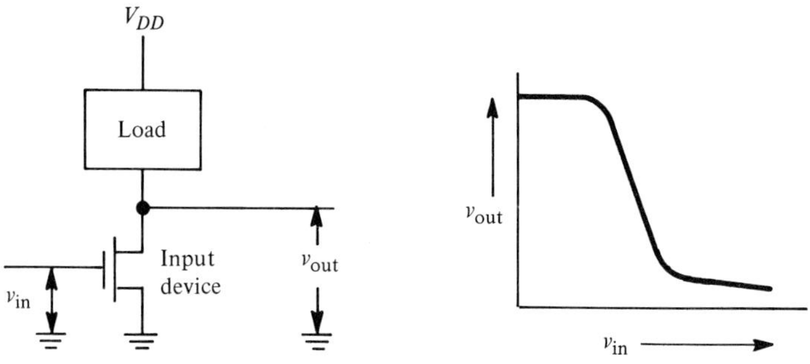

Figure 4–13 MOS inverter and its voltage transfer characteristics.

and will enter a nonsaturated condition as the input voltage increases, there will be two equations, one for each condition.

2. Write the equation for the current in the load.
3. Equate the current in the input device to the current in the load. Since these expressions will be functions of the device parameters, the bias conditions, input voltage, and output voltage, the resulting equation will contain the voltage transfer characteristics.
4. Normalize the output and input voltage in these equations to the maximum output voltage.
5. Calculate the values and plot the transfer curves.

The theoretical expressions for the current in an MOS device were derived in Chapter 2. The equations that define operation in the nonsaturation (Equation 2-73) and saturation (Equation 2-74) regions

$$\text{for} \quad |V_{GS} - V_T| \leqq |V_{DS}| \quad \text{(Saturation)} \qquad (2\text{-}74)$$

$$I_{DS} = -k(V_{GS} - V_T)^2$$

$$\text{for} \quad |V_{GS} - V_T| > |V_{DS}| \quad \text{(Nonsaturation)} \qquad (2\text{–}73)$$

$$I_{DS} = -k[2(V_{GS} - V_T)\, V_{DS} - (V_{DS})^2]$$

where $\quad k = k' \dfrac{W}{\ell} = \dfrac{\mu_p \varepsilon_{ox}}{2t_{ox}} \dfrac{W}{\ell}$

For the current in the input device, first consider a device with the substrate connected to the source as shown in Figure 4-15.

From the figure it is clear that

$$V_{GS} = v_{in} \qquad (4\text{-}1)$$

$$V_{DS} = v_{out} \qquad (4\text{-}2)$$

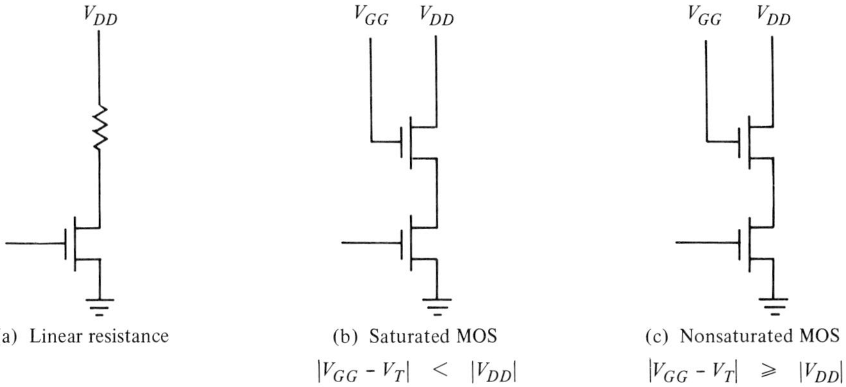

Figure 4–14 MOS inverter using three basic types of loads.

and

$$I_{DS} = i_I \tag{4-3}$$

The theoretical expressions for the current in the input device are obtained by substituting Equations 4-1 and 4-2 into Equations 2-74 and 2-73, ignoring the minus signs for convenience:

for $\quad |v_{\text{in}} - V_T| \leqq |v_{\text{out}}| \quad$ (Saturation)

$$i_I = k_I(v_{\text{in}} - V_T)^2 \tag{4-4}$$

and for $\quad |v_{\text{in}} - V_T| > |v_{\text{out}}| \quad$ (Nonsaturation)

$$i_I = k_I[2(v_{\text{in}} - V_T)\, v_{\text{out}} - v_{\text{out}}^2] \tag{4-5}$$

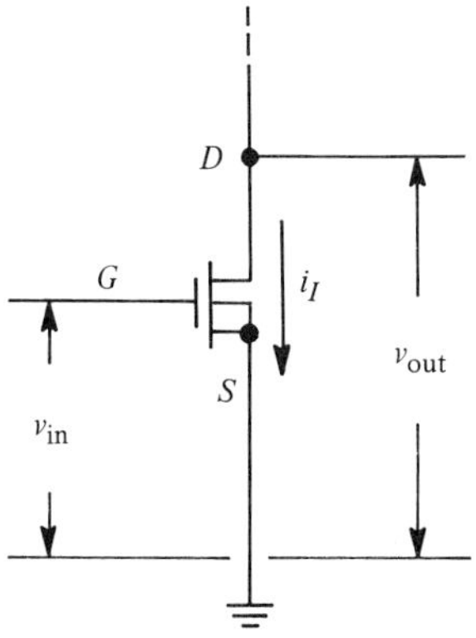

Figure 4–15 Input device with voltage and current definitions.

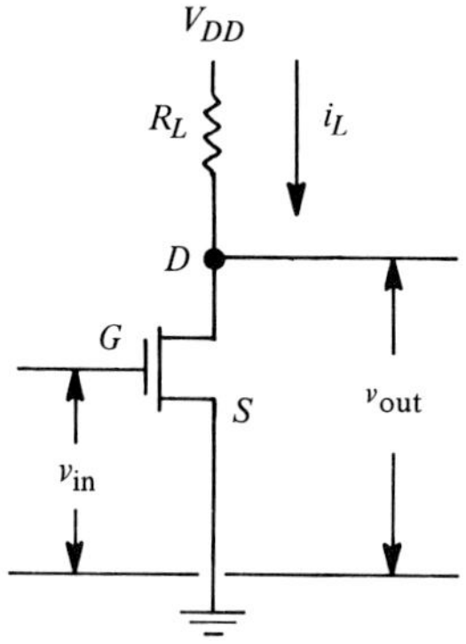

Figure 4–16 MOS inverter circuit with linear load resistor.

4.2.4 Linear Resistance Load

Consider the case where linear resistance is used as the load on the MOS input device as shown in Figure 4-16.

From Figure 4-16 it can be seen that

$$i_L = \frac{V_{DD} - v_{out}}{R_L} \tag{4-6}$$

If $i_L = i_I$, then

for $\quad |v_{in} - V_T| \leqq |v_{out}| \quad$ (Saturation)

$$\frac{V_{DD} - v_{out}}{R_L} = k_I(v_{in} - V_T)^2 \tag{4-7}$$

and for $\quad |v_{in} - V_T| > |v_{out}| \quad$ (Nonsaturation)

$$\frac{V_{DD} - v_{out}}{R_L} = k_I[2(v_{in} - V_T)\, v_{out} - v_{out}^2] \tag{4-8}$$

After v_{out} and $(v_{in} - V_T)$ are normalized to V_{DD} and the terms in Equations 4-7 and 4-8 are rearranged,

for $\quad \left|\frac{v_{in} - V_T}{V_{DD}}\right| \leqq \left|\frac{v_{out}}{V_{DD}}\right| \quad$ (Saturation)

$$1 - \frac{v_{out}}{V_{DD}} = k_I V_{DD} R_L \left(\frac{v_{in} - V_T}{V_{DD}}\right)^2 \tag{4-9}$$

and for $\quad \left|\frac{v_{in} - V_T}{V_{DD}}\right| > \left|\frac{v_{out}}{V_{DD}}\right| \quad$ (Nonsaturation)

$$1 - \frac{v_{out}}{V_{DD}} = k_I V_{DD} R_L \left[2 \left(\frac{v_{in} - V_T}{V_{DD}} \right) \left(\frac{v_{out}}{V_{DD}} \right) - \left(\frac{v_{out}}{V_{DD}} \right)^2 \right] \quad (4\text{-}10)$$

4.2.5 Saturated MOS Load

Next, consider the case where a saturated MOS device is used as the load for the MOS input device. In order to ensure saturation of the load device, the biasing must satisfy the condition that $|V_{GG} - V_T| \leqq |V_{DD}|$.

From Figure 4-17, it is clear that for the load device,

$$V_{GS} = V_{GG} - v_{out} \quad (4\text{-}11)$$

$$V_{DS} = V_{DD} - v_{out} \quad (4\text{-}12)$$

$$I_{DS} = i_L \quad (4\text{-}13)$$

The theoretical expression for the current in the saturated load device can be found by substituting Equations 4-11 and 4-13 into the expression for current, Equation 2-74 in Chapter 2.

$$i_L = k_L[(V_{GG} - v_{out}) - V_T]^2 \quad (4\text{-}14)$$

If $i_L = i_I$, then

for $\qquad |v_{in} - V_T| \leqq |v_{out}| \qquad$ (Saturation)

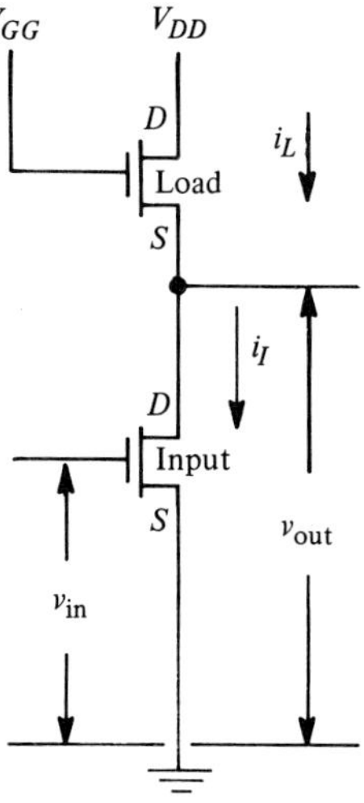

Figure 4–17 MOS inverter circuit with MOS load device.

$$k_L[(V_{GG} - v_{out}) - V_T]^2 = k_I(v_{in} - V_T)^2 \tag{4-15}$$

and for $|v_{in} - V_T| > |v_{out}|$ (Nonsaturation)

$$k_L[(V_{GG} - v_{out}) - V_T]^2 = k_I[2(v_{in} - V_T)\, v_{out} - v_{out}^2] \tag{4-16}$$

If v_{out} and $(v_{in} - V_T)$ are normalized to $(V_{GG} - V_T)$ and the terms in Equations 4-15 and 4-16 are rearranged,

for $\left|\dfrac{v_{in} - V_T}{V_{GG} - V_T}\right| \leq \left|\dfrac{v_{out}}{V_{GG} - V_T}\right|$

$$\left(1 - \frac{v_{out}}{V_{GG} - V_T}\right)^2 = \frac{k_I}{k_L}\left(\frac{v_{in} - V_T}{V_{GG} - V_T}\right)^2 \tag{4-17}$$

and for $\left|\dfrac{v_{in} - V_T}{V_{GG} - V_T}\right| > \left|\dfrac{v_{out}}{V_{GG} - V_T}\right|$

$$\left(1 - \frac{v_{out}}{V_{GG} - V_T}\right)^2 = \frac{k_I}{k_L}\left[2\left(\frac{v_{in} - V_T}{V_{GG} - V_T}\right)\left(\frac{v_{out}}{V_{GG} - V_T}\right) - \left(\frac{v_{out}}{V_{GG} - V_T}\right)^2\right] \tag{4-18}$$

For a given value of the coefficient k_I/k_L, the curve of normalized output voltage $v_{out}/(V_{GG} - V_T)$ vs. normalized input voltage $(v_{in} - V_T)/(V_{GG} - V_T)$ can be calculated for each condition using Equations 4-17 and 4-18.

The values of the parameter k_I/k_L are shown in Figure 4-18.

Note that by decreasing the W/ℓ ratio of the load, the load resistance is increased and the value of k_I/k_L increases, causing the transfer characteristics to have a steeper slope and a lower final value. It is also important to realize that the maximum output voltage $(V_{GG} - V_T)$ differs for different values of V_{GG}; however, the form of the normalized transfer curves is the same for all of the values. In the normal method of obtaining saturated load devices, the gate and drain are connected so that $V_{GG} = V_{DD}$.

4.2.6 Nonsaturated MOS Load

Finally, consider the case where a nonsaturated MOS device is used as the load for the MOS input device as previously shown in Figure 4-14c. To ensure that the load device is not saturated, the biasing must satisfy the condition that $|V_{GG} - V_T| > |V_{DD}|$.

From the discussions of the saturated load device, it is clear that for the nonsaturated load, Equations 4-11, 4-12, and 4-13 apply. As before, the theoretical expression for the current in the nonsaturated load device

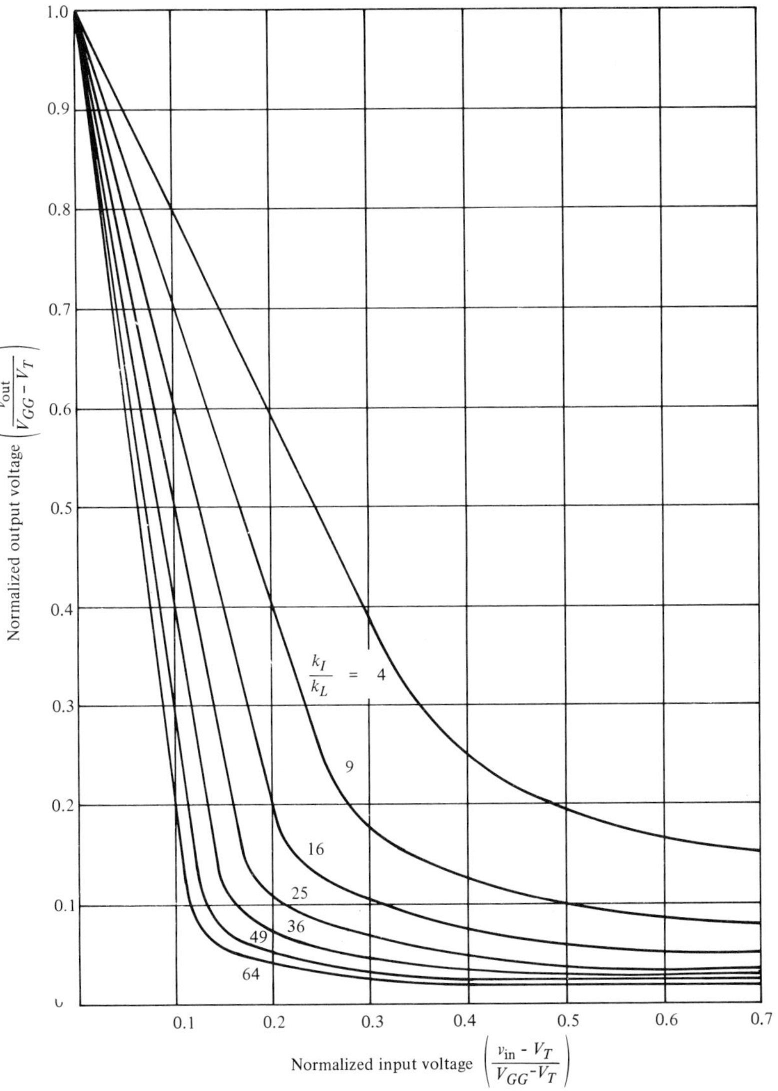

Figure 4–18 Normalized theoretical voltage transfer characteristics for a saturated MOS load device.

is obtained by substituting Equations 4-11, 4-12, and 4-13 into Equation 2-73.

$$i_L = k_L[2(V_{GG} - v_{out} - V_T)(V_{DD} - v_{out}) - (V_{DD} - v_{out})^2] \quad (4\text{-}19)$$

Equation 4-19 can be rewritten

$$i_L = k_L\{[V_{DD} - v_{out}][2(V_{GG} - V_T) - V_{DD} - v_{out}]\}$$

$$= k_L[2(V_{GG} - V_T) - V_{DD}]\left\{[V_{DD} - v_{out}]\left[1 - \frac{v_{out}}{2(V_{GG} - V_T) - V_{DD}}\right]\right\}$$

$$= k_L V_{DD}^2 \frac{[2(V_{GG} - V_T) - V_{DD}]}{V_{DD}}\left\{\left[1 - \frac{v_{out}}{V_{DD}}\right]\right.$$

$$\left.\left[1 - \frac{V_{DD}}{2(V_{GG} - V_T) - V_{DD}}\left(\frac{v_{out}}{V_{DD}}\right)\right]\right\} \quad (4\text{-}20)$$

It is now convenient to introduce into the equations an algebraic *biasing parameter* (m),

$$m = \frac{V_{DD}}{2(V_{GG} - V_T) - V_{DD}} \quad (4\text{-}21)$$

Equation 4-20 then becomes

$$i_L = \frac{k_L V_{DD}^2}{m}\left(1 - \frac{v_{out}}{V_{DD}}\right)\left(1 - m\frac{v_{out}}{V_{DD}}\right) \quad (4\text{-}22)$$

Since the biasing parameter has meaning only if the load device is nonsaturated, the voltages must satisfy the relation $|V_{GG} - V_T| > |V_{DD}|$. The maximum value of m occurs when $|V_{GG} - V_T| = |V_{DD}|$ where $m = V_{DD}/(2V_{DD} - V_{DD}) = 1.0$.

The value of m decreases as the gate voltage becomes greater and

$$m \to 0 \text{ as } V_{GG} \to \infty$$

Therefore, the range of m is $0 < m \leq 1$ when an MOS load device is used. A typical value of m for commonly used supply voltages and a high threshold process, where $V_{DD} = 15$, $V_{GG} = 27$ and $V_T = 4$ V, is

$$m = \frac{15}{2(27 - 4) - 15} = \frac{15}{31} = 0.484 \sim 0.5$$

If $i_L = i_I$,

for $\quad |v_{in} - V_T| \leq |v_{out}| \quad$ (Input device saturated)

$$k_L V_{DD}^2\left(1 - \frac{v_{out}}{V_{DD}}\right)\left(1 - m\frac{v_{out}}{V_{DD}}\right) = k_I(v_{in} - V_T)^2 \quad (4\text{-}23)$$

and for $\quad |v_{in} - V_T| > |v_{out}| \quad$ (Input device nonsaturated)

$$k_L V_{DD}^2\left(1 - \frac{v_{out}}{V_{DD}}\right)\left(1 - m\frac{v_{out}}{V_{DD}}\right) = k_I[2(v_{in} - V_T)\, v_{out} - v_{out}^2] \quad (4\text{-}24)$$

After v_{out} and $(v_{in} - V_T)$ are normalized to V_{DD} and the terms in Equations 4-23 and 4-24 are rearranged,

for $\left|\dfrac{v_{in} - V_T}{V_{DD}}\right| \leq \left|\dfrac{v_{out}}{V_{DD}}\right|$

$$\left(1 - \frac{v_{out}}{V_{DD}}\right)\left(1 - m\frac{v_{out}}{V_{DD}}\right) = m\frac{k_I}{k_L}\left(\frac{v_{in} - V_T}{V_{DD}}\right)^2 \tag{4-25}$$

and for $\left|\dfrac{v_{in} - V_T}{V_{DD}}\right| > \left|\dfrac{v_{out}}{V_{DD}}\right|$

$$\left(1 - \frac{v_{out}}{V_{DD}}\right)\left(1 - m\frac{v_{out}}{V_{DD}}\right) = m\frac{k_I}{k_L}\left[2\left(\frac{v_{in} - V_T}{V_{DD}}\right)\left(\frac{v_{out}}{V_{DD}}\right) - \left(\frac{v_{out}}{V_{DD}}\right)^2\right] \tag{4-26}$$

For a given value of coefficients m and k_I/k_L, the curve of normalized output voltage v_{out}/V_{DD} versus normalized input voltage $(v_{in} - V_T)/V_{DD}$ can be calculated for each condition using Equations 4-25 and 4-26. Curves for values of parameters m and k_I/k_L are shown in Figures 4-19 through 4-27. These voltage transfer curves can be used to design inverter circuits with specified transfer characteristics.

4.2.7 Generalized Equations

As discussed earlier, the equations for the load current for the three different types of loads follow.

Linear load resistance:

$$i_L = \frac{V_{DD} - v_{out}}{R_L} \tag{4-6}$$

saturated MOS load:

$$i_L = k_L[(V_{GG} - v_{out}) - V_T]^2 \tag{4-14}$$

nonsaturated MOS load:

$$i_L = \frac{k_L V_{DD}^2}{m}\left(1 - \frac{v_{out}}{V_{DD}}\right)\left(1 - m\frac{v_{out}}{V_{DD}}\right) \tag{4-22}$$

When slightly rearranged, these equations become

Linear load resistance:

$$i_L = \frac{V_{DD}}{R_L}\left(1 - \frac{v_{out}}{V_{DD}}\right) \tag{4-27}$$

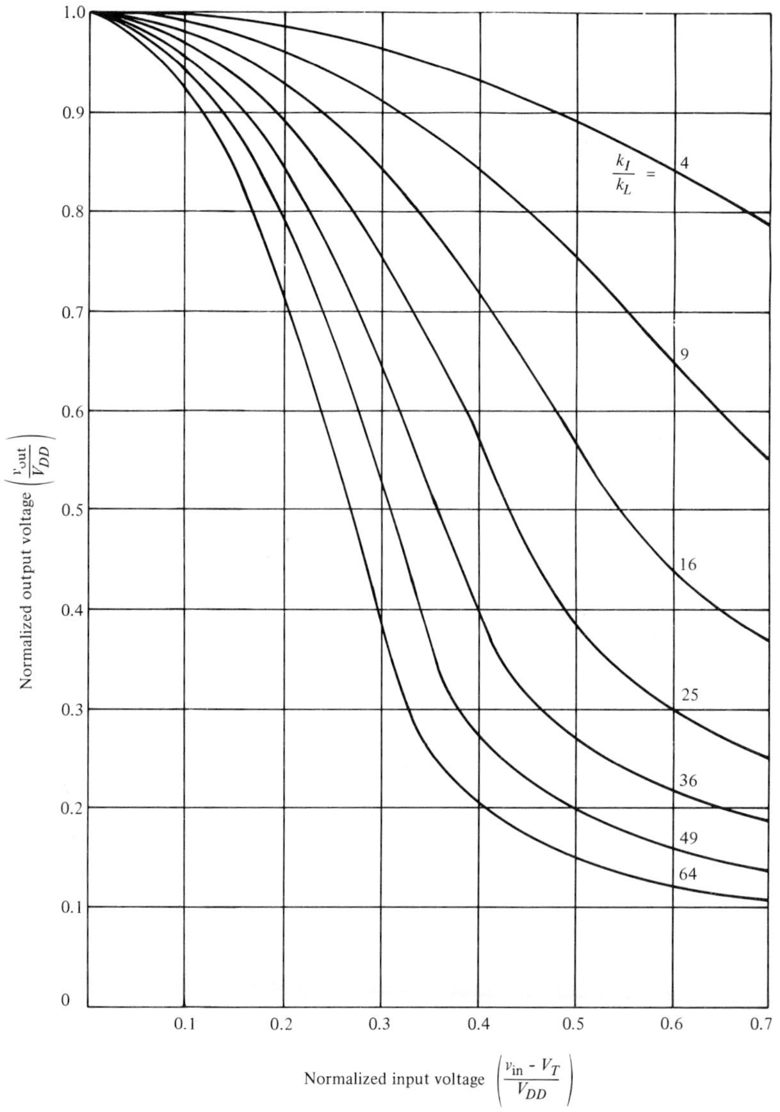

Figure 4–19 Normalized theoretical voltage transfer characteristics for a nonsaturated MOS load device. Biasing parameter $m = 0.1$.

saturated MOS load:

$$i_L = k_L(V_{GG} - V_T)^2 \left[1 - \frac{v_{out}}{V_{GG} - V_T}\right]^2 \qquad (4\text{-}28)$$

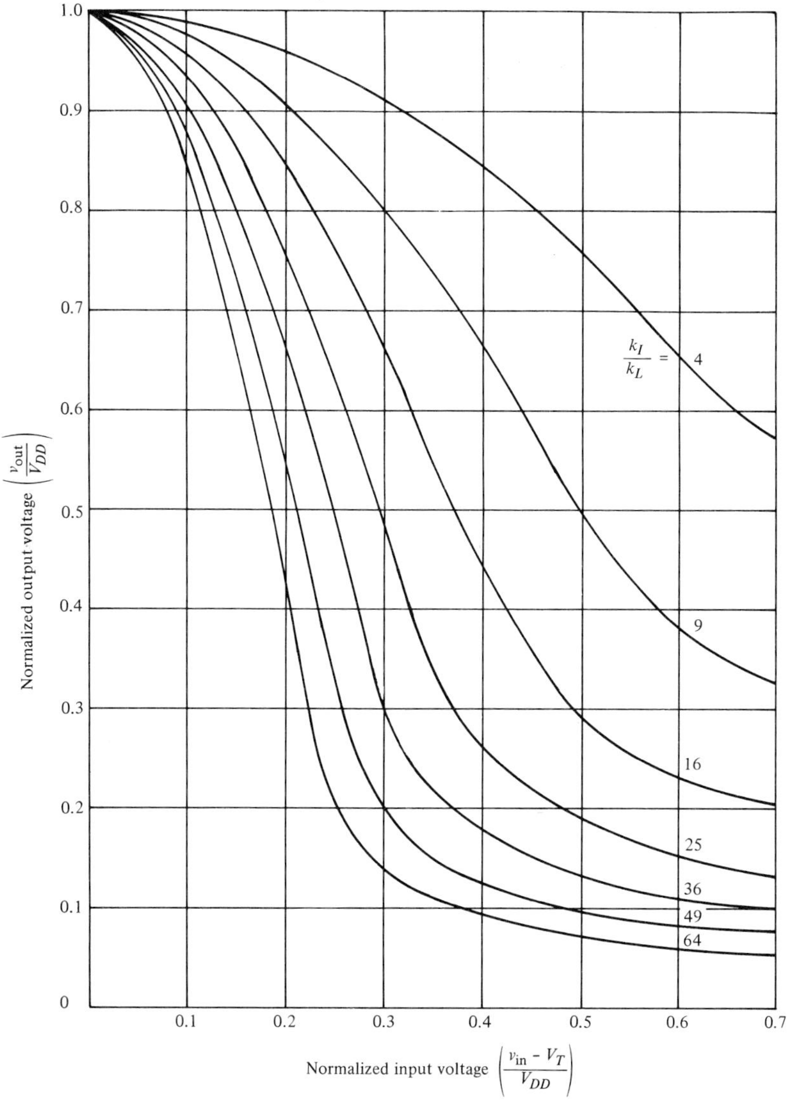

Figure 4–20 Normalized theoretical voltage transfer characteristics for a nonsaturated MOS Load Device. Biasing Parameter $m = 0.2$.

nonsaturated MOS load:

$$i_L = \frac{k_L V_{DD}^2}{m}\left(1 - \frac{v_{out}}{V_{DD}}\right)\left(1 - m\,\frac{v_{out}}{V_{DD}}\right) \tag{4-29}$$

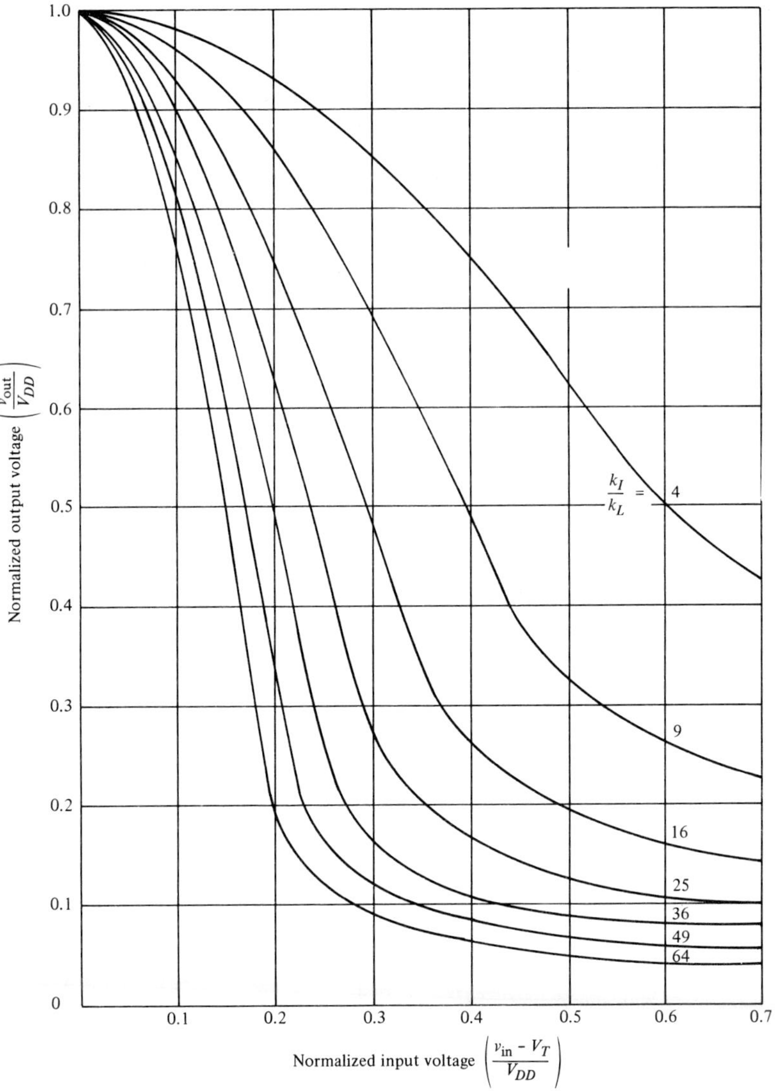

Figure 4–21 Normalized theoretical voltage transfer characteristics for a non-saturated MOS load device. Biasing parameter $m = 0.3$.

All three types of loads can be described by a single *general* equation.

$$i_L = I_{sc}\left(1 - \frac{v_{out}}{V_{OH\ max}}\right)\left(1 - m\,\frac{v_{out}}{V_{OH\ max}}\right) \tag{4-30}$$

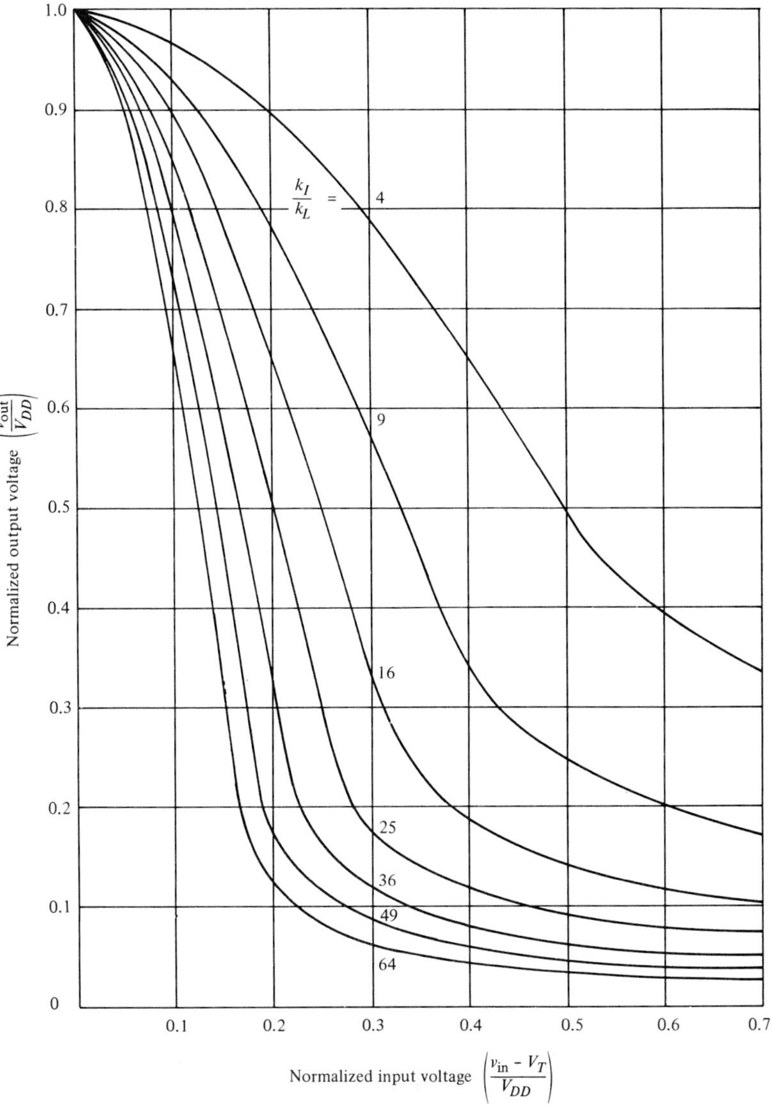

Figure 4–22 Normalized theoretical voltage transfer characteristics for a non-saturated MOS load device. Biasing parameter $m = 0.4$.

where I_{sc} is the *short-circuit current,* (i.e., $I_{sc} = i_L|_{v_{out} = 0}$) and $V_{OH\,max}$ is the *maximum output voltage,* (i.e., $V_{OH\,max} = v_{out}|_{i_L = 0}$). The values of I_{sc}, m, and $V_{OH\,max}$ for each case are defined in Table 4-2.

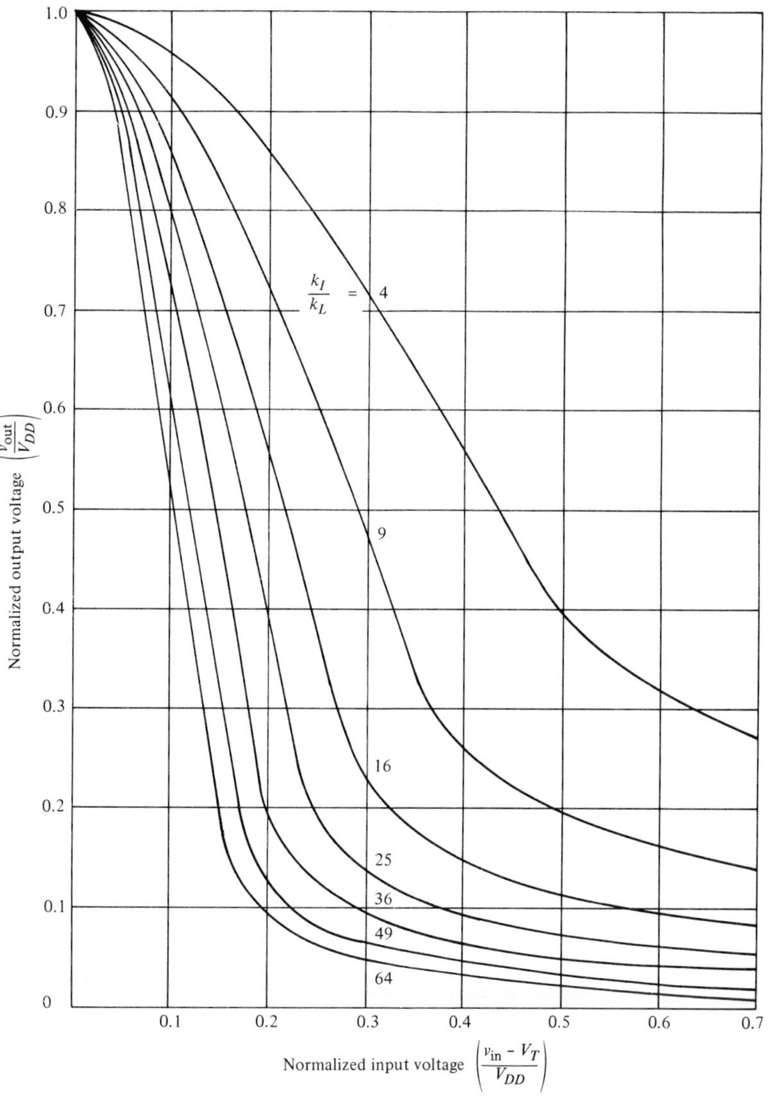

Figure 4–23 Normalized theoretical voltage transfer characteristics for a non-saturated MOS load device. Biasing parameter $m = 0.5$.

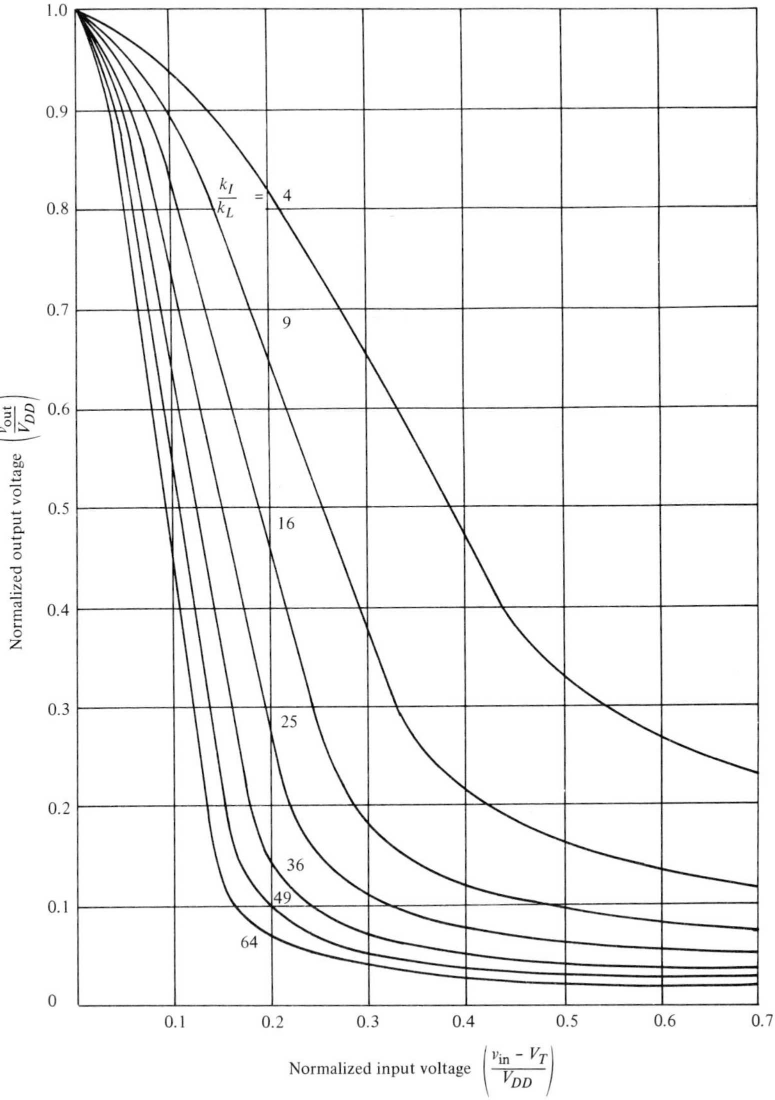

Figure 4–24 Normalized theoretical voltage transfer characteristics for a non-saturated MOS device. Biasing parameter $m = 0.6$.

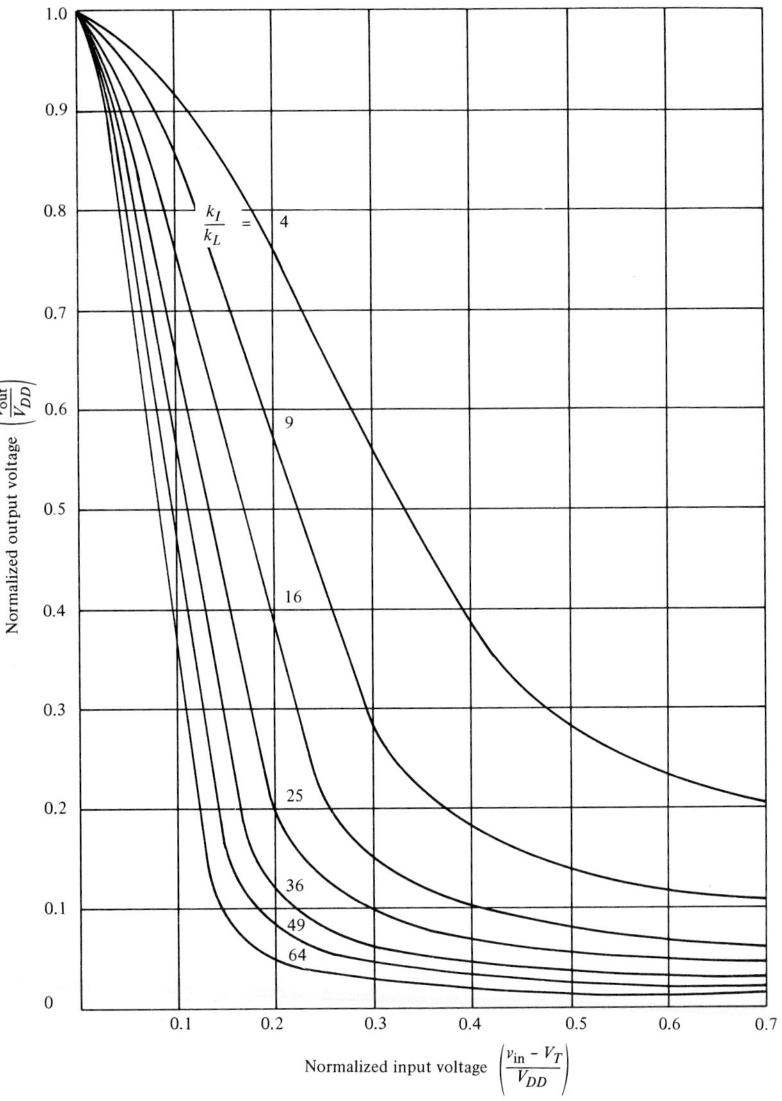

Figure 4–25 Normalized theoretical voltage transfer characteristics for a non-saturated MOS load device. Biasing parameter $m = 0.7$.

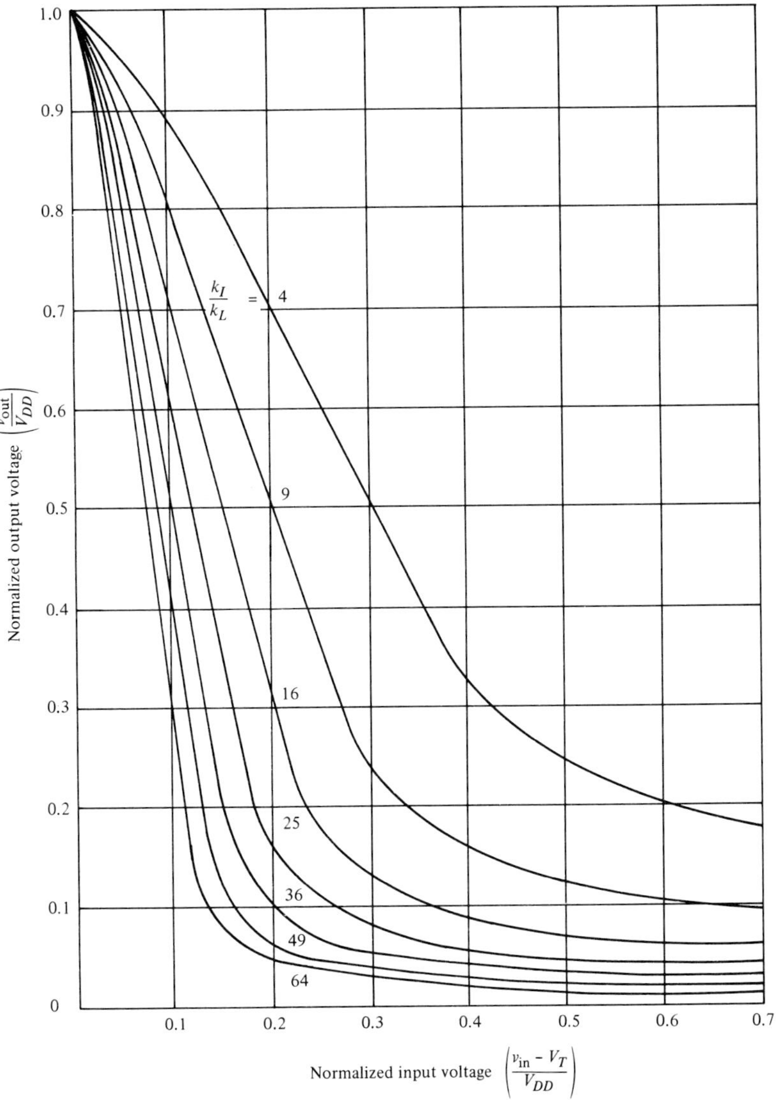

Figure 4–26 Normalized theoretical voltage transfer characteristics for a non-saturated MOS load device. Biasing parameter $m = 0.8$.

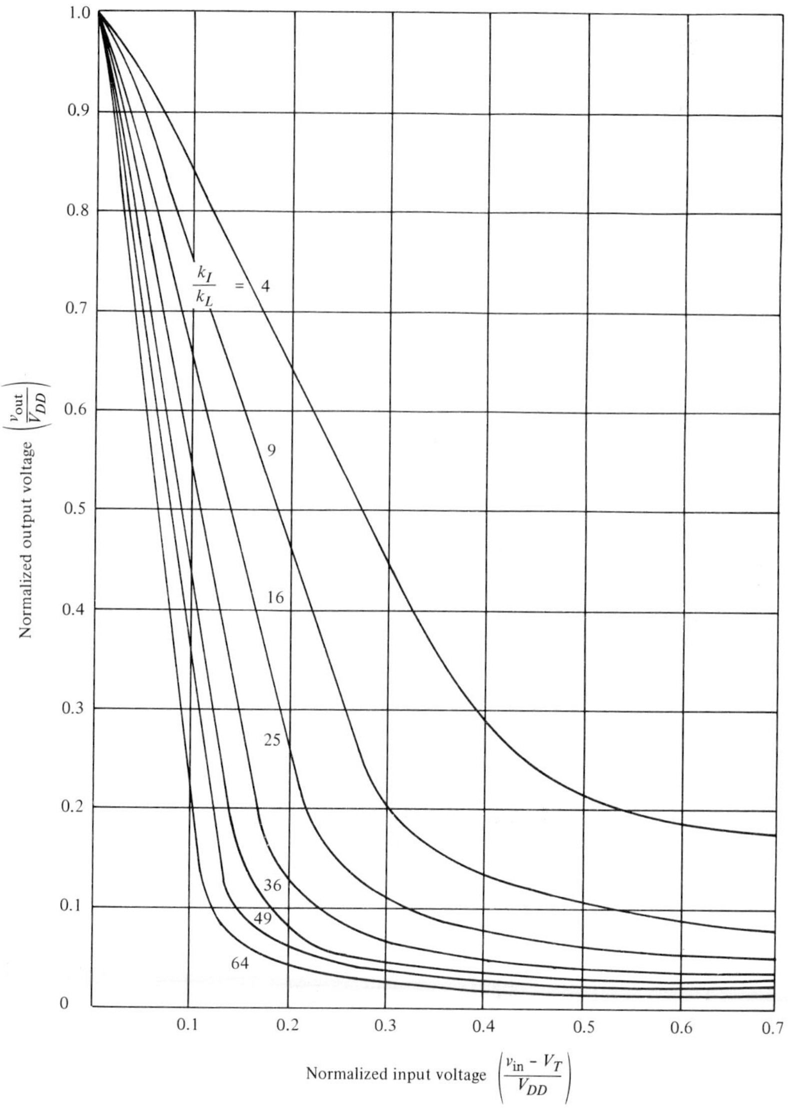

Figure 4–27 Normalized theoretical voltage transfer characteristics for a nonsaturated MOS load device. Biasing parameter $m = 0.9$.

Table 4–2 Circuit Parameter Relationships for Load Devices

Load device	*Circuit Parameter*		
	I_{sc}	m	$V_{OH\ \max}$
Linear load resistance	$\dfrac{V_{DD}}{R_L}$	0	V_{DD}
Saturated MOS load	$k_L\,(V_{GG} - V_T)^2$	1	$V_{GG} - V_T$
Nonsaturated MOS load	$\dfrac{k_L V_{DD}^2}{m}$	$\dfrac{V_{DD}}{2\,(V_{GG} - V_T) - V_{DD}}$	V_{DD}

4.2.8 Body Effect Considerations

As the output of an inverter circuit with an MOS load device and grounded substrate approaches V_{DD}, the source to substrate voltage of the load device increases. The apparent threshold voltage for the load device will therefore increase due to body effect (see Chapter 2).

Body effect has not been considered in the development of the voltage transfer characteristic curves, since a slightly more conservative approach to k_I/k_L ratio design is obtained when this effort is ignored.

If body effect is considered, it is significant to note that as the output voltage of an inverter approaches V_{DD} the effective load resistance increases causing the transfer characteristics to have a somewhat steeper slope. For actual designs, the distortion in the transfer curves is usually small. However, it is important to consider body effect in computing the maximum output voltage ($V_{OH\ \max}$) for a saturated load device. The correct expression for maximum output voltage becomes

$$V_{OH\ \max} = V_{GG} - V_T'$$

where

$$V_T' = V_T + av_{\text{out}}^x$$

Coefficients a and x are dependent on the process and the geometry of the load device.

4.2.9 Transient Response of the MOS Inverter Circuit

For an understanding of the transient behavior of the MOS inverter, the circuit shown in Figure 4-28 will be analyzed.

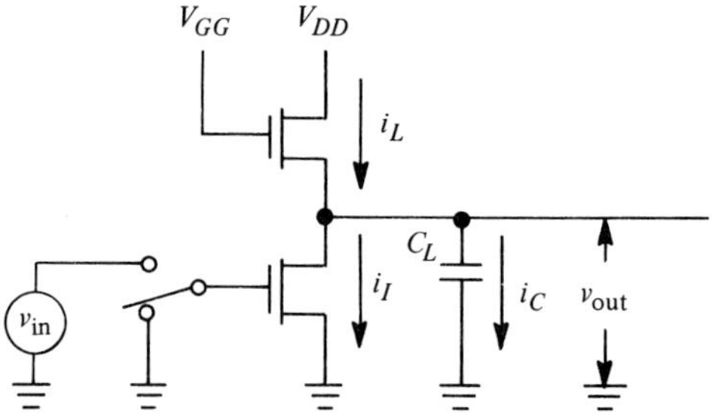

Figure 4–28 Transient model for an MOS inverter circuit.

To simplify the analysis, assume that

1. There are no storage or transit time effects in the device itself; the transient behavior of the circuit shown in Figure 4-28 is limited only by circuit capacitances and resistances.
2. All output capacitances are combined into a simple equivalent load capacitance (C_L). This simplification allows an analytical solution to be developed which is adequate for most MOS transient design problems. In the case of certain dynamic circuit configurations, a more complex equivalent circuit may be required, considering such factors as feedback and nonlinear capacitances.
3. The voltage input to the inverter is a step function for both the turn-on and the turn-off cases. The first case is assumed to have a value v_{in} where $|v_{\text{in}}| > |V_T|$, and the second case to have a value of zero. The simplest input waveform, a step, is used in the analysis to show more clearly the effects of the MOS device parameters.

Theoretical expressions will be obtained for turn-off and turn-on characteristics of the circuit.

4.2.10 Turn-Off or Rise Time

For the turn-off case, the inverter will be assumed to switch from the conducting state to a completely nonconducting state at the beginning of the transient. This is a direct consequence of two of the assumptions stated earlier, i.e., that the input is a step function and that the device itself has no inherent time delay.

As a result, in the circuit shown in Figure 4-28 for $t \geqq 0$, i_I is assumed equal to zero and the currents are given by:

$$i_L = I_{sc}\left(1 - \frac{v_{\text{out}}}{V_{OH\,\text{max}}}\right)\left(1 - m\,\frac{v_{\text{out}}}{V_{OH\,\text{max}}}\right) \tag{4-30}$$

$$i_C = C_L \frac{dv_{\text{out}}}{dt} \tag{4-31}$$

By equating the currents at the output node, the differential equation that describes the transient condition results.

$$I_{sc}\left(1 - \frac{v_{\text{out}}}{V_{OH\,\text{max}}}\right)\left(1 - m\,\frac{v_{\text{out}}}{V_{OH\,\text{max}}}\right) = C_L \frac{dv_{\text{out}}}{dt} \tag{4-32}$$

If we define the short-circuit resistance as

$$R_{sc} = \frac{V_{OH\,\text{max}}}{I_{sc}} \tag{4-33}$$

Equation 4-32 becomes

$$\left(1 - \frac{v_{\text{out}}}{V_{OH\,\text{max}}}\right)\left(1 - m\,\frac{v_{\text{out}}}{V_{OH\,\text{max}}}\right) = R_{sc}C_L \frac{d\left(\dfrac{v_{\text{out}}}{V_{OH\,\text{max}}}\right)}{dt} \tag{4-34}$$

By rearranging the equation for integration we obtain

$$\int dt = \int R_{sc}C_L \frac{d\left(\dfrac{v_{\text{out}}}{V_{OH\,\text{max}}}\right)}{\left(1 - \dfrac{v_{\text{out}}}{V_{OH\,\text{max}}}\right)\left(1 - m\,\dfrac{v_{\text{out}}}{V_{OH\,\text{max}}}\right)} \tag{4-35}$$

From a table of integrals we obtain the following general form:

$$\int \frac{dx}{(1-x)(1-mx)} = \frac{1}{-m+1}\,\ell\text{n}\left(\frac{1-mx}{1-x}\right) \tag{4-36}$$

The equation for turn-off or rise time (t_r) therefore becomes

$$t_r = R_{sc}C_L\left(\frac{1}{1-m}\right)\ell\text{n}\left[\frac{1 - m\left(\dfrac{v_{\text{out}}}{V_{OH\,\text{max}}}\right)}{1 - \left(\dfrac{v_{\text{out}}}{V_{OH\,\text{max}}}\right)}\right] \tag{4-37}$$

Hence

$$\varepsilon^{(1-m/R_{sc}C_L)t_r} = \frac{1 - m\left(\dfrac{v_{\text{out}}}{V_{OH\,\text{max}}}\right)}{1 - \left(\dfrac{v_{\text{out}}}{V_{OH\,\text{max}}}\right)} \tag{4-38}$$

Finally,

$$\frac{v_{\text{out}}}{V_{OH\,\text{max}}} = \frac{1 - \varepsilon^{-(1-m/R_{sc}C_L)t_r}}{1 - m\varepsilon^{-(1-m/R_{sc}C_L)t_r}} = \frac{1 - \varepsilon^{-(1-m)t/\tau_{sc}}}{1 - m\varepsilon^{-(1-m)t/\tau_{sc}}} \tag{4-39}$$

where

$$\tau_{sc} = R_{sc}C_L \tag{4-40}$$

The curve of $v_{out}/V_{OH\ max}$ as a function of t_r/τ_{sc} can now be calculated and plotted for various values of the biasing parameter (m). These results are shown in Figures 4-29a and 4-29b.

It is readily seen that Equation 4-40 is the familiar exponential function when $m = 0$, i.e., when the load device is linear.

For an understanding of the influence of body effect on turn-off time, consider the operating characteristics of an MOS load device with the substrate (body) at ground potential as it supplies current to charge a load capacitor (C_L). The schematic diagram is shown in Figure 4-30.

As the voltage across the capacitor (C_L) approaches V_{DD}, the value of V_{BS} increases; hence, the apparent threshold voltage increases by the body effect (see Chapter 2). The threshold voltage increase is given by $V_T' = V_T + aV_{BS}^x$, where typically $a = x = 1/2$. As the output voltage approaches V_{DD}, the biasing parameter (m) increases, causing the output

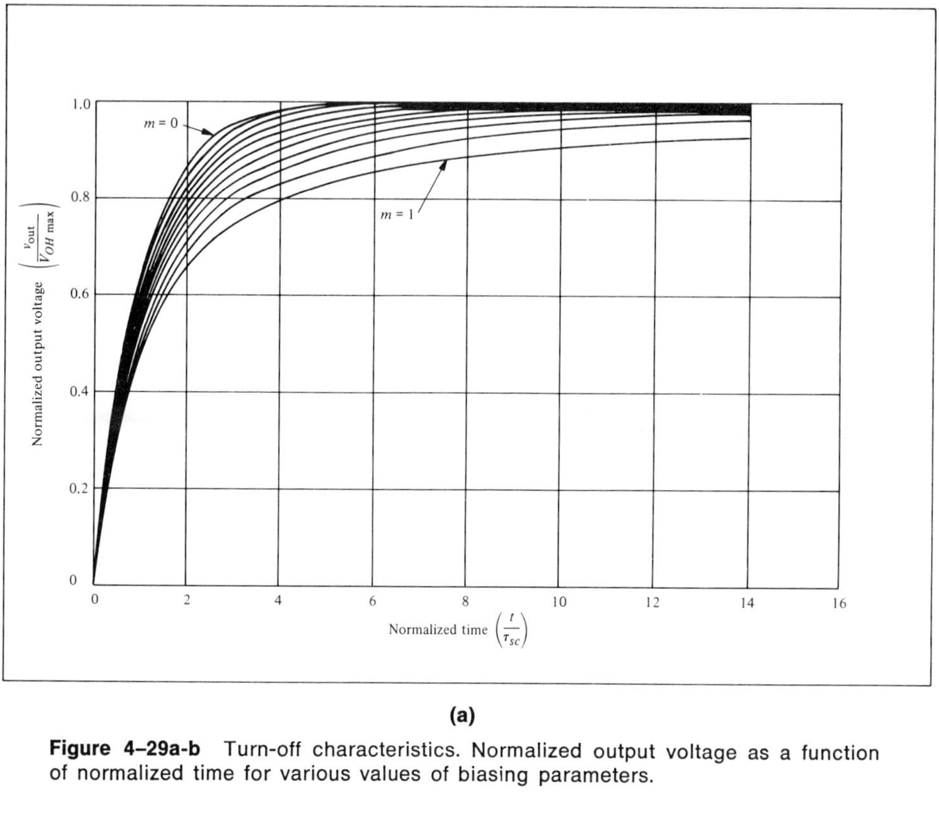

(a)

Figure 4–29a-b Turn-off characteristics. Normalized output voltage as a function of normalized time for various values of biasing parameters.

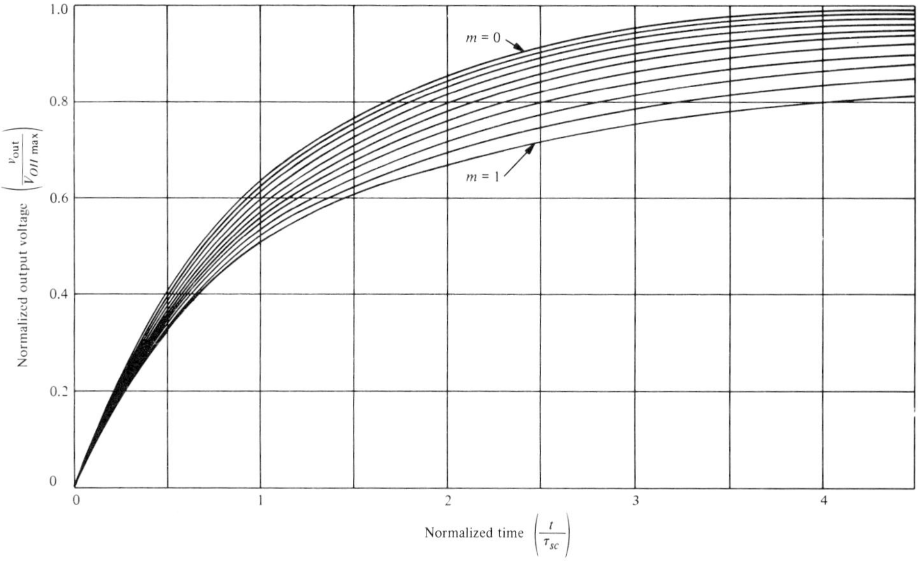

(b)

Figure 4–29 Continued.

to respond more slowly than indicated by the constant m turn-off characteristics shown in Figures 4-29a and 4-29b. The biasing parameter may be calculated at $v_{out} = V_{OH\ max}$ as well as at the beginning of turn-off when $v_{out} = 0$ V. The change in the value of m indicates the range of biasing that the load device is subjected to during the turn-off transient. A conservative approach to design, taking body effect into account, would indicate the use of the maximum value of m (the value determined at $v_{out} = V_{OH\ max}$); however, a more accurate design technique involves the use of some mid-range value of m. Empirical turn-off characteristics closely match the theoretical results for m calculated at $v_{out} = V_{OH\ max}/2$. The effect discussed here is much more pronounced on devices manufactured with a low threshold voltage process. As a typical example consider the following conditions and calculations of m for a load device with grounded substrate, where $V_{OH\ max} = 5$ V, $V_{GG} = 10$ V and $V_T = 1.5$ V. At $v_{out} = 0$ V, m is calculated as follows:

$$m = \frac{V_{DD}}{2(V_{GG} - V_T) - V_{DD}} = \frac{5}{2(10 - 1.5) - 5} = 0.416$$

At $v_{out} = V_{OH\ max}/2$ the biasing parameter (m') is given by

$$m' = \frac{V_{DD}}{2(V_{GG} - V_T') - V_{DD}}$$

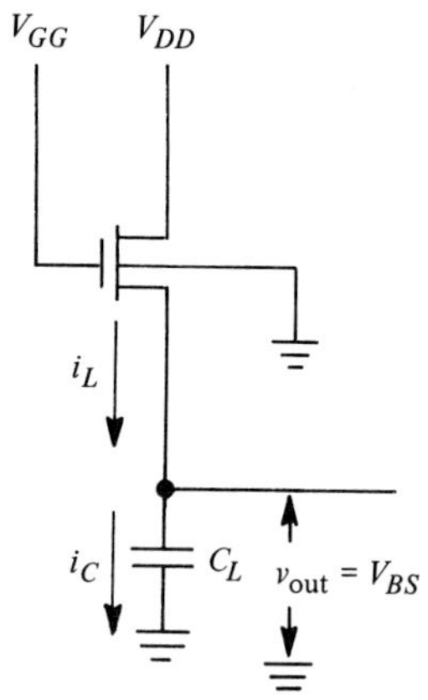

Figure 4–30 Schematic of an MOS load device with substrate grounded charging a load capacitance.

where

$$V_T' = V_T + \tfrac{1}{2}\sqrt{V_{BS}} = V_T + \tfrac{1}{2}\sqrt{V_{OH\,\max}}$$

Therefore, m' at $V_{OH\,\max}/2$ is

$$m' = \frac{5}{2(10 - 1.5 - \frac{1}{2}\sqrt{5/2}) - 5} = 0.479$$

From the calculations and Equation 4-37, it can be seen that the turn-off characteristics for the load device, including body effect, would be somewhat slower than the results of calculations in which the body effect is ignored.

4.2.11 Turn-On or Fall Time

For the turn-on case, the inverter will be assumed to switch from the nonconducting state to a completely conducting state at the beginning of the transient. This is a direct consequence of the assumptions made earlier that the input is a step function and that the device itself has no inherent time delay. A schematic diagram of the circuit for $t \geq 0$ is shown in Figure 4-31.

It will be assumed that the MOS load device is biased for operation in the nonsaturated mode. The solution of the problem for the linear

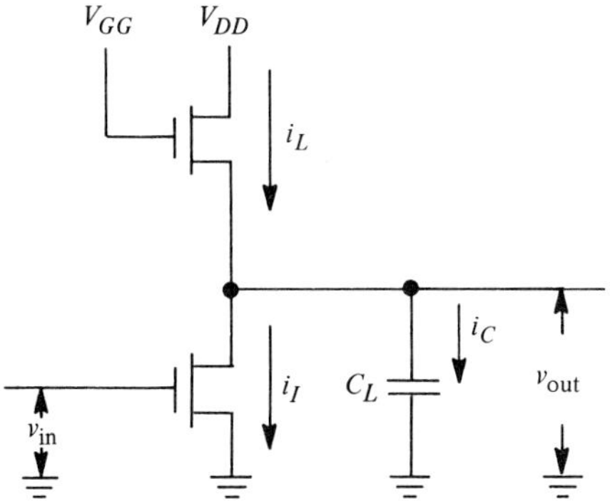

Figure 4–31 Schematic diagram for the turn-on transient analysis case.

load resistance and the saturated MOS load device can be obtained as limiting cases of the solution for the nonsaturated MOS load device.

The current through the load device is given by

$$i_L = \frac{k_L V_{OH\,\max}^2}{m}\left(1 - \frac{v_{out}}{V_{OH\,\max}}\right)\left(1 - m\frac{v_{out}}{V_{OH\,\max}}\right) \tag{4-30}$$

The current in the capacitor (i_C) is

$$i_C = C_L \frac{dv_{out}}{dt} \tag{4-31}$$

Since the input device is initially saturated and then when $|v_{in} - V_T| > |v_{out}|$ it passes into the nonsaturation region, two equations must be used to describe the current in the device.

For $\quad |v_{in} - V_T| \leqq |v_{out}|$

$$i_I = k_I[v_{in} - V_T]^2 \tag{4-4}$$

and for $\quad |v_{in} - V_T| > |v_{out}|$

$$i_I = k_I[2(v_{in} - V_T)\,v_{out} - v_{out}^2] \tag{4-5}$$

When the currents at the output node are added, two differential equations that describe the transient condition are obtained.

$$i_L = i_I + i_C$$

For $\quad |v_{in} - V_T| \leqq |v_{out}|$

$$\frac{k_L V_{OH\,\max}^2}{m}\left(1 - \frac{v_{out}}{V_{OH\,\max}}\right)\left(1 - m\frac{v_{out}}{V_{OH\,\max}}\right) = k_I(v_{in} - V_T)^2 + C_L\frac{dv_{out}}{dt} \tag{4-41}$$

and for $|v_{\text{in}} - V_T| \geqq |v_{\text{out}}|$

$$\frac{k_L V_{OH\,\max}{}^2}{m}\left(1 - \frac{v_{\text{out}}}{V_{OH\,\max}}\right)\left(1 - m\frac{v_{\text{out}}}{V_{OH\,\max}}\right) = k_I[2(v_{\text{in}} - V_T)\, v_{\text{out}} - v_{\text{out}}{}^2] + C_L\frac{dv_{\text{out}}}{dt} \tag{4-42}$$

By rearranging terms, normalizing $(v_{\text{in}} - V_T)$ to $V_{OH\,\max}$ and substituting the expression, we obtain

$$R_{\text{sc}} = \frac{m}{k_L V_{OH\,\max}} \tag{4-43}$$

Equations 4-43 and 4-44 reduce to

$$dt = \frac{R_{sc}C_L\, d\left(\dfrac{v_{\text{out}}}{V_{OH\,\max}}\right)}{a + b\dfrac{v_{\text{out}}}{V_{OH\,\max}} + c\left(\dfrac{v_{\text{out}}}{V_{OH\,\max}}\right)^2} \tag{4-44}$$

This equation can be integrated to obtain the equation for $(v_{\text{out}}/V_{OH\,\max})$ vs. fall time (t_f).

The integral in the equation can be obtained from a table of integrals as follows:

$$\int \frac{dx}{a + bx + cx^2} = \frac{-2}{\sqrt{-q}}\tanh^{-1}\frac{2cx + b}{\sqrt{-q}} \tag{4-45}$$

where $q = 4ac - b^2$ and $q < 0$.

By applying Equation 4-45, the solution to Equation 4-44 becomes

$$t_f = R_{sc}C_L\left\{\frac{-2}{\sqrt{-q}}\left[\tanh^{-1}\left(\frac{2c\dfrac{v_{\text{out}}}{V_{OH\,\max}} + b}{\sqrt{-q}}\right) - K\right]\right\} \tag{4-46}$$

where K is the constant of integration.

By solving for $v_{\text{out}}/V_{OH\,\max}$ from Equation 4-46,

$$\frac{v_{\text{out}}}{V_{OH\,\max}} = \frac{1}{2c}\left[-b + \sqrt{-q}\tanh\left(\frac{-t_f}{R_{sc}C_L}\frac{\sqrt{-q}}{2} + K\right)\right] \tag{4-47}$$

By comparing Equations 4-41 and 4-44, we obtain values for the constants in Equation 4-45 during the time interval when the inverter is saturated, that is, for $$\left|\frac{v_{\text{in}} - V_T}{V_{OH\,\max}}\right| \leqq \left|\frac{v_{\text{out}}}{V_{OH\,\max}}\right|$$

By using the subscript s for the saturated case, the constants are:

$$a_s = \left[1 - m\,\frac{k_I}{k_L}\left(\frac{v_{\text{in}} - V_T}{V_{OH\,\max}}\right)^2\right] \tag{4-48}$$

$$b_s = -\,(1 + m) \tag{4-49}$$

$$c_s = m \tag{4-50}$$

$$q_s = 4\left[1 - m\,\frac{k_I}{k_L}\left(\frac{v_{\text{in}} - V_T}{V_{OH\,\max}}\right)^2\right] m - (1 + m)^2 \tag{4-51}$$

$$\sqrt{-q_s} = \sqrt{(1 + m)^2 - 4m\left[1 - m\,\frac{k_I}{k_L}\left(\frac{v_{\text{in}} - V_T}{V_{OH\,\max}}\right)^2\right]}$$

$$= \sqrt{1 + 2m + m^2 - 4m + 4m^2\,\frac{k_I}{k_L}\left(\frac{v_{\text{in}} - V_T}{V_{OH\,\max}}\right)^2}$$

$$\sqrt{-q_s} = \sqrt{(1 - m)^2 + 4m^2\,\frac{k_I}{k_L}\left(\frac{v_{\text{in}} - V_T}{V_{OH\,\max}}\right)^2} \tag{4-52}$$

The constant of integration K_s in Equation 4-47 can be determined since $v_{\text{out}} = V_{OH\,\max}$ at $t = 0$ or, by substitution into Equation 4-47, the following is obtained

$$1 = \frac{1}{2c_s}\,[-b_s + -q_s \tanh\,(0 + K_s)]$$

Therefore,

$$K_s = \tanh^{-1}\left(\frac{2c_s + b_s}{\sqrt{-q_s}}\right) \tag{4-53}$$

These parameter values can be calculated and substituted into Equation 4-47 to obtain the expression for $v_{\text{out}}/V_{OH\,\max}$ vs. time. Further study of Equation 4-47 shows that the minus sign is the correct sign for the radical $\sqrt{-q}$. By rewriting Equation 4-47 for the saturation region,

$$\frac{v_{\text{out}}}{V_{OH\,\max}} = \frac{1}{2c_s}\left[-b_s + \sqrt{-q_s}\tanh\left(-\frac{t_f}{R_{sc}C_L}\,\frac{\sqrt{-q_s}}{2} + K_s\right)\right] \tag{4-54}$$

The output voltage will follow the resulting equation (Equation 4-54) until it has reached the value $v_{\text{out}}/V_{OH\,\max} = (v_{\text{in}} - V_T)/V_{OH\,\max}$. Then the inverter passes from saturated into nonsaturated operation, and the current equation must be changed.

The time at which $v_{\text{out}}/V_{OH\,\max} = (v_{\text{in}} - V_T)/V_{OH\,\max}$ is defined as t_f'.

For $t \geqq t_f'$ during the time interval when the inverter is nonsaturated, Equation 4-47 is applied to Equation 4-42 for

$$\left|\frac{v_{\text{in}} - V_T}{V_{OH\,\text{max}}}\right| > \left|\frac{v_{\text{out}}}{V_{OH\,\text{max}}}\right|$$

The constants from Equation 4-47 are obtained by comparing Equations 4-42 and 4-44. Using the subscript n for the nonsaturated case, the constants are

$$a_n = 1 \tag{4-55}$$

$$b_n = -\left[1 + m + 2m\frac{k_I}{k_L}\left(\frac{v_{\text{in}} - V_T}{V_{OH\,\text{max}}}\right)\right] \tag{4-56}$$

$$c_n = m\left(1 + \frac{k_I}{k_L}\right) \tag{4-57}$$

$$q_n = (4)(1)m\left(1 + \frac{k_I}{k_L}\right) - \left[1 + m + 2m\,\frac{k_I}{k_L}\left(\frac{v_{\text{in}} - V_T}{V_{OH\,\text{max}}}\right)\right]^2 \tag{4-58}$$

$$\sqrt{-q_n} = \sqrt{(1+m)^2 + 4m\,(1+m)\,\frac{k_I}{k_L}\left(\frac{v_{\text{in}} - v_T}{V_{OH\,\text{max}}}\right) + \left[2m\,\frac{k_I}{k_L}\left(\frac{v_{\text{in}} - V_T}{V_{OH\,\text{max}}}\right)\right]^2 - 4m\left(1 + \frac{k_I}{k_L}\right)} \tag{4-59}$$

When Equation 4-49 is rewritten for the nonsaturated case,

$$\frac{v_{\text{out}}}{V_{OH\,\text{max}}} = \frac{1}{2c_n}\left\{-b_n + \sqrt{-q_n}\,\tanh\left[-\frac{t_f}{R_{sc}\,C_L}\,\frac{\sqrt{-q_n}}{2} + K_n\right]\right\} \tag{4-60}$$

It is convenient to rewrite Equation 4-60 in the following form:

$$\frac{v_{\text{out}}}{V_{OH\,\text{max}}} = \frac{1}{2c_n}\left\{-b_n + \sqrt{-q_n}\,\tanh\left[-\frac{(t_f - t_f')}{R_{sc}C_L}\,\frac{\sqrt{-q_n}}{2} + K_n'\right]\right\} \tag{4-61}$$

where t_f' is the time at which the inverter enters the nonsaturation region of operation.

No attempt will be made to substitute the general expressions for b_n, c_n, q_n. Normally these parameters would be calculated separately and then substituted into Equation 4-61.

The constant of integration K_n' in Equation 4-61 can be determined, since

$$\frac{v_{\text{out}}}{V_{OH\,\max}} = \frac{v_{\text{in}} - V_T}{V_{OH\,\max}} \qquad \text{at } (t - t_f') = 0$$

$$\frac{v_{\text{in}} - V_T}{V_{OH\,\max}} = \frac{1}{2c_n} \{[-b_n + \sqrt{-q_n} \tanh (K'_n)]\}$$

therefore,

$$K'_n = \tanh^{-1} \left\{\frac{1}{\sqrt{-q_n}} \left[2c_n \left(\frac{v_{\text{in}} - V_T}{V_{OH\,\max}}\right) + b_n\right]\right\} \tag{4-62}$$

To clarify these equations let us consider the following example. Determine the turn-on characteristics for an MOS inverter where

$V_{GG} = 27$ V; $V_{DD} = V_{OH\,\max} = 15$ V; $V_{T\,\max} = 5$ V; and $v_{\text{in}} = 12$ V

$$m = \frac{15}{2(27 - 5) - 15} = 0.517$$

Let $m = 0.50$ and assume that $v_{\text{out}}/V_{OH\,\max} = 0.1$ from dc considerations. The normalized input voltage is

$$\frac{v_{\text{in}} - V_T}{V_{OH\,\max}} = \frac{12 - 5}{15} = 0.466$$

Then, from the curves of normalized output voltage versus normalized input voltage, Figure 4-23, obtain the value of $k_I/k_L = 20$.

At $t = 0$, $v_{\text{out}} = V_{DD}$ and the inverter is saturated; therefore, when

$$\left|\frac{v_{\text{in}} - V_T}{V_{OH\,\max}}\right| \leq \left|\frac{v_{\text{out}}}{V_{OH\,\max}}\right|$$

from Equations 4-48, 4-49, and 4-50,

$$a_s = [1 = (0.50)\,(20)\,(0.466)^2] = -1.18$$

$$b_s = -(1 + 0.50) = -1.50$$

$$c_s = 0.50$$

and from Equation 4-51,

$$\begin{aligned} q_s &= 4a_s c_s - (b_s)^2 \\ &= 4\,(1.18)\,(0.50) - (-1.50)^2 \\ &= -2.36 - 2.25 = -4.61 \end{aligned}$$

$$\sqrt{-q_s} = \sqrt{4.61} = 2.15$$

From Equation 4-53,

$$K_s = \tanh^{-1} \frac{2(0.50) + (-1.50)}{2.15}$$
$$= \tanh^{-1} (-0.233)$$
$$= -\tanh^{-1} (0.233)$$
$$K_s = -0.238$$

From Equation 4-54 the normalized output voltage is

$$\frac{v_{\text{out}}}{V_{OH\,\text{max}}} = \frac{1}{2(0.50)} \left[-(-1.50) + 2.15 \tanh \left(-\frac{t_f}{\tau_{sc}} \frac{2.15}{2} - 0.238 \right) \right]$$

$$\frac{v_{\text{out}}}{V_{OH\,\text{max}}} = 1.50 - 2.15 \tanh \left(1.075 \frac{t_f}{\tau_{sc}} + 0.238 \right)$$

Values for $v_{\text{out}}/V_{OH\,\text{max}}$ as a function of t_f/τ_{sc} are shown below.

$\frac{t_f}{\tau_{sc}}$	$\frac{v_{\text{out}}}{V_{OH\,\text{max}}}$
0.00	1.000
0.10	0.786
0.20	0.589
0.25	0.498
0.27	0.466

As shown in the derivation, the inverter passes from the saturation to the nonsaturation region when

$$\left| \frac{v_{\text{out}}}{V_{OH\,\text{max}}} \right| = \left| \frac{v_{\text{in}} - V_T}{V_{OH\,\text{max}}} \right| \qquad \text{at } t_f = t_f'$$

In this case,

$$\frac{v_{\text{out}}}{V_{OH\,\text{max}}} = \frac{v_{\text{in}} - V_T}{V_{OH\,\text{max}}} = 0.466$$

This transition point occurs approximately at

$$\frac{t_f}{\tau_{sc}} = 0.27$$

This represents zero time for the nonsaturated case. Therefore we must use Equation 4-61 for the remaining calculations in the nonsaturated condition when

$$\left|\frac{v_{\text{in}} - V_T}{V_{OH\,\max}}\right| > \left|\frac{v_{\text{out}}}{V_{OH\,\max}}\right|$$

and $t_f > t_f'$. By using Equations 4-55, 4-56, 4-57, 4-58, and 4-59,

$$a_n = 1.00$$

$$b_n = -[1 + 0.5 + 20\ (0.466)] = -\ 10.82$$

$$c_n = 0.5\ (1 + 20) = 10.5$$

$$q_n = 4(1)\ c_n - (-b_n^2)$$

$$q_n = (4)\ (1.0)\ (10.5) - (10.82)^2 = -75.07$$

$$\sqrt{-q_n} = \sqrt{75.07} = 8.67$$

Then, from Equation (4-62),

$$K_n' = \tanh^{-1}\left\{\frac{1}{8.67}\ [2(10.5)\ (0.466) + (-10.82)]\right\}$$

$$= \tanh^{-1} -0.119$$

$$= -0.119$$

From Equation 4-61,

$$\frac{v_{\text{out}}}{V_{OH\,\max}} = \frac{1}{2(10.5)}\left\{-(-10.82) + 8.67 \tanh\left[-\frac{(t_f - t_f')(8.67)}{2\tau_{sc}} + (-0.119)\right]\right\}$$

$$\frac{v_{\text{out}}}{V_{OH\,\max}} = 0.516 - 0.413 \tanh\left[4.33\ \frac{(t_f - t_f')}{\tau_{sc}} + 0.119\right]$$

Values for $v_{\text{out}}/V_{OH\,\max}$ are calculated below and are plotted with the values calculated for the saturated region in Figure 4-32.

$\frac{t_f - t_f'}{\tau_{sc}}$	$\frac{v_{\text{out}}}{V_{OH\,\max}}$
0.00	0.466
0.10	0.301
0.20	0.204
0.30	0.149
0.40	0.123
∞*	0.103

*tanh ∞ = 1.

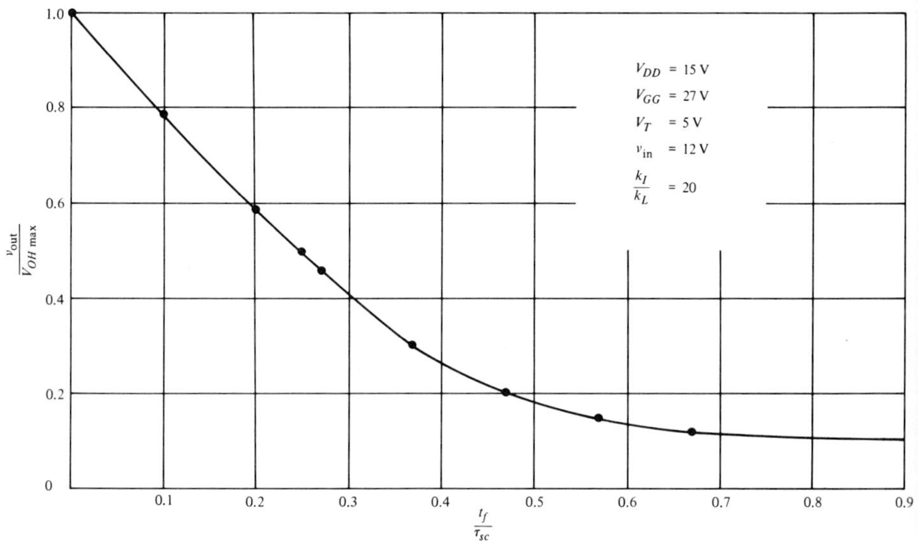

Figure 4–32 Normalized turn-on or fall time for an MOS inverter circuit.

In practice, frequently the load device current (i_L) is small compared to the current in the capacitor (i_C) over the transient voltage range of interest and can therefore be ignored. This simplifies the turn-on equivalent circuit and analysis. Assuming that the input is a step function and the input device switches from the nonconducting to conducting state at the beginning of the transient, a transient equivalent circuit can be constructed as shown in Figure 4-33.

At time $t = 0^+$ it is assumed that the capacitor (C_L) is charged to a voltage ($v_{out} = V_{OH\ max}$) the input (v_{in}) is equal to a dc level greater than threshold, and the equation for the current in the input device is described by the saturated current expression.

For $\quad |v_{in} - V_T| \leq |v_{out}|$

$$i_I = k_I(v_{in} - V_T)^2$$

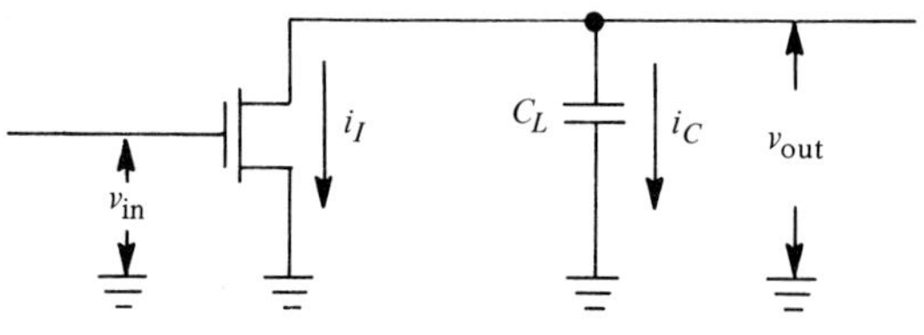

Figure 4–33 Simplified Turn-on equivalent circuit.

The current in the capacitor may be equated to the current of the input device.

$$i_C = -i_I \tag{4-63}$$

$$i_C = -C_L \frac{dv_{\text{out}}}{dt} \tag{4-31}$$

Therefore,

$$-C_L \frac{dv_{\text{out}}}{dt} = k_I(v_{\text{in}} - V_T)^2 \tag{4-64}$$

By rearranging terms and integrating, we obtain

$$\int dt = -\int \frac{C_L}{k_I(v_{\text{in}} - V_T)^2}\, dv_{\text{out}} \tag{4-65}$$

$$t_{\text{sat}} = -\frac{C_L v_{\text{out}}}{k_I(v_{\text{in}} - V_T)^2} + \text{constant } C_1 \tag{4-66}$$

The constant C_1 can be easily obtained by noting that for $t = 0^+$, $v_{\text{out}} = V_{OH\,\max}$. Hence,

$$\text{constant } C_1 = \frac{C_L V_{OH\,\max}}{k_I(v_{\text{in}} - V_T)^2} \tag{4-67}$$

and for $|v_{\text{in}} - V_T| \leqq |v_{\text{out}}|$

$$t_{\text{sat}} = \frac{C_L(v_{OH\,\max} - v_{\text{out}})}{k_I(v_{\text{in}} - V_T)^2} \tag{4-68}$$

For $|v_{\text{out}}| < |v_{\text{in}} - V_T|$, the current in the input device is described by the nonsaturated current expression which can be equated to the capacitor current as in the previous case. For some actual bias conditions and parameter values, the turn-on transient may be described adequately by Equation 4-68 alone.

For the discharge of the load capacitor in the nonsaturated region, the current in the capacitor is again set equal to the current of the input device

$$i_C = -i_I \tag{4-63}$$

and again

$$i_C = -C_L \frac{dv_{\text{out}}}{dt} \tag{4-31}$$

However, when $|v_{\text{in}} - V_T| > |V_{OH\,\max}|$ the current in the input device is expressed by

$$i_I = k_I[2(v_{\text{in}} - V_T)v_{\text{out}} - v_{\text{out}}^2] \tag{4-69}$$

By substituting into Equation 4-63,

$$-C_L \frac{dv_{\text{out}}}{dt} = k_I[2(v_{\text{in}} - V_T)v_{\text{out}} - v_{\text{out}}^2] \tag{4-70}$$

or

$$-\frac{dv_{\text{out}}}{2(v_{\text{in}} - V_T)v_{\text{out}} - v_{\text{out}}^2} = \frac{k_I}{C_L}\,dt \tag{4-71}$$

This may be rewritten and integrated to solve for v_{out} as a function of t.

$$-\int \frac{dv_{\text{out}}}{v_{\text{out}}[2(v_{\text{in}} - V_T) - v_{\text{out}}]} = \frac{k_I}{C_L}\int dt \tag{4-72}$$

From a table of integrals,

$$\int \frac{dx}{x(a + bx)} = -\frac{1}{a}\,\ell\text{n}\left(\frac{a + bx}{x}\right) \tag{4-73}$$

Examination of Equation 4-72 reveals that we can substitute v_{out} for x, $a = 2(v_{\text{in}} - V_T)$, and $b = -1$. Therefore,

$$-1\left\{\frac{-1}{2(v_{\text{in}} - V_T)}\,\ell\text{n}\left[\frac{2(v_{\text{in}} - V_T) - v_{\text{out}}}{v_{\text{out}}}\right]\right\} = \frac{k_I}{C_L}\,t + \text{constant } C_2 \tag{4-74}$$

where constant C_2 is a constant of integration which may be found by noting at $t = 0$, $|v_{\text{in}} - V_T| = |v_{\text{out}}|$. This corresponds to the switch-over point from the saturated to the nonsaturated equation for i_I.

By substituting $v_{\text{in}} - V_T = v_{\text{out}}$ into Equation 4-74,

$$\frac{1}{2(v_{in} - V_T)}\,\ell\text{n}\left[\frac{2(v_{\text{out}}) - v_{\text{out}}}{v_{\text{out}}}\right] = 0 + \text{constant } C_2 \tag{4-75}$$

Since $\ell\text{n}\ [1] = 0$, constant $C_2 = 0$, when the terms are rearranged, for $|v_{\text{in}} - V_T| > |v_{\text{out}}|$

$$t_{\text{nonsat}} = \frac{C_L}{k_I}\,\frac{1}{2(v_{\text{in}} - V_T)}\,\ell\text{n}\left[\frac{2(v_{\text{in}} - V_T) - v_{\text{out}}}{v_{\text{out}}}\right] \tag{4-76}$$

To solve for the total time to discharge the load capacitor (C_L) from $V_{OH\,\max}$ to a final value of V_F where device operation passes through the saturation and nonsaturation regions, the value obtained for t_{sat} (Equation 4-68) to reach the limit $v_{\text{out}} = v_{\text{in}} - V_T$ must be added to the value of t_{nonsat} to reach $v_{\text{out}} = V_F$. The expression for the total discharge time then becomes

$$t_{\text{total}} = \frac{C_L}{k_I(v_{\text{in}} - V_T)} \left\{ \frac{V_{OH\,\max} - (v_{\text{in}} - V_T)}{v_{\text{in}} - V_T} + \frac{1}{2}\,\ell\text{n}\left[\frac{2(v_{\text{in}} - V_T) - V_F}{V_F}\right]\right\} \tag{4-77}$$

For calculations using the simplified turn-on analysis technique, consider the following specifications and assumptions

$$\begin{aligned} V_{GG} &= 27\text{ V} & V_F &= 4.0\text{ V} \\ V_{DD} &= 15\text{ V} & \frac{k_I}{k_L} &= 20 \\ V_{T\,\max} &= 5\text{ V} & & \\ v_{\text{in}} &= 12\text{ V} & m &= 0.5 \end{aligned}$$

In this case, it is assumed that $i_L = 0$ and Equation 4-77 is used to determine t_{total} from Equation 4-40, $\tau_{sc} = R_{sc}C_L$

and from Equation 4-33, $$R_{sc} = \frac{V_{OH\,\max}}{I_{sc}} = \frac{V_{DD}}{I_{sc}} \quad \text{(Nonsaturation)}$$

or

$$R_{sc} = \frac{m}{k_L V_{DD}} \tag{4-78}$$

Since, for the example inverter

$$\frac{k_I}{k_L} = 20$$

substitution into Equation 4-78 gives

$$R_{sc} = \frac{20m}{k_I V_{DD}}$$

or by rearranging terms

$$k_I = \frac{20m}{R_{sc}V_{DD}}$$

By substituting given values into Equation 4-77,

$$t_{\text{total}} = \frac{\tau_{sc}V_{DD}}{20(m)(v_{\text{in}} - V_T)} \left\{ \frac{V_{DD} - (v_{\text{in}} - V_T)}{v_{\text{in}} - V_T} + \frac{1}{2}\,\ell\text{n}\left[\frac{2(v_{\text{in}} - V_T) - V_F}{V_F}\right]\right\}$$

$$t_{\text{total}} = \frac{\tau_{sc}(15)}{20(0.5)(12 - 5)} \left\{ \frac{15 - (12 - 5)}{12 - 5} + \frac{1}{2}\,\ell\text{n}\left[\frac{2(12 - 5) - 4}{4}\right]\right\}$$

$$t_{\text{total}} = \frac{15\tau_{sc}}{70}\left\{\frac{8}{7} + \frac{1}{2}\,\ell\text{n}\ 2.5\right\}$$

$$t_{\text{total}} = 0.22\tau_{sc}\left\{1.14 + \frac{1}{2}(0.916)\right\} = 0.35\tau_{sc}$$

The validity of this answer can be checked by comparing it with the results shown in Figure 4-32, a plot of the turn-on time for the same example, and considering i_L in the equation. By using the value

$$\frac{v_{\text{out}}}{V_{OH\,\text{max}}} = \frac{v_{\text{out}}}{V_{DD}} = \frac{4.0}{15} = 0.267$$

from the plot in Figure 4-30,

$$t/\tau_{sc} \simeq 0.39 \qquad \text{or} \qquad t \simeq 0.39\tau_{sc}$$

This result agrees reasonably well with that obtained by the simplified method using Equation 4-77. For most practical problems, Equation 4-77 will provide turn-on transient time results with adequate accuracy and eliminate the lengthy calculations required in a rigorous solution.

4.2.12 Rise and Fall Times

A frequent application of the MOS inverter circuit is in a series connection of several inverters or logic stages. Usually the propagation delay time must be determined for an input level transition through the series of stages. A conservative approach to the problem is to calculate the rise and fall times of each stage and simply add alternate values of rise and fall times for each stage in the series connection. The approach is conservative in that, although inputs to every stage are not step functions as assumed in the rise and fall time calculations, the output of an inverter will begin its transition before the input achieves the assumed step function value. In addition, an input to an inverter turning on will continue to rise beyond the assumed input voltage, causing a more rapid turn-on transition than calculated.

A practical approach to a more complex transient analysis would involve the use of a computer program that can take into account the basic device characteristics together with a complex input waveform. A number of programs have been developed by both MOS suppliers and users to perform this type of analysis. By using such a program, detailed transient performance may be predicted for an entire MOS LSI array, although for economic reasons it is more common to examine only critical signal paths.

4.3 INVERTER DESIGN EXAMPLES

Two examples of inverter design will be considered in this section: an inverter internal to the LSI chip that drives an internal capacitance load, and an output inverter that drives an external load. Tables 4-3 and 4-4

Table 4–3 Inverter Design Specifications

Parameter	*Minimum*	*Typical*	*Maximum*	*Units*	*Conditions/Comments*
V_{DD}	−13.5	−15.0	−16.5	V	
V_{GG}	−24.3	−27.0	−29.7	V	
V_{OL}	—	—	− 2.5	V	Low level (magnitude) output voltage
V_{OH}	−10.0	—	V_{DD}	V	High level (magnitude) output voltage
V_{IL}	—	—	− 3.5	V	Low level (magnitude) input voltage
V_{IH}	− 9.0	—	—	V	High level (magnitude) input voltage
T_A	0	+25	+70	°C	Operating ambient air temperature
C_L	—	—	20	pF	Load external to package
t_r (0 to −10.0 V)	—	—	1.0	μs	Output rise time with $C_L = 20$ pF

Table 4–4 *P*-Channel MOS Process Parameters (Design Range)

Parameter	*Minimum*	*Typical*	*Maximum*	*Units*	*Conditions/Comments*
V_T	− 3.0	− 4.0	− 5.0	V	$I_{DS} = 0$
k'	1.3	1.7	2.1	$\mu A/V^2$	25°C
ρ_P	—	—	100	Ω/□	*P*-region resistivity

list the specifications of the chip and process parameters that are essential to design. Included in the tables are actual values and ranges that are used in the example. In an inverter circuit the speed requirements and worst-case parameter conditions determine the size of the load transistor. The size of the input transistor may then be determined from the dc level specifications and worst-case parameter values that affect these levels.

4.3.1 Output Inverter Design

First consider the design of an output inverter using the high threshold *P*-channel MOS process. The basic circuit of an output inverter is shown in Figure 4-34. The output load consists of two other MOS chips and the interconnecting wiring which, for convenience, may be considered as a total capacitance load (C_L). Actual dimensions must be determined for the load and the input device.

4.3.2 Output Inverter Load Device

From the chip specifications in Table 4-3 and the process parameters in Table 4-4, parameters that reflect worst-case design conditions for speed are selected.

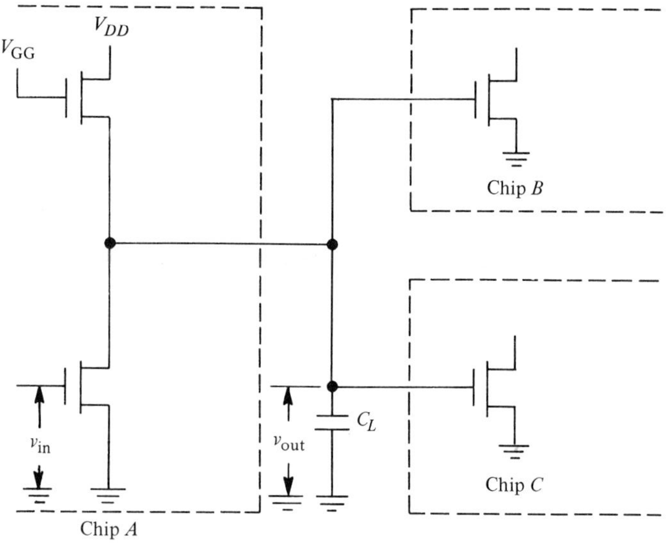

Figure 4–34 Basic output inverter circuit.

Low V_{DD}	−13.5 V
Low V_{GG}	−24.3 V
High V_T	− 5.0 V
Low k' @ 25°C	1.3 μA/V^2
High T_A	+70°C

In addition, temperature rise in the package must be considered. Since no specific package has been designated and the total power dissipation of the chip is not known, a temperature rise of 20°C will be assumed, in which case the junction temperature $T_j = 70°C + 20°C = 90°C$.

Following is a brief explanation of why each of the selected conditions is a worst-case consideration for the design of the load device.

- Low V_{DD}— With a lower V_{DD}, C_L charges to a lower final voltage; therefore, it takes longer to reach the required V_{OH} level.
- Low V_{GG}—A low V_{GG} means a higher value of biasing parameter (m); therefore, switching is slower (see turn-off characteristics curves, Figures 4-29a and 4-29b).
- High V_T—Since the load device is operating in the nonsaturation region, the expression for current is given by (Equation 2-73)

$$I_{DS} = -k[2(V_{GS} - V_T)V_{DS} - V_{DS}^2] \qquad (2\text{-}73)$$

If $v_{out} = 0$ and $I_{DS} = I_{sc}$, then, the equation for current in the load device becomes

$$i_L = I_{sc} = k_L[2(V_{GG} - V_T)V_{DD} - V_{DD}^2] \qquad (4\text{-}79)$$

Therefore, high V_T means less load current and slower speed.

- Low k'—Referring to the expression for current, it is evident that low k' means less load current and slower speed.
- High Temperature—Since k' decreases with increasing temperature, i_L also decreases with increasing temperature and the speed is again slower.

To determine the characteristics of the load device, first the biasing parameter (m') is calculated for an output voltage of ($V_{OH\,max}/2$). It is assumed that the body or substrate is grounded.

$$m = \frac{V_{DD}}{2(V_{GG} - V_T') - V_{DD}}$$

where V_T' is given by the expression

$$V_T' = V_T + \tfrac{1}{2}\sqrt{V_{BS}}$$

Substitution of values into the expression for m' gives

$$m' = \frac{13.5}{2\left[24.3 - \left(5.0 + \frac{1}{2}\sqrt{\frac{13.5}{2}}\right)\right] - 13.5} = 0.6$$

Next, the minimum value of output voltage (V_{OH}) is normalized to V_{DD}

$$\frac{V_{OH}}{V_{DD}} = \frac{-10.0}{-13.5} = 0.74$$

From the turn-off characteristic curves Figure 4-29b for a value of $m = 0.6$ and $v_{out}/V_{OH\ max} = 0.74$, find $t_r/\tau_{sc} = 1.9$. The value for C_L is obtained from the specifications. Since C_L is specified as capacitance external to the package, some value of capacitance must be added for the internal connection from the output inverter to the package as well as for the package itself. A reasonable value to assume for this is 5 pF; thus, the total capacitance that the output inverter must drive is 25 pF. The value of t_r in the specifications is 1.0 μs. From the expression for R_{sc}, Equation 4-33, and current in the load, Equation 4-79,

$$R_{sc} = \frac{V_{OH\ max}}{I_{sc}} = \frac{V_{DD}}{k_L[2(V_{GG} - V_T)V_{DD} - V_{DD}^2]} \tag{4-33}$$

By rearranging terms and simplifying,

$$k_L = \frac{1}{R_{sc}[2(V_{GG} - V_T) - V_{DD}]} \tag{4-80}$$

since

$$\tau_{sc} = R_{sc}C_L \tag{4-40}$$

By substituting the value of τ_{sc} found from the curves (Figure 4-29b) and rearranging Equation 4-40,

$$R_{sc} = \frac{t_r}{1.9\ C_L} = \frac{1\ \mu\text{s}}{1.9 \times 25\ \text{pF}} = 21 \times 10^3\ \Omega$$

By substituting R_{sc} into the expression for k_L in Equation 4-80,

$$k_L = \frac{1}{21 \times 10^3[2(24.3 - 5.0) - 13.5]} = 1.9\ \mu\text{A/V}^2$$

This is the minimum value of k_L at high temperature. For the actual dimensions of the load device, k' must be determined at high temperature.

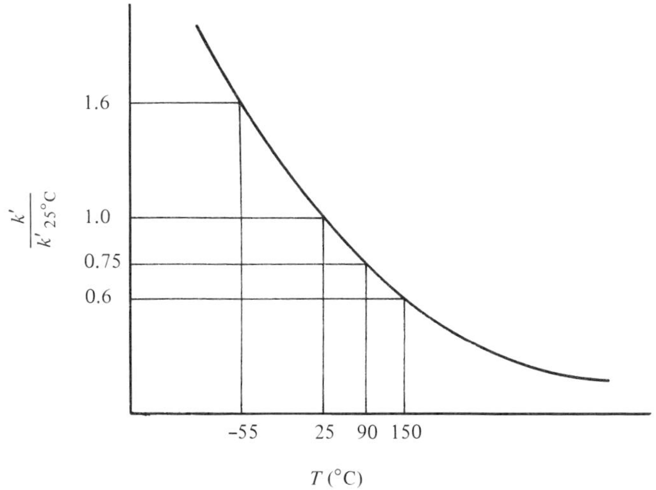

Figure 4–35 Variation of k' with temperature.

A curve of the variation of k' with temperature is shown in Figure 4-35. By using the minimum value of k' from the process specifications and multiplying by the temperature factor for 90° C,

$$k'_{90^\circ\,\mathrm{C}} = 0.75\,(k'_{25^\circ\,\mathrm{C}})$$

$$\min k'_{90^\circ\,\mathrm{C}} = 0.75\,(1.3\ \mu\mathrm{A/V^2}) = 0.975\ \mu\mathrm{A/V^2}$$

It is now possible to determine the size of the load device, using

$$k_L = k'\left(\frac{W}{\ell}\right)_L$$

or, rearranging terms,

$$\left(\frac{W}{\ell}\right)_L = \frac{k_L}{k'} = \frac{1.9}{0.975} = 1.95$$

To construct the smallest possible device with this $(W/\ell)_L$, we will assume an absolute minimum value of ℓ_L limited by the process at 0.2 mil and

$$W_L = \ell_L\,(1.95) = 0.2\,(1.95) = 0.39 \text{ mil}$$

4.3.3 Output Inverter Input Device

The input device must be designed using the worst-case conditions from the specifications for the low level output (V_{OL}). The worst-case parameters are as follows:

High V_{DD}	−16.5 V
High V_{GG}	−29.7 V
High V_T	− 5.0 V
k'	No effect
Temperature	No effect

Following is a brief explanation of why each of these considerations is a worst-case consideration.

- Temperature and k'—Temperature and k' have no effect on the output level with the input device in the on-state, since k' values always appear as a ratio in the equations; therefore, variations in k' cancel.
- High V_{DD}—This means a higher voltage which must be divided to provide the near ground output (−2 V) when the input device is in the on-state. It also means a higher current in the load device, which again means a lower impedance input device.
- High V_{GG}—This means a lower value of m; therefore, the current of the load device is higher which again means a lower impedance input device.
- High V_T—This is worst-case for the example because of the fixed input voltage of −9 V; i.e., the overdrive voltage beyond the threshold is low. For some bias conditions, it is possible that low V_T is worst-case since at low V_T there is more load current. A calculation must be made for both cases to determine the worst-case for each particular bias condition.

The design of the input device is approached in the following manner.

Calculate m for the worst-case conditions,

$$m = \frac{V_{DD}}{2(V_{GG} - V_T) - V_{DD}} = \frac{16.5}{2(29.7 - 5) - 16.5} = 0.49 \simeq 0.50$$

Calculate the ratio of V_{OL}/V_{DD}

$$\frac{V_{OL}}{V_{DD}} = \frac{-2.5}{-16.5} = 0.152$$

With the ratio design curves, Figure 4-19 through 4-27, the input is determined as follows:

$$\frac{V_{IH} - V_T}{V_{DD}} = \frac{-9 - (-5)}{-16.5} = 0.242$$

From the ratio design curves for $m = 0.5$, Figure 4-23, a minimum value of k_I/k_L is obtained

$$\frac{k_I}{k_L} \geq 35$$

or

$$k_I = 35k_L \qquad \text{or} \qquad k'\left(\frac{W}{\ell}\right)_L = 35k'\left(\frac{W}{\ell}\right)_L$$

Since k' cancels from the expression

$$\left(\frac{W}{\ell}\right)_I = 35\left(\frac{W}{\ell}\right)_L$$

To construct the smallest possible device, assume that the minimum value of ℓ is limited by the process to 0.2 mil.

$$\ell_I = 0.2 \text{ and } \left(\frac{W}{\ell}\right)_L = 1.95$$

therefore

$$W_I = (35)\,(0.2)\,(1.95) = 13.6 \text{ mils}$$

It should be noted that turn-on or fall time was not considered in the calculations. The low level output specification usually ensures that the impedance of the input device is low compared to the load device and therefore the resulting fall time is very much faster than the rise time. However, if fall time were specified, it may be calculated using Equation 4-77.

4.3.4 Output Inverter Power Considerations

Worst-case design considerations for power dissipation are as follows:

High temperature	90°C
High V_{DD}, V_{GG}	−16.5 V, −29.7 V
Low V_T	− 3.0 V
High k'	2.1 μA/V^2 @ 25°C

- High Temperature—The value shown is usually the worst-case since the highest junction temperature is reached at high ambient and care must be taken to ensure that the junction temperature does not rise to an excessive value. A junction temperature of +150°C is considered safe for an MOS chip.

- V_{DD}, V_{GG}, V_T, k'—The values listed are all in the direction to cause the maximum current through the inverter.

Power calculations are:

Power = $V_{DD}I_{DS}$, this assumes $v_{out}(0) = 0$ V

$$\text{Power} = V_{DD}k'\left(\frac{W}{\ell}\right)_L [2(V_{GG} - V_T)V_{DD} - V_{DD}^2]$$

For the example, values are substituted into the power equation. First, k' at 90°C is obtained from the graph of $k'/k_{25°C}'$ (Figure 4-34).

$$k_{90°C}' = k_{25°C}' \times 0.75 = 2.1\ (0.75) = 1.57\ \mu A/V^2$$

$$\text{Power} = (16.5)(1.57)(1.95)\ [2(29.7 - 3.0)\ 16.5 - 16.5^2]$$

$$\text{Power} = 30.8 \text{ mW}$$

It should be noted that this power is for a 100 percent duty cycle.

In actual array design, the power calculated for this inverter would be summed up with the power calculated for all other inverters and logic elements that are dissipating power in the array. Utilizing the thermal characteristics of the package and the maximum ambient temperature, the silicon or junction temperature would be calculated and compared with the assumed temperature. If the calculated temperature is close to the assumed value, adjustments may not be necessary in the power or speed calculations. If there is a significant discrepancy, especially if the temperature was assumed lower than the calculated value, one or more iterations may be necessary in the power as well as speed calculations.

4.3.5 Open Drain Considerations

It is frequently desirable to provide an open drain output that can be connected as shown in Figure 4-36. Chip *A* contains an output inverter circuit similar to the design example in Section 4.3.1, "output inverter design." Chip *B* contains an output transistor that must be designed to function in conjunction with the output inverter to provide acceptable inputs to Chip *C*. It is worthwhile to examine the design approach for an output transistor for this open drain configuration.

Since the output inverter and open drain transistors are located on different chips, the assumption must be made that process parameters for each of the two chips may be at opposite extremes. That is, if any process parameter is assumed at a maximum value for one chip, it is possible that it is at a minimum value on the other chip. Considering this

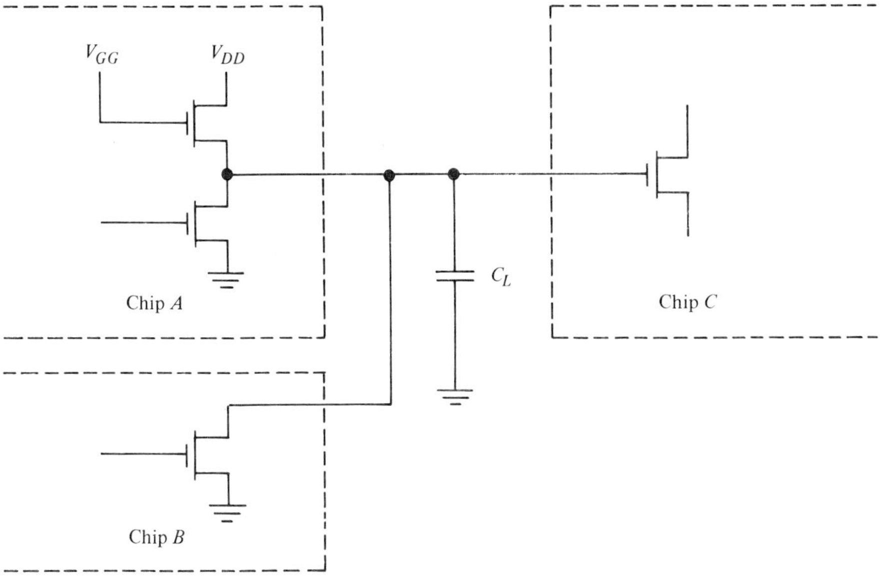

Figure 4–36 Open drain output circuit.

possibility, worst-case conditions must be determined and used in the design of the circuit.

In designing an output circuit that consists of an inverter and open drain transistor, the low output level is of major concern. Worst-case conditions are those that cause the highest load current; i.e., high k' and low V_T, on Chip A combined with the highest impedance, i.e., low k' and high V_T for the open drain transistor on Chip B. The design procedure involves a determination of the output inverter device sizes as in the example in Section 4.3.1. Then considering the worst-case conditions for low output level, the size of the open drain transistor on Chip B is calculated. This open drain transistor must be larger, i.e., have a wider gate region, and hence will occupy a larger area than the output inverter.

4.3.6 Internal Inverter

The design procedure for an inverter or logic gate that drives a load internal to the chip is basically the same as that for an output inverter. There are some differences in the details of the design procedure; these will be noted in the following example. Assuming the circuit and process specifications given in Tables 4-3 and 4-4, worst-case conditions must be selected and device dimensions determined.

4.3.7 Internal Inverter Load Device

The same worst-case conditions apply as in the case for the output inverter; therefore, $m' = 0.6$ and $V_{OH}/V_{DD} = 0.74$ apply to the internal inverter as well.

From the turn-off characteristic curves, the value $t_r/\tau_{sc} = 1.9$ was obtained. The load capacitance must now be specified. Typically, an internal stage with a fanout of 3 or 4 has a load capacitance of approximately 2 pF. This value will be assumed for the example.

The speed requirement for an internal stage is usually determined from the specified maximum clock frequency. For this example it will be assumed that a typical internal stage must have a rise or fall time of 200 ns or less for the chip to meet the maximum clock frequency requirements. Using the expression R_{sc} and rise time, $R_{sc} = t_r/1.9C_L = 200\text{ ns}/1.9 \times 2\text{pF} = 52.6 \times 10^3\ \Omega$

By substituting terms in Equation 4-80,

$$k_L = \frac{1}{R_{sc}[2(V_{GG} - V_T) - V_{DD}]} = \frac{1}{52 \times 10^3[2(24.3 - 5.0) - 13.5]}$$
$$= 0.76\ \mu\text{A/V}^2$$

The value for W/ℓ is obtained using the minimum value of k' at 90° C.

$$\left(\frac{W}{\ell}\right)_L = \frac{k_L}{k'} = \frac{0.76}{0.975} = 0.78$$

To determine the smallest device allowed by the process, a minimum value of $W_L = 0.3$ mil will be assumed.

$$\ell_L = \frac{W_L}{0.78} = \frac{0.3}{0.78} = 0.38\text{ mil}$$

4.3.8 Internal Inverter Input Device

The input device for the internal inverter is designed using the same worst-case conditions as those for the output inverter. The value of m is 0.5. The output low level (V_{OL}) is required to be at the threshold voltage or below for proper operation of succeeding stages, but it is advisable to allow for some noise on the output. Assuming +1 V of noise is permitted, the expression for the output voltage is

$$V_{OL} = V_T + V_N = -5.0 + 1.0 = -4.0\text{ V}$$

This means the V_{OL} of the inverter must not exceed -4.0 V when $V_T = -5.0$ V. When normalized to V_{DD} for the ratio design curves,

$$\frac{V_{OL}}{V_{DD}} = \frac{-4.0}{-16.5} = 0.24$$

To obtain this output, assume the input is at its minimum value or $V_{IH} = -9$ V. The normalized input expression becomes

$$\frac{V_{IH} - V_T}{V_{DD}} = \frac{-9 - (-5)}{-16.5} = 0.24$$

From the ratio design curves for $m = 0.5$,

$$\frac{k_I}{k_L} \geqq 24$$

$$\left(\frac{W}{\ell}\right)_I = 24 \left(\frac{W}{\ell}\right)_L$$

Assuming that the smallest possible inverter is to be constructed, the smallest value of ℓ allowed by the process is $\ell_I = 0.2$ mil. Then

$$\begin{aligned} W_I &= 24\ (\ell_I)\ (k_L) \\ &= 24\ (0.2)\ (0.78) \\ &= 3.7 \text{ mils} \end{aligned}$$

The size of the input device has been determined considering the low output level (V_{OL}) requirements. It is now necessary to verify that the turn-on or fall time meets the specified 200 ns maximum. By using the simplified expression for turn-on in Equation 4-77 and substituting the appropriate worst-case values,

$$t_{\text{total}} = \frac{C_L}{k_I(v_{\text{in}} - V_T)} \left\{\frac{V_{OH\,\max} - (v_{\text{in}} - V_T)}{v_{\text{in}} - V_T} + \frac{1}{2}\,\ell\text{n}\left[\frac{2(v_{\text{in}} - V_T) - V_F}{V_F}\right]\right\} \tag{4-77}$$

$$\begin{aligned} C_L &= 2 \text{ pF} \\ k_I &= 24k_L = 24(0.76)\ \mu\text{A/V}^2 \\ v_{\text{in}} &= -9 \text{ V} \\ V_T &= -5 \text{ V} \\ V_{OH\,\max} &= V_{DD} = -16.5 \text{ V} \\ V_F &= V_{OL} \text{ (internal stage)} = -4.0 \text{ V} \end{aligned}$$

$$t_{\text{total}} = \frac{2\ \text{pF}}{24(0.76)\ \mu\text{A/V}^2\ (9.0 - 5.0)} \left\{ \frac{16.5 - (9.0 - 5.0)}{9.0 - 5.0} + \frac{1}{2} \ell\text{n} \left[\frac{2(9.0 - 5.0) - 4.0}{4.0} \right] \right\}$$

$$t_{\text{total}} = \frac{2\ \text{pF}}{24(0.76)\ \mu\text{A/V}^2\ (9.0 - 5.0)} \left[\frac{16.5 - (9.0 - 5.0)}{9.0 - 5.0} + 0 \right]$$

$$t_{\text{total}} = 86.1\ \text{ns}$$

which is less than the specified 200 ns maximum.

It is interesting to note that the entire fall time for this example occurs while the input device is in the saturation region. That is, the final value of output voltage (V_F) is just equal to $v_{\text{in}} - V_T$.

4.3.9 Noise Considerations for an Internal Inverter Stage

In specifying performance requirements for an internal stage, it is important to note that the low output level (V_{OL}) need only drop to the threshold voltage less some noise margin and not to a fixed level. This means that when the threshold voltage for the chip is at the high end of the parameter limits -5 V in the example, V_{OL} is not required to be -2.5 V as in the case of the output inverter but $-5.0\ \text{V} - (-V_N)$. A typical value for V_N is 1.0 V, in which case $V_{OL} = -4.0$ V, resulting in a considerably reduced minimum k_I/k_L ratio and a smaller input device.

4.3.10 Stacked Input Devices

Figure 4-37 shows a circuit configuration with two MOS input transistors stacked in series between the output and ground. Since configurations of this type are encountered in MOS arrays, some comments on special design considerations are in order.

The circuit may be looked at simply as an inverter. When high level inputs are applied simultaneously to the gates of Q_1 and Q_2, a resistance ratio must exist between the load and input devices as in the case of the conventional inverter. Since Q_1 and Q_2 are in series between the output terminal and ground, the effective resistance of each device must be half that of the single input device, or the device (gate) width must be two times as wide. Not only is the area required by such a large circuit, but the capacitance at the input and output terminals is large, which adversely affects the transient response.

A secondary effect that must also be considered is that, with the circuit in the conducting state, the source of the upper input device Q_1 is

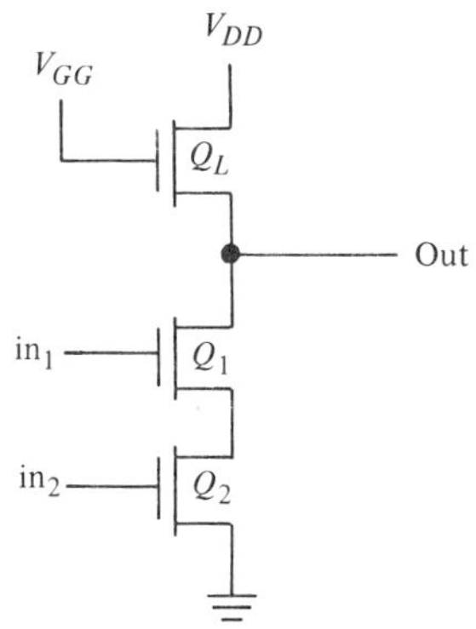

Figure 4–37 Typical schematic showing stacked input devices.

at a potential higher than ground, placing a somewhat higher level requirement on the input to maintain a specific on-resistance for the device. These considerations place a practical limit on the number of input devices that may be stacked between the output and ground. Typically this limit is two devices, although circuits with three stacked devices are sometimes used in MOS arrays.

4.4 SPECIAL CIRCUIT DESIGN PROBLEMS

4.4.1 MOS Bipolar Interface

The problem of interfacing MOS with bipolar circuits and a common approach to solve it are best illustrated by an example. Assume a typical bipolar (TTL) logic gate driven by an MOS output circuit manufactured by a low threshold process as shown in Figure 4-38. Input levels that are commonly specified for TTL circuits will be used. The problem is that of (a) providing an MOS transistor Q_B capable of sinking 1.6 mA of current with a maximum output voltage $V_{OL} = -0.4$ V and (b) providing an MOS transistor Q_A capable of supplying 100 μA of leakage current with a minimum output voltage (V_{OH}) of +2.4 V in the on-condition.

The internal inverter, with device Q_C and load R_L, is usually connected between V_{CC} and V_{GG}. V_{GG} is typically -10 to -12 V. R_L is a diffused resistor to allow maximum negative voltage on the gate of Q_B, since this device must be a very low impedance device. In order to reduce the size of Q_B, the device may be returned to a negative voltage; however, a diode is usually placed from the output to ground to avoid excessive negative out-

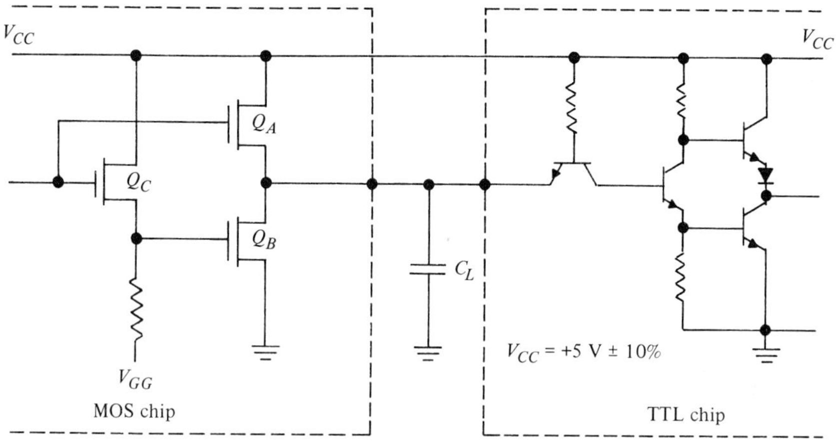

Figure 4–38 Typical MOS to TTL interface.

put voltage. The MOS device Q_A is usually designed to charge the external load capacitance C_L at a specific rate and the resulting device size must be checked against the leakage current and level requirements. Since the entire circuit will be required to respond within a specified time, the value of R_L is chosen from the transient considerations. The size of device Q_C must then be determined by the normal dc ratio design approach.

As with any MOS circuit design problem, worst-case design techniques must be employed. Frequently, when a diffused resistor such as R_L is used, careful consideration must be given to the production tolerances and the variations in the ohmic value of the resistor. An input circuit similar to that shown in Figure 4-39 provides a solution to the TTL-MOS interface problem.

Recall that the MOS circuit is one manufactured by a low threshold process. Typical output specifications for a TTL circuit indicate a "high" output level of +2.4 V minimum and a "low" output level of +0.4 V maximum, depending on the logic state of the output. Since a level of +2.4 V usually will not be sufficiently positive to ensure that Q_I is in the cutoff state, MOS device Q_D is added. The impedance of Q_D is normally not critical except that it must be sufficiently low to maintain the speed requirements. The TTL fanout will be reduced and power dissipation on the MOS chip will be increased as the impedance of Q_D is lowered.

4.4.2 Push-Pull Drivers

A useful circuit configuration for driving large capacitive loads is a push-pull device, a schematic diagram of which is shown in Figure 4-40. This circuit allows the use of low impedance devices to drive the output without

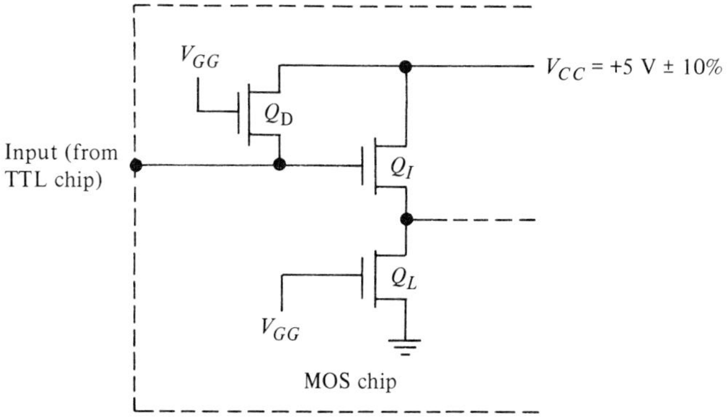

Figure 4–39 TTL to MOS circuit.

creating a power consuming dc path from V_{DD} to ground. In designing a push-pull circuit the saturated inverter circuit (Q_1, Q_2) is designed first. The load device Q_2 is designed in conjunction with Q_3 to provide the desired speed when the output load capacitance is charged toward V_{DD}. Normally, the rate at which Q_2 charges the gate of Q_3 is designed to be 60 to 80 percent of the overall circuit response. Device Q_1 is designed to discharge the gate of Q_3 rapidly to ensure that Q_3 is cut off before Q_4 starts conduction. Q_1, when on, must also provide a final low level output voltage to the gate of Q_3 which is below a threshold voltage. Finally, Q_4 is designed to discharge the output load capacitance in the specified time. If the push-pull circuit is designed to drive a dc load, these considerations must be accounted for and may modify the design approach to some extent.

A very important factor that must be considered when designing a push-pull driver is the body effect on the threshold voltage of both Q_2 and Q_3.

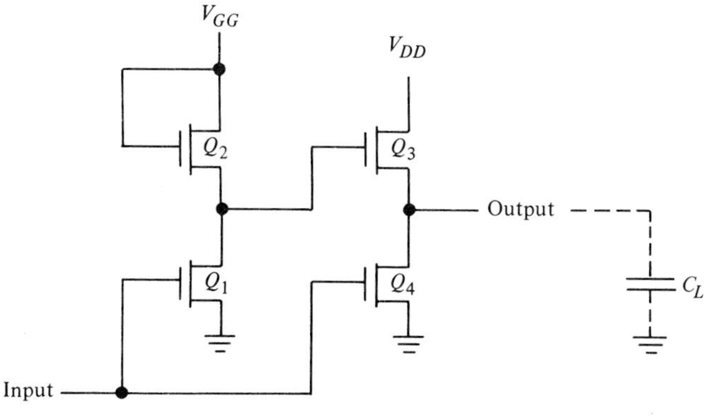

Figure 4–40 Typical push-pull driver circuit.

For certain bias conditions, it is possible that both Q_2 and Q_3 will be operating in the saturation region. This being the case, the final output voltage will be the result of the V_{GG} supply voltage less two threshold drops, both of which will be enhanced by the body effect phenomenon. It is possible that the output voltage will not be sufficient to meet the specified minimum value.

An approach to solving the level problem that results from the output device Q_3 operating in saturation is shown in the schematic diagram in Figure 4-41. Simply, the device Q_5 charges C_1 to $V_{GG} - V_T'$ when the input is high. A change in the state of the input level to the low condition causes point A to start charging toward V_{GG}. C_1 couples this negative-going transition to point B, causing Q_5 to cut off and placing a continuously increasing gate voltage on Q_2. The final output level available from Q_2 will be V_{GG}, allowing the output of Q_3 to achieve V_{DD}, or near V_{DD}, depending on actual power supply and threshold values. In addition to the improved output level characteristics, rise time characteristics are improved significantly with a boot strapped push-pull circuit, such as the one described.

In designing a boot strapped push-pull driver the procedure is basically the same as for the simple push-pull circuit. The output rise time specification determines the size of Q_2 and Q_3. Q_5 is usually a minimum size device, since its only function is to charge capacitor C_1 which is typically in the range of 1 pF. C_1 is designed to be much larger than the gate capacitance of Q_2 plus any stray capacitance to ground on point B. Q_1 and Q_4 are designed to achieve the same results as those for the simple push-pull circuit.

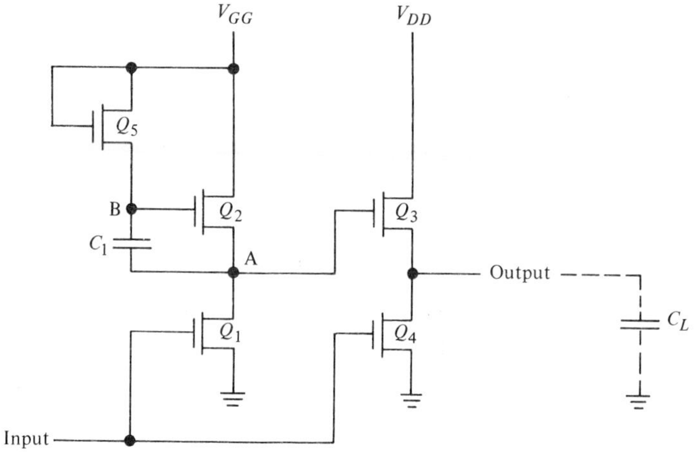

Figure 4–41 Typical boot strapped push-pull circuit.

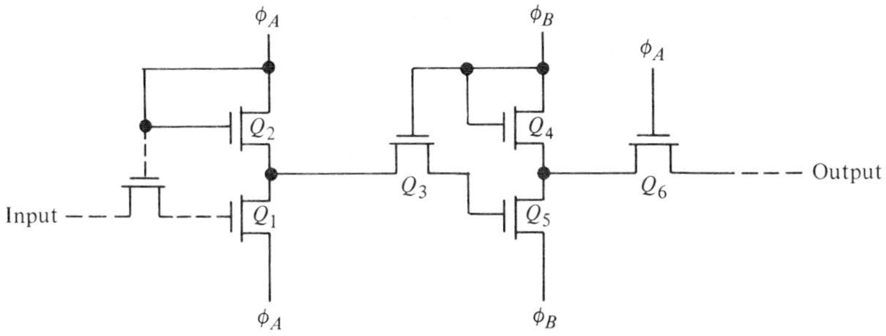

Figure 4–42 Typical schematic diagram of two 2-phase ratioless inverters.

4.4.3 Ratioless Circuits

The schematic diagram in Figure 4-42 is representative of a class of circuits in which operation does not depend on a resistance ratio between the load and input devices. The following discussion is concerned with the design considerations associated with one configuration of this class of circuit, the 2-phase ratioless inverter. Alternate ratioless mechanizations and descriptions of operation are discussed in Chapter 5.

Since there is no direct current path in this ratioless circuit, the design considerations are associated only with the transient characteristics. Figure 4-43 shows in greater detail the capacitances that are usually considered in ratioless design. These capacitances are all related very closely to the circuit topology and fabrication techniques. Some are desirable and others are not, but each must be considered in the design.

The parasitic capacitors C_4 and C_5 are the result of the gate overlapping

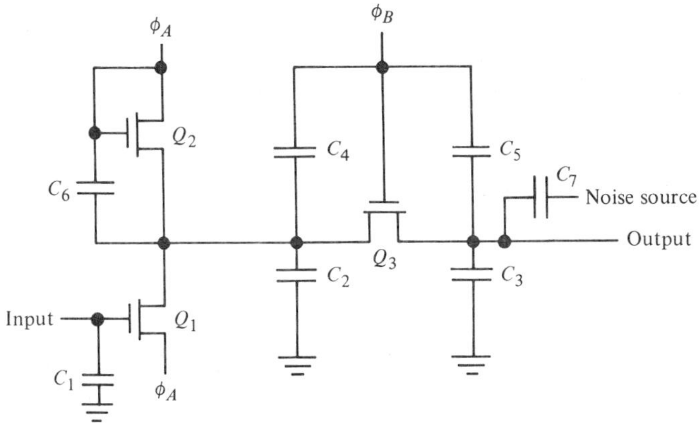

Figure 4–43 2-Phase ratioless inverter showing important capacitances.

the drain and source regions of device Q_3. These overlap capacitances are inherent in the fabrication process and are present on every MOS device. Their effect on device performance, however, is frequently so small that they are ignored. In the case of a ratioless stage, the effect of overlap capacitances is significant, especially on a series coupling device such as Q_3. Capacitor C_6 is the result of the gate to source capacitance of Q_2, which tends to remove any stored charge on capacitor C_2 when clock ϕ_A returns to zero. C_7 is intended to include any capacitance coupling to the storage node C_3. Such capacitance may result from interconnect line crossovers or from the overlap capacitances on stages being driven. The amount of noise coupled to the storage node must be estimated and accounted for in the design of any ratioless circuit.

The design of a ratioless circuit usually starts with an estimation of the load capacitance to be driven by the inverter. Referring to Figure 4-43, this load capacitance as represented by C_3 would consist of the sum of the fanout and interconnection capacitance. Next, the minimum value of the high logic level voltage to be required to cross C_3 must be selected based on the clock level and some reasonable noise considerations. The parasitic capacitor C_5, which will affect the voltage level across C_3 when ϕ_B returns to ground, must be estimated, together with any other coupling capacitance due to crossovers or adjacent circuits. If the magnitude of the effect of these capacitances is sufficiently great, C_3 may have to be enlarged. Capacitor C_2 shares its charge with C_3 during the time when Q_3 is in the conducting state; therefore, C_2 must be made large enough to ensure an adequate high logic level voltage on C_3.

Device Q_2 is designed to charge capacitor C_2. Since power drain is not a problem with this type of circuit, Q_2 may be a low impedance device if required by the speed specification. Q_1 and Q_3 must discharge the capacitors C_2 and C_3 if the input to Q_1 is a high logic level. This discharge time usually limits the maximum speed of operation of a 2-phase ratioless inverter. Since Q_3 has a high level gate drive from the clock (ϕ_B) and C_3 is significantly smaller than C_2, the problem of discharge may, as a simplifying measure, be reduced to that of discharging C_2 through device Q_1. The design of Q_1 is based on this discharge time, assuming a specific input voltage level on the gate of Q_1.

It is important to note that the actual design of a ratioless inverter involves a trade-off of input voltage levels and device sizes. It is also important to note the dependence of the design upon the actual circuit topology. Frequently, the design procedure involves one or more iterations. After the device sizes have been determined based on the original assumptions, the assumptions may require re-examination to see whether they can be improved. Any discrepancy between the actual designed value and the assumed value of a capacitance must be examined for its effect on circuit performance, and if necessary, device sizes may have to be adjusted. Since the design of ratioless circuits is so intimately related to device topology, it should be apparent that numerous specialized layout techniques are used to enhance circuit performance where desired. Various proprietary techniques are employed by different MOS suppliers. Design and layout of ratioless circuits has been greatly facilitated through computer aids. A number of programs are available in which electrical performance specifications, such as speed and voltage levels, are used to provide the circuit parameters, i.e., capacitances and device sizes that are required to achieve the desired performance. Once the basic layout of a ratioless cell has been developed, modification of a cell is a relatively simple matter using computer aids. The required adjustments in geometry as a result of new electrical specifications are placed into the layout program and a new cell can be generated automatically.

REFERENCES

1. R. H. Crawford, *MOSFET in Circuit Design,* McGraw-Hill, New York, 1967.
2. A. S. Grove, *Physics and Technology of Semiconductor Devices,* John Wiley & Sons, New York, 1967.
3. D. Frohman-Benchkowsky and L. Vadasz, "Computer-Aided Design and Characterization of Digital MOS Integrated Circuits," *IEEE J. Solid State Circuits,* **SC-4,** 57–64 (April 1969).

5
Logic Design with MOS

5.1 INTRODUCTION

This chapter discusses the theory and practice of implementing MOS static and dynamic logic. MOS static logic is similar to the logic employed in bipolar integrated circuits; whereas, dynamic logic is unique in that it uses multiphase clock circuits in ratio and ratioless configurations. The most widely used configurations, viz., static, 2-phase ratioed, 2-phase ratioless, and 4-phase ratioless, are discussed in this chapter to give the designer an understanding of the basic building blocks. These basic building blocks of complex combinatorial logic and the delay elements of each class are essentially identical in function, although the various characteristics of each may influence the relative ease of satisfying a specific system requirement. The designer may find that static logic is the simplest to implement; however, it should be remembered that the unique capabilities of MOS technology are not fully utilized in such applications.

5.2 STATIC LOGIC

MOS static logic timing is accomplished by shifting data with dc flip-flops and series coupling devices. A schematic representation of an MOS static inverter stage is shown in Figure 5-1. The relatively high impedance MOS transistor (Q_L) requires a smaller area than the equivalent diffused load resistor. This load transistor is held conducting at all

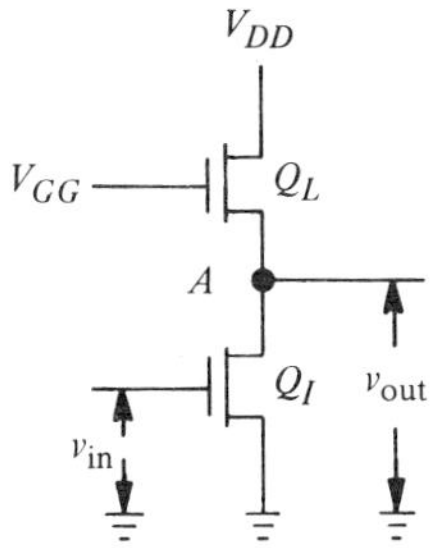

Figure 5–1 MOS inverter.

times by a negative bias supply (V_{GG}). As explained in Chapter 4, by selecting suitable geometries of the input transistor (Q_I), the load device (Q_L) and the power supply voltages (V_{DD} and V_{GG}), it is possible to switch v_{out} between V_{DD} and ground as a function of v_{in}.

For *P*-channel MOS devices with grounded substrates, a sufficiently negative voltage applied to the gate of the input transistor will cause conduction, driving the output node *A* (in Figure 5-1) close to ground potential as determined by the ratio of the on-resistances of MOS devices Q_I and Q_L.

If the voltage applied to the gate of the input device is at or near ground potential, conduction will not occur; consequently, output node *A* will be held at V_{DD}, a negative voltage. When the negative voltage and ground level are defined as opposite logical states, it is obvious that the input voltage (v_{in}) is "inverted" to the opposite logical state at the output node. Thus the circuit in Figure 5-1 satisfies the requirements for a logical inverter.

Advantages of static logic circuits are:

- No high voltage clock drivers are required as in dynamic logic circuits.
- Logic signal storage may be achieved at dc without capacitive storage memory.
- Functionally similar to bipolar logic circuits, which enables rapid conversion of existing systems to MOS LSI with minimum redesign.
- Generally good noise immunity.

Disadvantages of static logic include:

- Higher power dissipation than dynamic logic, since a dc path to ground exists whenever the input transistor is conducting.
- Flip-flops must be implemented using master-slave techniques which re-

quire a large number of devices, resulting in higher cost per logic function.

- Device geometries are larger than dynamic ratioless circuits, thus permitting fewer logic functions per array.

MOS static logic may be used in asynchronous systems; systems in which no central clock is available; systems with extremely slow clock rates; and systems for converting existing bipolar logic systems to MOS logic with minimum redesign.

5.2.1 Static Logic Elements

Two basic input device structures are used in the MOS implementation of digital logic. In Figure 5-2 the MOS devices are connected in series; whereas, in Figure 5-3 the MOS devices are connected in parallel. The logical operation performed by such gates is a function of the definition of a logic 1 and a logic 0 with respect to V_{DD} and ground; therefore, they can serve as NAND or NOR gates.

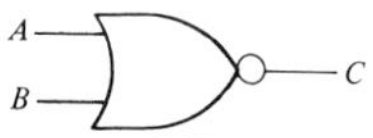

NOR

Positive true logic for *P*-MOS
Most positive voltage = 1
Most negative voltage = 0

Truth Table

A	*B*	*C*
0	0	1
0	1	0
1	0	0
1	1	0

$C = \overline{A + B}$

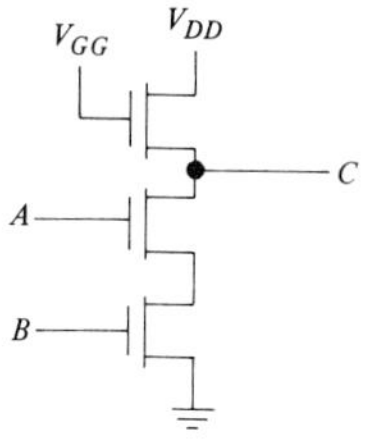

Truth Table for *N*-MOS or *P*-MOS

A	*B*	*C*
$\lvert V\rvert > \lvert V_T\rvert$	$\lvert V\rvert > \lvert V_T\rvert$	Ground
$\lvert V\rvert > \lvert V_T\rvert$	Ground	$\lvert V\rvert \doteq \lvert V_{DD}\rvert$
Ground	$\lvert V\rvert > \lvert V_T\rvert$	$\lvert V\rvert \doteq \lvert V_{DD}\rvert$
Ground	Ground	$\lvert V\rvert \doteq \lvert V_{DD}\rvert$

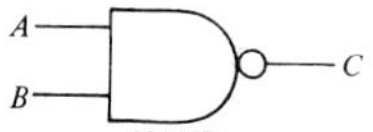

NAND

Negative true logic for *P*-MOS
Most negative voltage = 1
Most positive voltage = 0

Truth Table

A	*B*	*C*
1	1	0
1	0	1
0	1	1
0	0	1

$C = \overline{A \cdot B}$

Figure 5–2 Series implemented gate.

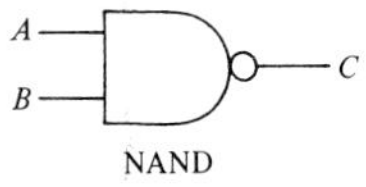

Positive true logic for *P*-MOS
Most positive voltage = 1
Most negative voltage = 0

Truth Table

A	*B*	*C*
0	0	1
0	1	1
1	0	1
1	1	0

$C = \overline{A \cdot B}$

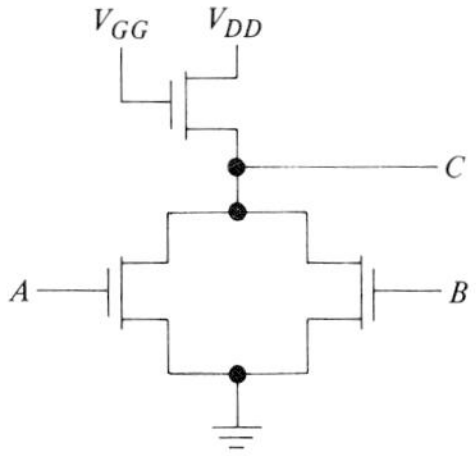

Truth Table for *N*-MOS or *P*-MOS

A	*B*	*C*
$\lvert V \rvert > \lvert V_T \rvert$	$\lvert V \rvert > \lvert V_T \rvert$	Ground
$\lvert V \rvert > \lvert V_T \rvert$	Ground	Ground
Ground	$\lvert V \rvert > \lvert V_T \rvert$	Ground
Ground	Ground	$\lvert V \rvert \doteq \lvert V_{DD} \rvert$

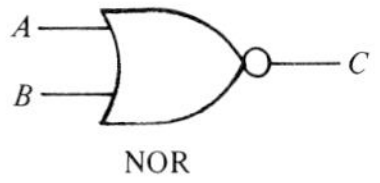

Negative true logic for *P*-MOS
Most negative voltage = 1
Most positive voltage = 0

Truth Table

A	*B*	*C*
1	1	0
1	0	0
0	1	0
0	0	1

$C = \overline{A + B}$

Figure 5–3 Parallel implemented gate.

Since the v_{out} of the series implemented gate is required to approach closely ground potential when both input devices are conducting, the total resistance of the series input devices must be equal to the resistance of the input device of the simple inverter. To lower the resistance, each series input device must be physically enlarged. For this reason, series input devices are avoided wherever possible. Parallel implementation allows much more flexibility, since the addition of parallel gates lowers the total output resistance, bringing v_{out} closer to ground potential. Combinations of these series and parallel forms greatly increase the flexibility of MOS logic circuits. Several possible configurations are shown in Figure 5-4.

5.2.2 Static Memory Elements

The three most widely applied static MOS memory elements are the cross-coupled latch (*RS* flip-flop), the master-slave flip-flop, and the quasi-static sample and hold flip-flop.

The MOS cross-coupled latch implementation illustrated in Figure 5-5 uses parallel input devices in each gate to minimize device geometries. The

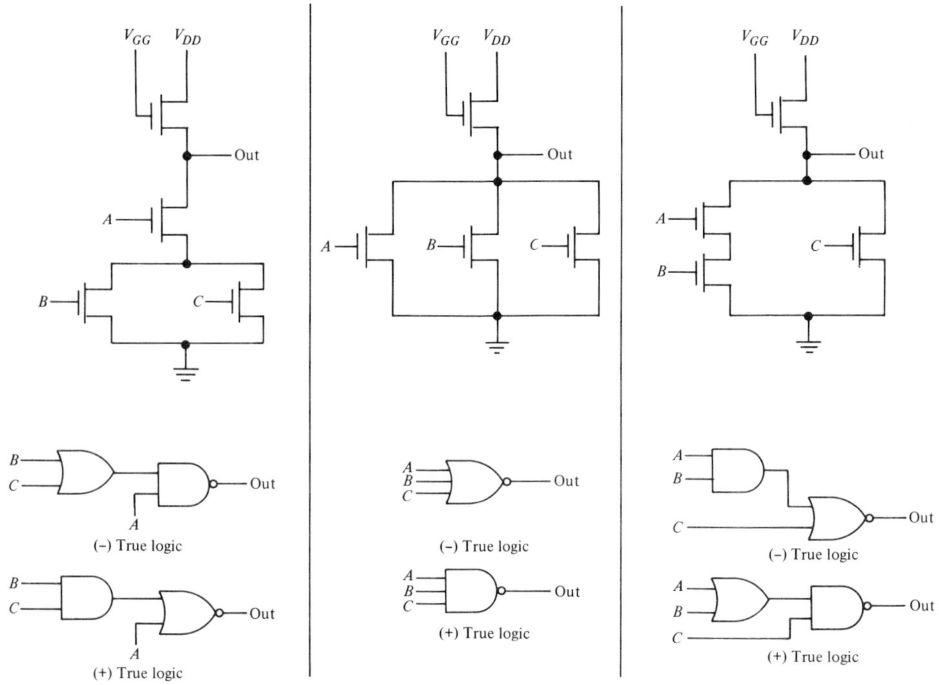

Figure 5–4 Static *P*-MOS complex mechanizations.

operation of this flip-flop is described by the truth table of Figure 5-5. Like bipolar *RS* flip-flops, the MOS version exhibits an indeterminate output state when $R = S =$ logic 1.

The master-slave flip-flop principle, as shown in Figure 5-6, is commonly applied in both bipolar and static MOS circuits. Isolation between the master and slave sections of the flip-flop is obtained by alternately enabling each section with timing signals (load and $\overline{\text{load}}$) which are mutually exclusive and are usually complements, thus resulting in a delay function. By adding complex logic to the inputs, the master-slave flip-flop may be used in shift registers, counters, and a number of applications that require memory. The prime deterrent to extensive use of the master-slave flip-flop in MOS dc logic circuits is its complexity. The large number of devices that are required (Figure 5-7) consume a considerable amount of chip area. As a result of an effort to minimize the transistor count, another isolation type flip-flop, the quasi-static sample and hold, has been developed.

In this form of flip-flop, unlike other dc memory elements, two capacitors (C_1 and C_2) are used to store temporarily a logic input. The circuit schematic and special timing waveforms (sample, transfer, and transfer de-

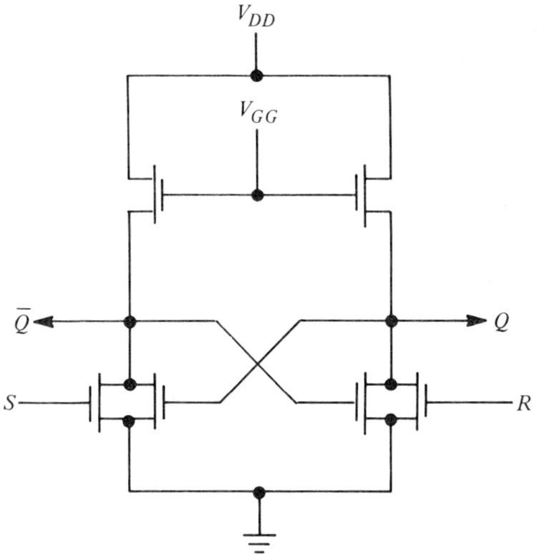

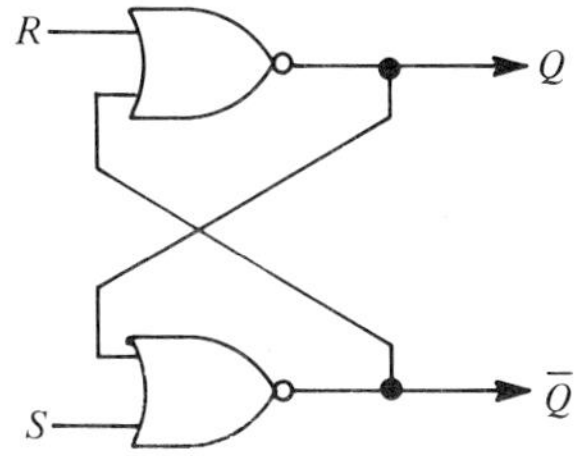

Inputs		Outputs
R_n	S_n	Q_{n+t}
0	0	Q_n
1	0	0
0	1	1
1	1	Indeterminate

t = propagation delay of circuit

Figure 5–5 *RS* flip-flop using cross-coupled NOR gates (negative true logic).

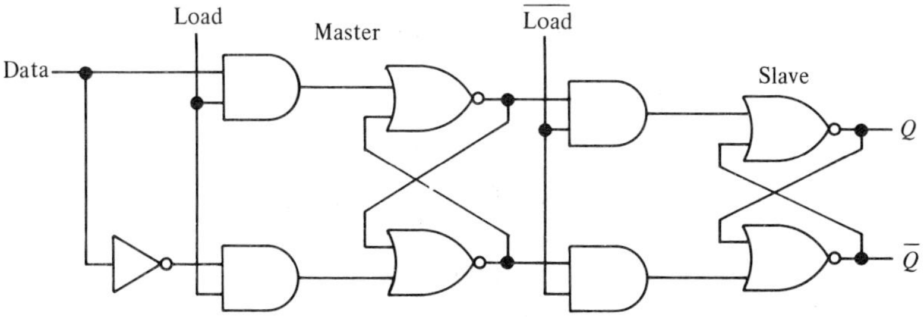

Figure 5–6 Master-slave flip-flop logic diagram.

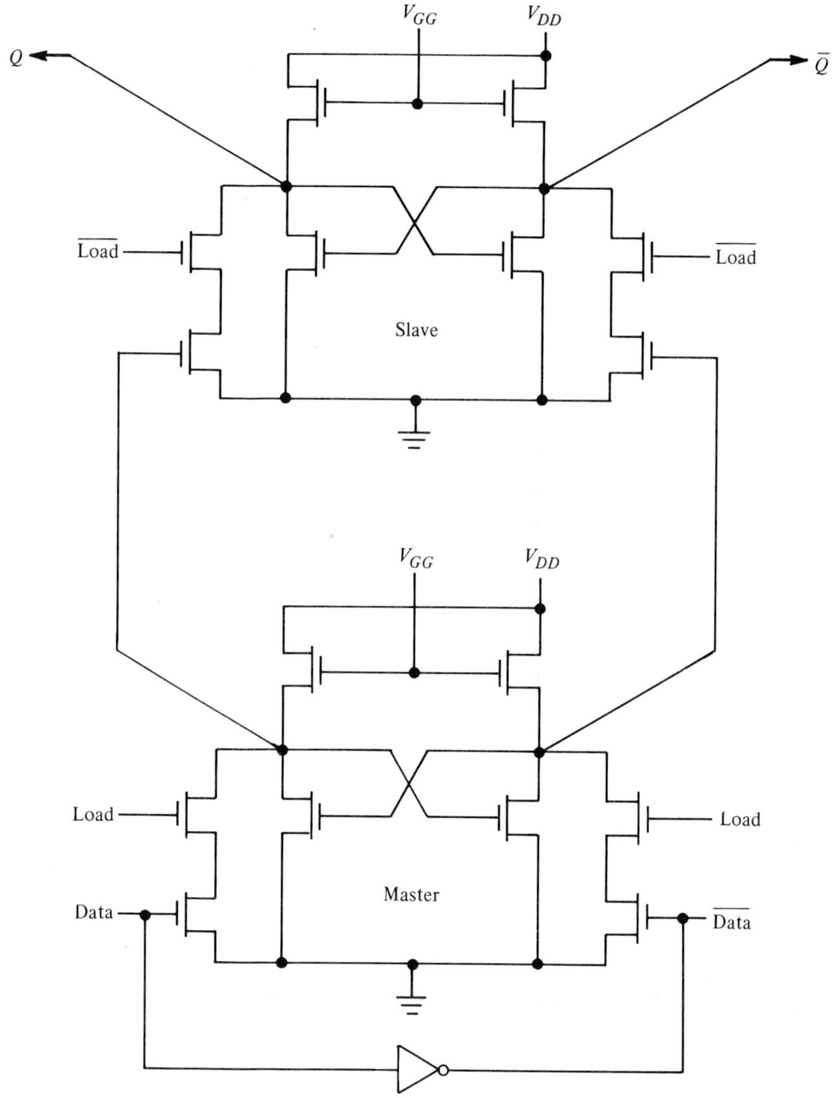

Figure 5–7 MOS master-slave flip-flop schematic.

layed) are shown in Figure 5-8. The negative portion of the sample and the positive portion of the transfer signal should be in the microsecond range. The transfer delayed waveform may be generated from the transfer signal such that its transitions occur as shown in the figure.

If v_{in} is negative, the sample signal enables C_1 to charge negatively. Q_2

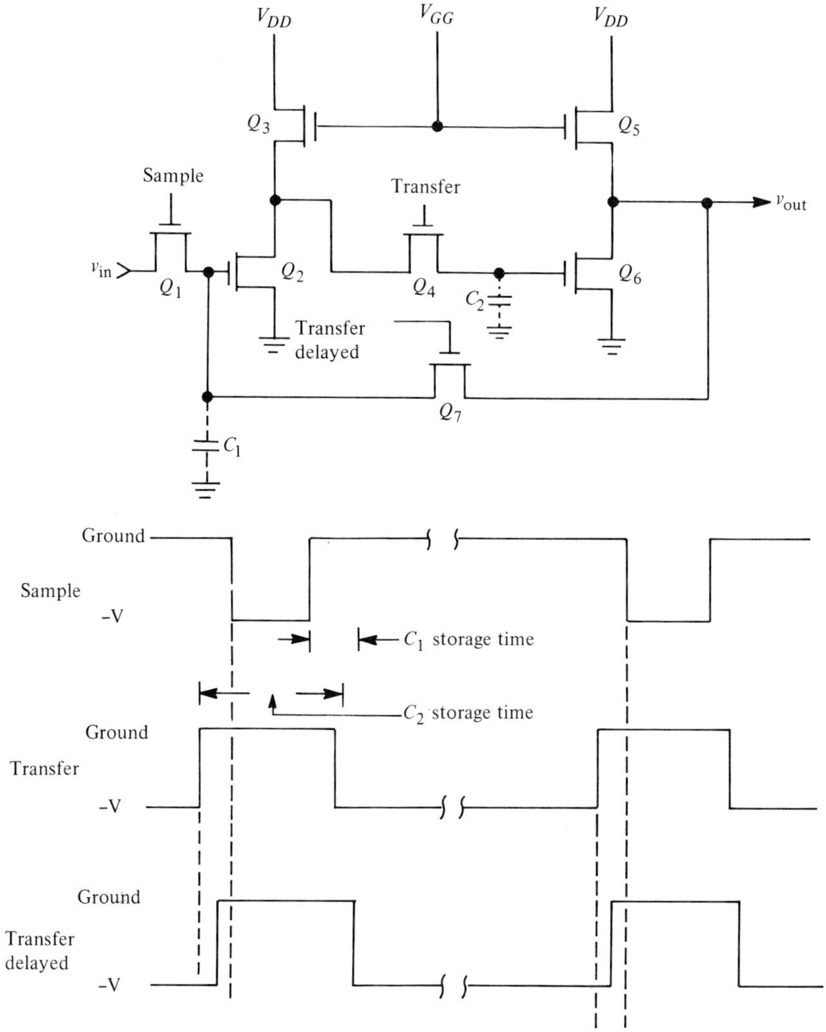

Figure 5–8 Quasi-static sample and hold.

and Q_3 invert v_{in} and when transfer goes negative, C_2 is discharged, causing v_{out} to go negative. When transfer delayed goes negative, dc conditions resume and C_1 is held charged through Q_5 and Q_7. Thus C_1 must be capable of holding its charge during the period between negative portions of the sample and the transfer delayed signals. When transfer returns to ground, Q_4 turns off, holding C_2 discharged, thus v_{out} remains negative.

When transfer delayed returns to ground, Q_7 turns off, allowing C_1 to be controlled by the next input.

If the next v_{in} is at ground during sample, C_1 will be discharged, Q_2 and Q_3 invert v_{in} and, when transfer goes negative, C_2 is charged, bringing v_{out} to ground. When transfer delayed goes negative, C_1 is held discharged through Q_6 and Q_7, thus latching the circuit in a dc mode. When transfer returns to ground, Q_4 turns off, holding C_2 negatively charged until the next negative transition of transfer. Q_6 is held on by C_2, thus holding v_{out} at ground. During the period that transfer is at ground, C_2 must be capable of holding a sufficiently negative charge to keep Q_6 conducting.

5.3 DYNAMIC LOGIC

Dynamic or clocked logic circuits evolved from static circuits as a means of reducing power dissipation. Several distinct circuit types have been developed, including 2-phase ratioed, 2-phase ratioless, and 4-phase ratioless.

The fundamental characteristic of dynamic logic is that it uses clocked load devices that are turned on and off synchronously as a function of system or subsystem timing. Temporary capacitive charge storage is used to retain information between clocking periods.

Dynamic logic utilizes the MOS device characteristics to best advantage. Some advantages of dynamic logic over static logic are:

- The flip-flop, or delay function, can be mechanized with fewer MOS devices.
- System timing problems are simplified.
- Less chip area is required per logic function, resulting in a lower cost per function.
- Since power is dissipated only when the load device is on, the power consumption is lower.

5.3.1 2-Phase Ratioed Logic

A characteristic of ratioed circuits is that when both the load and input devices are conducting, there is a dc current path from the V_{DD} power or clock supply to ground. Proper operation of a ratioed logic gate depends on the ratio of on-resistance of the load device to that of the input device. Intrinsic gate capacitance is used for temporary storage of charge.

Some of the advantages of 2-phase ratioed logic circuits are:

- The maximum number of logic levels per bit time is limited only by the required operating frequency.

- Lower noise and fanout sensitivity than ratioless circuits.
- Low clock line loading.
- Well suited for automated design and artwork generation.
- Fewer layout problems than ratioless circuits.

Disadvantages include:

- Generally higher power dissipation than ratioless circuits.
- Larger device geometries required to implement a given logic function than ratioless.

2-Phase Ratioed Memory Element

The basic operation of dynamic logic can be understood by considering two cascaded inverters that generate a double inversion as well as one clock period of delay due to the time required for charging and discharging nodal capacitance. Each inverter displays a ½-bit (half a clock period) delay. Figure 5-9 shows the logic diagram, circuit schematic, and clock timing. Typical input, output, and internal waveforms are shown in Figure 5-10. Capacitors C_1 through C_5 represent the total capacity associated with each node. The following paragraph describes the operation of this circuit for cases of v_{in} below and above threshold voltage.

If v_{in} is at ground when ϕ_2 becomes negative, Q_1 is turned on and C_1 is grounded. Q_3 is biased off. When ϕ_2 returns to ground, C_1 remains at ground. As ϕ_1 becomes negative, devices Q_2 and Q_4 turn on and, with Q_3 off, C_2 and C_3 are charged negatively to approximately V_{DD}. The negative voltage on C_3 causes Q_6 to turn on. As ϕ_1 returns to ground, the charge on C_2 and C_3 remains, holding Q_6 on. As ϕ_2 becomes negative, Q_5 and Q_7 begin to conduct, holding C_4 at and discharging C_5 to ground potential. If v_{in} is negative when ϕ_2 goes negative, Q_1 turns on and C_1 is charged negatively. When this voltage reaches V_T, Q_3 is turned on, discharging C_2. C_3 remains charged since Q_4 is off. As ϕ_2 returns to ground, C_1 remains negatively charged and Q_3 remains on. As ϕ_1 becomes negative, Q_4 turns on and discharges C_3, thus turning Q_6 off. As ϕ_2 becomes negative, Q_5 and Q_7 are turned on, and C_4 and C_5 charge to approximately V_{DD}.

The speed determining factor in this circuit is the rise time (t_r) required to sufficiently charge the output node capacitors. ϕ_1 must be negative long enough to charge capacitors C_2 and C_3, and ϕ_2 must be negative long enough to charge C_1 and C_4. Rise time is basically a function of the load resistance and output node capacitance. The fall time (t_f) is primarily a function of the output node capacitance and the on-resistance of the input

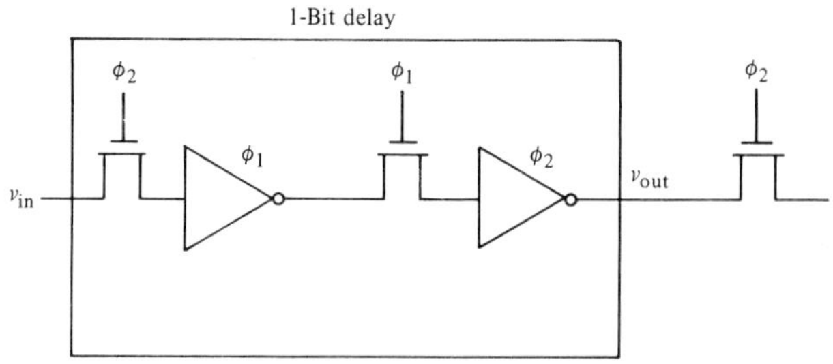

(a) Logic diagram

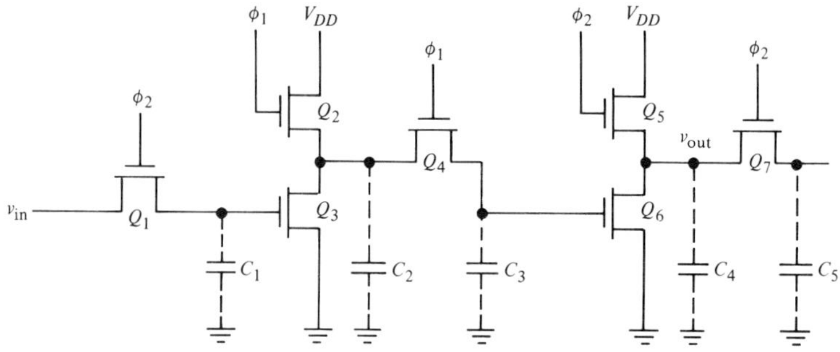

(b) Schematic

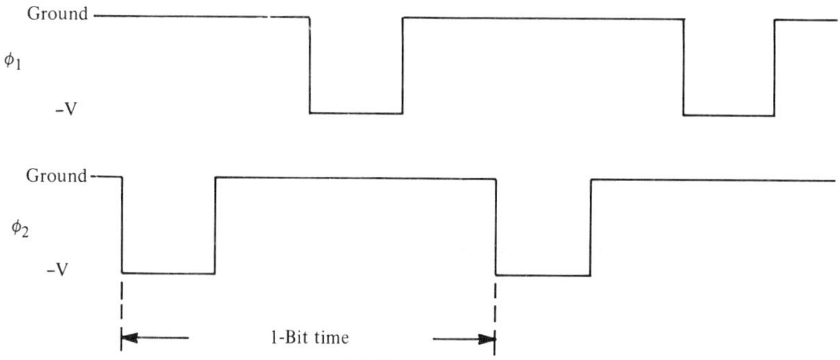

(c) Clock relationships

Figure 5–9 2-Phase ratioed MOS bit of delay with clocked inverters.

transistor. Since the input transistor must have a lower resistance than the load transistor, the rise time will be longer than the fall time.

Specifically, since the 2ϕ-bit delay requires nonoverlapping clocks to prevent race conditions, let us arbitrarily assume that the width of each clock is slightly less than half the total clock period. If the $t_r = 90$ ns and

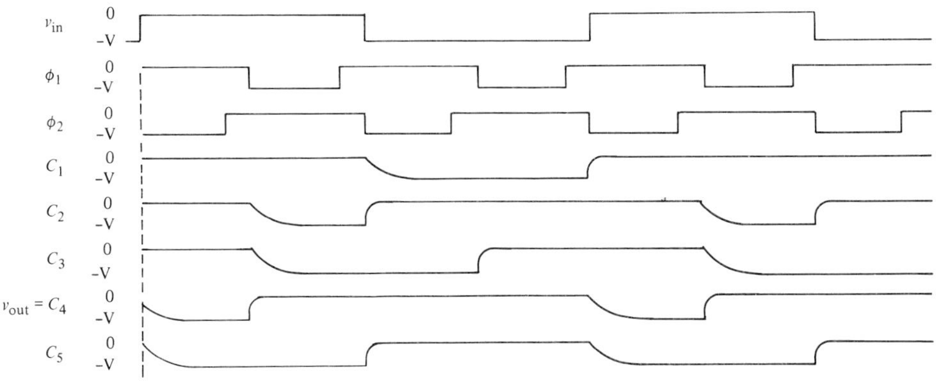

Figure 5–10 2-Phase ratioed timing diagram—1 bit delay (circuit shown in Figure 5–9).

t_f = 16 ns for a specific temperature, clock voltage swing, and power supply voltage, the minimum width of each clock phase should be greater than 90 ns. Assuming also that 200 ns are required for sufficient propagation time through the two inverter stages, the maximum operating speed would be 5 MHz.

If the system operating speed permits longer bit delays than the minimum required by the logic circuits, additional logic may be performed during each clock cycle, as shown in Figure 5-11. The additional inversions shown in Figure 5-11a and 5-11b are redundant; however, Figure 5-11c illustrates how more complex levels of logic may be implemented.

In any implementation, a coupling or series device must be included between logic elements that have load transistors clocked with different clock phases.

2-Phase Ratioed Complex Mechanizations

Typical combinatorial (complex) logic gating structures are shown in Figure 5-12. Either clock phase may be applied to the gate of the load device. In the series input device implementation, all devices must be conducting to drive the output node to ground. In the parallel input device implementation, any input device in the conducting state drives the output to ground. The propagation delay of cascaded logic elements with a common load device clock phase is additive and is summed from the input of one type of coupling gate (i.e., ϕ_1) to the input of the other type (i.e., ϕ_2). The worst-case propagation time determines the minimum clock phase width and, thus, the maximum operating speed.

Memory elements in 2-phase applications are usually integrated into the

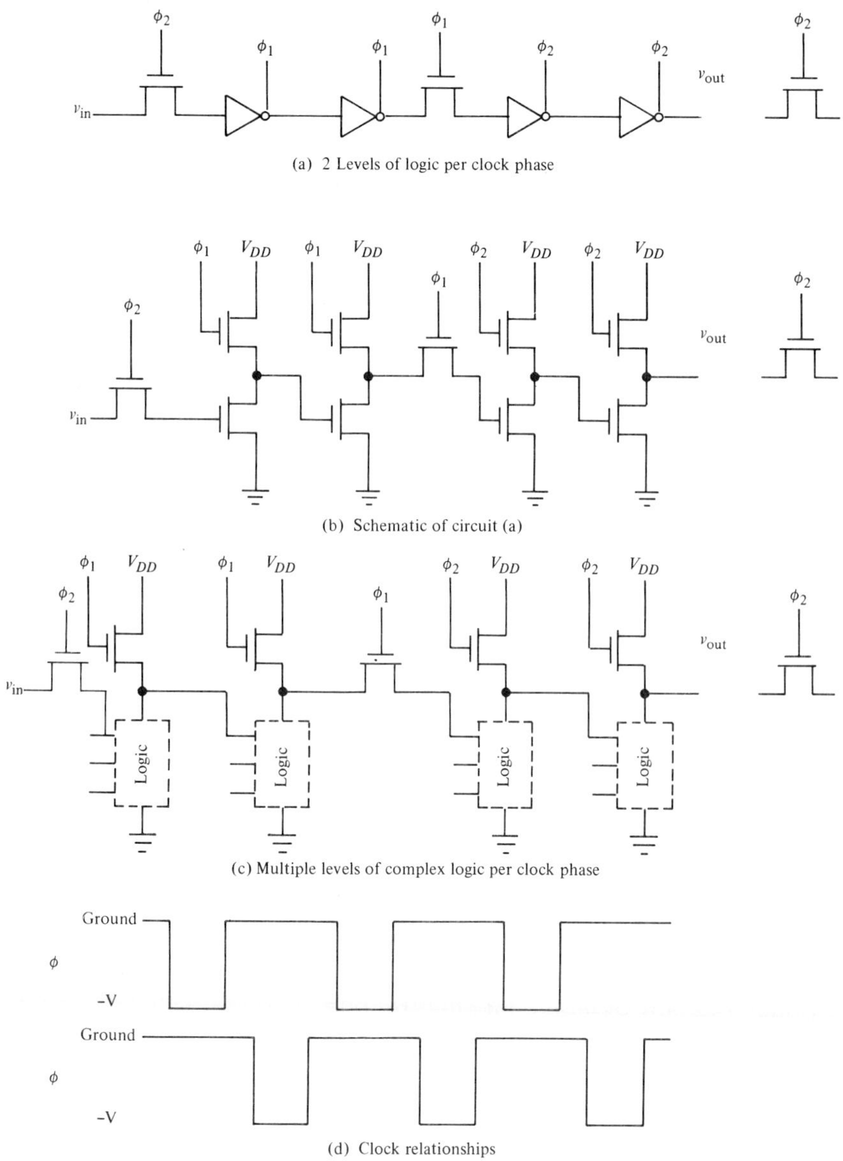

Figure 5–11 2-Phase ratioed logic elements.

basic combinatorial logic matrix, utilizing the inherent delay of MOS gates and series devices. Examples include the *n*-stage shift register and sample and hold register of Figures 5-13 and 5-14, respectively. This technique will be demonstrated with a design example later in this chapter.

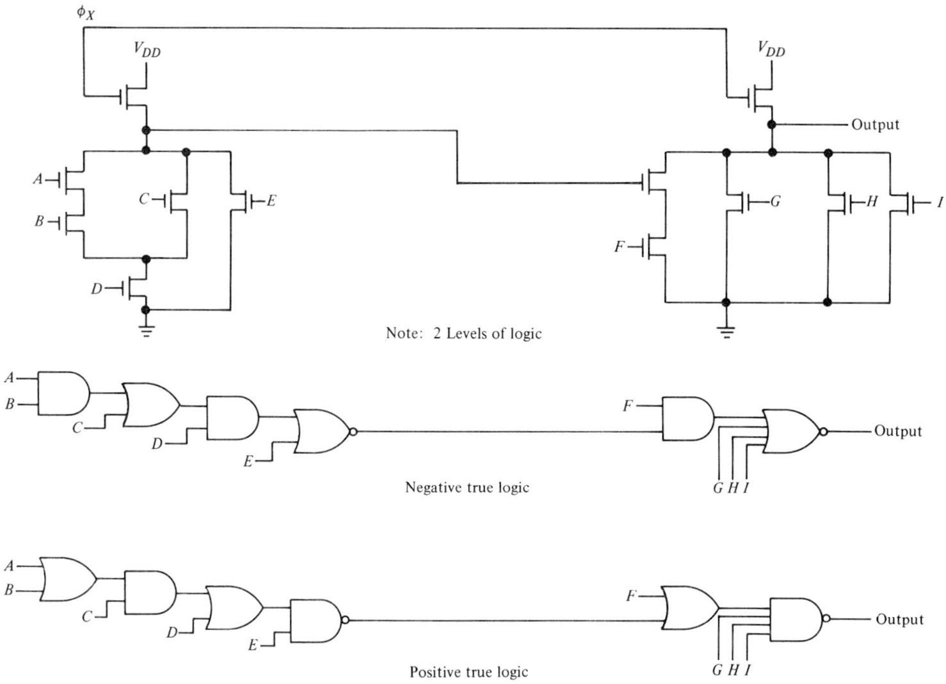

Figure 5–12 2-Phase ratioed complex mechanization.

5.3.2 2-Phase Ratioless Logic

Ratioless circuits do not maintain the resistance ratio relationship between the evaluating (input) and the load transistors as described in the ratioed implementation. Whereas, ratioed circuits use the voltage divider principle to provide the ground output level, ratioless circuits employ a conditional capacitor discharge to provide the ground output level. This enables the use of minimum-size load and input devices, thus permitting more logic within a given area.

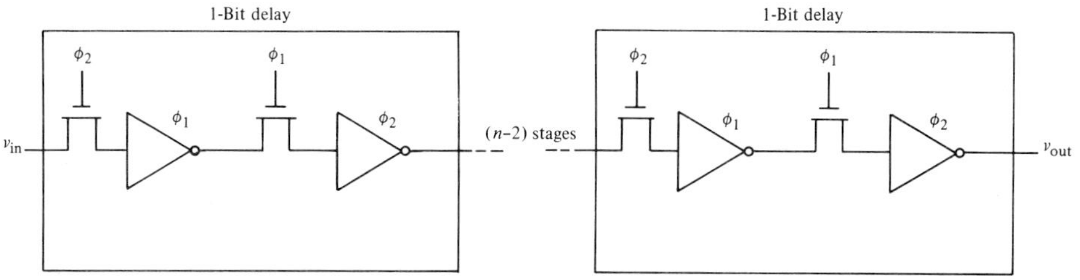

Figure 5–13 *n*-Stage shift register.

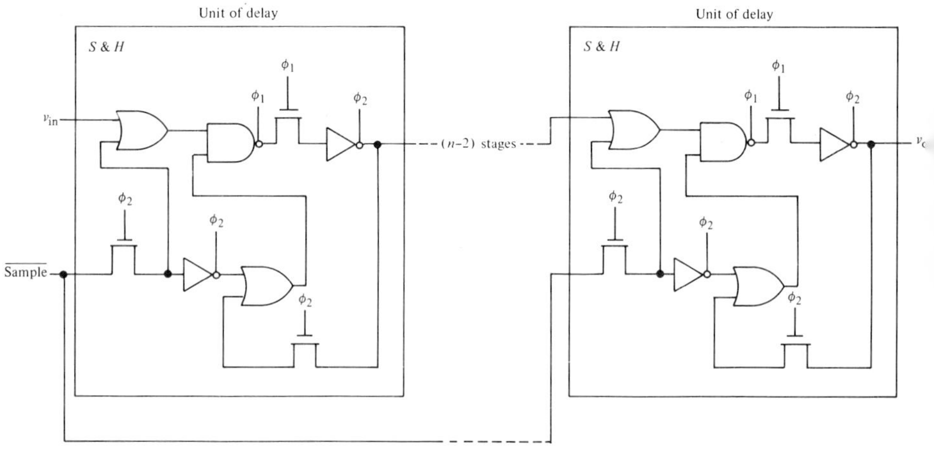

Figure 5-14 *n*-Stage shift register positive true logic (shifting determined by $\overline{\text{sample}}$ duration and rate).

The 2-phase ratioless implementation normally provides only two levels of logic per bit time. There are a number of 2-phase ratioless circuit techniques; several will be discussed here. The capacitors that are shown on circuit schematics represent total nodal capacitance, including interconnect and intrinsic device capacitance.

2-Phase Ratioless Configuration 1

Configuration 1 of 2-phase ratioless logic is illustrated in Figure 5-15. It uses four bus lines: V_{DD}, ϕ_1, ϕ_2, and ground. The major drawback to this circuit is that a dc path exists between V_{DD} and ground while the respective clock phases are negative, thus consuming dc power. In this particular circuit, the clock lines have minimum capacitance loading. Power is supplied primarily through V_{DD} and ground and is directly proportional to the clock widths. The sequence of events which results in a bit of delay is described in the following text and illustrated by the timing diagram of Figure 5-15c.

When the ϕ_1 clock goes negative, the input voltage is coupled onto capacitor C_1 through the series coupling transistor Q_1. Transistor Q_2 is also turned on by the negative ϕ_1 clock. If the input signal voltage v_{in} is negative, C_1 will be negatively charged and this negative voltage on the gate of

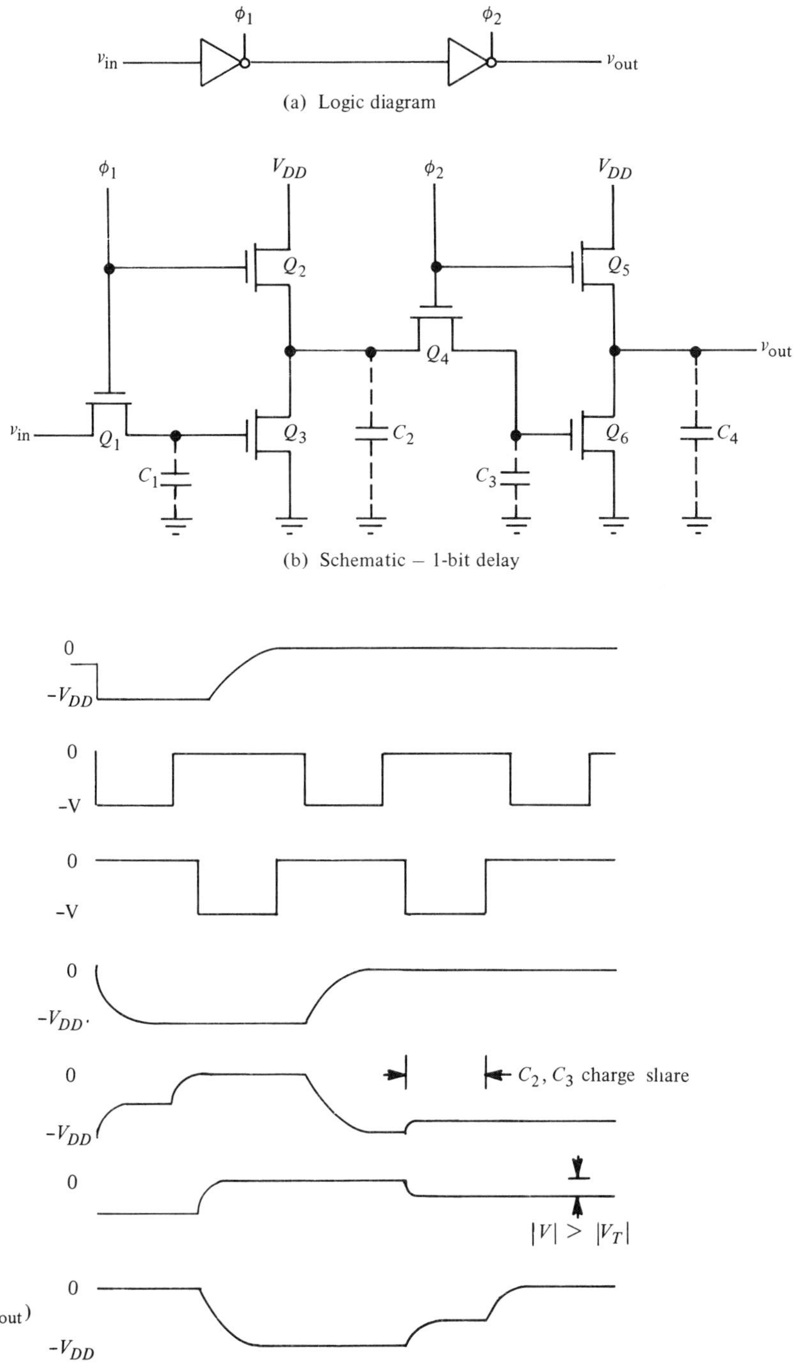

(a) Logic diagram

(b) Schematic – 1-bit delay

(c) Timing waveforms

Figure 5–15 Configuration 1 (2-phase ratioless logic).

Q_3 will turn it on. When Q_2 and Q_3 are both conducting, their common node forms a voltage divider between V_{DD} and ground. C_2 will be charged to the divided voltage level. When the ϕ_1 clock returns to ground, transistors Q_1 and Q_2 are biased off; thus, the charge will remain on C_1, keeping Q_3 on. Since the negative voltage is no longer applied to C_2 through Q_2, C_2 is discharged to ground through Q_3. When clock ϕ_2 goes negative, Q_4 and Q_5 are turned on. Because the stored charge on C_1 keeps Q_3 conducting, a path exists between C_3 and ground through Q_4 and Q_3. C_3 will therefore be discharged to ground potential. With the gate of Q_6 at ground potential, Q_6 is biased off. Since Q_5 is on, C_4 will charge to approximately V_{DD} through Q_5. This charge will remain on C_4 when ϕ_2 returns to ground.

If v_{in} is at ground when ϕ_1 goes negative, C_1 is grounded by Q_1, which is conducting. Therefore, the gate of Q_3 is at ground, biasing Q_3 off. Since Q_2 is on, C_2 is charged to approximately V_{DD}. When ϕ_1 goes to ground, Q_1 turns off, keeping Q_3 off, which causes C_2 to remain negatively charged. When ϕ_2 goes negative, Q_4 and Q_5 are biased on. The charge on C_2 distributes itself such that a negative voltage appears on capacitor C_3. By appropriately selecting the ratio of capacitors C_2 and C_3, the voltage transferred to capacitor C_3 is sufficiently negative to bias transistor Q_6 on. A voltage divider now exists between V_{DD} and ground through Q_5 and Q_6. C_4 is charged to this voltage while ϕ_2 is negative. When ϕ_2 returns to ground, Q_4 and Q_5 are biased off. However, the negative charge stored on C_3 keeps Q_6 conducting. Since the negative voltage is no longer applied to C_4 through Q_5, C_4 is discharged to ground through Q_6.

2-Phase Ratioless Configuration 2

Configuration 2 of 2-phase ratioless logic is illustrated in Figure 5-16. This is similar to the first version in that it also uses four bus lines. Unlike the first version, there is never a dc path from V_{DD} to ground, thus eliminating dc power consumption. However, the capacitance loading on the clock lines is higher since more devices are required. Power consumption is equal to CV^2f, where C is the clock line capacitance, V is the clock voltage, and f is the clock frequency. The sequence of events which results in a bit of delay is described in the following text and illustrated in Figure 5-16c for the cases of v_{in} greater and less than threshold voltage.

When ϕ_1 goes negative, transistors Q_1 and Q_2 are turned on, precharging C_2 negative. The input voltage v_{in}, is transferred to C_1 through Q_1. When v_{in} is negative, C_1 charges negative and turns on Q_4. Next, ϕ_1 returns to ground and ϕ_2 goes negative, turning on Q_3 and Q_5. Since Q_3, Q_4, and Q_5 are all conducting, a path exists to ground from C_3 through Q_4; from C_2 through Q_3 and Q_4; and from C_4 through Q_3, Q_4, and Q_5. Therefore, C_2, C_3, and C_4 all assume ground potential. Q_6 is also turned on as

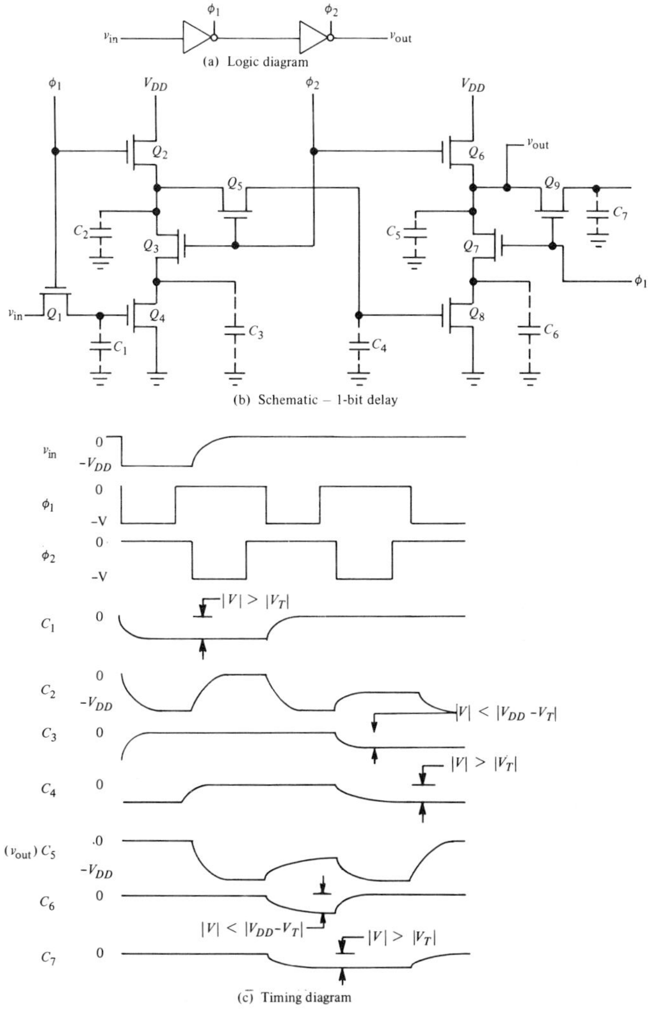

Figure 5–16 Configuration 2 (2-phase ratioless logic).

ϕ_2 goes negative, precharging C_5. As ϕ_2 returns to ground, Q_5 turns off and a ground level remains on the gate of Q_8. When ϕ_1 again goes negative, transistor Q_7 and Q_9 turn on and the charge on C_5 is distributed such that a negative voltage appears on C_6 and C_7. By appropriately selecting the ratios of capacitors C_5, C_6, and C_7, the voltage transferred to C_7 will be sufficiently negative to bias a transistor on.

If the v_{in} is at ground when ϕ_1 goes negative, a ground potential is applied to C_1 and the gate of Q_4. C_2 is precharged negatively through the conducting Q_2. As ϕ_1 goes to ground, an effective ground remains on C_1 biasing Q_4 off. When ϕ_2 goes negative, Q_3 and Q_5 turn on and the charge on C_2 is distributed such that a negative voltage appears on C_3 and C_4. By appropriately selecting the ratios of capacitors C_2, C_3, and C_4, the voltage transferred to C_4 is sufficient to turn Q_8 on. Also, as ϕ_2 goes negative, C_5 is precharged negatively through Q_6. As ϕ_2 returns to ground, Q_5 turns off and the negative potential is retained on C_4, keeping Q_8 turned on. When ϕ_1 goes negative, Q_7 and Q_9 turn on. Since Q_7, Q_8, and Q_9 are all conducting, a path exists to ground from C_6 through Q_8, from C_5 through Q_7 and Q_8, and from C_7 through Q_7, Q_8, and Q_9. Therefore, C_5, C_6, and C_7 are discharged to ground potential.

2-Phase Ratioless Configuration 3

Configuration 3 of 2-phase ratioless logic, illustrated in Figure 5-17, uses only two bus lines, viz., ϕ_1 and ϕ_2 clocks, which is advantageous from the standpoint of layout. Power and ground are supplied by these clock lines. Although a minimum number of devices is required, capacitance loading of the clock lines is increased, since these lines are not only connected to transistor gates but also to transistor sources and drains. Power consumption is equal to CV^2f, where C is the clock loading capacitance, V is the clock voltage, and f is the clock frequency. The sequence of events, which results in a bit of delay, is described in the following text and illustrated in the timing diagram of Figure 5-17c.

When the ϕ_1 clock goes negative, the input voltage is coupled onto C_1 through Q_1. Also, transistor Q_2 is turned on since its gate and drain terminals are returned to the ϕ_1 clock line. This causes C_2 to charge negatively at this time. If v_{in} is negative, C_1 is charged negatively. When ϕ_1 returns to ground, the source of Q_3 is grounded and Q_1 is biased off, thus retaining the negative charge on C_1. Because its source is at ground and its gate is negative, Q_3 is turned on discharging C_2. As ϕ_2 goes negative, Q_4 and Q_5 are turned on. A path to ground through Q_3 and Q_4 is created and C_3 is discharged to ground. C_4 is charged negatively through Q_5. As ϕ_2 returns to ground, Q_4 and Q_5 turn off. Since C_3 was discharged to ground, Q_6 remains off, thereby retaining the negative charge on C_4.

2-Phase Ratioless Complex Mechanizations

The circuits of 2-phase ratioless configuration 3 are used in Figure 5-18 to illustrate the application of complex input device structures to achieve complex logic functions using a single load device. The flexibility of ratio-

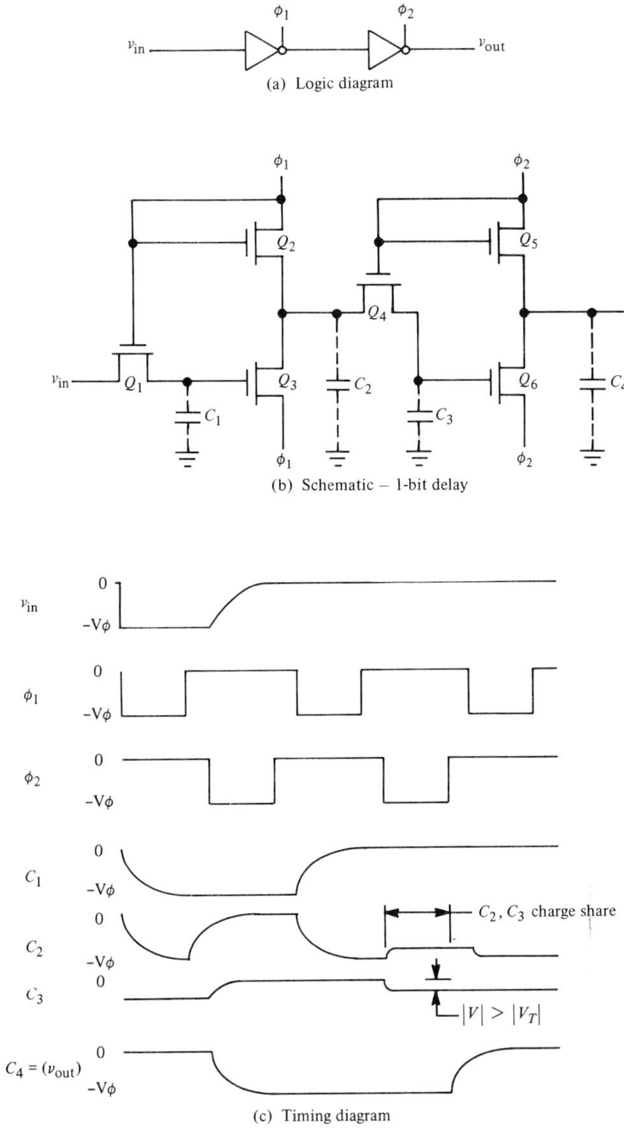

Figure 5–17 Configuration 3 (2-phase ratioless logic).

less logic is enhanced by the use of minimum sized devices for both load and input devices, since there is no need to maintain a resistance ratio. The restriction that must then be considered is desired operating speed. The clock durations must be adequate to charge fully and discharge condi-

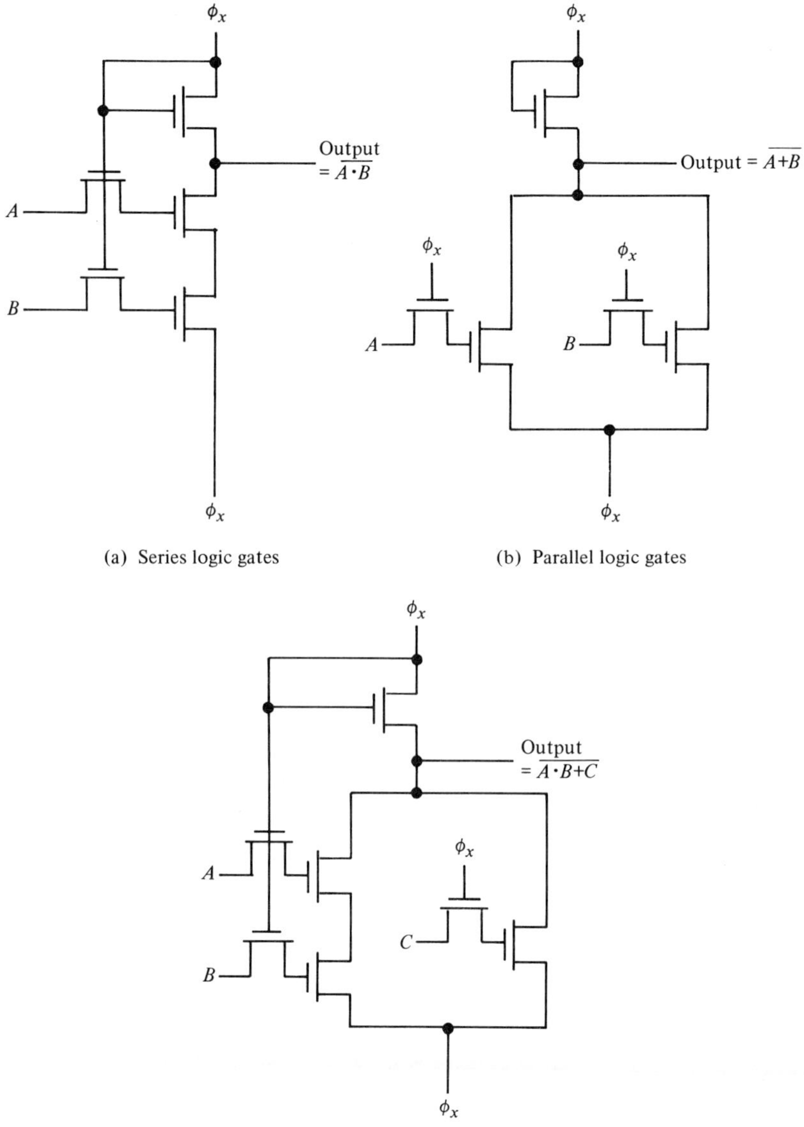

(a) Series logic gates

(b) Parallel logic gates

(c) Series-parallel logic gates

Figure 5–18 2-Phase ratioless complex logic gates (negative true logic).

tionally the load capacitance. The maximum speed is usually limited by the discharge time (t_f) when two or more series input devices are used, as the total input structure resistance is greater than the load device resistance. This is a function of the bias conditions on the load and input devices as well as their geometries, however.

It should be noted that each input to the logic gates must include a coupling transistor or series device. These coupling transistors are clocked with the same phase that clocks the load transistor of the logic element.

2-Phase Ratioless Minimum Delay Gates

Logic complexity may frequently be reduced if more than two levels of logic can be performed per bit time. 2-phase ratioless minimum delay gates enable two additional logic levels per bit time, as illustrated in Figure 5-19.

The operation of a ϕ_1 minimum delay gate (Figure 5-19b) which is driven by a ϕ_2 logic gate and drives a ϕ_1 logic gate, is basically the same as that of a ϕ_2 minimum delay gate, which is driven by a ϕ_1 logic gate and drives a ϕ_2 logic gate (Figure 5-19a). The ϕ_1 minimum delay gate output is precharged and the gate of the input device is discharged during clock ϕ_2. During clock ϕ_1, input data are sampled and the output node is conditionally discharged. Since the series device of both the minimum delay gate and the subsequent logic gate are clocked by the same phase, the evaluated output of the minimum delay gate is immediately transferred to the input of the subsequent logic gate.

The complexity of the minimum delay gate is restricted. On standard 2-phase ratioless logic circuits, the clock widths are determined by the discharge time of the output nodes. When using 2-phase minimum delay gates, each clock phase must be on sufficiently long to discharge the output nodes of the minimum delay elements and conditionally charge or discharge the input node on the following gate. Figure 5-20 illustrates the implementation of an "exclusive OR" logic circuit with only $\frac{1}{2}$ bit time delay enabled by the use of minimum delay gates.

5.3.3 4-Phase Ratioless Logic

4-phase ratioless dynamic logic, an extension of the 2-phase concept, employs four recycling clock signals and may be implemented in a number of configurations. Five of the most common configurations are analyzed, and the relative advantages and disadvantages are discussed in the following text.

In general, the main advantages of 4-phase ratioless logic over 2-phase ratioless logic are:

- It enables four levels of logic evaluation to be performed per bit time. As will be discussed later, up to eight levels per bit time are possible utilizing minimum delay gates.
- The problem of charge-sharing between output and evaluation nodes may be eliminated to allow higher system operating speeds.

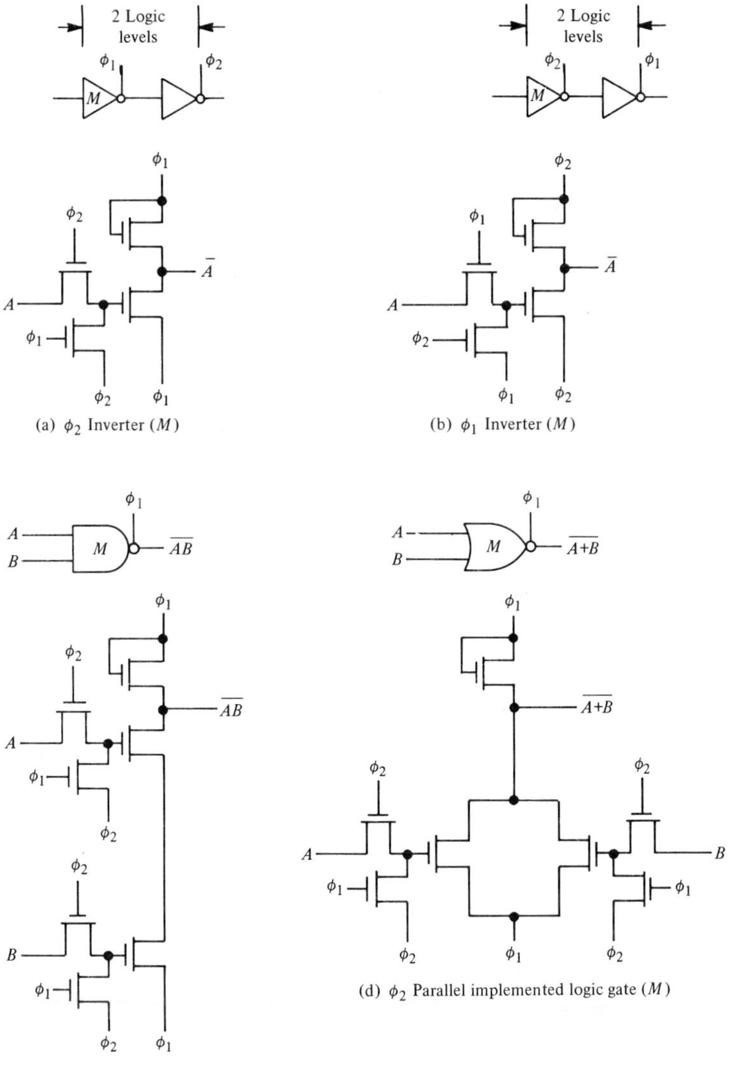

(a) ϕ_2 Inverter (M)

(b) ϕ_1 Inverter (M)

(c) ϕ_2 Series implemented logic gate (M)

(d) ϕ_2 Parallel implemented logic gate (M)

Figure 5–19 2-Phase ratioless minimum delay logic gates (negative true logic).

- Coupling devices are not usually required on inputs to each gate.

Some disadvantages of 4-phase logic are:

- Chip layout restrictions are greater due to the increased number of noise generating clock signals.

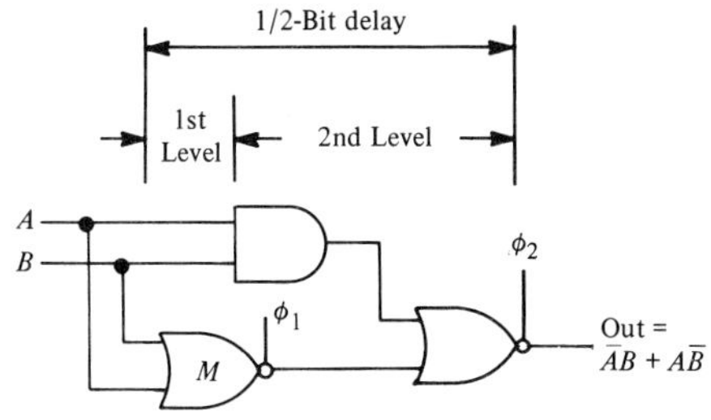

(a) Logic diagram (negative true logic)

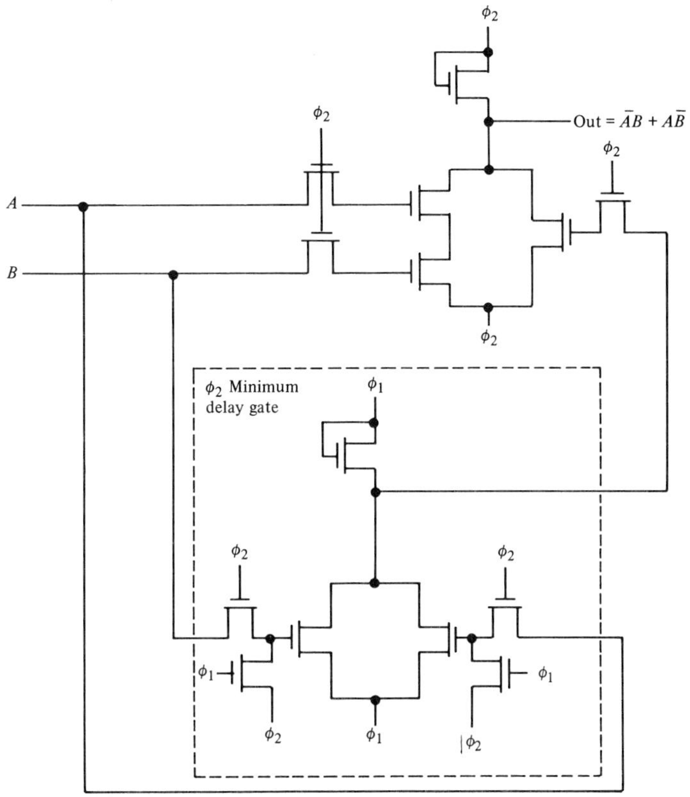

(b) Schematic

Figure 5–20 2-Phase ratioless minimum delay gate application.

- Information is stored on interconnect capacitance between gates, making all signals more susceptible to noise and complicating the layout.
- More clock/power connections to package are required, thus reducing the number of connections available for input/output signals.

Power consumption is equal to CV^2f, where C is the total clock line capacitance loading, V is the clock voltage, and f is the clock frequency. As might be anticipated, in 4-phase logic there are four gate types that correspond to the clock subcycle during which evaluation occurs. Due to the operational characteristics of each 4-phase configuration, the interconnection of gate types becomes somewhat more complex inasmuch as only certain combinations of gate types can be directly interfaced. These combinations are usually categorized as either "major-major" or "major-minor."

Major-major 4-phase is characterized by each gate type having the capability of driving two other gate types. Such gates are termed "major" gates. As shown in Figure 5-21, a type 1 may drive a type 2 or 3, a type 2 may drive a 3 or 4, a type 3 may drive a 4 or 1, and a type 4 may drive a 1 or 2.

Major-minor 4-phase is characterized by two major gates (types 2 and 4) and two minor gates (types 1 and 3). The minor gate is capable of driving only one other gate type. As shown in Figure 5-22, a type 1 may only drive a type 2, a type 2 may drive a 3 or 4, a type 3 may only drive a 4, and a type 4 may drive a 1 or 2.

In the following paragraphs, typical 4-phase configurations are analyzed

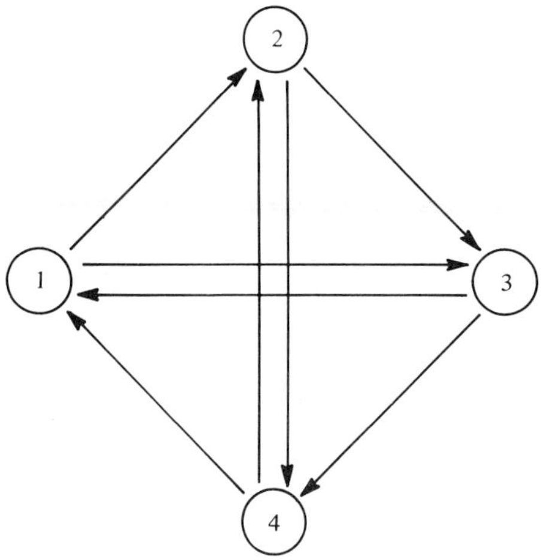

Figure 5–21 Major-major gate interface diagram.

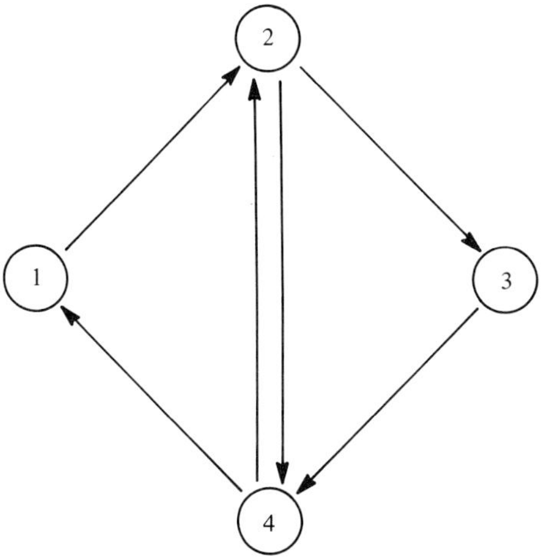

Figure 5–22 Major-minor gate interface diagram.

to clarify the major-major and major-minor concepts. Each gate type is defined as having three sequential modes of operation as follows:

- *Precharge*—the load device is conducting and the output node of the gate is charged to a negative voltage.
- *Evaluate* (or Sample)—the output node capacitance is conditionally discharged as a function of the input voltage(s).
- *Hold*—the gate holds the output node capacitance either charged or discharged as a result of the evaluation.

Given two gates, *A* and *B*, in order for gate *A* to drive gate *B*, whenever gate *B* is in the evaluate mode, gate *A* must be in a hold mode to enable correct evaluation by gate *B*.

4-Phase Ratioless Configuration 1

Configuration 1 of 4-phase ratioless logic falls in the major-major category in which a nonoverlapping 4-phase scheme is used, as shown in Figure 5-23. All gate outputs are precharged during the clock phase n, inputs are evaluated at phase $n + 1$, and outputs are held at phases $n + 2$ and $n + 3$.

From an analysis of the schematic for each gate type and clock relationships (Figure 5-23b–5-53d), a clock program table may be constructed (Figure 5-23c) to illustrate the major-major characteristics. In this table,

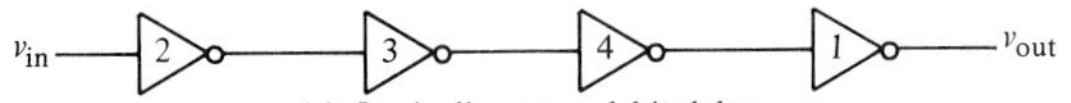

(a) Logic diagram – 1-bit delay

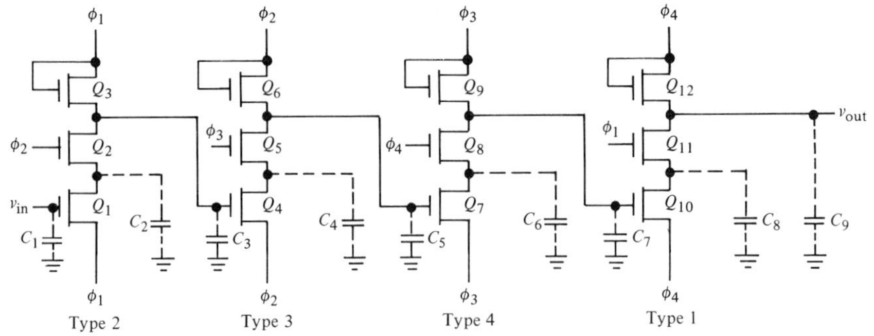

(b) Schematic

Gate type \ Clock ϕ	1	2	3	4
1	*E*	*H*	*H*	*P*
2	*P*	*E*	*H*	*H*
3	*H*	*P*	*E*	*H*
4	*H*	*H*	*P*	*E*

P = Precharge mode
E = Evaluation mode
H = Hold mode

(c) Major-major clock program

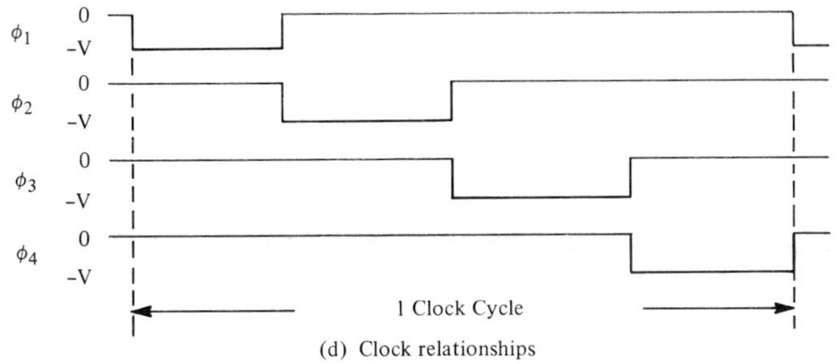

(d) Clock relationships

Figure 5–23 Configuration 1 (4-phase ratioless logic).

each gate type is characterized as having three modes (i.e., precharge, evaluate, and hold) which are tabulated as a function of clock phase. The type 2 gate is in the precharge mode when ϕ_1 is negative on, thus turning Q_3 on and Q_2 off, and allowing C_3 to charge through Q_3. The evaluate mode occurs when ϕ_2 goes negative and Q_3 is turned off, Q_2 is turned on, and C_3 either discharges through Q_2 and Q_1 (if the input voltage causes C_1 to charge negatively) or distributes between C_3 and C_2 (if the input voltage is at ground). The hold mode occurs during the time when ϕ_3 and ϕ_4 are negative on, retaining the charge on C_3 that results from the evaluation mode.

It should be emphasized that any gate in the evaluate mode must be fed by a gate type in a hold mode to prevent erroneous evaluation. A comparison of Figures 5-21 and 5-23 verifies that this condition is met for each gate type.

For example, when a type 3 gate is in the evaluate mode (ϕ_3 clock on), type 1 and 2 gates are in the hold mode, thus enabling correct evaluation. To illustrate a disallowed gate connection, consider the case where a type 4 gate is driving a type 3 gate. This combination cannot operate properly, since the type 4 gate is precharging the input node of the type 3 gate when the type 3 gate is evaluating, causing the type 3 gate output to always be at ground when in the hold mode.

The example circuit of Figure 5-23a and 5-23b illustrates the capability of achieving four levels of logic per bit time using 4-phase logic. This circuit, as shown, is redundant, since either the types 1 and 3 or types 2 and 4 may be removed and still maintain a noninverting bit of delay. The usefulness becomes apparent when complex input gate structures are implemented within each gate type, as is demonstrated later in this chapter.

Assuming a negative v_{in}, the mechanism for a bit of delay is as follows:

During ϕ_1 on: *Type 2 gate* (precharge mode) Q_2 is off, Q_3 is on, and C_3 is charged negatively through Q_3.

During ϕ_2 on: *Type 2 gate* (evaluate mode) Q_1 and Q_2 are on, Q_3 is off, and C_3 is discharged to ground through Q_1 and Q_2.

Type 3 gate (precharge mode) C_5 is charged negatively since Q_4 and Q_5 are off, and Q_6 is biased on. The negative potential on the gate of Q_7 turns Q_7 on and discharges C_6 to ground potential.

During ϕ_3 on: *Type 2 gate* (hold mode) Q_2 and Q_3 are not conducting, thus holding C_3 discharged.

Type 3 gate (evaluate mode) The ground level (C_3 discharged) on the gate of Q_4 holds Q_4 off, Q_6 is off, and Q_5 is turned on. The charge on C_5 is shared between C_4 and C_5 with the capacitance ratios designed such that a negative voltage sufficient to turn on Q_7 remains on C_5.

Type 4 gate (precharge mode) Q_8 is off, Q_9 is on, thus charging C_7 through Q_9. As C_7 charges negatively, Q_{10} is turned on and C_8 is discharged.

During ϕ_4 on: *Type 2 gate* (hold mode) Q_2 and Q_3 remain off, holding C_3 at ground potential.

Type 3 gate (hold mode) Q_5 and Q_6 remain off, holding the negative voltage on C_5.

Type 4 gate (evaluate mode) The negative voltage on C_5 turns Q_7 on and ϕ_4 turns Q_8 on, thus discharging C_7 to ground potential.

Type 1 gate (precharge mode) Q_{10} and Q_{11} are held off, Q_{12} is biased on, charging C_9 negatively.

During ϕ_1 on: *Type 4 gate* (hold mode) With Q_8 and Q_9 off, C_7 is held at ground potential.

Type 1 gate (evaluate mode) Q_{12} is off, Q_{11} is turned on, C_7 holds the gate of Q_{10} at ground, thus holding it off.

The charge on C_9 is shared between C_8 and C_9 which are designed with a ratio such that v_{out} remains sufficiently negative to turn on a subsequent device.

4-Phase Ratioless Configuration 2

The example of configuration 2 of the 4-phase ratioless logic shown in Figure 5-24 is very similar to the 4-phase configuration 1 circuit.

The category is major-major and the clock relationships are identical to configuration 1. The difference is the addition of a device which precharges the evaluation device nodes (C_2, C_4, C_6, and C_8) in addition to the output node, thus eliminating the charge sharing characteristic of the previous example. This increases the noise immunity of the circuits and simplifies the circuit design in that the ratio of the output to the evaluating node capacitances need not be considered.

Since clock lines provide all power and ground connections to the devices, extreme care must be exercised during layout to minimize the noise which may be coupled onto signal lines and appear as logic level changes.

4-Phase Ratioless Configuration 3

Figure 5-25 shows another configuration in the major-major category. The clocking scheme utilizes overlapping clocks. The mechanization is unique in that the gate output and output node are separated by a series coupling device.

This configuration also eliminates the charge sharing characteristic displayed by the 4-phase configuration 1 example and uses one less device per gate than configuration 2.

If v_{in}, shown in Figure 5-25b, is assumed to be a negative voltage, the

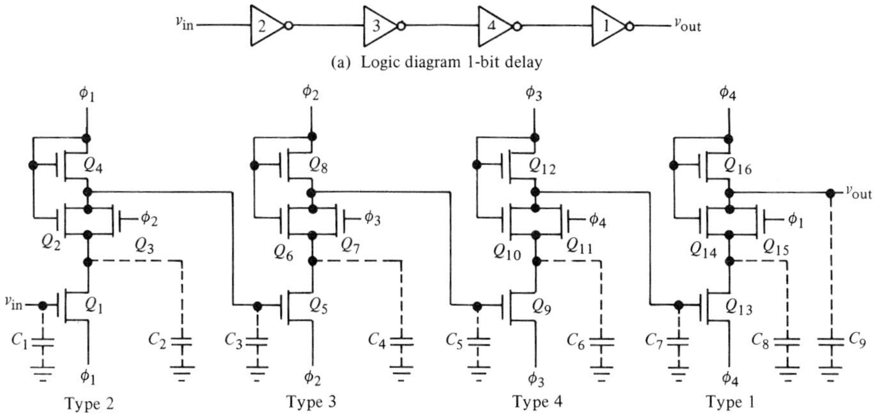

(a) Logic diagram 1-bit delay

(b) Schematic

Gate type \ Clock ϕ	1	2	3	4
1	*E*	*H*	*H*	*P*
2	*P*	*E*	*H*	*H*
3	*H*	*P*	*E*	*H*
4	*H*	*H*	*P*	*E*

P = Gate in precharge mode
E = Gate in evaluate mode
H = Gate in hold mode

(c) Major-major clock program

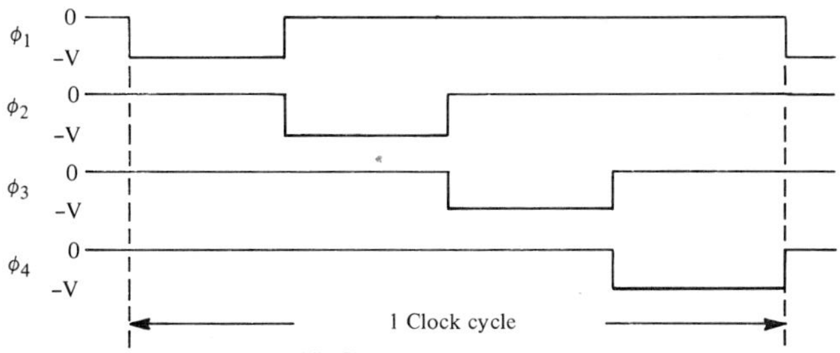

(d) Clock relationships

Figure 5–24 Configuration 2 (4-phase ratioless logic).

sequence that produces a bit of delay is given below. (Note: Each clock period is divided into four subcycles as shown in Figure 5-25d).

Subcycle 2: Type 3 gate (precharge mode) Q_2 and Q_3 are turned on, precharging C_2 and C_3 to a negative voltage.

Subcycle 3: Type 3 gate (evaluate mode) The negative voltage on C_1

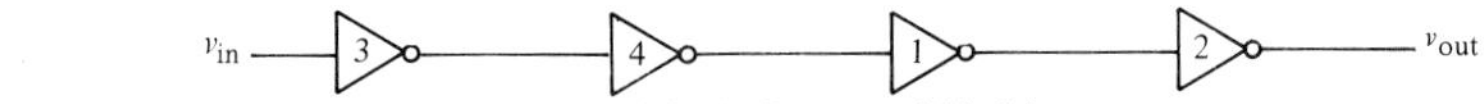

(a) Logic diagram – 1-bit delay

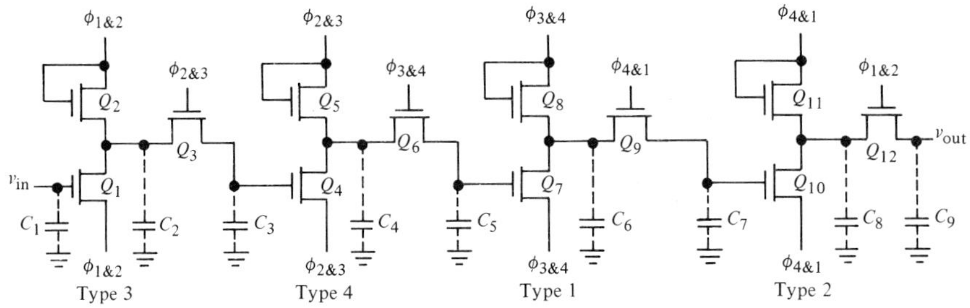

(b) Schematic

Gate type \ Clock subcycle	1	2	3	4
1	*E*	*H*	*H*	*P*
2	*P*	*E*	*H*	*H*
3	*H*	*P*	*E*	*H*
4	*H*	*H*	*P*	*E*

P = Gate in precharge mode
E = Gate in evaluate mode
H = Gate in hold mode

(c) Major-major clock program

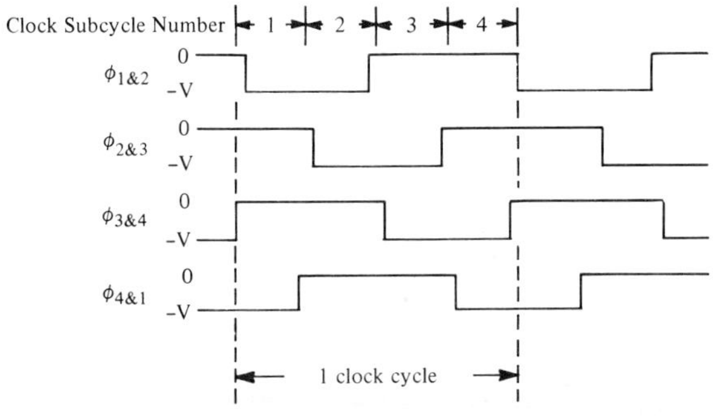

(d) Clock relationships

Figure 5–25 Configuration 3 (4-phase ratioless logic).

turns on Q_1, Q_2 is biased off, and Q_3 remains on, thus discharging C_2 and C_3 through Q_1 and presenting a ground level as input to Q_4 of the type 4 gate.

Type 4 gate (precharge mode) Q_4 is off, Q_5 and Q_6 are biased on, thus precharging C_4 and C_5 negatively.

Subcycle 4: Type 3 gate (hold mode) Q_1, Q_2, and Q_3 are off and C_3 is discharged to ground.

Type 4 gate (evaluate mode) As $\phi_{2\&3}$ returns to ground, the ground on C_3 holds Q_4 off, Q_5 is turned off, Q_6 remains on, and C_4, C_5 remain negatively charged.

Type 1 gate (precharge mode) Q_7 is held off by $\phi_{3\&4}$, Q_8 and Q_9 are turned on, and C_6 and C_7 are precharged to a negative voltage.

Subcycle 1: Type 4 gate (hold mode) As $\phi_{3\&4}$ returns to ground, Q_6 is turned off, holding the negative charge on C_5.

Type 1 gate (evaluate mode) With $\phi_{3\&4}$ returned to ground, the negative voltage on C_5 keeps Q_7 on, Q_8 is off, and Q_9 is on; therefore, C_6 and C_7 discharge to ground through Q_7 and Q_9.

Subcycle 2: Type 1 gate (hold mode) As $\phi_{4\&1}$ returns to ground, Q_9 is turned off, and C_7 remains discharged.

Type 2 gate (evaluate mode) With $\phi_{4\&1}$ at ground, the ground on C_7 holds Q_{10} off, Q_{11} is off, Q_{12} remains on, and the negative charge on C_8 and C_9 is retained, thus making v_{out} negative.

4-Phase Ratioless Configuration 4

The circuit shown in Figure 5-26 is an example of 4-phase major-minor gates. This configuration utilizes two overlapping and two nonoverlapping clock signals. The minor gates do not have evaluating devices and thus the outputs are only valid during the subcycle (hold mode) following the evaluation subcycle.

Assuming a negative voltage input (v_{in}) at C_1, the mechanism for the bit of delay is as follows:

Subcycle 1: Type 2 gate (precharge mode) Q_2 and Q_3 are conducting, thus precharging C_2 and C_3 negatively.

Type 3 gate (precharge mode) Q_5 is on, and C_4 is precharged negatively.

Subcycle 2: Type 2 gate (evaluate mode) As ϕ_1 returns to ground, Q_1 is turned on, Q_2 remains on, thus discharging C_2 and C_3 to ground.

Type 3 gate (precharge mode) Conditions remain unchanged from subcycle 1.

Subcycle 3: Type 2 gate (hold mode) As $\phi_{1\&2}$ returns to ground, Q_2 is biased off and C_3 is held discharged.

Type 3 gate (evaluate mode) Since C_3 is discharged to ground, Q_4 is off. Q_5 is off and C_4 remains negatively charged.

Type 4 gate (precharge mode) Q_7 and Q_8 are turned on, and C_5 and C_6 are precharged negatively.

Type 1 gate (precharge mode) Q_{10} is turned on, thus precharging C_7.

Subcycle 4: Type 2 gate (hold mode) Q_2 and Q_3 remain off, holding C_3 discharged.

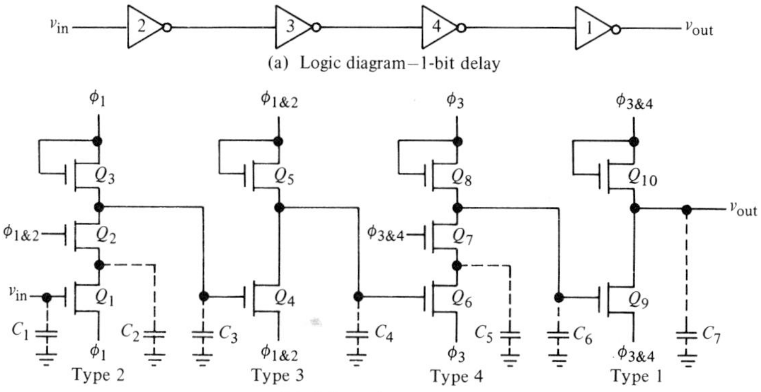

(a) Logic diagram—1-bit delay

(b) Schematic

Gate type \ Clock subcycle	1	2	3	4
1	*E*	*H*	*P*	*P*
2	*P*	*E*	*H*	*H*
3	*P*	*P*	*E*	*H*
4	*H*	*H*	*P*	*E*

P = Precharge mode
E = Evaluate mode
H = Hold mode

(c) Major-minor clock program

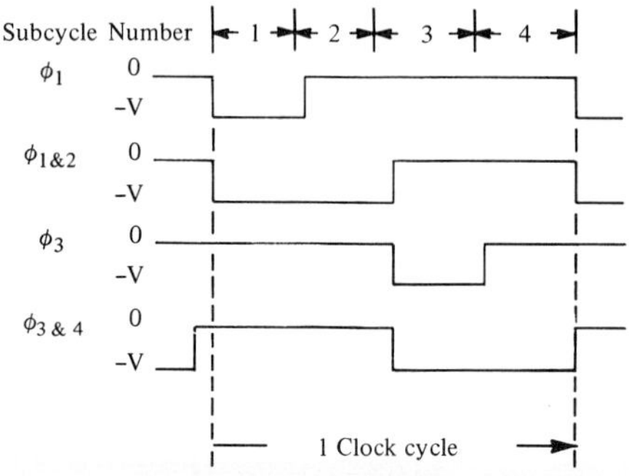

(d) Clock relationship

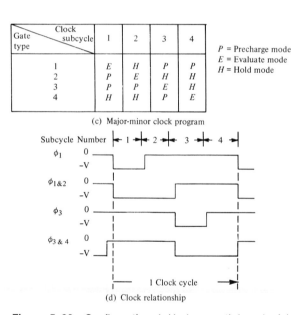

Figure 5–26 Configuration 4 (4-phase ratioless logic).

Type 3 gate (hold mode) Since C_3 is discharged, Q_4 remains off, Q_5 remains off, and C_4 is held at a negative potential.

Type 4 gate (evaluate mode) As ϕ_3 returns to ground, the negative voltage on C_4 turns Q_6 on, Q_7 is on, Q_8 is off, thus discharging C_5 and C_6 to ground.

Type 1 gate (percharge mode) Conditions remain unchanged from subcycle 3.

Subcycle 1: Type 4 gate (hold mode) Since Q_7 and Q_8 are not conducting, C_6 is held at ground potential.
Type 1 gate (evaluate mode) As $\phi_{3\&4}$ returns to ground, the ground level on C_6 holds Q_9 off, thus retaining the negative charge on C_7.

Assuming a ground potential input (v_{in}) at C_1, the bit delay is accomplished as follows:

Subcycle 1: Type 2 gate (precharge mode) Q_1 is biased off Q_2 and Q_3 are conducting, thus precharging C_2 and C_3
Type 3 gate (precharge mode) Since $\phi_{1\&2}$ is negative, Q_5 is on, and C_4 is precharged negatively.
Subcycle 2: Type 2 gate (evaluate mode) As ϕ_1 returns to ground, Q_1 is held off by the ground potential on C_1. Q_3 is off and Q_2 is on, thus retaining the negative charges on C_2 and C_3
Type 3 gate (precharge mode) Q_4 is off, Q_5 is conducting and precharges C_4 negatively.
Subcycle 3: Type 2 gate (hold mode) As $\phi_{1\&2}$ goes to ground, Q_2 is turned off. With Q_2 and Q_3 not conducting, the negative charge is held on C_3.
Type 3 gate (evaluate mode) Since $\phi_{1\&2}$ is now at ground, Q_5 is biased off. The negative voltage on C_3 turns Q_4 on, discharging C_4.
Type 4 gate (precharge mode) As ϕ_3 and $\phi_{3\&4}$ go negative, Q_7 and Q_8 turn on. Capacitors C_5 and C_6 are charged negatively.
Type 1 gate (precharge mode) Q_{10} is turned on. Thus C_7 is precharged negatively.
Subcycle 4: Type 2 gate (hold mode) Q_2 and Q_3 are off, thus holding C_3 fully charged.
Type 3 gate (hold mode) Since C_3 is storing a negative voltage, Q_4 remains on, thus holding C_4 discharged.
Type 4 gate (evaluate mode) Q_6 remains off since C_4 is at ground. Q_7 remains on and C_5 and C_6 remain negatively charged.
Type 1 gate (precharge mode) Conditions remain unchanged from subcycle 3.
Subcycle 1: Type 4 gate (hold mode) Q_7 and Q_8 are not conducting, holding C_6 negatively charged.
Type 1 gate (evaluate mode) When the source of Q_9 is returned to ground, Q_9 turns on due to negative charge on C_6. Q_{10} turns off. C_7 is discharged to ground through Q_9, and v_{out} represents the ground on C_1 delayed by one bit time.

The major-minor clock program of Figure 5-26c is verified by the preceding analysis. A comparison of Figure 5–26c to the major-minor gate interface diagram of Figure 5–22 will demonstrate the major-minor concept.

4-Phase Ratioless Configuration 5

This major-minor configuration (Figure 5-27) is very similar to 4-phase configuration 4 (Figure 5-26). The major gates (types 2 and 4) are identical as is the clocking scheme. The minor gates have an evaluating device

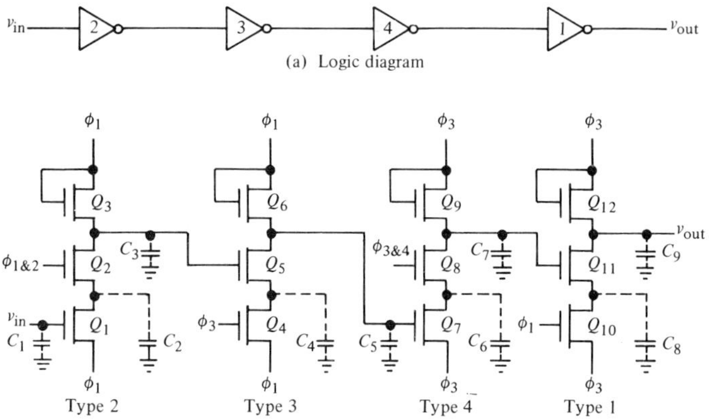

(b) Schematic

Gate type \ Clock subcycle	1	2	3	4
1	*E*	*H*	*P*	*P*
2	*P*	*E*	*H*	*H*
3	*P*	*P*	*E*	*H*
4	*H*	*H*	*P*	*E*

P = Precharge mode
E = Evaluate mode
H = Hold mode

(c) Major-minor clock program

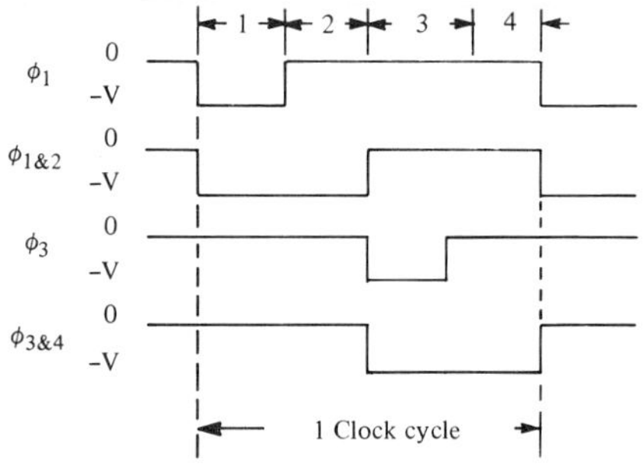

(d) Clock relationships

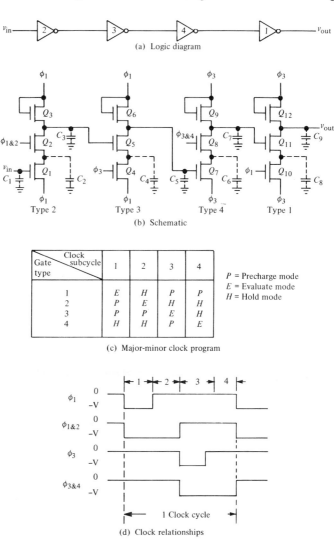

Figure 5–27 Configuration 5 (4-phase ratioless logic).

placed below the input device. This enables the use of only two clock phases to supply power, source, and sink, whereas the other two clock phases control the gates of evaluating devices. Due to the resulting lower power requirements on two clock phases ($\phi_{1\&2}$, $\phi_{3\&4}$), in some cases it is beneficial to bring only the ϕ_1 and ϕ_3 clocks onto the chip and generate the $\phi_{1\&2}$ and $\phi_{3\&4}$ clocks internally, although the internal clock generators consume a large amount of chip area.

Assuming a negative voltage level v_{in} on C_1, the mechanism for a bit of delay is as follows:

Subcycle 1: Type 2 gate (precharge mode) The negative ϕ_1 clock holds Q_3 on. $\phi_{1\&2}$ turns Q_2 on, thus charging C_2 and C_3 negatively.
Type 3 gate (precharge mode) ϕ_1 turns Q_6 on. As C_3 is precharged negatively, Q_5 will either turn on (if C_4 is discharged) or remain off (if C_4 is charged). In either case, C_4 will be charged negatively by the end of ϕ_1. C_5 is precharged through Q_6

Subcycle 2: Type 2 gate (evaluate mode) As ϕ_1 returns to ground, Q_3 is off, Q_2 is on, and the negative voltage on C_1 turns Q_1 on, discharging C_2 and C_3.
Type 3 gate (precharge mode) Q_4 and Q_6 are off. Q_5 is off; thus C_4 and C_5 remain charged.

Subcycle 3: Type 2 gate (hold mode) Q_2 and Q_3 are off, and C_3 is held discharged.
Type 3 gate (evaluate mode) Q_6 is off, ϕ_3 turns Q_4 on, discharging C_4. Q_5 is held off by ground potential on C_3, thus C_5 remains charged.
Type 4 gate (precharge mode) With ϕ_3 negative, Q_9 is on. When $\phi_{3\&4}$ goes negative, Q_8 can either turn on (if C_6 is at ground potential) or remain off (if C_6 is negatively charged). In either case, C_6 and C_7 are charged negatively at the end of ϕ_3.
Type 1 gate (precharge mode) ϕ_3 turns Q_{12} on. Q_{11} is conditionally turned on (if C_8 is discharged) by the negative voltage on C_7. C_8 and C_9 are then precharged.

Subcycle 4: Type 2 gate (hold mode) Q_2 and Q_3 are off, and C_3 is held discharged.
Type 3 gate (hold mode) Q_5 is held off by the ground level from C_3, and C_5 is held charged.
Type 4 gate (evaluate) ϕ_3 goes to ground turning Q_9 off, $\phi_{3\&4}$ turns Q_8 on, the negative charge on C_5 turns Q_7 on, and C_6 and C_7 are discharged through Q_7 and Q_8.
Type 1 gate (precharge mode) The precharge on C_8 and C_9 is retained as Q_{11} is held off by the ground on C_7.

Subcycle 1: Type 4 gate (hold mode) As $\phi_{3\&4}$ returns to ground, Q_8 turns off, the ground on ϕ_3 turns Q_9 off, and C_7 is held discharged.

Type 1 gate (evaluate mode) ϕ_1 turns Q_{10} on, and discharges C_8. Ground voltage on C_7 holds Q_{11} off; thus C_9 remains charged and represents the negative bit of information which was on C_1 one bit time earlier.

4-Phase Ratioless Complex Mechanizations

The 4-phase ratioless complex mechanizations may be combinations of series and parallel connected input devices that replace the single input device in the simple inverter. The series structure may not normally be stacked as high as the 2-phase ratioless, since only a quarter bit time is available for discharging the output node.

Theoretically, it is possible to build series/parallel stacks of devices to any logic complexity. Limitations are found in actual circuits, however, due to speed requirements and device size limitations.

4-Phase Ratioless Minimum Delay Elements

A minimum delay gate is possible within the 4-phase ratioless configuration 3 circuit implementation. Figure 5-28 shows two minimum delay mechanizations. The minimum delay gate must be simple implementation, since its output is evaluated during the off time between the precharge clock phase and the clocking phase to the evaluating device. The use of the minimum delay gate enables eight possible logic levels per bit time.

5.4 SYNCHRONOUS SEQUENTIAL MACHINES

The unique characteristics of dynamic MOS LSI circuits are used to best advantage when applied to synchronous sequential machine designs. A synchronous sequential machine is defined as a logic network with outputs that are a function of the present inputs and the prior sequence of inputs. Any such machine may be reduced to the classic form of Figure 5-29, consisting of static (or dc) combinatorial logic networks (without feedback) followed by clocked delay elements whose outputs feedback into the logic network. The machine may have any number of inputs, delay elements, or outputs depending upon the application. However, all delay elements must change state simultaneously.

The input to each delay element (P_1, $P_2 \ldots P_n$) and the machine outputs (O_1, $O_2, \ldots, O_n$) are dc logic functions of the control inputs (I_1, $I_2, \ldots, I_n$) and the present outputs of the delay elements (Q_1, $Q_2, \ldots, Q_n$). The present state of the machine is defined as the present state of the delay element outputs (Q_1, $Q_2, \ldots, Q_n$). The next state of the machine is defined as the present state inputs to the delay elements (P_1 $P_2, \ldots, P_n$). When the delay elements are clocked, their present inputs become their

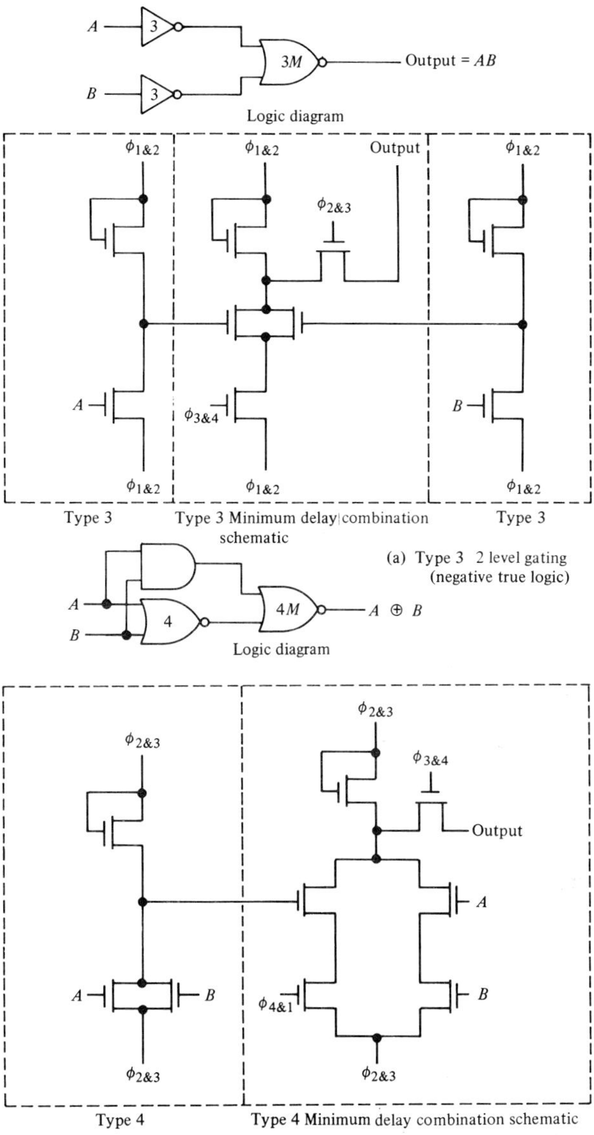

(b) Type 4 2 level gating (negative true logic)

Figure 5–28 Two mechanizations of 4-phase ratioless minimum delay gates.

$$\text{Inputs} = I_n$$
$$\text{Outputs} = O_n = f\,[\,I_1, I_2, \ldots, I_n), (Q_1, Q_2, \ldots, Q_n)]$$

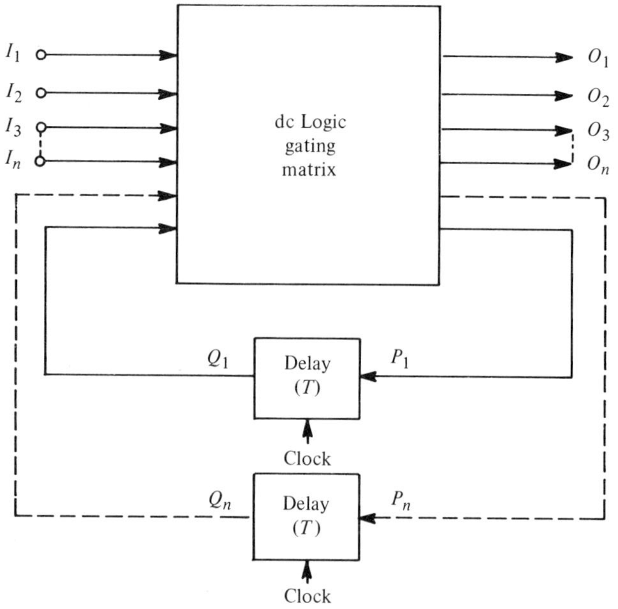

Figure 5–29 General synchronous sequential machine block diagram.

new outputs. The machine can change state only at clock time, cannot have any race problems, and has no transient unstable states.

Any sequential machine that does not meet the specifications for a synchronous sequential machine is referred to as an asynchronous machine. Although a given sequential machine may typically be mechanized with fewer static logic elements when using asynchronous rather than synchronous design techniques, the problems with asynchronous MOS designs become very complex. The primary problem is the difficulty of predicting gate propagation delay times because of their dependency on chip layout. This factor creates race and coincidence timing conditions which necessitate extremely careful design.

Synchronous logic design has numerous advantages when MOS LSI technology is used.

- Leads directly to the use of dynamic logic elements, which enables more functions and lower power dissipation per chip than static MOS logic elements.

- Reduces race and coincidence conditions, which greatly simplifies design.
- Enables the use of state flow charts to describe system operation which aids in preliminary design.

State Diagrams

The sequential operation of a machine of the form shown in Figure 5-29 may be completely described by a state diagram as shown in Figure 5-30. This diagram shows every state that the machine can be in (in this case, two delay elements allow four states). The states are numbered arbitrarily S_1 through S_4. The control inputs that cause state changes are shown beside each state change path. States always change at clock time unless a control input causes a state to hold, as shown by the small loops from a state back into itself. Output terms are difficult to show on the diagram and are tabulated below it in logic equation form, in terms of the states and control inputs. State diagrams of this type are difficult to interpret

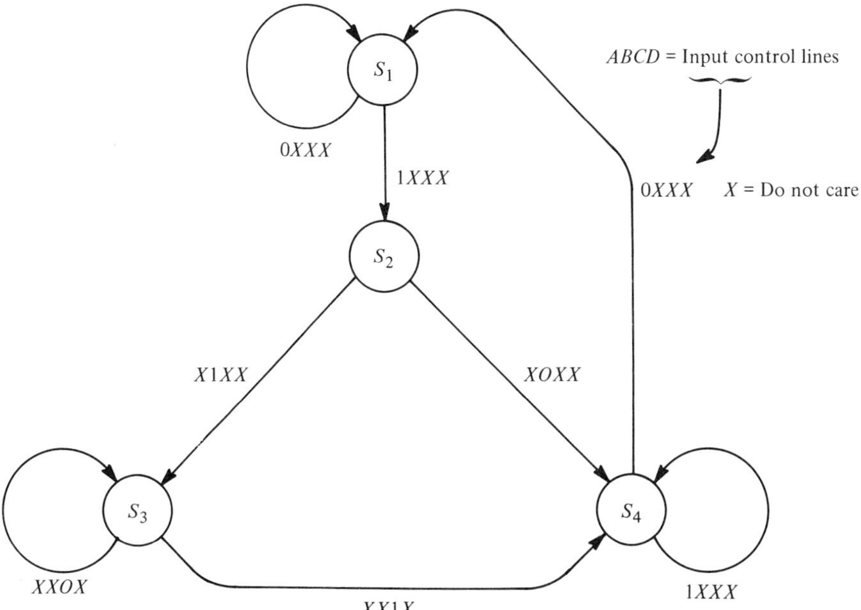

$O_1 = S_2$
$O_2 = D \cdot S_3$
$O_3 = \overline{D} \cdot S_3$

Figure 5–30 Sequential machine state diagram.

because logic decisions for state control are not obvious and large output terms must be tabulated separately.

State Flow Charts

The sequential machine operation described by the state diagram of Figure 5-30 may also be described with a state flow chart as shown in Figure 5-31. Although the chart contains the same information, the format

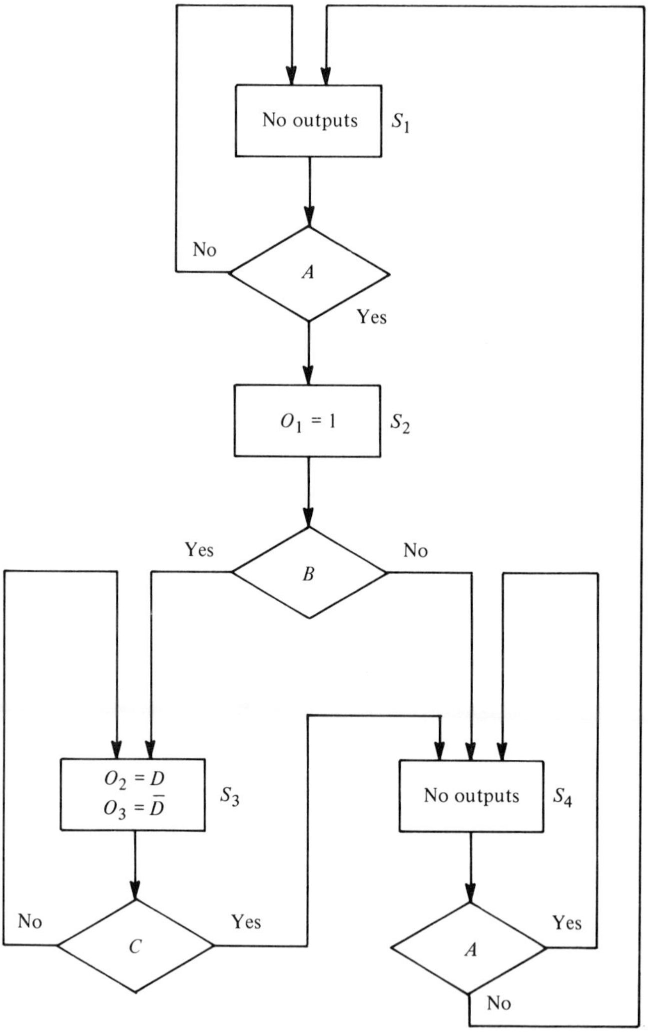

Figure 5–31 State flow chart.

is easier to interpret. This technique is analogous to computer software flow charts, and thus provides a common interface between hardware designers and programmers, which is extremely useful if the machine is to be either computer controlled or programmable.

Two symbols are required, the state (or action) block and the diamond-shaped decision block. Uniformity and clarity are achieved by keeping the flow from top to bottom, with the exception of feedback loops, and by locating the initialized state at the top. The output equations are noted within the state block by which they are controlled. Unlike some flow charts, the state flow chart must be continuous in that all logic paths must lead to a defined state.

For reasons that will become evident when system organization and partitioning are discussed in Chapter 6, if portions of the flow chart are used repeatedly or are basic functions that may be considered subroutines that can be time-shared with other machines, the flow chart may be divided as shown in Figures 5-32 and 5-33. In this case, the subroutine execution is controlled by start-finish signals, which may be bussed to additional machines sharing the subroutine chip.

Another example of the division of random control logic flow charts into separate sections is illustrated in Figure 5-34. Each control section can then be designed separately from the individual flow charts.

5.4.1 Design Procedure Example

In this example, a machine is used that eliminates switch bounce noise from a calculator keyboard signal and produces a synchronized output. A word description of a sequential machine is translated into a state flow chart, and a design procedure for preliminary logic synthesis is defined to provide a mechanization using MOS dynamic logic. An analysis of the completed machine is then made by reversing the synthesis procedure.

Machine Description

A network is supplied with clock pulses, timing pulses (T), and an asynchronous control signal A. Each of the timing pulses (T) are one clock period wide and occur regularly, many clock periods apart. The control signal (A) may bounce for an unknown length of time when changing from a logic 0 to a logic 1, or vice versa. It is known, however, that the control signal will not bounce again after existing at either zero or one continuously without bounce for a time equal to the period of T.

It is required that the network produce one output pulse equal to the period of timing pulse (T), only after A has been true (logic 1) without

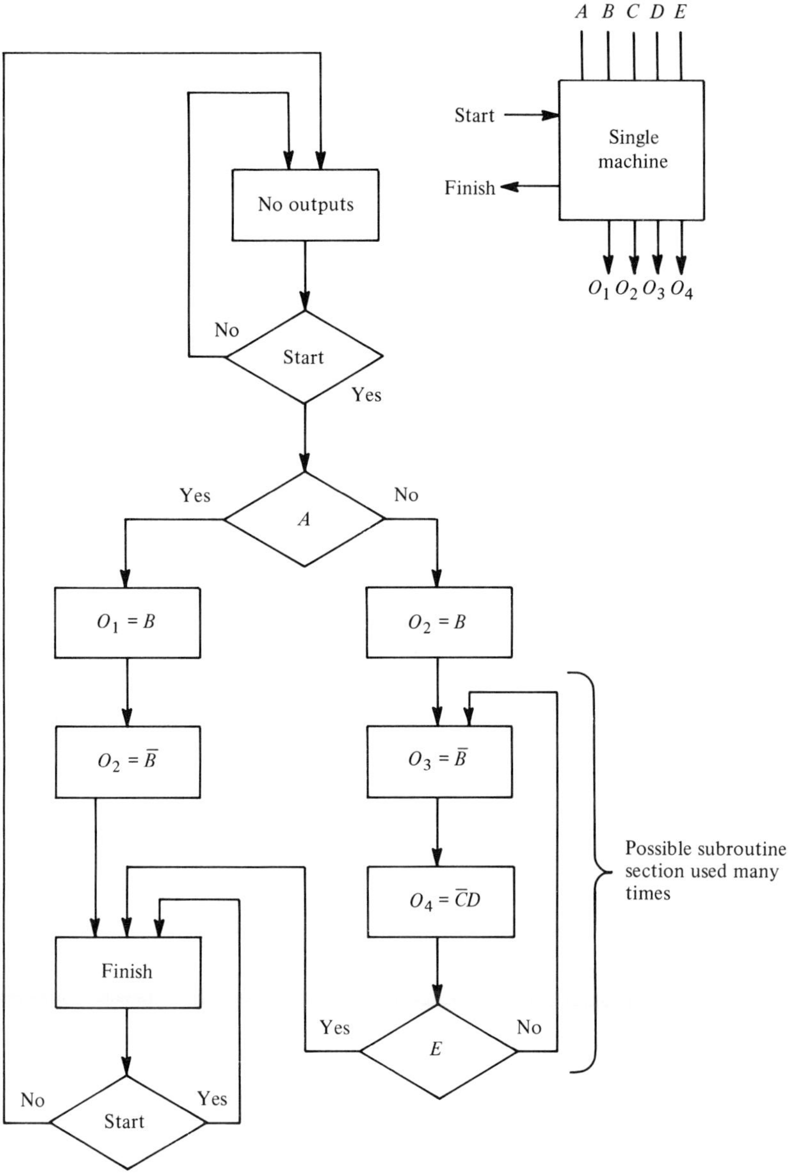

Figure 5–32 Large flow chart with subroutine section.

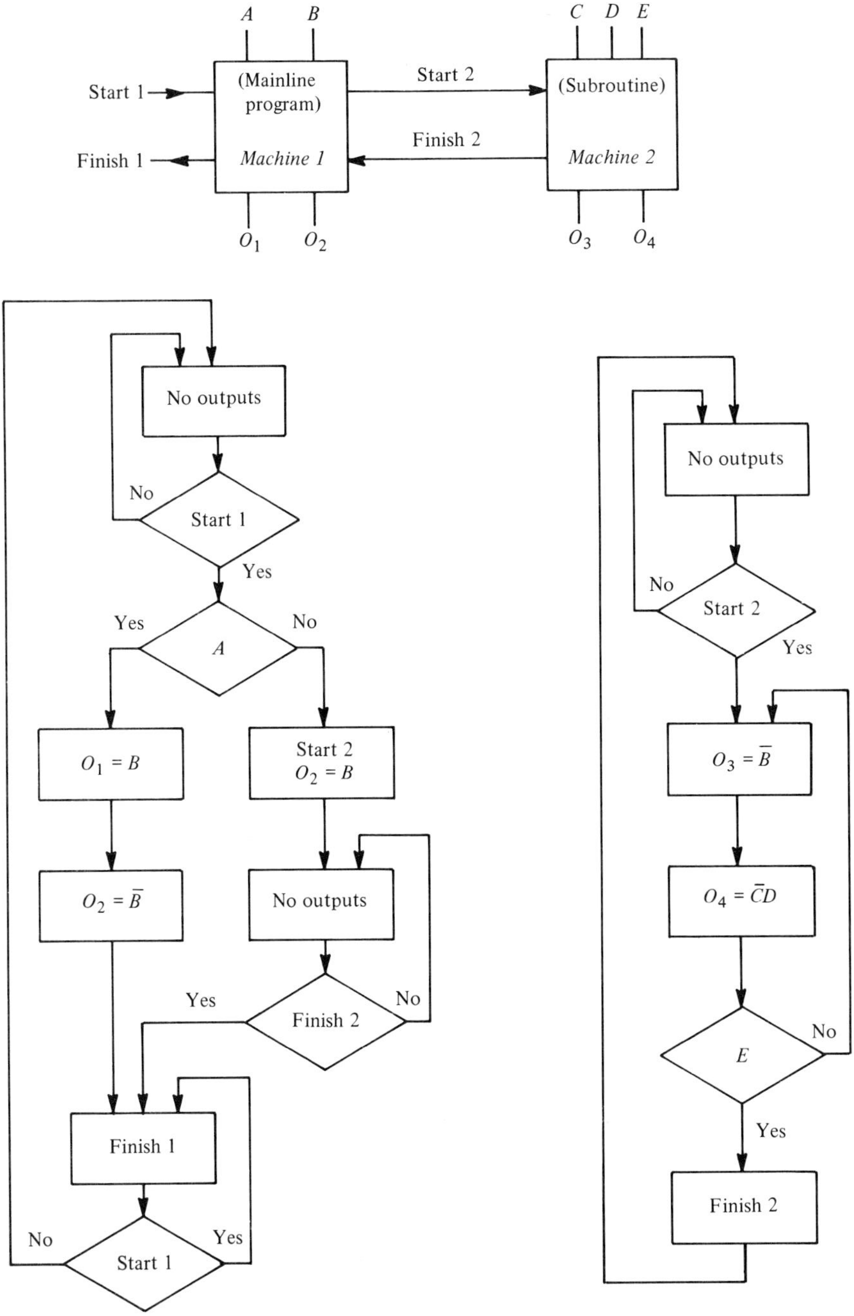

Figure 5–33 Machine of Figure 5–32 with subroutine separated.

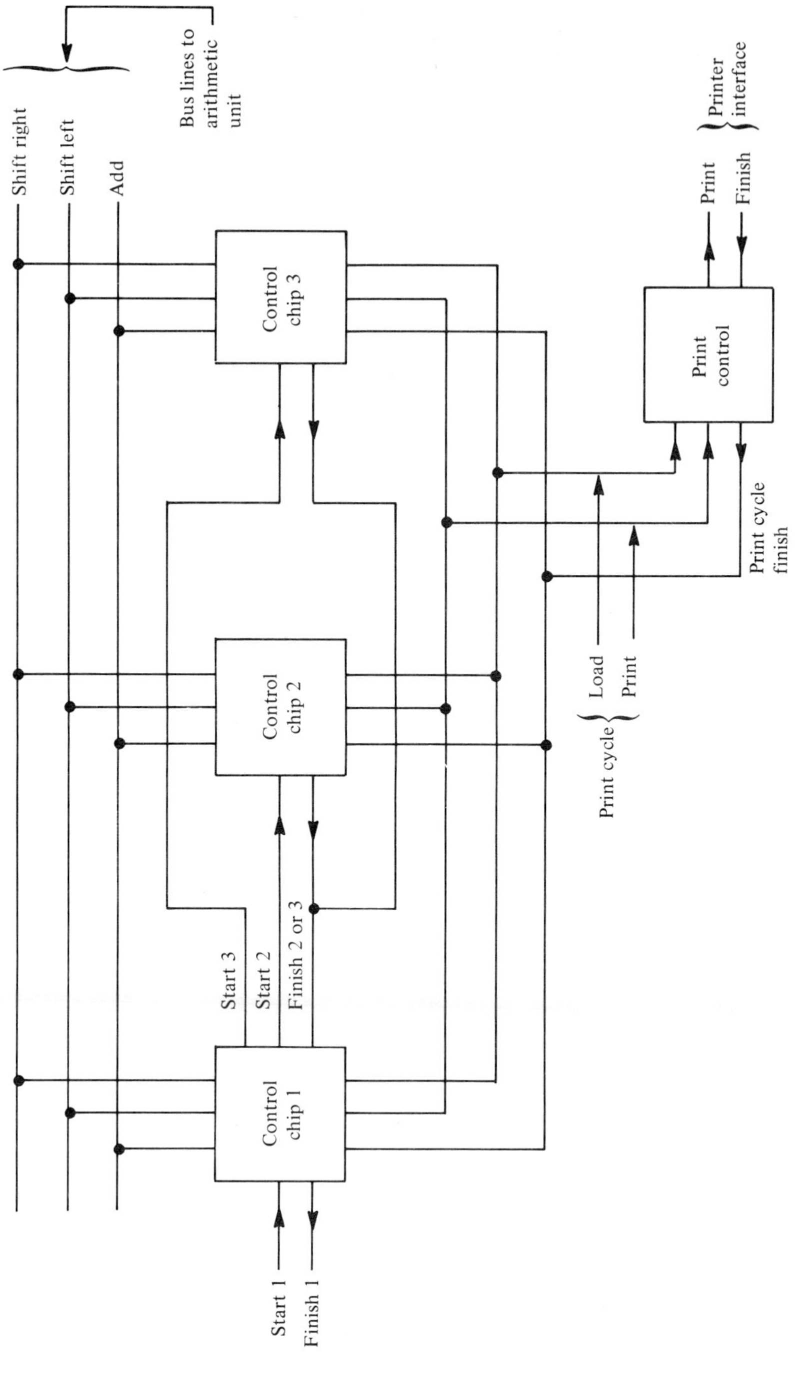

Figure 5–34 Control chips with bus lines.

bounce for the previous interval between T pulses. A new output will not be produced unless A first returns to logic 0 for a continuous bounce-free period equal to the period of T, and then again satisfies the requirements for a logic 1.

Synthesis Procedure

Numerous design procedures have been developed to provide the designer with a systematic approach to obtaining minimum flip-flop input logic equations.[1-3] These approaches usually involve the algebraic manipulation of the application equations and the characteristic (or difference) equations for flip-flops. The procedure in this example enables the designer to proceed from the state flow chart to the flip-flop logic equations using tabular and logic mapping techniques alone.[4] The result is a process that is adaptable to computer software implementation and automated logic design.

State Flow Chart Construction—The first step is to construct a state flow chart that completely describes the required machine operation as previously defined. This is the most difficult part of the design procedure. Important considerations include questions such as: Will the machine initialize correctly upon application of power? Are there undefined states that may be entered due to system noise? If so, a path must be provided for exit to a defined state. Also, entry into undefined states must be provided as a test condition to verify exit path logic.

The final flow chart for this example is given in Figure 5-35. In constructing the chart, the machine is assumed to be in its initialized state S_1. Since input A must exist without bounce for a full T period, a timing reference is established when A and T occur simultaneously and the machine advances to state S_2. Each clock period during the next T period input A is tested for possible bounce. If A goes to zero during this period, the machine returns to state S_1 and starts over. If input A does not bounce before the next T pulse, state S_3 is achieved and an output is generated during the next T period. Since the machine cannot assume the initial state S_1 until A remains at a logic 0 for one T period, it now advances to state S_4. State S_4 is similar to state S_1 except that the test is now for $\bar{A}$ instead of A. When $\bar{A}$ and T occur simultaneously, a timing reference is established, and the machine advances to state S_5. If A should bounce (to a logic 1) during S_5, the state returns to S_4. If $\bar{A}$ remains for one T period, the switch-opening bounce requirement is satisfied and the machine is returned to state S_1 to await the next input A.

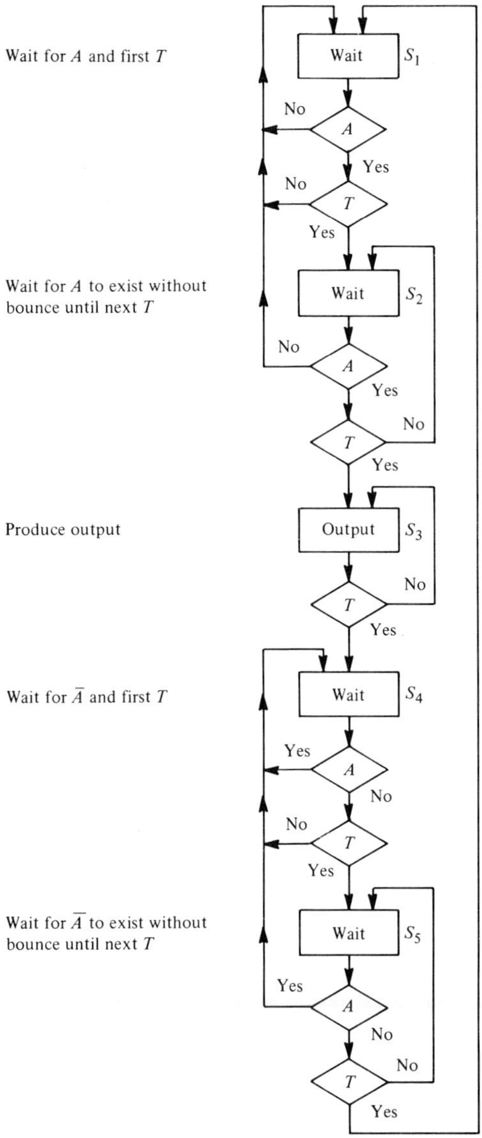

Figure 5–35 State flow chart example.

State Table Generation—This table (Figure 5-36) summarizes the logic of the state flow chart. It is constructed by first listing all eight possible states in numerical order. The next state and output conditions are then tabulated as a function of the present state and inputs A and T. Next

Present state	Next state				Output			
	Inputs (A,T)				Inputs (A,T)			
	(0,0)	(0,1)	(1,0)	(1,1)	(0,0)	(0,1)	(1,0)	(1,1)
1	1	1	1	2	0	0	0	0
2	1	1	2	3	0	0	0	0
3	3	4	3	4	1	1	1	1
4	4	5	4	4	0	0	0	0
5	5	1	4	4	0	0	0	0
6	X	X	X	X	0	0	0	0
7	X	X	X	X	0	0	0	0
8	X	X	X	X	0	0	0	0

Figure 5–36 State table for example.

state conditions of unused states are initially assumed to be "don't cares" (X).

For instance, if the machine is in state S_1, its next state is also state S_1, as long as inputs (A, T) remain (0, 0), (0, 1), or (1, 0). However, if (A, T) go to (1, 1), the next state is S_2. Similarly, the machine output is a zero in all states except S_3, where it is a logic 1 regardless of the states of inputs (A, T).

The existence of redundant states would be obvious from the state table (i.e., two or more present states having identical next states and outputs). When this occurs, all redundant states may be merged into a single state. In this example, five states are required and cannot be reduced.

Secondary Assignments—Each state must be defined by a unique combinatorial state of delay elements (flip-flops). In general, m states require n delay elements, where $2^n \geq m$ (n and m are integers). In this example $m = 5$; therefore, $n = 3$. The assignment of codes to states is completely arbitrary, although some assignments result in less logic than others. Techniques have been developed to aid in this step; but the result is primarily a function of the experience and intuition of the designer.

Caution must be exercised in assigning state codes when any control inputs are asynchronous with respect to the machine clock. If the control signal is to cause a state change involving more than one flip-flop, and it occurs just before clock time, there is a possibility of an erroneous state transition. This is due to the different time delays associated with the input logic of each flip-flop. This problem may be eliminated either by synchronizing such control inputs or by assigning state codes in such a manner

Present state	State code		
	Q_A	Q_B	Q_C
1	0	0	0
2	0	1	0
3	0	1	1
4	1	1	0
5	1	0	0
6	0	0	1
7	1	0	1
8	1	1	1

Figure 5–37 Secondary assignment.

that state transitions that involve such inputs require changing only one flip-flop.

For this example, the latter solution is used, and the resultant assignment of codes to states is made in Figure 5-37. It should be noted that the transition from state S_3 and S_4 is a function only of the synchronous input T; therefore, two flip-flops are allowed to change.

Transition Table Generation—The transition table (Figure 5-38) is created by substituting the state numbers of the state table with the previously assigned secondary state codes. The table contains the next state (P) information for each flip-flop as a function of the present state (Q) and inputs A and T, and thus corresponds to a combined truth table for the application equations of each flip-flop.

$$Q_{n+1} = P = f(Q_{An}, Q_{Bn}, Q_{Cn}, A, T)$$

Present state (PS)	($Q = Q_n$) PS Code			Next state ($P = Q_{n+1}$) Inputs (A,T) (0,0)			(0,1)			(1,0)			(1,1)		
	Q_A	Q_B	Q_C	P_A	P_B	P_C	P_A	P_B	P_C	P_A	P_B	P_C	P_A	P_B	P_C
1	0	0	0	0	0	0	0	0	0	0	0	0	0	1	0
2	0	1	0	0	0	0	0	0	0	0	1	0	0	1	1
3	0	1	1	0	1	1	1	1	0	0	1	1	1	1	0
4	1	1	0	1	1	0	1	0	0	1	1	0	1	1	0
5	1	0	0	1	0	0	0	0	0	1	1	0	1	1	0
6	0	0	1	X	X	X	X	X	X	X	X	X	X	X	X
7	1	0	1	X	X	X	X	X	X	X	X	X	X	X	X
8	1	1	1	X	X	X	X	X	X	X	X	X	X	X	X

Figure 5–38 Transition table.

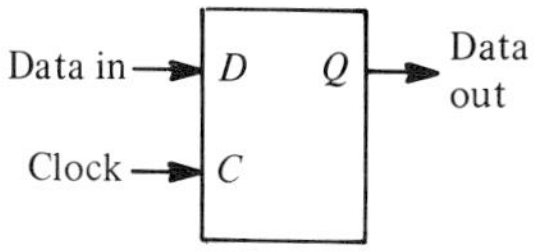

(a) Logic symbol

D	P
0	0
1	1

(b) Characteristic table

$Q \rightarrow P$		D
0	0	0
0	1	1
1	0	0
1	1	1

(c) Excitation table

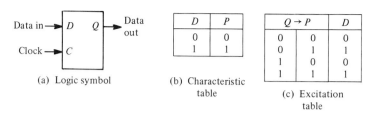

Figure 5–39 *D* flip-flop.

Flip-Flop Type Selection—Flip-Flop types considered here are assumed to be clocked devices, i.e., the device does not change state until the clock input has changed state.

All clocked flip-flops may be described by characteristic tables which define the next state output P (where $P = Q_{n+1}$) in terms of the inputs. The information in the characteristic table may be used to generate an excitation table which describes what the inputs to the flip-flop must be before the clock transition to change from one output state to another. Examples of the *D, RS* (reset dominant), *SR* (set dominant), *JK*, and *SH* (sample and hold) flip-flop types are given in Figures 5-39 through 5-43. These flip-flops differ from the classical types in that MOS dynamic logic implementation allows set (or reset) dominant set-reset flip-flops and eliminates the classical "undefined" state when set and reset terms occur simultaneously. Also, the sample and hold type flip-flop is used primarily with dynamic

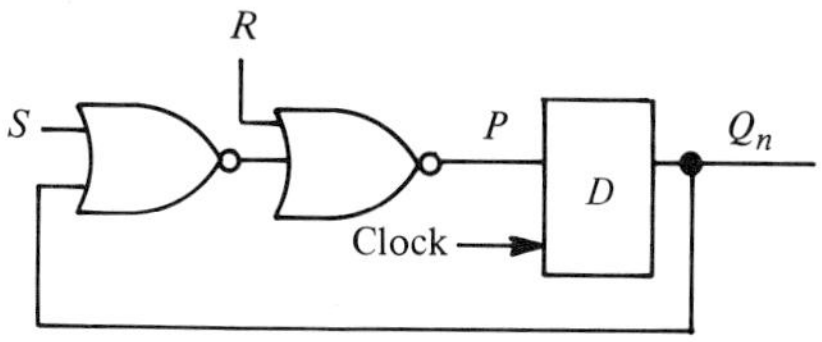

(a) Logic diagram

S	R	P
1	1	0
1	0	1
0	1	0
0	0	Q_n

(b) Characteristic table

$Q \rightarrow P$		S	R
0	0	1 or 0	1 or X
0	1	1	0
1	0	X	1
1	1	X	0

(c) Excitation table

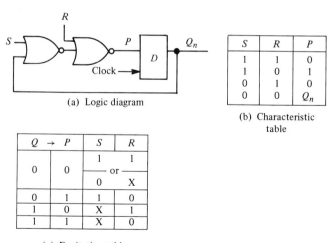

Figure 5–40 *RS* flip-flop (reset dominant).

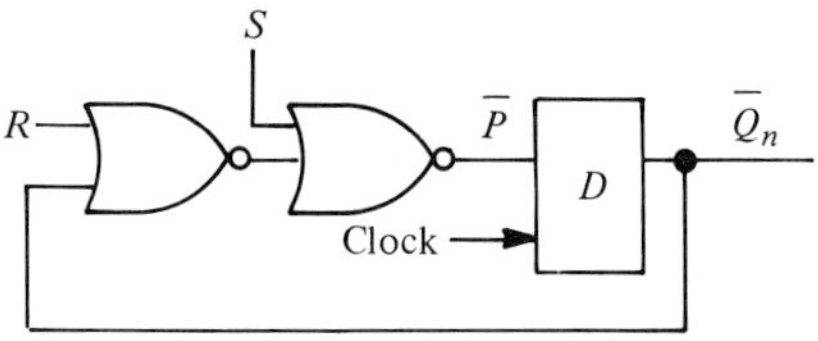

(a) Logic diagram

S	R	P
1	1	1
1	0	1
0	1	0
0	0	Q_n

(b) Characteristic table

Q	→ P	S	R
0	0	0	X
0	1	1	X
1	0	0	1
1	1	1	1
		or	
		X	0

(c) Excitation table

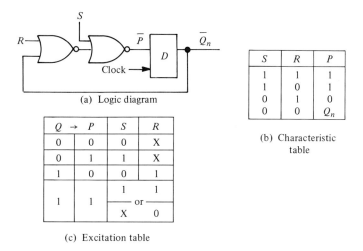

Figure 5–41 *SR* flip-flop (set dominant).

MOS logic and has not been presented in this manner before. For this design example, flip-flop *A* and *B* will be defined as *SR* (*set dominant*) and *C* will be a *JK* flip-flop.

Flip-Flop Control Matrices Construction—The next step in the synthesis procedure is to create a map that completely describes the logic required

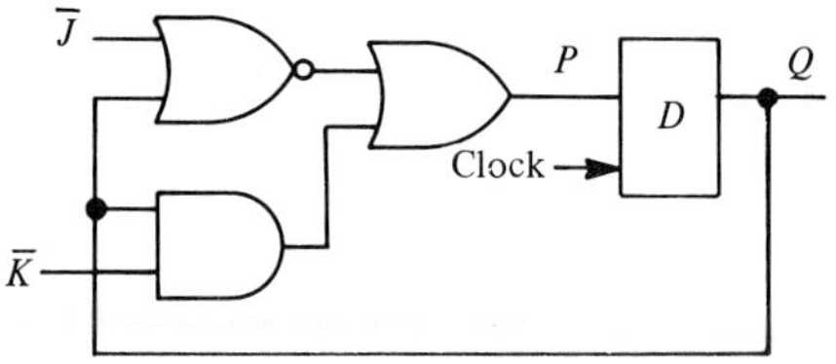

(a) Logic diagram

J	K	P
1	1	$\overline{Q}_n$
1	0	1
0	1	0
0	0	Q_n

(b) Characteristic table

Q →	P	J	K
0	0	0	X
0	1	1	X
1	0	X	1
1	1	X	0

(c) Excitation table

Figure 5–42 *JK* flip-flop.

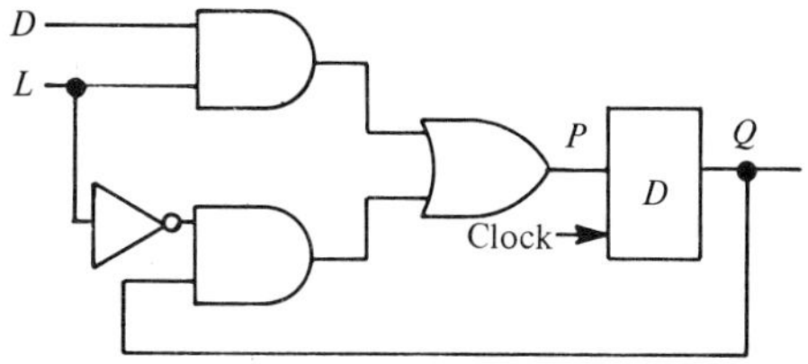

D = data input
L = sample data input

(a) Logic diagram

D	L	P
1	1	1
1	0	Q_n
0	1	0
0	0	Q_n

$Q \rightarrow P$		L	D
0	0	0	X
		or	
		1	0
0	1	1	1
1	0	1	0
1	1	0	X
		or	
		1	1

Figure 5–43 Sample and hold flip-flop.

on each input of the flip-flop to perform the desired function. This is accomplished with the transition table (Figure 5-38), the excitation table of the desired flip-flop type (Figures 5-39 through 5-43) and a five variable map (i.e., Vietch, Karnaugh, etc.) for each flip-flop in the machine. The procedure for generating the control matrices is as follows:

First, consider one five variable map at a time. For example, the map for flip-flop A, an SR type (Figure 5-41).

Second, from the transition table (Figure 5-38) for each combination of Q_A, Q_B, Q_C, A, T, locate the corresponding P_A and note the state transition of flip-flop A from $Q_A \rightarrow P_A$. (i.e., for combination 00000, $P_A = 0$ and $Q_A \rightarrow P_A = 0 \rightarrow 0$; for combination 01111, $P_A = 1$, and $Q_A \rightarrow P_A = 0 \rightarrow 1$).

Third, after each $Q_A \rightarrow P_A$ transition has been determined, examine the excitation table (Figure 5-41c) for the SR flip-flop and extract the S, R terms that correspond to the appropriate $Q_A \rightarrow P_A$ transition. Place the resulting terms into the five variable map in the cell defined by the Q_A, Q_B, Q_C, A, T combination. (i.e., for combination 00000, $P_A = 0$, $Q_A \rightarrow P_A = 0 \rightarrow 0$; $(S, R) = (0, X)$; therefore, $(0, X)$ is placed in cell 00000 of flip-flop A control matrix (Figure 5-44).

This process is repeated until the control matrices for flip-flops A, B, and C are completed (Figures 5-44 through 5-46). The input equations for S_A, R_A, S_B, R_B, J_C, and K_C are then written directly from the control ma-

Q_A Q_B Q_C \ A,T	00	01	11	10
0 0 0	0X	0X	0X	0X
0 0 1	XX	XX	XX	XX
0 1 1	0X	1X	1X	0X
0 1 0	0X	0X	0X	0X
1 1 0	X0	X0	X0	X0
1 1 1	XX	XX	XX	XX
1 0 1	XX	XX	XX	XX
1 0 0	X0	01	X0	X0

$S_A - R_A$

Karnaugh map

Input equations

$S_A = Q_C T$

$R_A = \overline{(Q_B + A + \overline{T})}$

Figure 5–44 Flip-flop *A*: $S_A - R_A$ control matrix.

trices. If negative true logic is used, the equations should be written in the form of the product of sums that corresponds to NOR logic implementation. If positive true logic is used, equations should be in the form of the sum of products that correspond to NAND logic implementation.

Output Logic Mapping—Referring to the state table (Figure 5-36), the output of the machine is tabulated as a function of the inputs. A five variable map may be used to plot the resulting function with the output equation being written from that map. In the example, however, it is obvious that the output is a logic 1 whenever the machine is in state S_3; thus, the output $= S_3 = Q_C$.

Preliminary Logic Diagram—At this point a decision must be made whether the resulting logic design is to be input to an automated MOS chip design process, or is to be a custom design.

An integral part of an automated design system is a family of standard logic functions (standard cells) which must be used by the logic designer

Q_A Q_B Q_C \ A,T	00	01	11	10
0 0 0	0X	0X	1X	0X
0 0 1	XX	XX	XX	XX
0 1 1	11	11	X0	X0
0 1 0	01	01	X0	X0
1 1 0	X0	01	X0	X0
1 1 1	XX	XX	XX	XX
1 0 1	XX	XX	XX	XX
1 0 0	0X	0X	1X	1X

$S_B - R_B$

Karnaugh map

Input equations

$S_B = AT + AQ_A + Q_C$

$R_B = \overline{(A + Q_A \overline{T})}$

Figure 5–45 Flip-flop *B*: $S_B - R_B$ control matrix.

Q_A	Q_B	Q_C	A,T: 00	01	11	10
0	0	0	0X	0X	0X	0X
0	0	1	XX	XX	XX	XX
0	1	1	X0	X1	X1	X0
0	1	0	0X	0X	1X	0X
1	1	0	0X	0X	0X	0X
1	1	1	XX	XX	XX	XX
1	0	1	XX	XX	XX	XX
1	0	0	0X	0X	0X	0X

J_C - K_C

Karnaugh map

Input equations

$J_C = \overline{(\overline{Q_A} + \overline{Q_B} + \overline{A} + \overline{T})}$

$K_C = T$

Figure 5–46 Flip-flop C: J_C – K_C control matrix.

much like conventional bipolar logic families. The logic diagram is thus defined as the interconnection of these standard cells to perform the desired logic function.

If the system is to be custom designed, the logic designer may create complex logic gates, the complexity of which is constrained only by the practical considerations of speed, power, and size.

When factoring input equations the following guidelines should be kept in mind:

- If 2-phase ratioed logic is to be used, the maximum number of logic levels between delay blocks must not exceed a limit that is dependent upon operating speed, temperature, and voltages as described in Section 5.3.1.
- If 2-phase ratioless logic is to be used, the number of logic levels between delay blocks is restricted to two levels. Two additional simple logic levels (i.e., inverters) may be used if necessary, as described in Section 5.3.2.
- If 4-phase ratioless logic is to be used, the number of logic levels should be restricted to four. Four additional simple logic levels (i.e., inverters) may be used if necessary, as described in Section 5.3.3.

Figure 5-47 illustrates the preliminary logic diagram for this example (assuming negative true logic), which uses a maximum of four logic levels.

If a preliminary partitioning of the system into chips has been made, it is helpful to allow one logic drawing for each chip, identify all chip input and output logic signals, provide a chip interrconnect list, and identify the longest (containing the most levels of logic) paths between delay elements.

Mechanization With D *Type Flip-Flops*—Since the *D* type flip-flop is relatively simple to implement with MOS dynamic logic, the example problem in "Machine Description" (Section 5.4.1) is reworked and mechaniza-

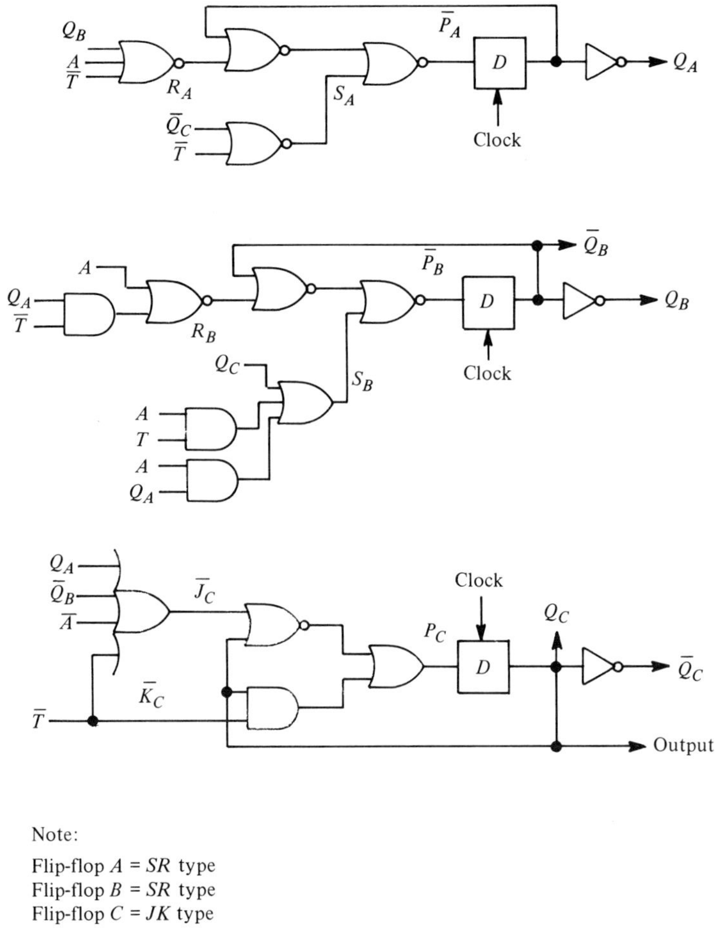

Figure 5–47 Preliminary logic diagram of flow chart example.

tion accomplished with *D* flip-flops. All design steps 1 through step 5 of the "Synthesis Procedure" remain the same. Control matrices for flip-flops *A*, *B*, and *C* (Figures 5-48 through 5-50) are generated by following a procedure similar to that defined in step.[6] The resulting preliminary logic diagram is shown in Figure 5-51.

5.4.2 Design Analysis

It is sometimes necessary to analyze an existing circuit either to define its operation or to check the synthesis procedure. The analysis procedure is exactly the reverse of the synthesis procedure.

Q_A	Q_B	Q_C \ A,T	00	01	11	10
0	0	0	0	0	0	0
0	0	1	X	X	X	X
0	1	1	0	1	1	0
0	1	0	0	0	0	0
1	1	0	1	1	1	1
1	1	1	X	X	X	X
1	0	1	X	X	X	X
1	0	0	1	0	1	1

Karnaugh map

Input equation

$$D_A = (Q_B + A + \overline{T})(Q_C + Q_A)(\overline{Q}_C + T)$$

Figure 5–48 Flip-flop *A* (*D* type) control matrix.

Q_A	Q_B	Q_C \ A,T	00	01	11	10
0	0	0	0	0	1	0
0	0	1	X	X	X	X
0	1	1	1	1	1	1
0	1	0	0	0	1	1
1	1	0	1	0	1	1
1	1	1	X	X	X	X
1	0	1	X	X	X	X
1	0	0	0	0	1	1

Karnaugh map

Input equation

$$D_B = (Q_C + A + \overline{T})(Q_A + Q_B + T)(A + Q_B)(Q_C + A + Q_A)$$

Figure 5–49 Flip-flop *B* (*D* type) control matrix.

Q_A	Q_B	Q_C \ A, T	00	01	11	10
0	0	0	0	0	0	0
0	0	1	X	X	X	X
0	1	1	1	0	0	1
0	1	0	0	0	1	0
1	1	0	0	0	0	0
1	1	1	X	X	X	X
1	0	1	X	X	X	X
1	0	0	0	0	0	0

Karnaugh map

Input equation:

$$D_C = Q_B\overline{Q}_A\,(Q_C + T)(\overline{T} + \overline{Q}_C A)$$

Figure 5–50 Flip-flop *C* (*D* type) control matrix.

- *Input Equations* Write input equations for each flip-flop (Figure 5-52a) from logic diagram.
- *Flip-Flop Control Matrices Construction* Plot control matrix for each flip-flop (Figures 5-52b–5-52d). Plot D_A, D_B, and D_C on separate five variable maps.

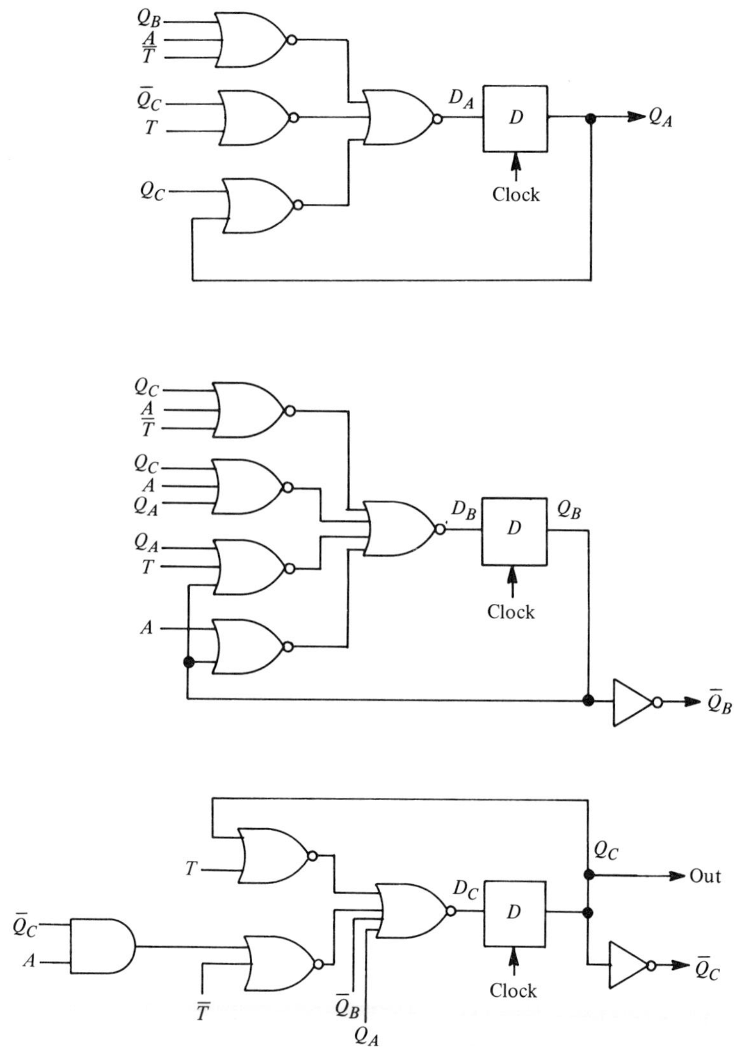

Figure 5–51 Preliminary logic diagram of flow chart example using *D* type flip-flops.

- *Transition Table Generation* Assuming that there is no prior knowledge of the previous assignment of codes to states, the present state codes may be listed in any order in the table (Figure 5-53). The next state (*P*) columns are then easily filled in by referring to the respective control matrix (Figure 5-52). Inspect the D_A, D_B, or D_C terms in each

$$D_A = (Q_B + A + \overline{T})(Q_C + Q_A)(\overline{Q}_C + T)$$

$$D_B = (Q_C + A + \overline{T})(Q_A + Q_B + T)$$
$$(A + Q_B)(Q_C + A + Q_A)$$

$$D_C = \overline{Q}_A Q_B (Q_C + T)(\overline{T} + \overline{Q}_C A)$$

(a) Input equations

Q_A	Q_B	Q_C \ A,T	00	01	11	10
0	0	0	0	0	0	0
0	0	1	0	0	1	0
0	1	1	0	1	1	0
0	1	0	0	0	0	0
1	1	0	1	1	1	1
1	1	1	0	1	1	0
1	0	1	0	0	1	0
1	0	0	1	0	1	1

(b) D_A Control matrix (example analysis)

Q_A	Q_B	Q_C \ A,T	00	01	11	10
0	0	0	0	0	1	0
0	0	1	0	0	1	0
0	1	1	1	1	1	1
0	1	0	0	0	1	1
1	1	0	1	0	1	1
1	1	1	1	1	1	1
1	0	1	0	0	1	1
1	0	0	0	0	1	1

(c) D_B Control matrix (example analysis)

Q_A	Q_B	Q_C \ A,T	00	01	11	10
0	0	0	0	0	0	0
0	0	1	0	0	0	0
0	1	1	1	0	0	1
0	1	0	0	0	1	0
1	1	0	0	0	0	0
1	1	1	0	0	0	0
1	0	1	0	0	0	0
1	0	0	0	0	0	0

(d) D_C Control matrix (example analysis)

Figure 5–52 Analysis of Figure 5–51.

Present state	Present state code			Next state, Inputs (A,T): (0,0)			(0,1)			(1,1)			(1,0)		
	Q_A	Q_B	Q_C	P_A	P_B	P_C	P_A	P_B	P_C	P_A	P_B	P_C	P_A	P_B	P_C
(1)	0	0	0	0	0	0	0	0	0	0	1	0	0	0	0
(6)	0	0	1	0	0	0	0	0	0	1	1	0	0	0	0
(3)	0	1	1	0	1	1	1	1	0	1	1	0	0	1	1
(2)	0	1	0	0	0	0	0	0	0	0	1	1	0	1	0
(4)	1	1	0	1	1	0	1	0	0	1	1	0	1	1	0
(8)	1	1	1	0	1	0	1	1	0	1	1	0	0	1	0
(7)	1	0	1	0	0	0	0	0	0	1	1	0	0	1	0
(5)	1	0	0	1	0	0	0	0	0	1	1	0	1	1	0

Figure 5–53 Transition table (analysis of Figure 5–51).

cell and extract, from the corresponding characteristic table, the value of P_A, P_B, or P_C.

- *Secondary Assignments* The assignments may be made arbitrarily but, for the sake of clarity in this example, assignments are assumed to be the same as those used in the "Synthesis Procedure" (i.e., present state number 1 corresponds to $(Q_A, Q_B, Q_C) = (0, 0, 0)$ etc.).

- *State Table Generation* The transfer of information, from the transition table to the state table (Figure 5-54) is straightforward and yields additional information not apparent when the original state table was constructed (Figure 5-36). A comparison of the two tables indicates that the next state conditions for the defined present states are identical. Furthermore, it can be seen that there are exit paths from undefined states ($S_6 - S_8$).

Present state	Next state, Inputs (A,T): 0,0	0,1	1,1	1,0
1	1	1	2	1
6	1	1	4	1
3	3	4	4	3
2	1	1	3	2
4	4	5	4	4
8	2	4	4	2
7	1	1	4	2
5	5	1	4	4

Figure 5–54 State table (analysis of Figure 5–51).

5.5 CONVERSION OF PRELIMINARY DESIGN TO 2-PHASE RATIOED LOGIC

Upon completion of the preliminary design procedures as outlined in the "Synthesis Procedure" (Section 5.4.1) the resulting preliminary logic diagram may be easily converted to a form which may be implemented with 2-phase ratioed logic elements.

5.5.1 Delay Manipulation

One of the most important design considerations is the effective utilization of the inherent delay characteristics of MOS LSI. Figures 5-55a and 5-55b

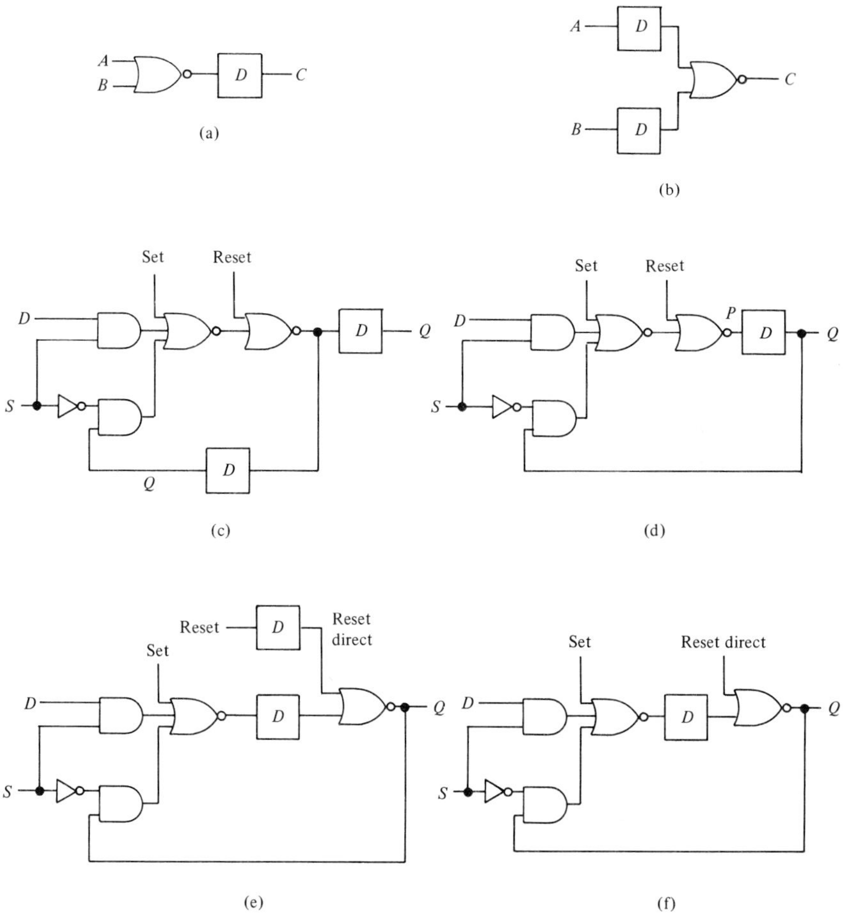

Figure 5–55 Delay element manipulation.

illustrate how the delay blocks which resulted from the preliminary design may be moved through a dc logic gating matrix without altering the input-output characteristics of the circuit. Figure 5-55c is a typical circuit that might result from the preliminary design procedures, and Figure 5-55d shows how the two delay blocks may be combined. Figure 5-55d is a sample and hold flip-flop with synchronous set and reset capabilities. To work this into a smaller, more standard flip-flop the delay block is moved to the left as shown in Figure 5-55e. This results in a new reset signal (reset direct) which is delayed by one clock time from reset and is synchronous because it passes through a delay block. The result is the circuit of Figure 5-55f.

5.5.2 Substitution

After the number of delay blocks have been minimized and the locations defined, each delay block is replaced with the 2-phase ratioed delay element shown in Figure 5-56. The delay element is characterized by two cascaded dc inverters, each with a ½-bit delay block that designates the clock phase of the preceding node.

5.5.3 Minimization

After substitution, logic reduction may be possible by moving the ½-bit delay blocks through the logic as was previously demonstrated. Figure 5-57 illustrates the reduction accomplished by moving the ϕ_2 delay block to the left and combining the two dc inverters.

The ϕ_1 and ϕ_2 delay blocks may be moved through the logic with two restrictions: (1) at least one node must always separate any ϕ_1 and ϕ_2 delay blocks; and (2) the maximum number of nodes between ϕ_1 and ϕ_2 delay blocks, which are a function of the required operating speed must be calculated and not exceeded.

For the 2-phase ratioed logic circuits, ½-bit delay elements are functionally identical to a series device having its gate clocked with the phase associated with the delay block. After substituting each ϕ_1 delay block with

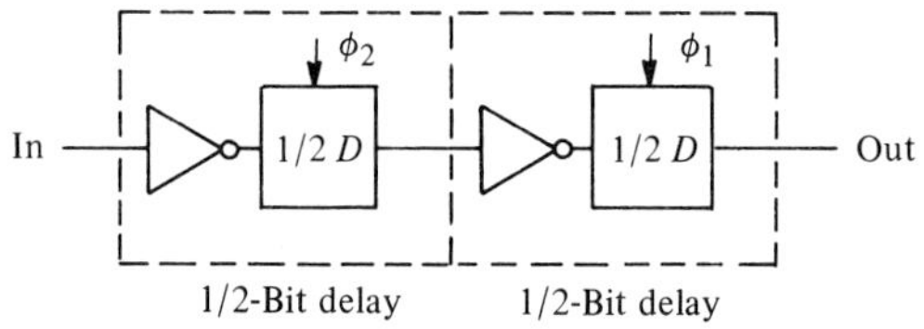

Figure 5–56 2-Phase ratioed delay element.

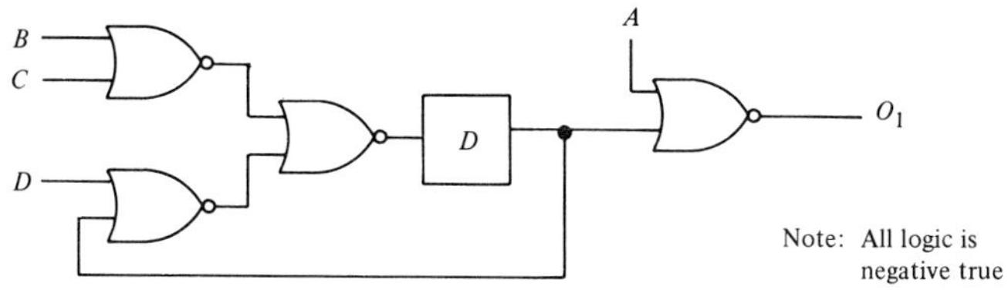

(a) Preliminary logic diagram

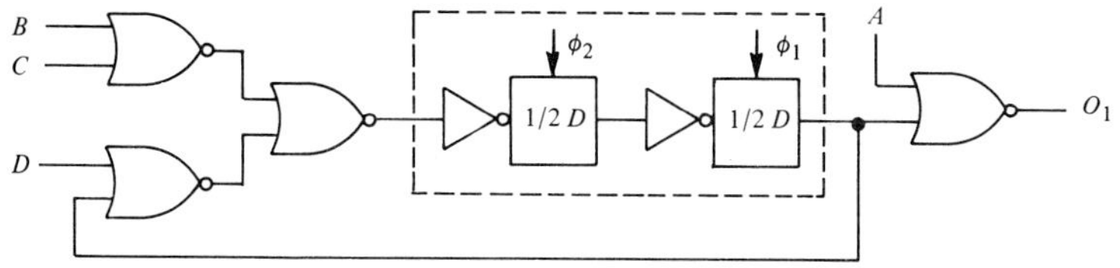

(b) 2-Phase ratioed delay element substitution

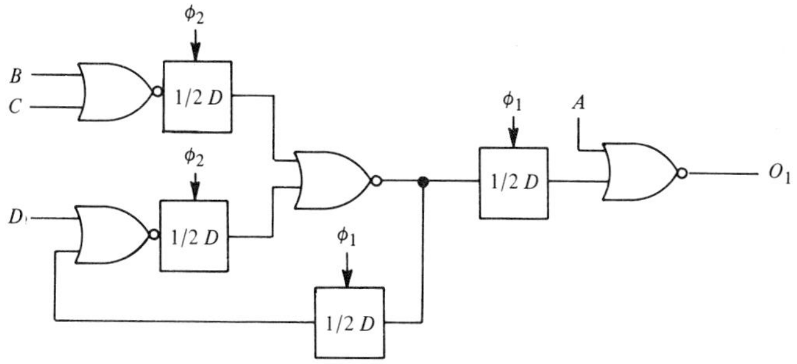

(c) Minimization

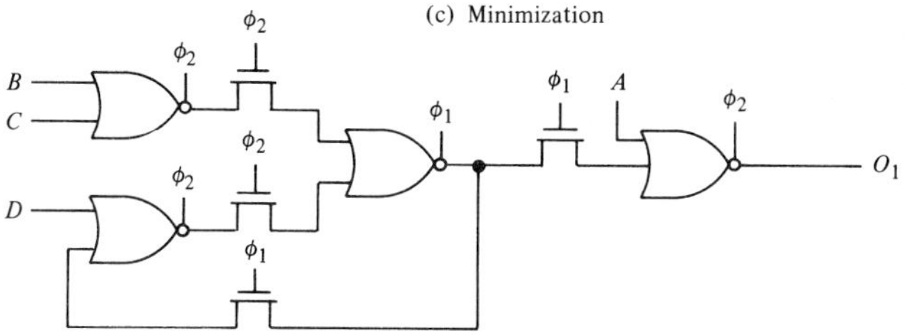

(d) Final logic diagram

Figure 5–57 Conversion to 2-phase ratioed logic.

a ϕ_1 clocked series gate and all ϕ_2 delay blocks with a ϕ_2 clocked series gate, all remaining nodes are clocked with the phase of the subsequent series gate. Figure 5-57d illustrates the resulting circuit. Figure 5-58 shows another development from preliminary logic diagram to the 2-phase ratioed logic diagram.

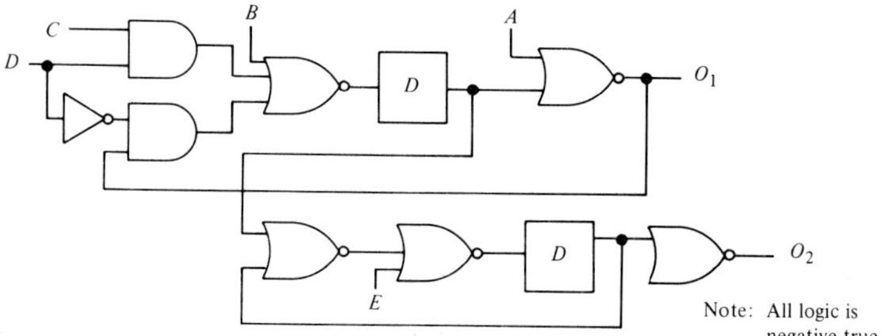

(a) Preliminary logic diagram

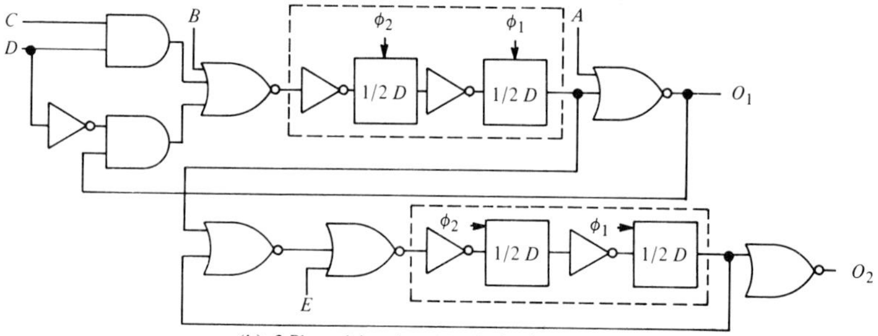

(b) 2-Phase delay element substitution

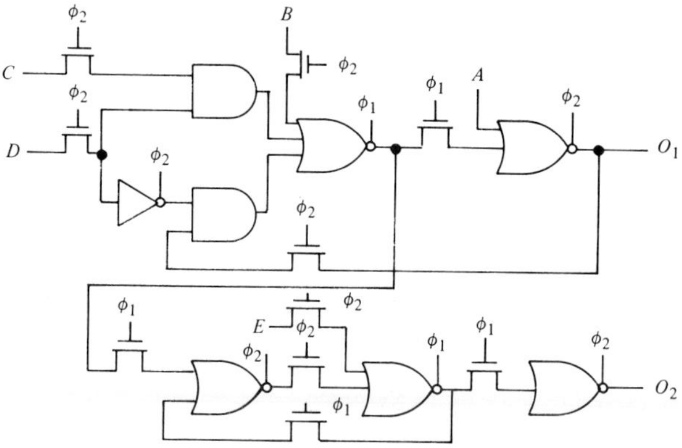

(c) Minimization and final logic diagram

Figure 5–58 Example conversion to 2-phase ratioed logic.

By using the techniques described above, it is possible to convert the preliminary logic diagram of the design example (Figure 5-47) to 2-phase ratioed logic as shown in Figure 5-59. By reversing the previous procedures it is possible to analyze a 2-phase ratioed logic diagram and work back to a state flow chart.

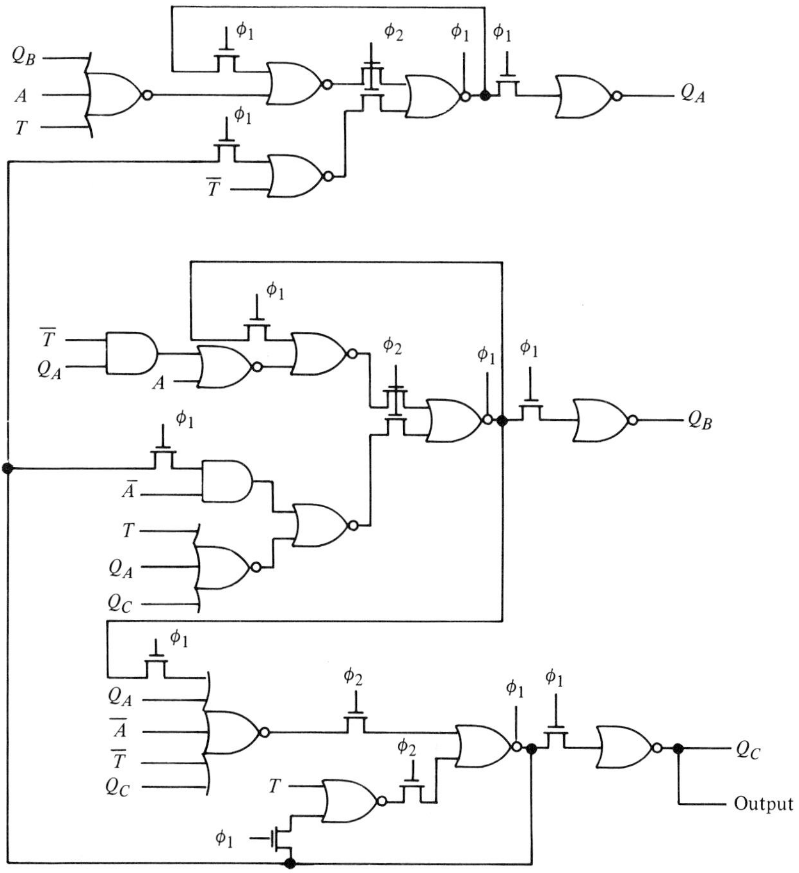

Notes: 1. Logic symbols negative true.
2. All unmarked nodes are clocked with ϕ_2.

Figure 5–59 Preliminary logic diagram of Figure 5–47 converted to 2-phase ratioed logic.

5.6 CONVERSION OF PRELIMINARY DESIGN TO 2-PHASE RATIOLESS LOGIC

Once a preliminary logic diagram is generated in the form defined in the "Design Procedures" (Section 5.4.1.), the conversion to 2-phase ratioless logic is relatively straightforward and is very similar to the 2-phase ratioed design of "Conversion of Preliminary Design to 2-Phase Ratioed Logic" (Section 5.5).

5.6.1 Delay Manipulation

This step is identical to the procedure described in the 2-phase ratioed case in "Delay Manipulation" (Section 5.5.1). The objective is to minimize the number of delay blocks as well as work the logic into a standard form which may simplify the chip layout by enabling the use of standard logic cells.

5.6.2 Substitution

After the number of delay blocks is minimized, and the locations determined, each is replaced by the 2-phase ratioless delay element as shown in Figure 5-60.

The delay element is characterized by two cascaded dc inverters, each with a ½-bit of delay block that designates the clock phase of the preceding node.

5.6.3 Minimization

After substitution, further reduction may be possible by moving the ϕ_1 and ϕ_2 delay blocks through the logic, until a minimum configuration is obtained. Since normal 2-phase ratioless logic allows only one level of logic per clock phase, every node on the logic diagram must have an associated ϕ_1 or ϕ_2 delay block following minimization. In addition, consecutive nodes must have alternating ϕ_1, ϕ_2 delay blocks.

If unassigned nodes exist, either the logic must be modified or, for simple logic functions, minimum delay gates may be used as shown in the 2-phase ratioless minimum delay gate section.

Since 2-phase ratioless circuits exhibit a ½-bit delay, each dc logic gate and associated delay block may be replaced with a 2-phase circuit of the

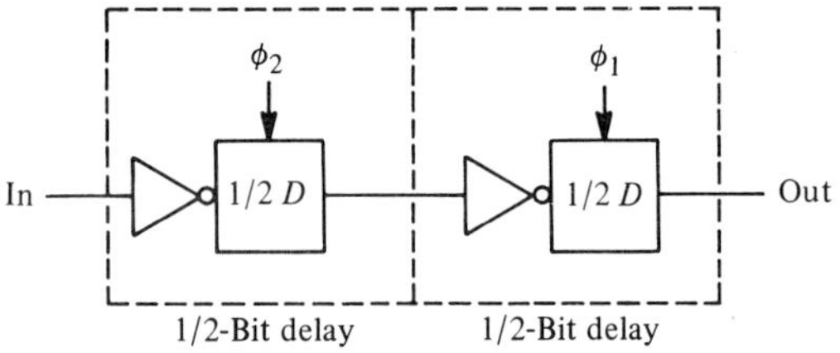

Figure 5–60 2-Phase ratioless delay element.

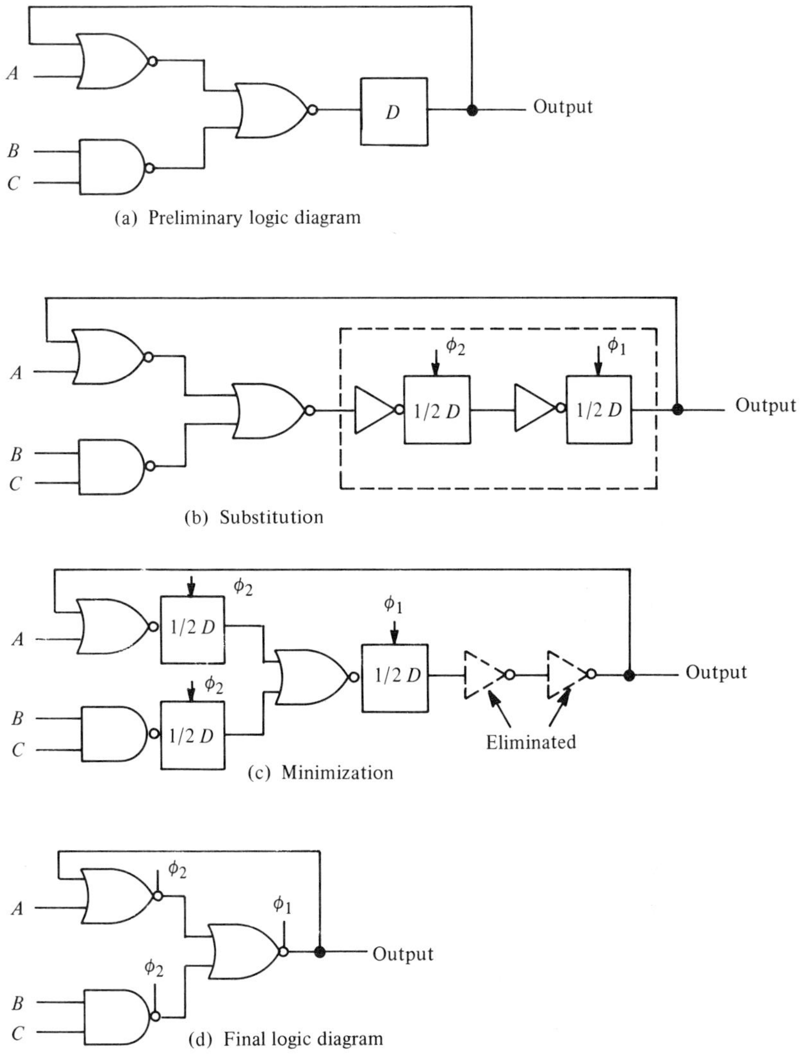

Figure 5–61 Development of circuit from preliminary logic diagram to 2-phase logic diagram.

type designated by the clock phase of the delay block. Figure 5-61 illustrates the development of an example circuit from preliminary logic diagram to a 2-phase ratioless diagram. Figure 5-62 shows a schematic of Figure 5-61 implemented with 2-phase ratioless configuration 3 circuits (Figure 5-17).

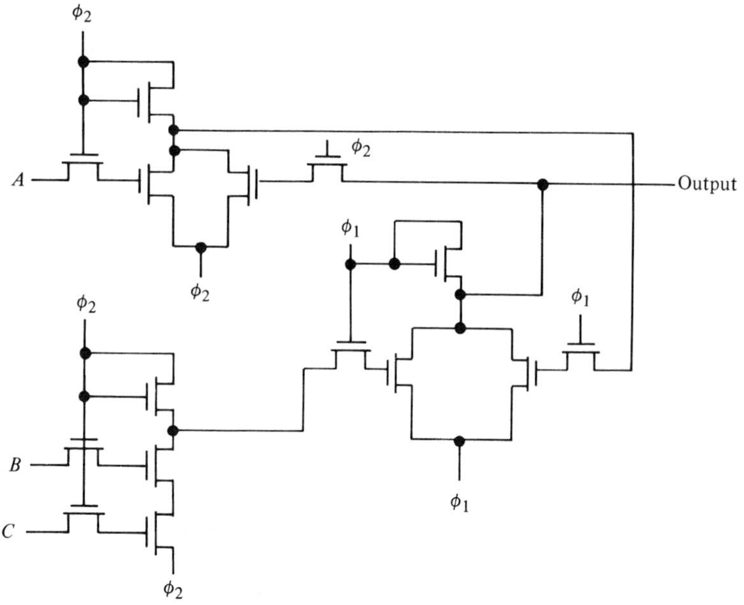

Figure 5–62 Schematic of circuit from Figure 5–61 mechanized with 2-phase ratioless configuration 3.

The application of minimum delay gates to achieve additional logic levels is illustrated in the conversion of the preliminary logic diagram of Figure 5-51 to 2-phase ratioless logic as shown in Figure 5-63.

5.7 CONVERSION OF PRELIMINARY DESIGN TO 4-PHASE LOGIC

Upon completion of the preliminary design procedures, the resulting preliminary logic diagram can be easily converted into a form using 4-phase dynamic logic elements. The preliminary logic diagram must be in the form of a generalized sequential machine, having dc combinatorial logic without feedback and delay blocks that are clocked synchronously.

5.7.1 Delay Manipulation

This procedure is identical to that described previously for 2-phase logic. The objective is to minimize the number of delay elements either by combining two or more or by altering the timing of input signals.

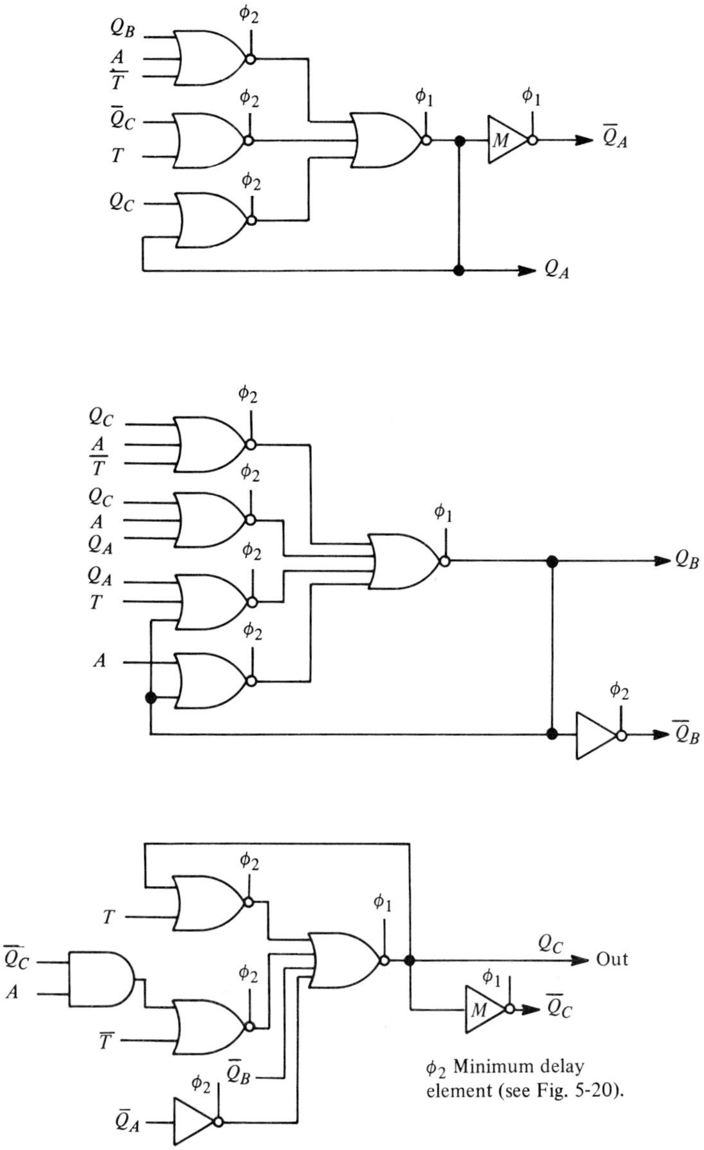

Figure 5–63 Preliminary logic diagram of Figure 5–51 implemented with 2-phase ratioless logic.

5.7.2 Substitution

When the number of delay blocks has been minimized, a 4-phase delay element representation is substituted for each delay block. Figure 5-64 shows the 4-phase delay element representation.

The delay block consists of two cascaded dc inverters, each with a delay block that represents ½-bit time of delay and the gate type associated with the node evaluation clock phase (ϕ_A, ϕ_B). Major gate types that satisfy the associated gate interface diagram must be used. Therefore, for major-major gates, types 2 and 4 or types 1 and 3 may be used. For major-minor gates, types 2 and 4 must be used. The order is important only to the extent that the output gate should be the desired circuit output node type, and that the assignment is consistent with the requirements of any interfacing logic.

5.7.3 Minimization

As with the 2-phase delay elements, the 4-phase ½-bit of delay blocks may be moved through the logic as shown in Figure 5-55. The objective is to integrate the delay into the dc logic matrix. The two dc inverters associated with the delay element should be eliminated if possible.

The delay blocks are then arranged such that no more than one unassigned node (i.e., node without associated delay block) exists between two assigned nodes. Since 4-phase ratioless major gate types exhibit ½-bit time delay, all dc gates with associated delay blocks may be replaced with a 4-phase circuit of the type indicated by the delay block. Unassigned nodes must then be replaced with 4-phase circuits that satisfy the appropriate gate interface diagram (i.e., major-major or major-minor), and do not increase the total delay between previously assigned nodes. This may be accomplished by replacing unassigned nodes with 4-phase circuit types either one type number higher than the gate type from which it is driven, or one

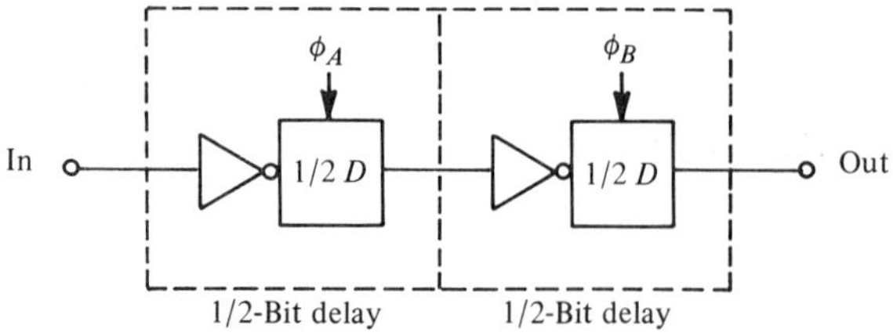

Figure 5–64 4-Phase delay element.

type number lower than the gate type that it is driving. Therefore, if an unassigned dc node is driven from a type 4 gate, it is replaced with a type 1 gate.

In order to reduce the number of different gate types required and simplify the chip circuit design, it is usually desirable to manipulate the logic such that unassigned nodes are simple, commonly used logic functions such as inverters.

Occasionally, it is necessary to utilize minimum delay gates in order to satisfy a circuit requirement. These special circuits are described in "4-Phase Ratioless Minimum Delay Elements" (Section 5.3.3) and allow an increase in maximum possible number of logic levels to eight per bit time.

5.7.4 Conversion Procedure Examples

- *Example No. 1* Figure 5-65 shows the development of a 4-phase logic diagram from a preliminary logic diagram. The final logic diagram has the gate type labeled inside each logic symbol.

- *Example No. 2* Another example is shown in Figure 5-66, which illustrates the case where the dc logic matrix of the preliminary design does not contain a sufficient number of nodes to allow full integration of the delay function without altering input timing relationships.

- *Example No. 3: Calculator Keyboard* The design example of Figure 5-51 may be converted using the previous procedure. Figure 5-67 shows the resulting 4-phase logic diagram.

- *Example No. 4: Serial Full Adder* Another example is shown in Figure 5-68. This is a serial full adder with carry flip-flop, and one flip-flop for storage of the sum. *D* type flip-flops with synchronous reset capability are used. The preliminary logic diagram is shown with the delay elements substituted (type 2 inverters and type 4 NOR gates). The type 2 delay blocks are moved back through the logic until a minimum configuration is obtained. Note that the inverter output constitutes a node between type 2 and type 4 major gates; therefore, the gate must be a type 3. As a result the two delay flip-flops are completely integrated into the full adder logic.

5.8 MOS LSI TEST CONSIDERATIONS

One of the more important logic design considerations is the test requirements of the final product. The following guidelines should be followed when analyzing the test requirements of a logic function.

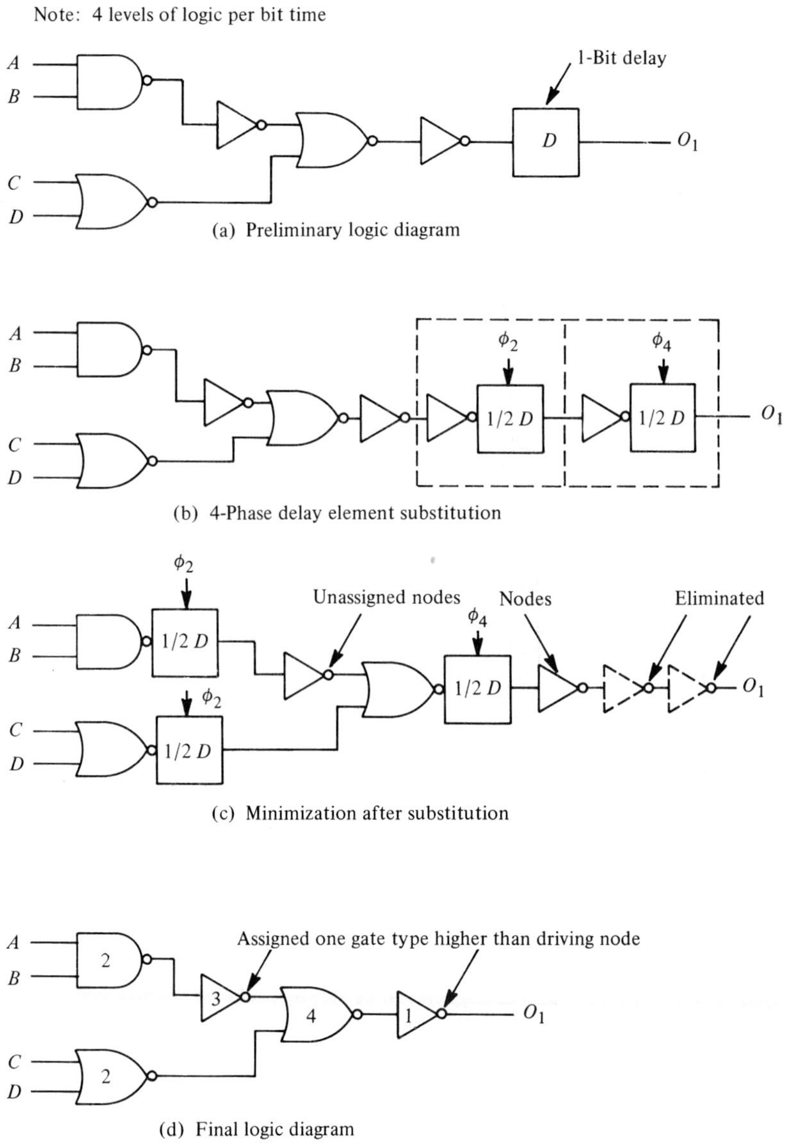

Figure 5–65 Example 1 of 4-phase logic design procedure.

1. Verify that all logic elements can be tested.
2. Minimize test pattern length.
3. Minimize test terminals.
4. Insure tester-chip synchronization.

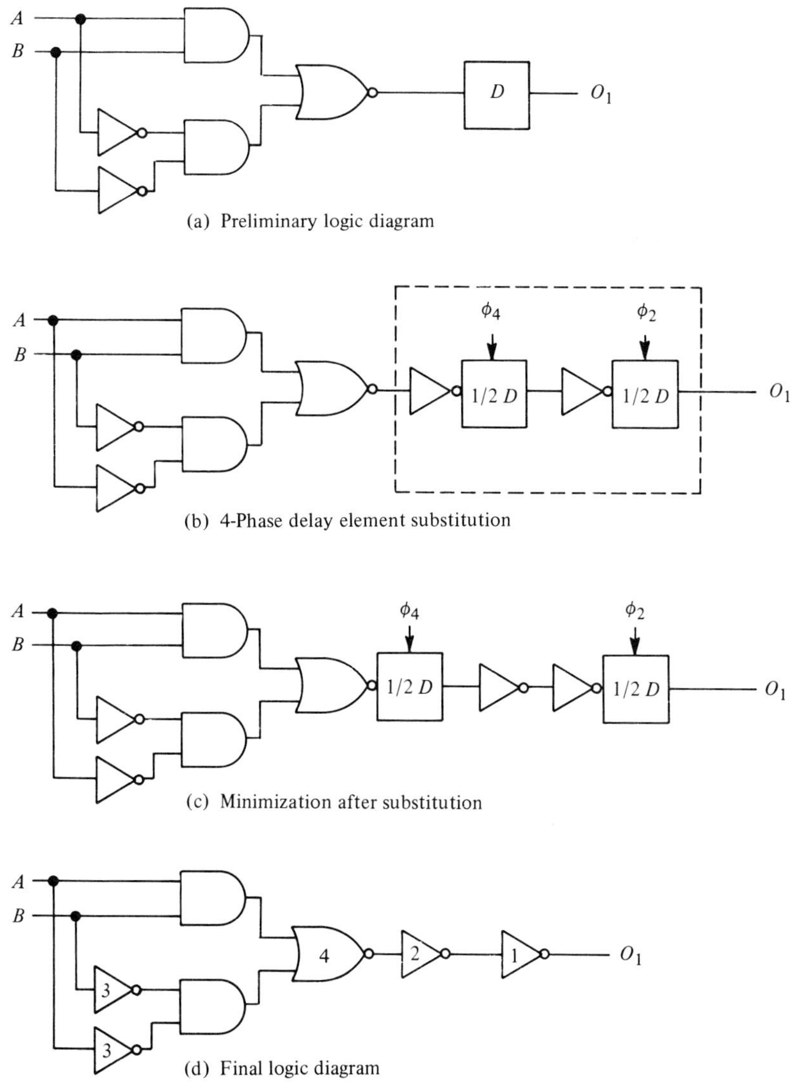

(a) Preliminary logic diagram

(b) 4-Phase delay element substitution

(c) Minimization after substitution

(d) Final logic diagram

Figure 5–66 Example 2 of 4-phase logic design procedure.

5.8.1 Undetectable Failures

There are numerous logic configurations which cannot be completely tested. One example of this is the majority voting circuit of Figure 5-69. The three logic blocks [$f(x)$] are identical and feed the majority voting block that performs the logic function OUT $= C(A\bar{B} + \bar{A}B) + AB$, thus

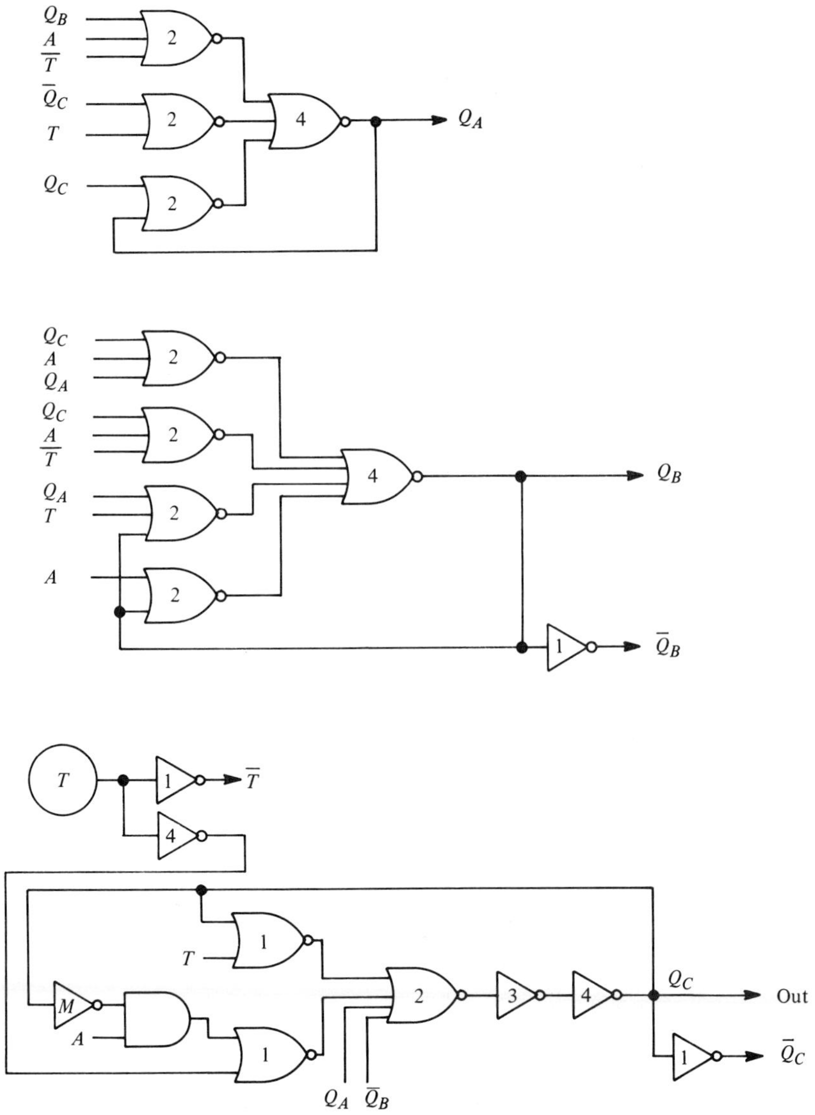

Note: M = Minimum delay element

Figure 5–67 Example of Figure 5–51 converted to 4-phase ratioless MOS logic.

generating an output whenever two or more $f(x)$ logic blocks yield the same outputs. Since a single failure can only disable one of the three $f(x)$ logic blocks, it will never be detected. This problem can be solved by providing additional test inputs into the voting block for selection of one $f(x)$ block at a time, allowing each section to be tested separately.

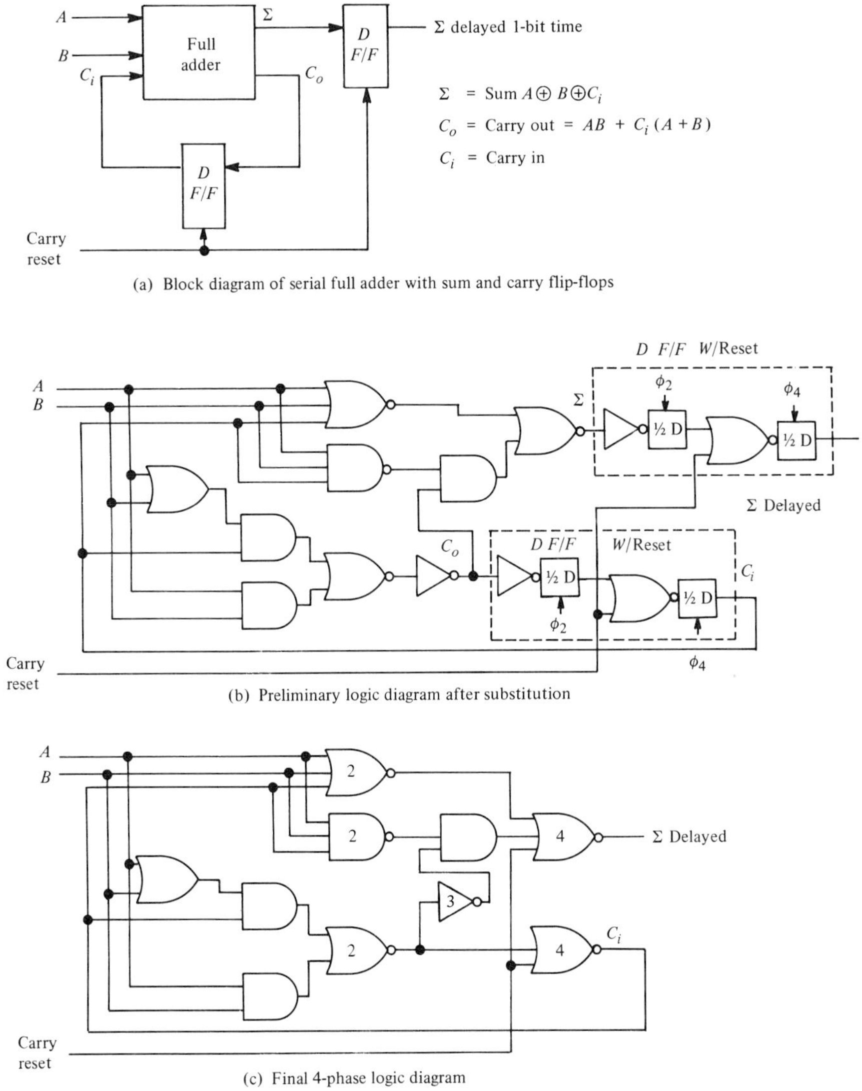

(a) Block diagram of serial full adder with sum and carry flip-flops

(b) Preliminary logic diagram after substitution

(c) Final 4-phase logic diagram

Figure 5–68 Example of 4-phase conversion.

Another type of logic that cannot be tested is the reset from unused states. An example of this is the twisted ring counter of Figure 5-70, although the principle applies to any counter or sequential machine with unused states. In this example, the counter has eight unused states and extra logic (resets) must be added to insure that the circuit has a way of exiting

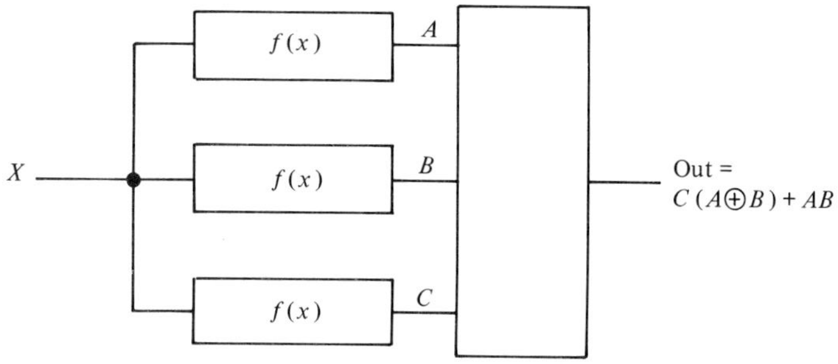

Figure 5–69 Majority voting logic.

from an unused state to a used state if it is erroneously entered during power turn-on or by system noise.

This ability to exit unused states and the logic that causes it to exit cannot be tested unless the tester can force the machine being tested into an unused state. This requires the addition of more logic to force the unused states to occur during testing. In this case (Figure 5-70), a test signal OR-ed into the first bit of delay could force the unused states. Another approach is to use three flip-flops to count to eight; thus, eliminating any unused states.

Figure 5-71 is an example of the unused state problem shown in the form of a flow chart of a sequential machine with three states (S_0, S_1, S_2). The unused state (S_3) must also be shown on the flow chart and an escape path provided to return the circuit to normal operation in case the unused state is accidentally entered due to power turn-on or system noise. This escape route cannot be tested unless logic is added to forcibly enter the unused states during testing.

Other testing problems which occur are mostly concerned with long test programs that exceed the tester shift register length, test pad pin limitations, and tester logic synchronizing which is covered in Chapter 6.

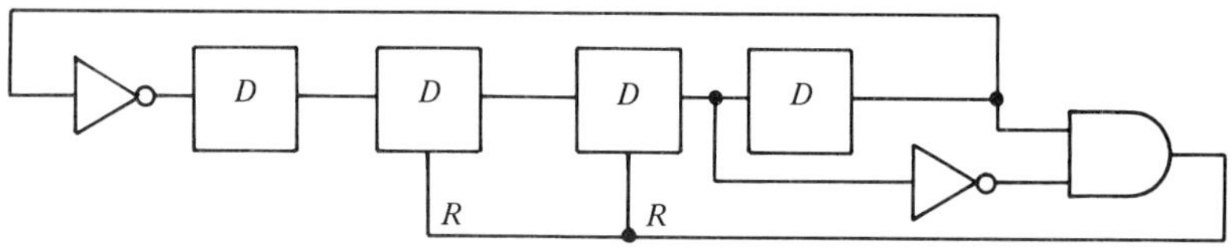

Figure 5–70 Twisted ring counter.

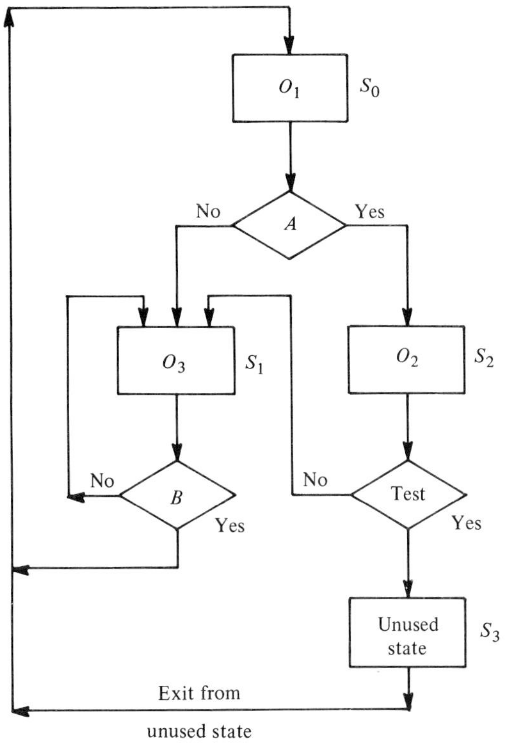

Figure 5–71 Flow charting unused states for testing.

5.8.2 Reduction of Test Program Length

Lengthy test programs can usually be shortened by proper partitioning. It is particularly advantageous for the person who does the logic design and partitioning to also write the test program. As a minimum, the logic designer should be aware of test equipment limitations and have some familiarity with testing problems before he does the logic design and partitioning. The ramifications on system cost can be very significant.

Long test programs can also be reduced by adding test inputs to an existing design as shown in the counter of Figure 5-72. Pulses on this test input will cause the second half of the counter to advance faster than the normal rate which results in shortening the required test pattern length. Programs may also be shortened by adding output test points.

5.8.3 Test Terminal Limitations

The impact on system cost of chip connections (pads), as emphasized in Chapter 6, dictates the minimization of connections that are used only for test purposes.

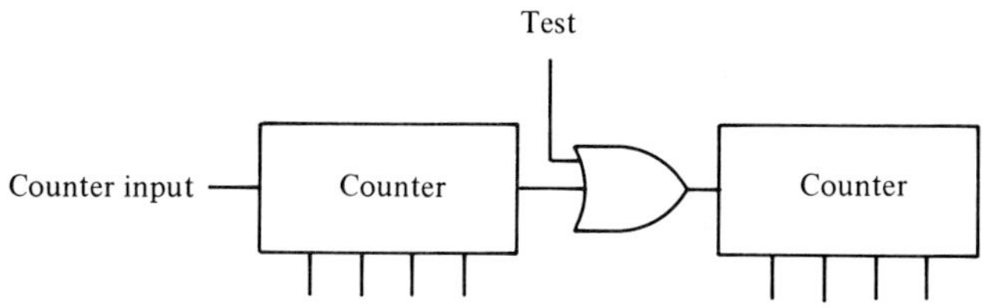

Figure 5–72 Long counters shortened by test pad.

As previously shown, test requirements may be reduced by functionally separating the logic during the test procedure. Several techniques that enable additional test functions while minimizing chip connections have been developed.

Figure 5-73 shows a chip with two control inputs (or timing inputs) that are mutually exclusive during normal system operation. An unused condition has been decoded on the chip for an extra test function. This principle may be used whenever there are input pin logic combinations that are not used during normal system operation. Obviously four input test functions may be coded on two test pins.

Figure 5-74 shows how test functions may be serially coded. A single pad feeds a serial to parallel shift register. If, for example, 4 bits are stored and decoded, 16 test functions may be generated. This circuit need not be used exclusively for testing. The same serial to parallel converter used for saving pins on control inputs can generate test functions if there are any unused serial input combinations during normal system operation.

Output test points may be multiplexed as shown in the example in Figure 5-75. If the four test points were brought out directly, it would require four pads and four output drivers, which consume large amounts of valuable chip area. The addition of test point select gates reduces this to one

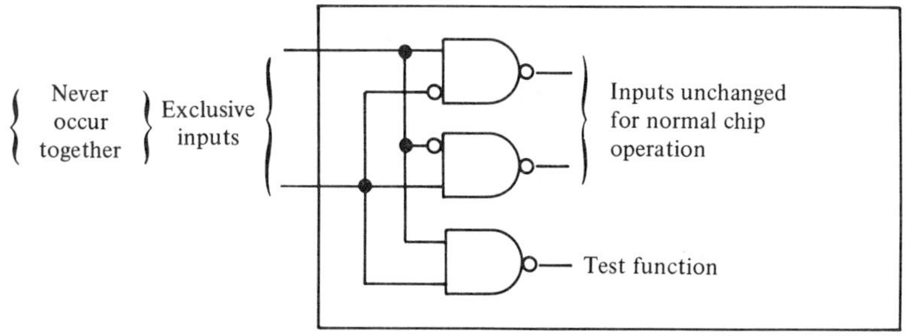

Figure 5–73 Test functions from exclusive control signals.

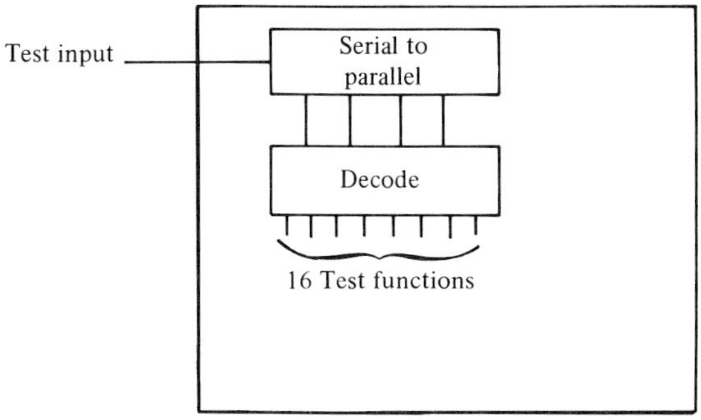

Figure 5–74 Serially coded test function.

output driver and three pads. The more test points that are required, the larger the savings realized by multiplexing.

5.8.4 Tester-Chip Synchronizing

Many testers cannot be reset or synchronized to chip outputs. The test pattern registers are recirculating continuously and the chip logic must be synchronized to the tester. Chip logic that continuously recirculates such as counters or recirculating shift registers must therefore have some sort of reset or sync input signals so that the tester can initialize them to known states. This reset line may be used to force the logic into the unused states which will synchronize it and test the escape routes from the unused states at the same time.

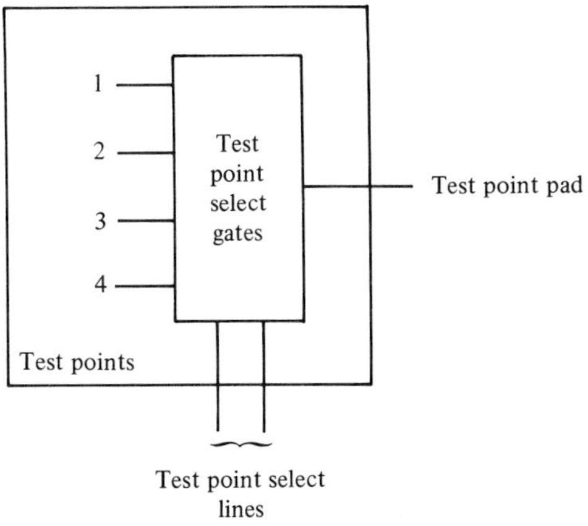

Figure 5–75 Multiplexed test points.

REFERENCES

1. M. J. Phister, *Logical Design of Digital Computers,* John Wiley & Sons, New York, 1958.
2. R. S. Ledley, *Digital Computer and Control Engineering,* McGraw-Hill, New York, 1960.
3. T. C. Bartee, I. L. Lebaw, and I. S. Reed, *Theory of Design of Digital Machines,* Lincoln Laboratory Publications, McGraw-Hill, New York, 1962.
4. W. E. Wickes, *Logic Design With Integrated Circuits,* John Wiley & Sons, New York, 1968.

6

System Design with MOS Arrays

6.1 INTRODUCTION

When a systems manufacturer considers the use of MOS LSI technology in either an existing system or a new system design, the factors to be analyzed and the trade-offs that must be weighed are numerous and quite different from those encountered in designing systems with conventional SSI (small-scale integration) or MSI (medium-scale integration) circuits.

This chapter covers the important aspects of the multilevel interface between the systems manufacturer and the MOS LSI supplier. Also discussed are the system organization and design considerations that yield the most efficient MOS arrays in terms of cost, performance, reliability, and testability.

6.2 CUSTOMER VENDOR INTERFACE

The clear dividing line that once separated the system manufacturing business from the semiconducter suppliers' business has been virtually eliminated with the development of custom LSI circuits. In the past, most systems were built by assembling standard off-the-shelf parts or components, which resulted in a distinct division of responsibilities between the supplier and customer. Now that the customer has recognized the advantages of LSI and has begun using custom LSI circuits, his interface with the LSI supplier can occur at a number of points. Some of the possible interface points are:

- System specification
- System algorithm
- System block diagram
- Preliminary logic diagram (equations) of system
- Partitioned preliminary logic diagrams of chips
- MOS logic diagrams
- Chip specifications
- Chip composites
- Chip rubylith or mask masters

In view of these many interface points, the LSI supplier's interface organization must be staffed with personnel in fields such as systems engineering, logic design, circuit design, and mechanical design to work effectively with any potential customer. Similarly, customers electing to use LSI must acquire sufficient knowledge of their LSI suppliers' business to use the MOS technology to best advantage. This means that customers must acquire talent and understanding in the fields of semiconductor processes, process control, device design, design limits and rules, LSI layout techniques and constraints, and LSI testing. The degree to which the supplier and customer must understand each other's business depends greatly upon the interface point at which they mutually agree to operate.

Since no single interface level is best for all customer-vendor relationships, the nine levels listed will be examined along with the advantages and disadvantages of each.

6.2.1 System Performance Specifications

This is normally the highest level of interface. Here the customer supplies the design requirements and acceptance criteria to his LSI supplier, who is given the freedom to select what he considers the best design approach to implement the system functions within the specified budget. Thus the LSI supplier must assume complete responsibility for successful system performance as defined by the specification. If the LSI supplier totally understands the functions and application of the system, this interface can result in optimizing the implementation. However, the LSI supplier's chip development costs are likely to be high because of the extensive engineering required. Also, there is a danger that some important system parameters may not be specified or that design or test requirements may be misunderstood. Inevitably, there are changes that occur as the system design progresses and these are likely to cause program delays and increase costs.

6.2.2 System Algorithm

The next interface level usually consists of system flow charts that show the steps used in solving the system problem. Although this provides more detailed definition of what the customer expects, it may result in less than an optimum system implementation with LSI technology. Again, as with the system specification interface, the LSI circuit development costs are likely to be quite high and important system parameters may be overlooked.

6.2.3 System Block Diagram

This interface level consists of a diagram that divides the system into functional blocks, such as an adder-subtractor, multiplexer or register, which the customer believes are necessary to solve the system problem. The LSI supplier is given the freedom to optimize the implementation of each functional block with his LSI technology. However, although the LSI circuits may perform the functions described, the system may be inadequate in some respects. In this interface level, the LSI supplier's development costs may be reduced somewhat since little system engineering is required.

6.2.4 Preliminary Logic Diagram (or Logic Equations) of System

This interface point allows the LSI supplier freedom to implement the system logic using dynamic (2-phase or 4-phase) or dc logic techniques as appropriate. Also, if the LSI supplier has an automated design capability, he usually can interface well with preliminary logic diagrams or logic equation inputs. This results in lower chip development costs and reduced development time spans. The LSI vendor usually simulates the logic of each chip and verifies that the individual chips can be thoroughly tested. This interface level often provides the best trade-off between development cost and responsibility for proper performance.

6.2.5 Partitioned Preliminary Logic Diagrams of Chips

At this interface level, the customer provides the LSI vendor with a set of logic diagrams which he has already partitioned into individual LSI chips. Although this clarifies the chip to chip interfaces, it somewhat reduces the LSI supplier's flexibility to use his process to maximum advantage. For instance, the LSI supplier may be able to partition the system into smaller, less costly chips than the customer. The responsibility for specifying the chip to chip performance parameters, power supplies, clocks, etc., must be given to the LSI supplier, but the customer must guarantee that the logic on each chip is testable. As in the system logic diagram interface, this level of interface does allow maximum use of the LSI supplier's capabilities.

6.2.6 MOS Logic Diagrams

In this case, the customer converts the system logic into detailed MOS logic diagrams using static logic, 2-phase ratioed or ratioless logic, 4-phase ratioless logic, or combinations of these. The use of the vendor's standard MOS cells for the conversion provides an excellent basis for communication between customer and vendor and results in an effective interface.

6.2.7 Chip Specification

In this interface, the customer generates specifications that detail performance requirements and test criteria for each LSI chip and then negotiates price and delivery with perhaps several LSI manufacturers. If correctly prepared, these specifications should eliminate any design or test interpretation difficulties. On the other hand, they may severely limit the vendor from taking full advantage of his particular LSI technology. In this interface, the customer must have an in-depth knowledge of his supplier's process capabilities, design techniques, design limits, packages, and test equipment.

6.2.8 Chip Composite

At this interface point, the customer supplies a composite drawing, usually 400x or 500x, of the physical layout of each LSI chip. From the composite, the LSI supplier prepares rubylith artwork for each masking operation required by his process. This interface gives the customer the tightest control over the design of the LSI chip, but requires that he have thorough knowledge of his supplier's process. In addition, he must know how to design devices correctly and then translate this information into a physical layout of the chip using the supplier's design rules. The advantage of tight design control should be weighed against the many disadvantages. First, it is very unlikely that a composite drawn using design rules from one LSI supplier will be directly usable by another vendor because of differences in their processes. Thus, the customer may be locking himself into a single source situation. Second, the LSI supplier's design rules often change as he develops methods to improve his yield. These design rule changes must be incorporated into the composite by the customer to achieve the best yield. Third, the LSI supplier has no responsibility for the proper functioning of the LSI circuits, but only for proper processing to the design limits. Testing responsibility usually rests with the customer, and the LSI supplier only performs device parameter measurements. If the customer desires, he may provide functional test criteria for good chips and have the LSI vendor test and package the good dice. This interface level minimizes the LSI supplier's development costs but may raise the overall costs, since the customer has to duplicate some of his LSI supplier's skills and facilities. The LSI supplier is likely to amortize these skills and facilities over higher volume, thus lowering the costs of a given job. This approach does have an advantage in this it should shorten the LSI supplier's chip delivery time span. However, it does not necessarily shorten the overall system development time span unless the customer can design the circuits faster than his LSI supplier can do the same job.

6.2.9 Chip Rubylith or Mask Masters

This interface consists of the customer providing his LSI supplier with either rubylith artwork for each photo mask or mask master plates for use in preparing mask working plates. Except for the customer taking responsibility for the preparation of the rubyliths or mask master plates, the advantages and disadvantages cited previously for the chip composite interface point are also valid here. However, the customer now must have additional capabilities in rubylith cutting. If mask master plates are to be supplied by the customer, he must also have or have access to a photo facility with accurate, high resolution photoreduction and step and repeat capability. The additional facilities required are costly and difficult to justify unless they can be amortized over large volume. The semiconductor vendor is more likely to amortize such costs than the customer.

From the foregoing, optimum customer interfaces with LSI suppliers are most probable at the points from system block diagram to chip specification. These interface levels take best advantage of the strong points of both sophisticated systems manufacturers and semiconductor suppliers and result in least duplication of expensive facilities and skills.

6.3 SYSTEM PARTITIONING CONSIDERATIONS

In implementing a system with MOS LSI arrays, a great deal of attention must be given to partitioning. Partitioning is the procedure of dividing system logic diagrams into large-scale arrays in the most efficient manner possible. The results of this procedure drastically affect system cost, performance, and reliability.

In this section, the objectives of system partitioning are first defined; then the limitations which the technology imposes on partitioning are considered. Next, guidelines to effective partitioning are presented and their interrelationships analyzed. Finally, the effects of each guideline on the objectives of system partitioning and some criteria for making rational trade-offs between diametrically opposed guidelines are discussed.

6.3.1 Objectives of System Partitioning

The objectives of system partitioning are to reduce system cost, increase system performance, and increase system reliability.

System cost is a function of the costs of the LSI chip processing, packaging, assembly, testing, and system maintenance. The partitioning phase directly influences each of these factors to some extent.

System performance can be adversely affected by indiscriminate partitioning of a system. As is true for most systems, the interface between subsystem units is a potential problem area and must be carefully analyzed.

System reliability is a function of the system configuration, the reliability of the individual LSI circuits, the system assembly and packaging methods, the system testing procedures, and the normal operating environment of the system.

6.3.2 Limitations on System Partitioning

Before the partitioning phase can be initiated, it is necessary to understand the limitations or constraints that are imposed on partitioning by current MOS LSI technologies. The primary considerations are the number of MOSFETs per chip and interface restrictions between chips.

Factors that influence the maximum number of MOS transistors per chip are:

- Maximum allowable chip size
  1. photolithographic restrictions
  2. package restrictions
- Desirable chip size
- Type of logic element
- Speed requirements

- Power dissipation
  1. package dissipation
  2. required ambient operating temperature range
- Number of outputs per chip

Much interrelationship exists within these factors; therefore, although the results of their effects are known, the degree to which each contributes toward these results is somewhat less defined.

The maximum allowable die size that can be achieved today is approximately 300 × 300 mils; however, this is an adventure in photolithography. Therefore, it is necessary to consider the desirable die size, which is a function of the state-of-the-art in MOS LSI processing and the cost of the package. When the die is relatively small, 100 × 100 mils or less, the package cost is predominant. For large chips, i.e., 140 × 140 mils or greater, the cost is determined primarily by process yield. Current studies place the desirable chip sizes between 100 × 100 mils and 150 × 150 mils. For shift registers, the optimum size falls somewhere between 500 and 1000 bits. In the case of read only memories (ROMs), the optimum size is somewhere between 2000 and 4000 bits of storage. Random access memories (RAMs) vary between 128 and 1024 bits for the dynamic type and between 100 and 200 bits for the static type. It is extremely difficult to estimate the amount of combinatorial logic that can be placed on a chip of optimum size, since the logic and interconnect complexities vary widely. However, a very general estimate is about 400 to 700 devices or 100 to 130 gates per chip.

A fairly accurate technique for die size estimation is presented in Section 6.3.10, "Estimation of Chip Area." The estimates for shift registers, ROMs and RAMs are fairly accurate inasmuch as the interconnect patterns are simple and easy to predict.

Type of logic and speed requirements are closely related to chip size and power limitations. Most systems may be implemented in MOS LSI with either dc (static) or ac (dynamic) logic. However, each system normally contains some inherent characteristics that can take better advantage of one particular type. In general, if a choice exists, dynamic logic is preferable, if only to reduce chip size. It is generally agreed that increased operating speeds require increased chip area whether dynamic or static logic is used.

The chip power requirement is the sum of the individual device power dissipations. Power limitations of a chip must be evaluated with a knowledge of available voltages, the type of process, type of logic, and package. When these are established, the power limitations are essentially the power dissipation rating of the package in its predicted environment. Chips that exceed the rated dissipation of the package, although rare, require either functional capability or logic design modifications to reduce power requirements.

On the other hand, when the various factors that influence power limitations of a chip are not fully established, there are several alternatives. For instance, the change from the standard threshold process to a low voltage process provides approximately a 75 percent power savings. When dc logic is used, power can be saved by increasing the on-resistance of the MOS load devices, assuming speed is not reduced below requirements. When ac ratioed logic is used, a power saving can be achieved by lowering the duty cycles of the clock phases. The last and probably the most obvious alternative is to change to another package with a higher power dissipation rating.

Another important factor that limits chip complexity concerns the output buffer circuits which are required to drive the capacitance between chips. The physical size and power requirements of the drivers may be an order of magnitude larger than the average internal logic function. Chip outputs must also have a pad for connection to the package. This pad is usually a probe point during the wafer testing and a bonding point during final assembly. Each of these pads requires an equivalent area of two or three average logic functions. The number of inputs and outputs on a chip also affect the chip in two other areas, chip complexity, and packaging considerations. A large number of input or output leads complicates chip design because the signals must be physically routed to the appropriate locations on the chip. This necessitates a package with many leads and, although an exact correlation does not exist, a package with more leads is normally more expensive and additional labor is required to bond each additional lead.

Other factors that can limit system partitioning flexibility are problems associated with the interface between chips. Capacity storage on the interface between chips should be avoided for two reasons. First, the equivalent leakage resistance that parallels the storage capacitance is several orders of magnitude lower than the leakage resistance internal to a chip. This reduces impedance, necessitates greater storage capacitance, and places more stringent requirements on output driver circuits. Second, capacitive storage is usually associated with either serial or flip-flop data storage. Since the most noise-sensitive portions of any system are the inputs to flip-flops in storage, capacity storage on the interface between chips should be avoided for optimum noise immunity.

Also, for medium to high speed systems that use ac ratioed logic, it is difficult, because of the slow response of output stages, to try to propagate a signal between more than two chips during a single clock phase. This holds true principally for systems designed for clock frequencies of approximately 500 kHz or more.

6.3.3 Guidelines to System Partitioning

The parameters that directly influence system cost, performance and reliability include:

- Number of chips
- Number of chip types
- Test requirements
- Interconnections between chips
- Package cost

A set of guidelines may be drawn from an examination of the effects of each parameter on the partitioning objectives. These guidelines are conflicting; however, not all can be optimized simultaneously. In most instances, these guidelines are mutually exclusive, since the optimization of one guideline opposes optimization of one or more of the other guidelines. It is necessary, therefore, to make judicious trade-offs or compromises between these guidelines in a manner that best satisfies the partitioning objectives.

The interrelationships of these guidelines are complex and depend to a great extent upon the individual system configuration. However, in general, the following relationships give a fair approximation of the results obtained by optimizing one guideline to the exclusion of all others.

6.3.4 Guideline Number 1—Reduce the Number of Chips

Partitioning to minimize the total number of chips will, at the limit, result in placing the entire system on a single chip. From the previous discussions on chip size limitations, a system designer usually finds it necessary to partition a system into a number of chips. This number is a function of economics and engineering judgment.

Optimization of this guideline characteristically results in larger chips as more system logic is included within each chip. Package costs will decrease as fewer packages are required per system. Testing requirements will decrease as less wafer probe test hardware, test programs, and associated documents are needed.

A reduction in the number of chips will also, most likely, increase the number of chip types. It is generally difficult to partition around repetitive areas in the system logic. Additional logic must be included with each repetitive section to achieve the optimum chip size and minimize the number of chips. This reduces the possibility of using one chip type more than once in a system and increases the number of chip types required.

6.3.5 Guideline Number 2—Reduce the Number of Chip Types

To minimize the number of chip types, repetitive functional areas are used in the system logic. The system is partitioned such that each chip type is used several times in each system, thus increasing the production volume of these chips, and lowering their costs. An example is a 16-bit parallel arithmetic-logic unit for a minicomputer which is partitioned into four identical chips of 4 bits per chip, thus enabling one chip type to be used four times per system. Repetition can frequently be forced by the logic designer through the addition of redundant logic even though chip size is often increased. An example of this is a system which requires serial shift registers of 128-, 256-, and 512-bit lengths. Instead of generating three different chips, one 512-bit register with outputs tapped off of bits 128, 256, and 512, will satisfy all requirements. This technique results in increasing interconnection and package costs since more package connections are required.

Optimization of this guideline typically results in a larger number of chips, since the repetitive functions seldom result in chips that are optimum in terms of chip complexity and size. Thus more chips are required to implement the remainder of the system logic. One advantage is that lower inventories are required for system maintenance.

6.3.6 Guideline Number 3—Reduce the Test Requirements

The reduction of test requirements may have a great impact on system cost. The primary consideration is to partition each array such that it may be tested on existing test equipment.

Optimization of this guideline frequently increases interconnect and chip size through the addition of test points and special test functions as described in Chapter 5 under Section 5.8, "MOS LSI Test Considerations." This guideline not only ensures that a chip is testable, but also reduces the time and equipment involved in testing a chip.

6.3.7 Guideline Number 4—Reduce the Number of Interconnections Between Chips

The reduction of interconnect between chips, when carried to an extreme, will increase the number of chip types. For instance, if a system requires shift registers that are 128-, 256-, and 512-bits long, this guideline dictates the development of three chip types rather than one 512-bit chip type with extra output taps, as previously discussed.

This guideline becomes particularly important when package pin requirements for a chip slightly exceeds the capacity of a desirable, low cost, standard package. In Section 6.4.6, "New Systems," several techniques for reducing interconnects are discussed, including wired-OR drivers, time-shared drivers, and coded control signals.

6.3.8 Guideline Number 5—Reduce Package Cost

Reduction of package cost must be considered here because it can be a major factor in the price of production parts. It can drastically affect any of the other guidelines. Obviously, standard packages are desirable due to their generally lower cost and better availability.

6.3.9 Guideline Trade-offs

For any given system each of the guidelines previously discussed is rated in the light of system requirements and characteristics. Several criteria that are used to trade-off between guidelines are: the number of systems to be produced; the desired cost of each system; the desired system reliability; and the desired system performance.

System cost trade-offs can be simplified by considering the cost per data processing function. This is a measure of the total costs, including packaging, assembly, processing, and systems costs for a section of system logic containing one or more functions. Figure 6-1 illustrates the marginal cost relationships of these factors as a function of the number of chips into which a system is partitioned.

As the number of chips increases, more packages are required and packaging costs increase. Similarly, the costs associated with assembling each system increase. Testing costs also increase as a greater number of test programs and associated documentation are generated. Maintenance costs are increased as more spare parts are required. Chip costs decrease

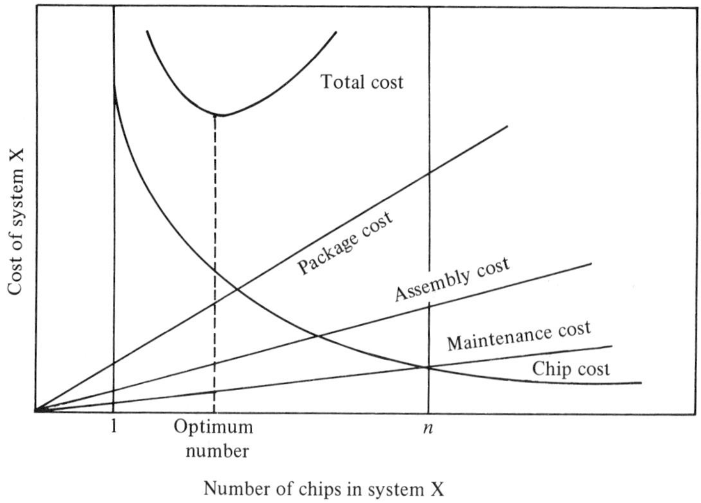

Figure 6–1 System X marginal costs.

as the number of chips increases, since decreased chip size results in greater yields. However, the lowest cost per function, and thus the lowest system cost, occurs when the system is partitioned into the number of chips that corresponds to the minimum inflection point of the total cost curve. Development costs must also be considered in trading off these guidelines. High system production quantities dictate the reduction of the number of chips since the development costs can be effectively amortized. On the other hand, small system production quantities may dictate the reduction of the number of chip types. This will increase the production volume of these chips and allow more effective amortization of the development cost, although the cost per function will increase.

System reliability is improved as the number of interconnections between chips decreases. This is due primarily to the reduced number of packaging bonds and solder connections between packages.

System noise immunity is safeguarded by reducing the number of interconnections, since fewer signals are exposed to the relatively noisy environment external to the package. System operating speeds can also be increased because of the reduction of high capacity loading present in the interconnects.

6.3.10 Estimation of Chip Area

In order to make predictions concerning chip feasibility and cost, it is necessary to develop an accurate procedure for estimating chip area and power dissipation. This is not a simple task, since chip size is heavily dependent upon the type and complexity of the logic and the efficiency of the layout technique. Attempts have been made to create complex algebraic formulas, but experience has shown that use of a standard cell family in conjunction with an automated design system is by far the most accurate technique.

The Die Size Estimate Work Sheets shown in Tables 6-1 and 6-2

Table 6–1 Die Size Estimate Work Sheet

		Typical power dissipation		*Totals*		
Standard cell logic function (negative true logic)	*Typical area (mil^2)*	*High V_T* (mW)*	*Low V_T** (mW)*	*Number used*	*Area (mil^2)*	*Power diss. (mW)*
Inverter	24	8	1.4			
2-Input NOR	36	8	1.4			
3-Input NOR	48	8	1.4			
4-Input NOR	60	8	1.4			
5-Input NOR	72	8	1.4			
6-Input NOR	84	8	1.4			
2-Input NOR expander	48	—	—			
2-Input NAND	36	8	1.4			
3-Input AND-NOR	48	8	1.4			
2-Input NAND expander	48	—	—			
Exclusive OR	59	16	2.9			
RS binary flip-flop	130	16	2.9			
Sample and hold flip-flop	98	11.5	2.2			
Delay flip-flop	46	8	1.4			
Full adder, serial	192	16	2.9			
Push-pull output driver	54	14.6	2.7			
Push-pull output, boot strapped	82	14.6	—			
Open drain output driver	42	—	—			
DTL or TTL output driver	348	—	25***			
Inverter output driver	42	37	6.8			
				Totals		

*$V_{DD} = -15$ V, $V_{GG} = V\phi = -27$ V (typical)
**$V_{DD} = -7.5$ V, $V_{GG} = V\phi = -15$ V (typical)
***$V_{DD} = -0$ V, $V_{SS} = +5$ V, $V_{GG} = -10$ V (typical)

illustrate this standard cell approach to die size estimation. Table 6–1 shows the parameters associated with a limited subset of 2ϕ ratioed standard cells. Although the parameters vary from one vendor to another, typical values of area per function and power dissipation for both high and low threshold processes are listed. The number of cell types, areas, and power are then totaled from an inspection of the partitioned logic diagram.

The total area is then carried over to line 1 of Table 6-2, a continua-

Table 6–2 Die Size Estimate Work Sheet

1. Total active area (from Table 6-1)		____ $mils^2$
2. Interconnect (IC) allowance (percent of line 1)	____%	____ $mils^2$
3. Total area (add line 2 to line 1)		____ $mils^2$
4. Square root of line 3 (active area plus IC area)	____x	____mils
5. Pad requirements: Inputs____Outputs____ [Total no. of pads] ____ V_{DD}____V_{GG}____V_{SS}____CP_1____CP_2____GND____		
6. Pad allowance (for ≤ 15 pads, use 10 x 10. For > 14 pads, use 20 x 20)	____x	____mils
7. Scribing allowance	____x	____mils
8. a. Circuit determined die size (sum of lines 4, 6, and 7)	____x	____mils
b. Pad-determined die size (multiply total no. of pads in line 5 by 1.8 and add 26)	____x	____mils
c. Final die size (entry in line 8a or 8b, whichever is larger)	____x	____mils
9. Die area (square entry in line 8c)		____ $mils^2$

tion of the Die Size Estimate Work Sheet. An interconnection allowance in the form of a percent of total active area is estimated and placed on line 2. This allowance varies appreciably, depending upon the complexity of the interconnect. Typical values may range between 25 percent for primarily serial interconnects to more than 200 percent for complex random logic functions.

The interconnect allowance figures are based primarily on prior experience. A new total is generated on line 3, and the square chip dimensions are placed on line 4. Pad requirements are noted on line 5. The pad allowance is defined by the total pad count being ≤14 or >14. This allowance is listed on line 6. Scribing allowance is included on line 7 and a new total is generated on line 8. The computation on line 8b gives the pad-determined chip size, thus providing an alternate estimate criteria for the rare case where a large number of pads dictates a larger chip than that determined by active area and pad allowance.

The chip size expected from a manual layout is estimated as a percentage of the standard cell estimate. This reduction ranges from 20 to 50 percent, depending upon the type of logic, complexity of the interconnect, and the skill of the layout design engineer.

6.4 SYSTEM DESIGN CONSIDERATIONS

The approach used in partitioning a system into subfunctions that would be feasible for integration into MOS arrays depends on whether the system already exists or a new design is required.

6.4.1 Existing Systems

In the conversion of an existing system to MOS arrays, the probability of taking full advantage of the technology in the redesign is not very high. Frequently, the cost and time schedule constraints rule out complete redesign. The basic problem then involves the analysis of the existing system logic and definition of suitably partitioned MOS arrays, using an approach such as that shown below.

1. Analyze system logic
   a. Can the design take advantage of inherent MOS capabilities?
   b. Is the design compatible with MOS logic?
   c. Is the speed (number of logic levels) compatible with MOS characteristics?
2. Redraw system in MOS logic taking advantage of its properties
   a. Use bit of delay in lieu of *D* flip-flops.
   b. Use complex combinatorial logic input gate structures.
3. Examine possible ways to partition logic (considering control as well as data paths)
   a. Cross-sectional
   b. Register organization
   c. Functional
   d. Mix of cross-sectional and functional
   e. Distributive
4. Evaluate speed, drive, and package considerations
5. Repeat procedure until satisfied with system design

6.4.2 System Logic Analysis

In analyzing the existing system logic, the first task is to determine how, or if, the present design can take advantage of inherent MOS capabilities. Factors to be considered include: Can the readily obtainable delay elements of MOS be utilized to best advantage? Can logic be simplified by using complex input gating structures? The logic must then be examined to determine the compatibility of the design with MOS logic. Some of the factors to be questioned are: What class of MOS logic (i.e., static, dynamic ratioed, dynamic ratioless, 2-phase, 4-phase, etc.) is most compatible with existing design in terms of speed, number of logic levels per bit time, size requirement, and performance? Do power supply limitations or interface requirements impose stringent restrictions on applications of MOS circuits? Is the drive capability of MOS sufficient to handle the interface requirement? Can logic be reduced to the general form of a synchronous sequential machine, or must asynchronous design techniques be used? The answers to these questions may indicate the necessity for a logic redesign to correct the problem.

6.4.3 Conversion of Logic Diagram to MOS Logic

Once the system requirements have been analyzed and problems resolved, the next step is to convert the existing logic diagram to MOS logic. The trade-offs between a custom and a standard cell design must be carefully weighed, keeping in mind the ramifications of each approach on cost, turnaround time, and the customer-vendor interface in general.

The existing logic diagram must be reduced to the form of a preliminary logic diagram and then converted to the desired MOS logic circuit type, using the procedures defined in Chapter 5, for 2-phase ratioed, 2-phase ratioless, or 4-phase ratioless logic. All logic functions must be clearly identified including output buffer circuits.

6.4.4 Partitioning of System Logic Diagram

Partitioning is probably the most important, and most difficult task of the design procedure. Using the guidelines presented in Section 6.3.1, "Objectives of System Partitioning," the system logic diagram is divided into smaller logic diagrams, each representing a single MOS chip.

Partitioning remains more of an art than a science, with success being largely a function of the engineer's experience. Several techniques (or combinations thereof) that find frequent application are functional, cross-sectional, and distributive partitioning.

Functional—This approach subdivides the system logic diagram on boundaries defined by one or more complete logical functions. For instance, a small digital system might be partitioned as shown in Figure 6-2. Each control chip performs a particular subroutine utilizing a common arithmetic logic unit (ALU) when given a "start" signal, and generates a "finish" signal when the algorithm is completed.

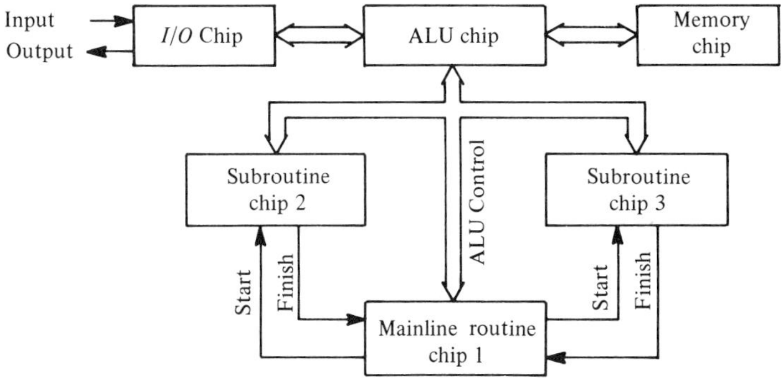

Figure 6–2 Functional partitioning.

Cross-Sectional—It is often possible to reduce chip interconnections of a parallel organized system by partitioning on a cross-sectional basis, as shown in Figure 6-3. In this case, a parallel adder is shown with the partition on a bit boundary. The partition may be expanded to include as many cross-sectional bits as permitted by chip size or package connections. For example, an 8-bit (byte) cross section would require 26 signal pads and occupy an area less than 100 × 100 mils.

Distributive—In general, due to the comparatively large chip area consumed by input/output connection (pads), it is advisable to trade logic complexity for a minimum number of leads. In other words, if duplica-

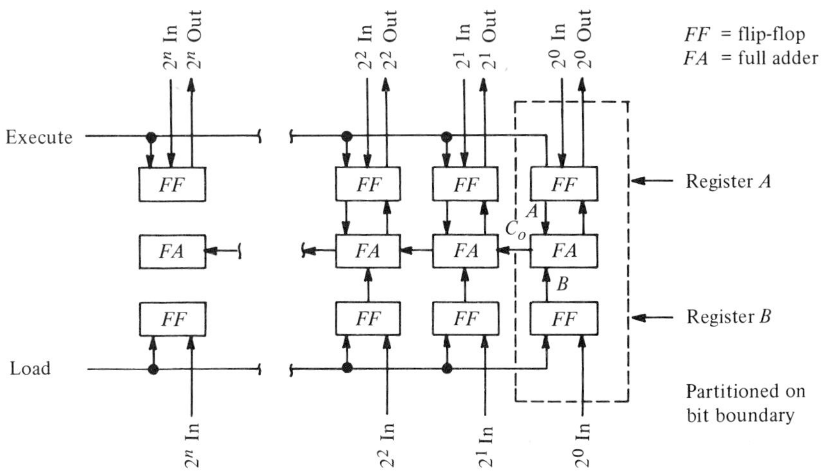

Figure 6–3 Cross-sectional partitioning.

tion of a function on two or more chips significantly reduces the number of interconnects required, the system cost may be lowered. For instance, in Figure 6-4a, a 4-bit decode counter (*X*) interfaces with a decoder (*Y*) which fans out 10 lines to functions *A*, *B*, *C*, and *D*. If this configuration is partitioned into three chips, as shown in Figure 6-4b, the chip interfaces consist of 10 signal lines. However, Figure 6-4c illustrates how duplication of the decoder (function *Y*) enables encoded information to be input to several arrays, reducing the interface to four signal lines.

6.4.5 Evaluation

Upon completion of the preliminary partitioning stage, each logic diagram is evaluated to verify absence of design errors.

In the case of 2-phase ratioed designs, all propagation delay paths (or

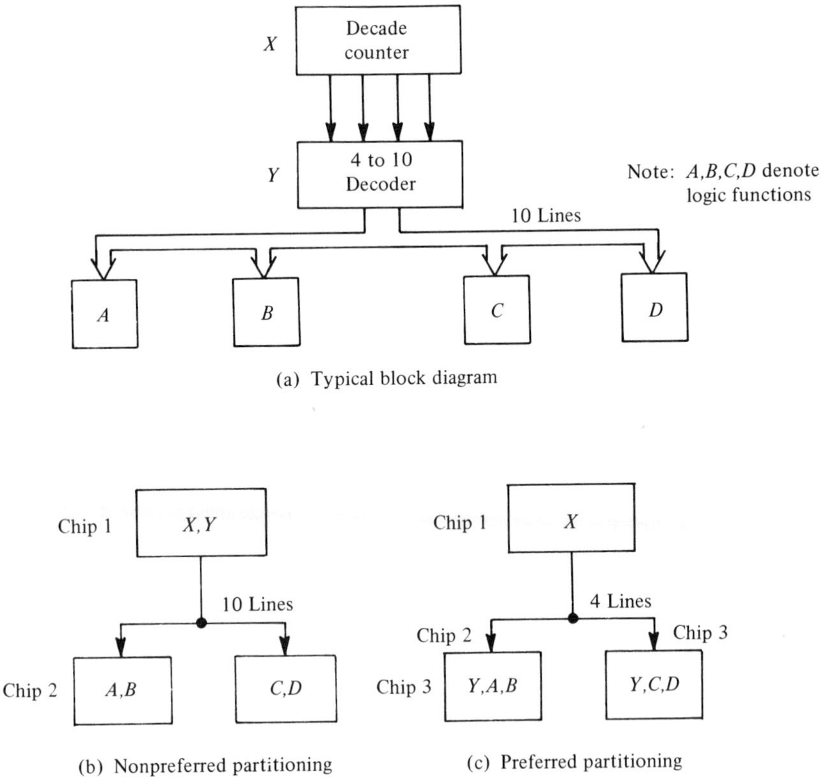

Figure 6–4 Distributive partitioning.

multiple logic levels per bit time) are analyzed to assure proper system operation during worst-case conditions. Signals between chips are examined for any problems that may arise due to output buffer characteristics including delay, fanout, and speed. The testability of each array is evaluated in the light of available production test capabilities and restrictions.

The estimated die sizes are re-examined to ensure that they are within desired limits and, finally, the total worst-case power dissipation is calculated and compared against the selected package rating.

If problems exist in any of these areas, logic modification or repartitioning is necessary. The prime considerations of system cost, reliability, and packaging requirements must be kept in mind throughout this procedure. The time spent during this early phase is usually well invested.

6.4.6 New Systems

When designing a new system, it is possible to take full advantage of MOS technology. In this case, there is freedom to design around the previously discussed limitations imposed by the technology. The degree to which this can be accomplished is directly related to the ingenuity, experience, and effort of the system logic designer. An approach to partitioning a new system is outlined below and key points are discussed in the following text.

1. Examine system peculiarities
   a. Input/output requirements
   b. Throughput requirements
2. Consider approaches to processing (functional diagram)
   a. Serial
   b. Parallel
   c. Serial/parallel
   d. Operations in parallel, data serial
   e. Distributive arithmetic
   f. Bus organization
3. Generate detailed logic diagram
4. Examine ways to partition logic
   a. Cross-sectional bit
   b. Register organization
   c. Functional

   d. Mix of cross-sectional and functional
   e. Distributive

5. Investigate pin-count reduction techniques
6. Evaluate speed, drive, and packaging considerations
7. Iterate procedure until satisfied with design

6.4.7 System Peculiarities

The first step is to examine any system peculiarities. Input/output requirements must be completely specified, including formats and all electrical and performance parameters. The system transfer or throughput characteristics must be examined, including allowable processing delays, operating frequency, and logical operations. A preliminary functional block diagram should be generated, describing the data flow and subsystem functional relationships. A careful examination of the functional block diagram must be made to gain insight into possible system organization. Factors to be considered are commonality and interaction of functions.

6.4.8 Approaches to Processing

Once the system requirements have been established, various approaches to organization of system processing must be considered at the detailed functional diagram level. Several approaches, or combinations thereof, that are conducive to implementation with MOS LSI are: serial, parallel, serial/parallel, operations in parallel, data serial, distributive arithmetic, and bus organization.

Serial processing defines the performance of all logical operations on a bit-by-bit basis, resulting in minimum hardware and interconnect. Although the instruction execution frequency is lower than with other system organizations, the obvious cost advantages that result from serial processing may dictate this approach.

Parallel processing allows increased instruction execution speed at the expense of logic complexity by performing logical operations on a multi-bit basis. The economical partitioning of a system of this type is generally difficult unless all parallel interfaces are contained within a single array.

A combination of serial and parallel processing can be utilized in applications where not all system sections require high operating speed.

Digital systems that require high operating frequency and perform several different operations on identical data may be efficiently organized such that logical operations are performed in parallel and data are kept in a serial format. This is illustrated in Figure 6-5, where an analog

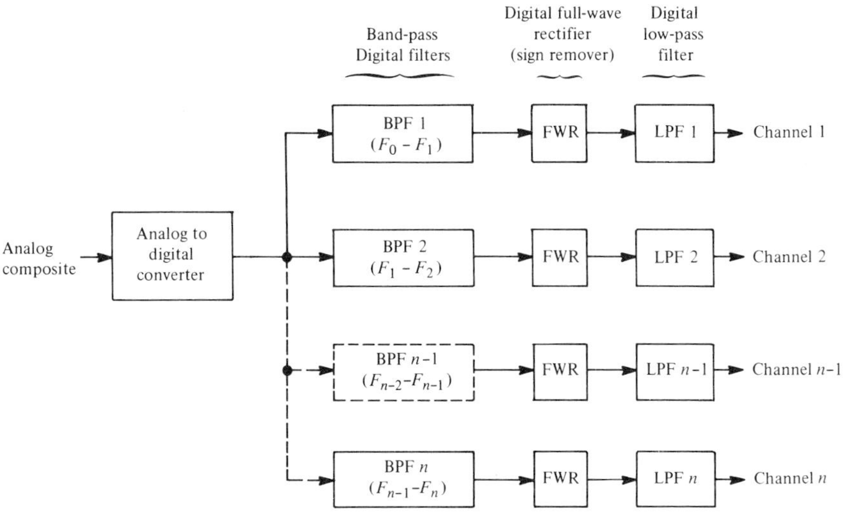

Figure 6–5 Operations in parallel, data serial.

signal is converted to serial format and fed into an N-channel spectrum analyzer that consists of digital filters, rectifiers, and low-pass filters. This enables efficient partitioning and can often reduce the number of chip types in a system.

The unique characteristics of MOS LSI technology are particularly amenable to the design of sampled-data systems. The capability of economically using digital building blocks such as multipliers, shift registers, and adders to implement complex z-transform expressions[1] has resulted in increasing applications in many areas, most notably that of digital data communications systems and servo-control systems. Typical uses include digital filters, phase locked loops, discriminators, and voltage controlled oscillators in systems such as modems (modulators-demodulators) and vocoders (voice-encoding and decoding). This organization technique is termed distributive arithmetic due to the resultant distribution of the arithmetic functions throughout the system logic, wherever required, by he system algorithm. An example of a section of a demodulator is shown in Figure 6-6. The arithmetic algorithms usually employ serial data interfaces, thus resulting in low interconnect requirements and economical partitioning.

Another approach particularly amenable to MOS LSI systems is the bus organization. This technique, utilizing the open drain and wired-OR capabilities of MOS, is described in a subsequent section. Figure 6-7 illustrates a simple bus organized computer which features serial data

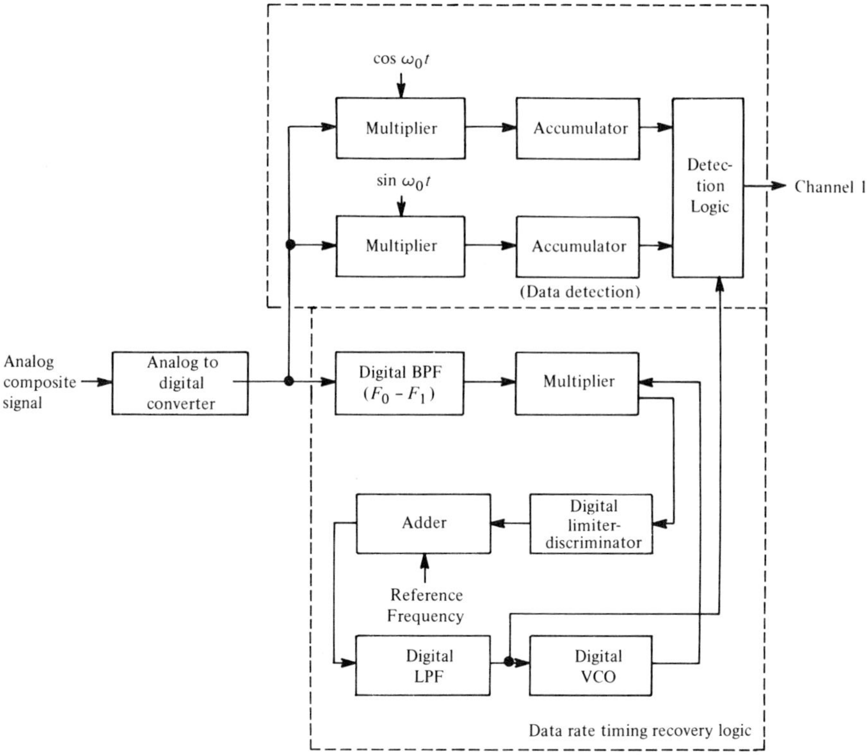

Figure 6–6 Distributive arithmetic.

transfers on a bus. An instruction bus results from paralleling up to N microprogrammed read only memories (ROMs), and controls the source and destination of information on the serial data bus.

6.4.9 Detailed Logic Diagram

After the system organization is defined, the functional block diagram must be mechanized with MOS logic elements, and a detailed logic diagram is generated. The design procedure is covered in detail in Chapter 5.

6.4.10 Partitioning Techniques

The logic diagram must then be partitioned into MOS arrays as described in Section 6.3, "System Partitioning Considerations." In addition to the general partitioning guidelines, such as cross-sectional, functional, and distributive, there are several other techniques for the reduction of inter-

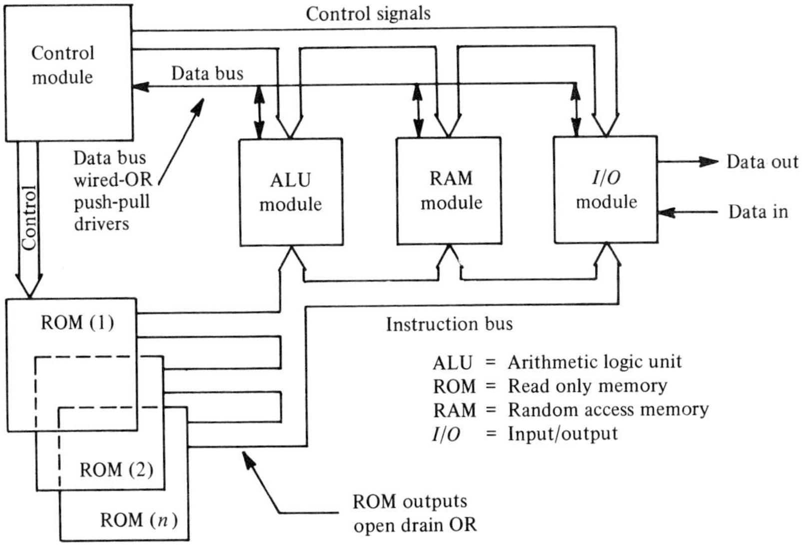

Figure 6–7 Bus organization (micro-programmable computer).

connects between chips: wired-OR, coded control, distributed time-base and time-shared drivers.

Figure 6-8 shows the application of the wired-OR technique to MOS. This use of a bus line normally at V_{DD} allows any chip to have an open drain MOSFET which pulls the bus line to ground. Only one pull-up (load) device or external resistor is used, and its location is unimportant. This bus line provides for an expandable system as other chips may subsequently be added to the bus. The pin requirements are reduced since the OR function is accomplished external to the chips.

The wired-OR technique may also be used with push-pull driver circuits as shown in Figure 6-9. This circuit requires two additional devices

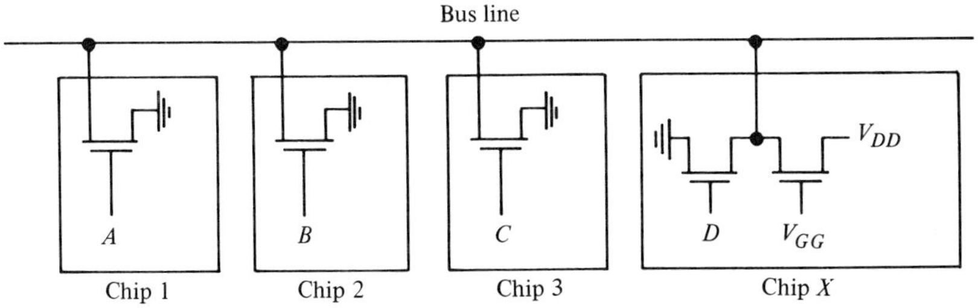

Figure 6–8 Open drain wired-OR.

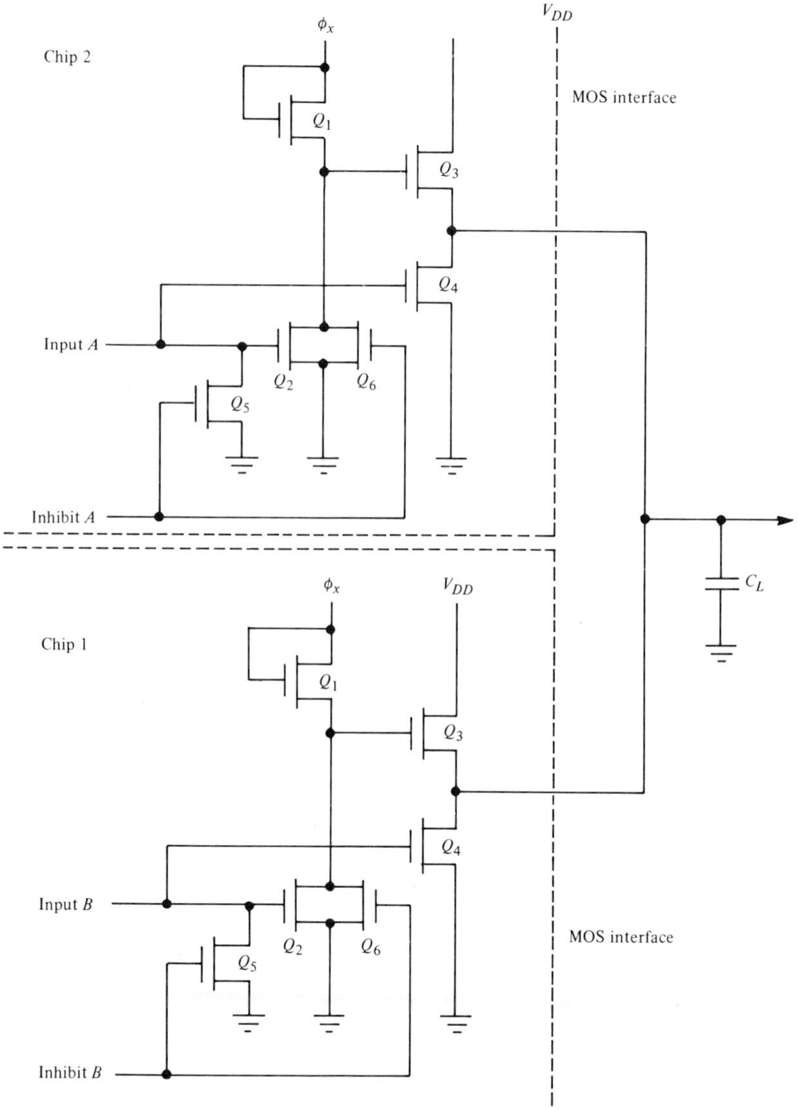

Figure 6–9 Wired-OR push-pull drivers.

(Q_5 and Q_6) plus a control signal (inhibit) that selectively enables one driver circuit to output data on the bus, while inhibiting all other drivers.

Figure 6–10 shows another use of the bus technique, the multiple function wired-OR. In this case a flip-flop feedback is brought back from the

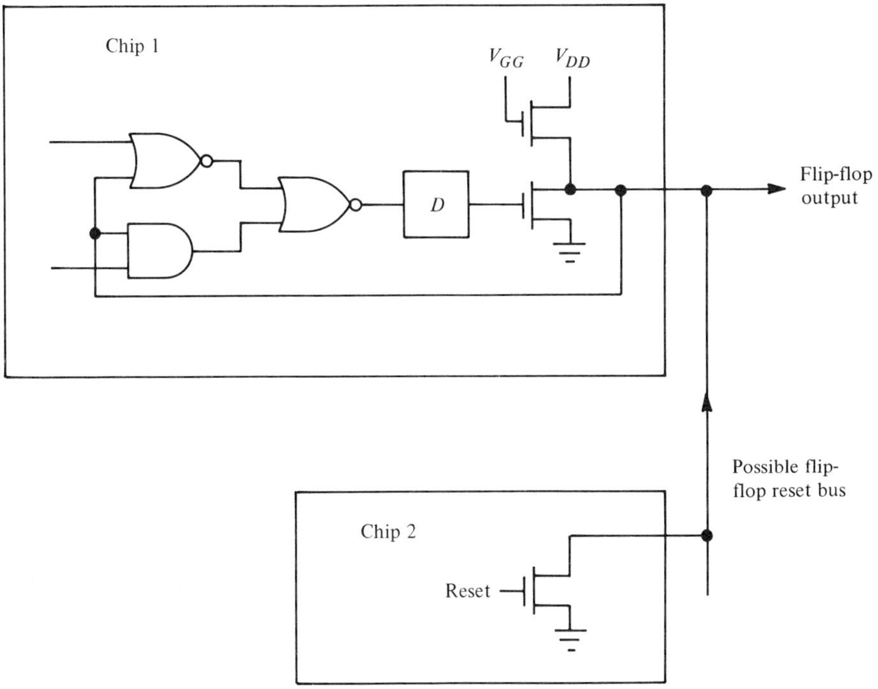

Figure 6–10 Alternate wired-OR technique.

output driver instead of directly from the delay element. This allows the use of a single output lead for flip-flop output and flip-flop reset. Although this technique reduces pad requirements, it should not be used unless absolutely necessary to keep pin-count under the package maximum, since it complicates testing. Normally, test equipment channels can be defined as either input or output, but not both.

In order to test chip 1 of Figure 6-10, it is necessary to both test for correct output logic and to input a reset signal to test that function. This requires one input and one output channel from the tester, plus additional hardware to switch the two tester channels between the output pad of chip 1 at appropriate times during the test sequence.

Figure 6-11 illustrates another method for reducing pin-count. Whenever mutually exclusive control signals are to be transferred from one chip to another, and system operating speed permits, they may be encoded on one chip and decoded on another. If the signals are not mutually exclusive, they may still be reduced as shown in Figure 6-12. Here the

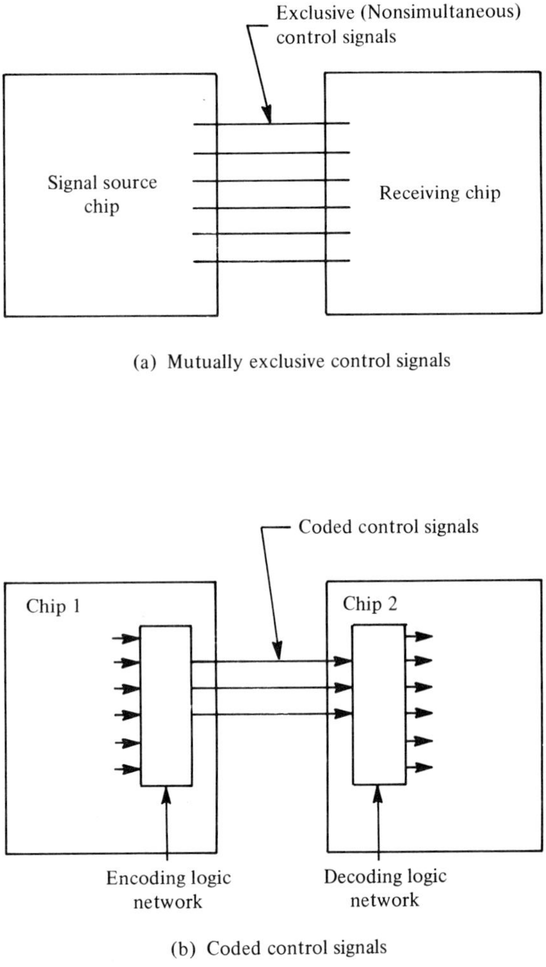

Figure 6–11 Coded control signals.

control signals are serially multiplexed using parallel to serial and serial to parallel converters.

Figure 6-13 shows how sequential timing pulses may be generated using a gated sync pulse and a shift register. Distributing the time-base functions throughout a system can reduce interconnect and pin-count. Each pin that is eliminated saves approximately 100 mils^2 in chip pad area and, since the delay bit requires only 50 mils^2, this is a very useful technique. Output pins that are eliminated create even greater savings, since the output drivers are also eliminated. Output drivers typically

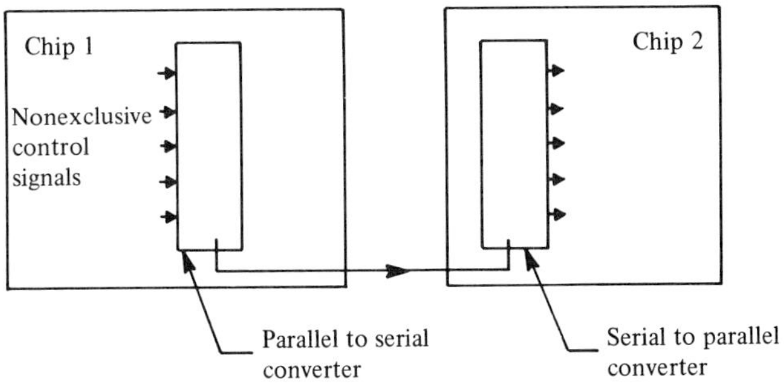

Figure 6–12 Serialized control signals.

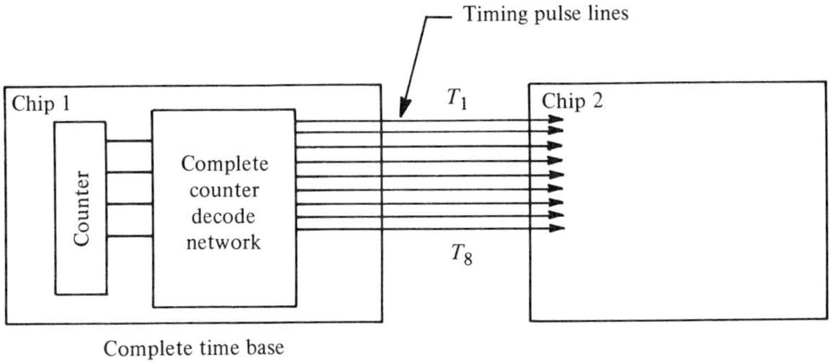

(a) Decoded timing signals

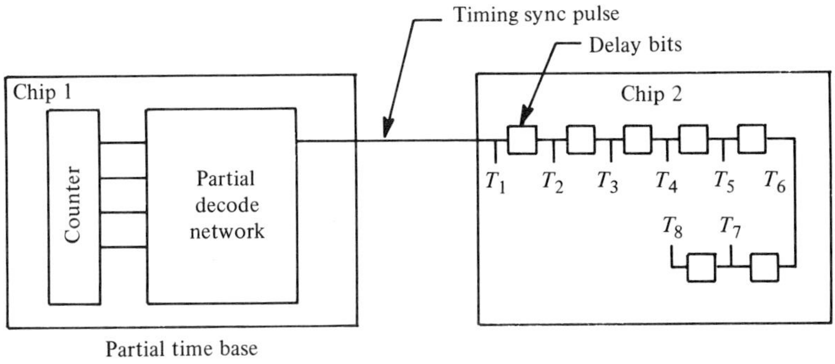

(b) Reduced interface using sync pulse

Figure 6–13 Distributed time-base.

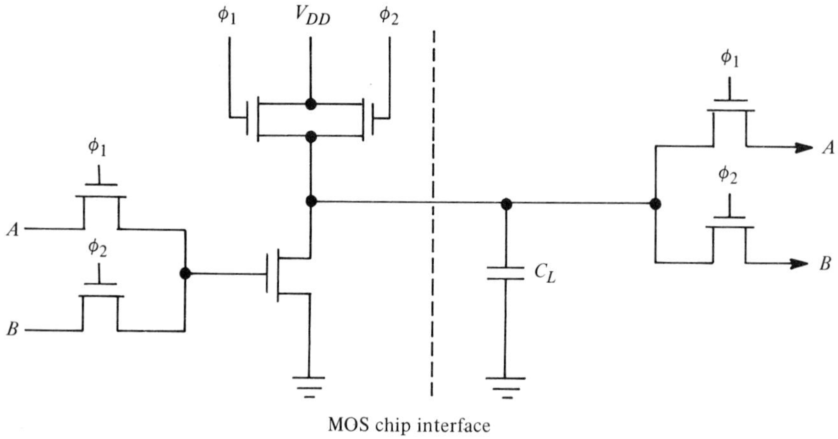

Figure 6–14 Time-shared inverter driver.

require 40 mils² or more each, resulting in a total area reduction of at least 90 mils² for each pin eliminated.

Yet another method for reducing pin-count is to time-share drivers. Figures 6-14, 6-15, and 6-16 illustrate time-shared inverter drivers, push-pull drivers, and open drain drivers, respectively. In each case the clocked series coupling devices control the data path between chips. When clock phase one (ϕ_1) is on, signal A is transferred across the interface, and during clock phase two (ϕ_2), signal B is transferred.

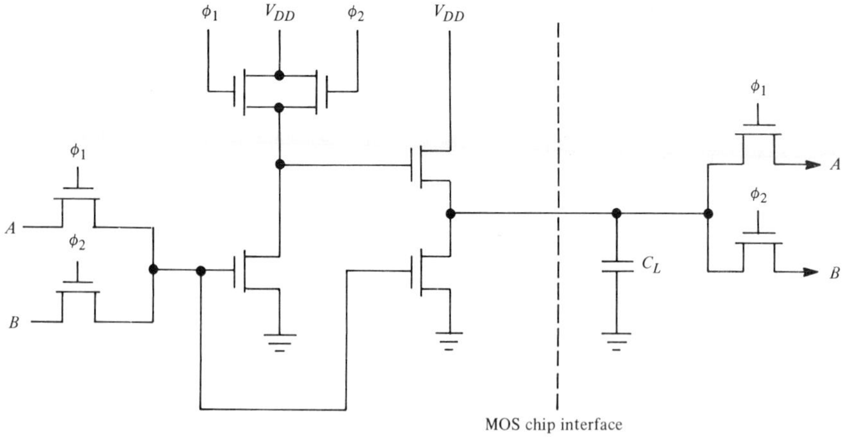

Figure 6–15 Time-shared push-pull driver.

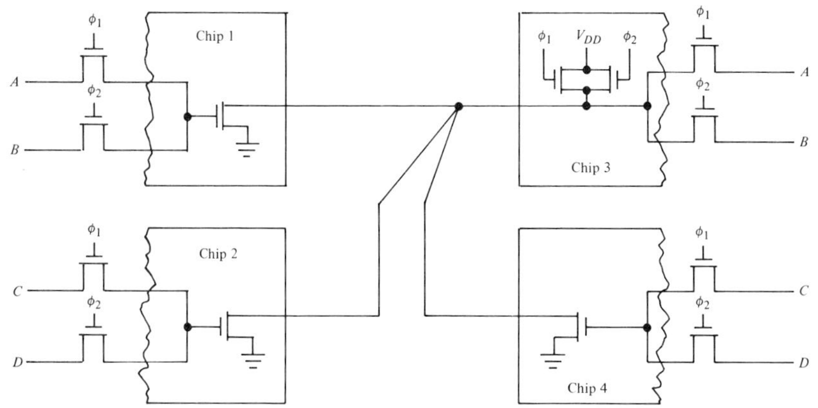

Figure 6–16 Time-shared open drain drivers.

6.5 LOGIC SIMULATION

The requirement for a means of verifying the logic design of a system prior to committing the chips to fabrication becomes obvious when considering the large development costs and long turnaround time of an MOS array. One approach to the problem is to fabricate a breadboard of the system using off-the-shelf components. A second approach, which is becoming increasingly important, is the use of computer software modeling techniques to simulate system logic.

6.5.1 Hardware Breadboard Approach

Frequently, breadboarding is an unnecessary step in MOS chip development. The cost of breadboarding includes extra breadboard logic diagrams as well as parts cost, assembly costs, checkout, and troubleshooting costs, and frequently is more expensive than the MOS chip development cost. Careful logic design with flow charts and the use of synchronous logic allows direct conversion to MOS arrays with high confidence without the need for breadboarding.

However, large systems with many chips such as desk calculators and minicomputers are almost impossible to design without some logic errors. They are also usually too complicated for easy computer simulation and, even if an effective simulation is accomplished, a test word of extreme length and complexity must be created to allow the computer to check the system for proper operation of all possible input combinations. In such cases, breadboarding allows direct real time testing and evaluation of the human-machine interface. Ideas for improvements are more obvious when an actual working breadboard is used. If the system is constructed such that each breadboard pc card represents one MOS LSI chip, each card may be replaced with a MOS LSI chip as it is completed.

Asynchronous logic designs sometimes are breadboarded because of the possibility of race conditions. This will increase confidence but will not guarantee a good design inasmuch as even MOS breadboard parts cannot exactly simulate the delays and logic of the final chip. This is why asynchronous logic should be avoided or be carefully designed and analyzed on paper to guarantee that no race or spike problems occur even if the logic works in breadboard form.

Breadboarding with MOS Components—Several MOS manufacturers produce a line of standard components suitable for mechanization of a prototype system. These parts include logic gates and shift registers. Even though the breadboard cannot exactly simulate the time delays of the final array, experience has shown that a close approximation is possible with the final chips usually being faster than the breadboard.

The usual procedure used to construct a breadboard is illustrated in Figure 6-17. Figure 6-17a shows a typical result of preliminary design with the translation to a final logic diagram implemented with standard cells shown in Figure 6-17b. Conversion of the final logic diagram to a

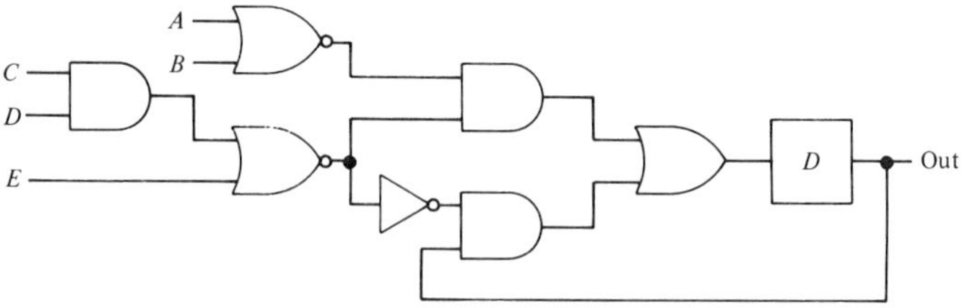

Figure 6–17a Preliminary design logic diagram.

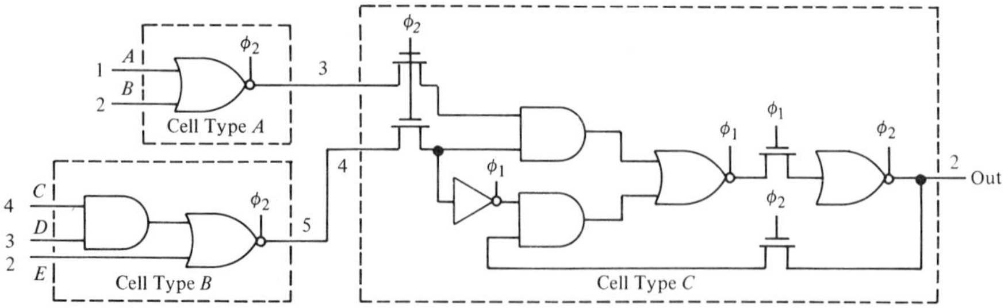

Figure 6–17b Final standard cell logic diagram.

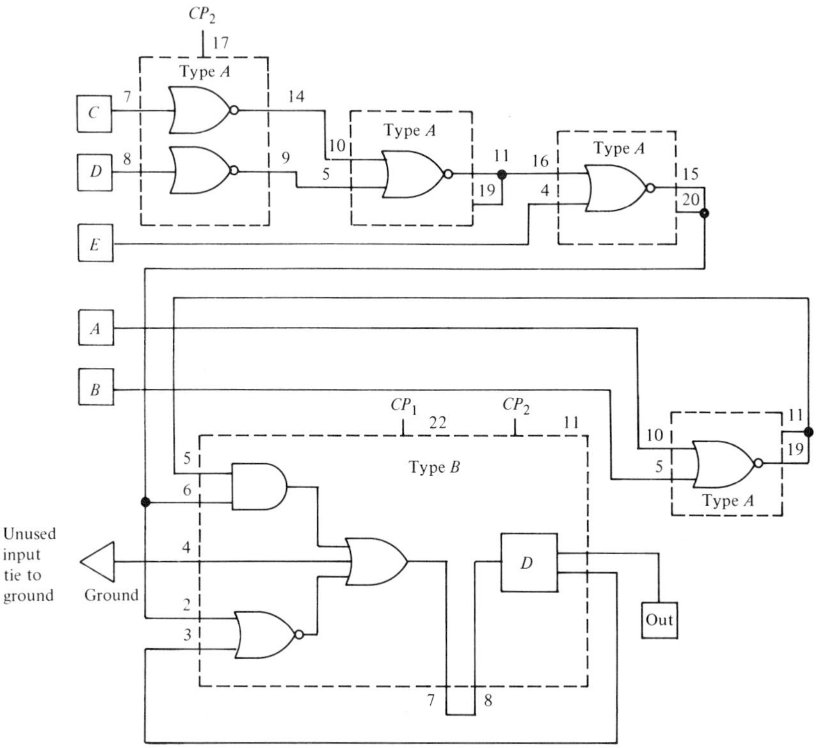

Figure 6–17c Breadboard logic diagram.

breadboard logic diagram is, of course, a function of the logic circuit types that are available. Assuming that only inverters, 2-input NOR gates, and sample and hold flip-flops are available, the conversion is as shown in Figure 6-17c. The breadboard logic diagram should define all interfaces between logic elements and any external connections. All breadboard package pin numbers and circuit types of Figure 6-17c are shown for illustration purposes and may not reflect available components. Great care must be exercised to ensure that no logic errors occur in the creation of the breadboard logic diagrams.

Breadboarding with DTL/TTL Components—Often it is desired that a system be built with DTL or TTL to demonstrate feasibility or to get into prototype production quickly with plans to later convert to MOS. In this case, one should design with (or convert to) synchronous logic. This can later be converted to MOS dynamic logic with few changes and no need for a new breadboard.

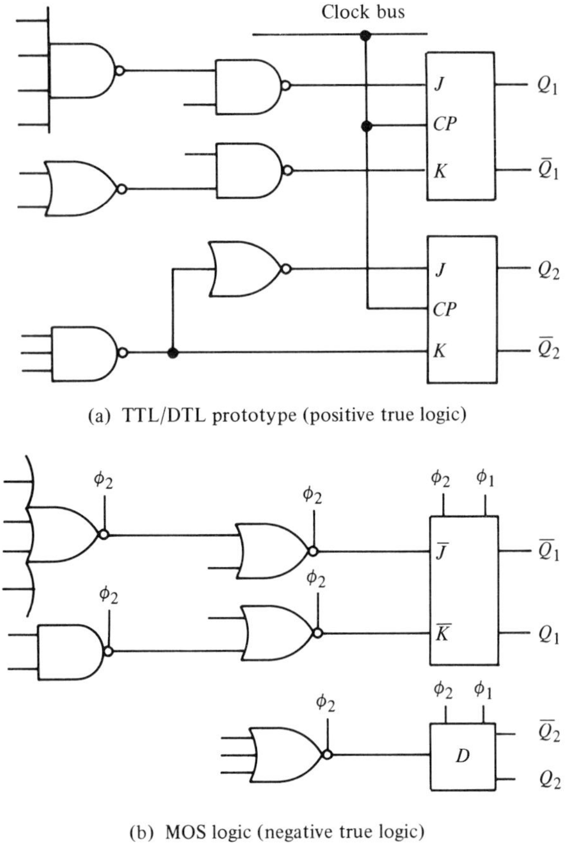

Figure 6–18 TTL/DTL MOS equivalents.

This method is illustrated in Figure 6-18. With positive true logic, dc NAND gates that have as many as eight inputs or NOR gates that have as many as two inputs may be replaced directly with ϕ_2 clocked 2-phase ratioed MOS gates. Synchronously clocked *JK* flip-flops and *D* flip-flops may be directly replaced with ϕ_1 and ϕ_2 clocked MOS equivalents.

Figure 6-19 shows how a 2-phase ratioed MOS standard cell can be simulated with TTL or DTL logic and a *D* flip-flop. This is done simply by replacing all series ϕ_1 gates with *D* flip-flops and changing the remainder of the logic to dc NAND logic.

6.5.2 Computer Software Simulation Approach

If the system is well-defined and contains fewer than 5000 logic elements, breadboarding may be supplemented or replaced by computer simulation.

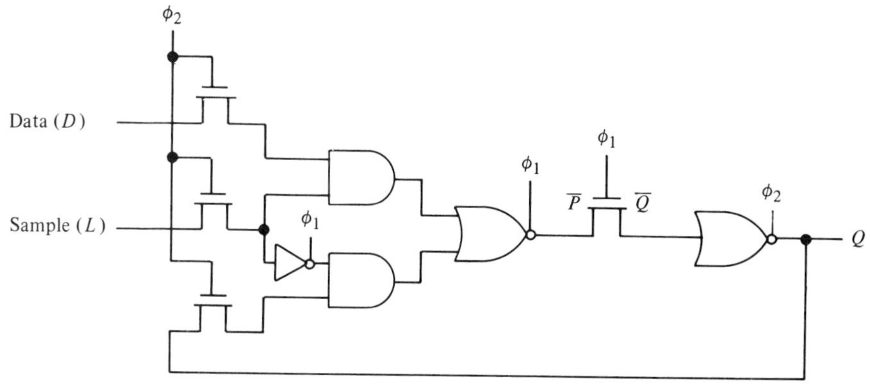

(a) MOS sample and hold flip-flop (negative true logic)

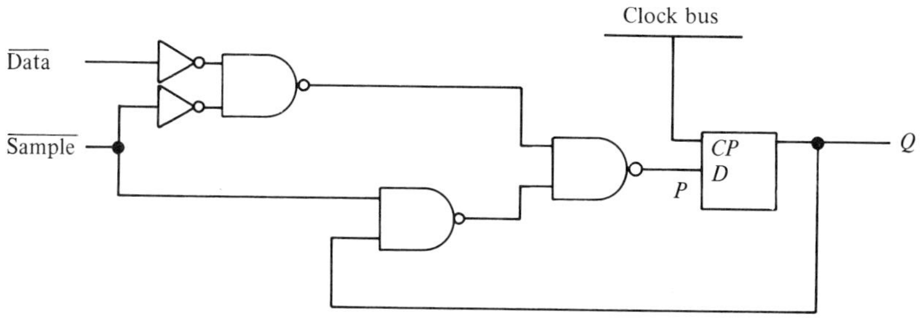

(b) DTL/TTL sample and hold (positive true logic)

Figure 6–19 TTL/DTL MOS equivalent circuits.

Many of the existing simulation programs are based on a program developed by the National Security Agency (referred to as Banning LOGBLOSIM (Logic Block Simulation). Many companies have generated proprietary simulation programs, each with its own format but most have several general points in common. These are listed below.

- Net list or logic equations
- Input test sequence
- Predicted output sequence

All simulation programs require the logic network to be coded in the form of a list that describes the logic. This is sometimes called a net list and is equivalent to a set of logic equations subdivided into simple functions or blocks such as NANDs, NORs, and delay bits. The net list could be made from the application equations. An example of a logic network and its net list is shown in Figure 6-20. All inputs and nodes are numbered arbitrarily, but without repetition. The net list shows the node number, its logic function, and its inputs. This list is coded on punched cards according to the format of the particular simulation program. More sophisticated programs will have delay bits and logic combined into larger preprogrammed functional blocks, such as *JK* flip-flops or exclusive ORs.

All simulation programs also require some sort of truth table or test sequence to apply to the logic network inputs. This must be generated by the program user and must be sufficiently long and complicated to reproduce closely the actual circuit operation or to try all possible input test combinations. This is the most difficult part of simulation. After describing the logic to the computer with the net list, the input test sequence will exercise the logic to see if it will actually perform the logic function for which it was designed.

The third item, predicted output sequence, is not absolutely necessary. All programs will print out the actual output sequence. Most programs will also accept a predicted output sequence, compare it with the actual, and flag any differences.

If the input test sequence proves sufficient and the logic net list has no errors, the resulting output sequence will indicate whether or not the logic was designed correctly.

Many programs will not accept a large functional block such as a *JK* flip-flop. In this case they must be replaced with smaller elements such as a delay bit preceded by NAND gates as shown in Figure 6-21. In this manner, the equations for any larger block may be written to subdivide it into smaller elements.

In summary, computer simulation is a good technique for evaluating a

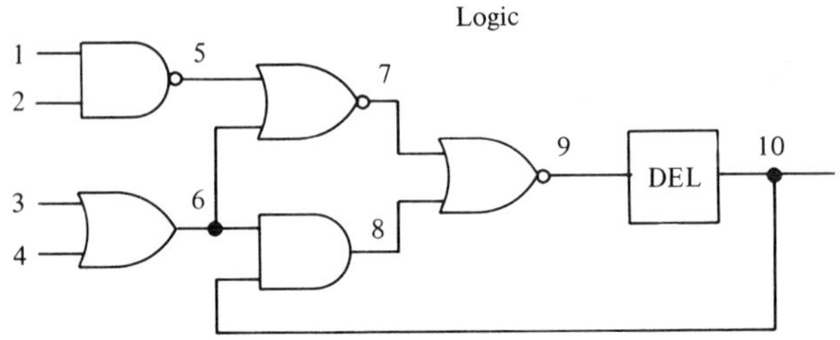

Net list

Node	Type	Inputs
5	NAND	1 2
6	OR	3 4
7	NOR	5 6
8	AND	6 10
9	NOR	7 8
10	DEL	9

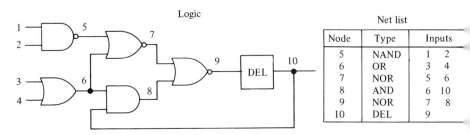

Figure 6–20 Example of logic net list.

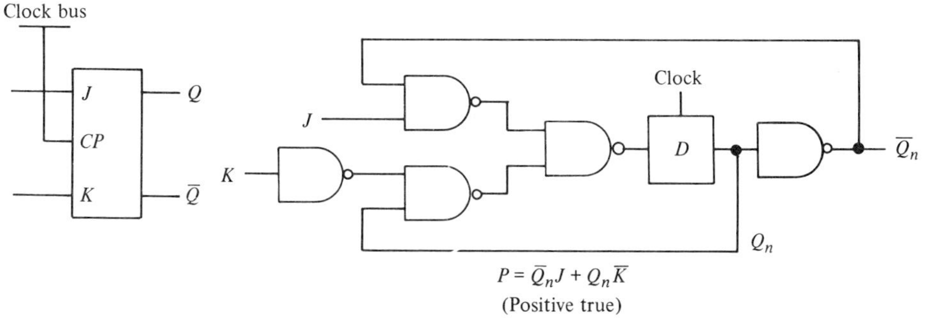

Figure 6–21 Breaking down a flip-flop for simulation.

logic design if it is applied to synchronous logic and if an input test sequence can be written that will check the logic design concept. Unlike production testing it is not sufficient to devise a test sequence that only functionally checks the logic gates. To check the design concepts one must create a test sequence that closely resembles the actual system operation.

6.6 MOS ARRAY TESTING

The production testing of complex MOS arrays presents a formidable problem to MOS suppliers and users. As a result of the system design and partitioning objectives, MOS arrays have a much lower pad to gate ratio than a design of equivalent logic complexity that uses first generation ICs or MSI circuits. This is quite unlike systems that use conventional ICs where the designers can provide numerous input and output test points from each circuit which enable functional separation and simplifies test procedures.

Numerous testing philosophies and techniques have been developed in an attempt to reduce the cost of testing. Computer aids in test pattern generation and automated production test equipment have greatly increased the efficiency of the MOS array test phase.

6.6.1 Test Pattern Generation

The magnitude of the testing problem becomes evident if a chip is viewed as a logic function that may be described and tested by a truth table approach. If n is the number of inputs to a chip, 2^n possible input combinations exist. Also, if the chip has m internal states, 2^m combinations of states exist. Therefore, for an exhaustive test on such a chip, 2^{n+m} test sets would be required. For example, a chip with 10 inputs and 10 internal states requires 2^{20} test sets. At a rate of 100,000 tests per second, the total test time would be about 9 hours per chip.

Fortunately, it is not necessary to test a chip by a complete truth table approach, since the test pattern required for chip testing needs only to determine whether there are any failures in the logic gates on the chip. The logic design is assumed to be correct following the simulation phase.

Sometimes the same pattern may be used for both logic design checking and chip testing if it can be kept within the maximum bit length. However, a test sequence that completely checks the logic gates on a chip will not necessarily give any assurance that the logic will perform the system function for which it was designed. This assurance may require a longer test that closely approximates the actual system operation. This problem is especially serious if the system breadboard or simulation phase is bypassed and simple tests are written only to test the logic on the chips, as would be the case with most computer generated test patterns.

The establishment of test requirements is the first problem to overcome in writing test programs. It is generally accepted practice to attempt to ensure that a test sequence or program will detect all single fault failures. A single fault failure is one which occurs alone and is assumed to be either a short-circuit or an open-circuit device or wire. It is possible, of course, to have any group of wires or devices fail simultaneously. Proof that a test sequence will detect all single fault failures does not prove that any multifault failure will also be detected. The number of possible multifault failures on a single MOS array is astronomical and all of them cannot be checked. Even the checking of all single fault failures is an enormous job that is best handled by a computer.

6.6.2 Computer Aids to Test Pattern Generation

Three types of computer programs are used in writing test sequences: simulation, which produces correct output patterns for any input test program; validation, which checks for detection of any single fault failure by trial input test program; and generation, which generates trial test programs.

The simulation program is identical to that used for logic design checking. It is also useful for manual test pattern operation, since it will verify the output patterns that result from any input test pattern created.

The validation program checks to see if all the logic on the chip is tested by a trial test program. It does this by first running through a regular simulation using the trial test pattern as inputs and stores the outputs created by this test pattern assuming a good chip. Then, one at a time, it fails every device on the simulated chip, first by shorting it to ground, then by open-circuiting it. The entire test pattern is completely repeated each time.

If the forced failure does not cause one of the outputs to differ from the output obtained with the fault-free model, this indicates that the test pattern was either insufficient to test the failed device, or the device was redundant and could have been omitted from the logic. The validation program is used to check a test written either by hand or by computer to see if it will detect all single fault failures.

The generation program is a way of creating test programs by computer. There are many different programs with various degrees of sophistication. Almost all of them can generate a test for simple combinatorial logic or simple shift registers. None of the generally available programs will produce a guaranteed minimum length test. Some of the more sophisticated programs can produce tests for simple sequential machine logic designs. Again, none of these can produce tests that are guaranteed to check all single fault failures on large sequential machines, such as those that are presently being built with MOS arrays, without considerable manual intervention and feedback.

6.6.3 Manual Test Pattern Generation

Trial test sequences that are manually generated have several advantages. If the sequence is written by the logic designer who created the logic, it is more likely to test the logic design concepts and system operation of the logic. This gives assurance that the logic has been designed correctly instead of just checking for device failures within the logic, as the computer generated program does. After the trial test program is written, it may be validated by computer to locate any logic gates that have not been checked and then reworked slightly to test those gates. Frequently, the designer will discover redundant logic during this validation process and can omit them from the final design. Sometimes, because of logic design errors, portions of the logic cannot be tested. By writing his own test pattern, the designer will discover this sooner and will be able to correct it more readily than if he had to troubleshoot an unfamiliar and almost random test program created by a computer. Computer test program generation is, of course, less costly and faster than a manually generated program.

6.6.4 Production Test Equipment

Almost all MOS LSI array production test requirements are satisfied by computer-controlled test systems similar to the Programmable Automatic Functional Tester (PAFT). A photograph and block diagram of this tester are shown in Figures 6-22 and 6-23, respectively.

Figure 6–22 PAFT II LSI tester.

The system is controlled by a general-purpose computer with a 4000 word (16 bits/word) memory and is capable of interfacing two test stations. The computer controls digital to analog converters that supply the voltage, clock, and power levels to the device(s) under test while control logic supplies multiphase clocks that have a programmable logic 1 level (0 to −32 V) and a logic 0 level fixed at less than −1.0 V. Clock frequency is programmable from 100 Hz to 2 MHz. A teletype unit serves as the primary operator interface for manual intervention or program modification. The paper tape reader provides the primary program/data input to the computer. The system provides 64 channels with a test pattern length of either 800, 400, or 200 bits per channel. Various parameter tests may be performed, including a stress test for breakdown of inputs, leakage tests, and a continuity test.

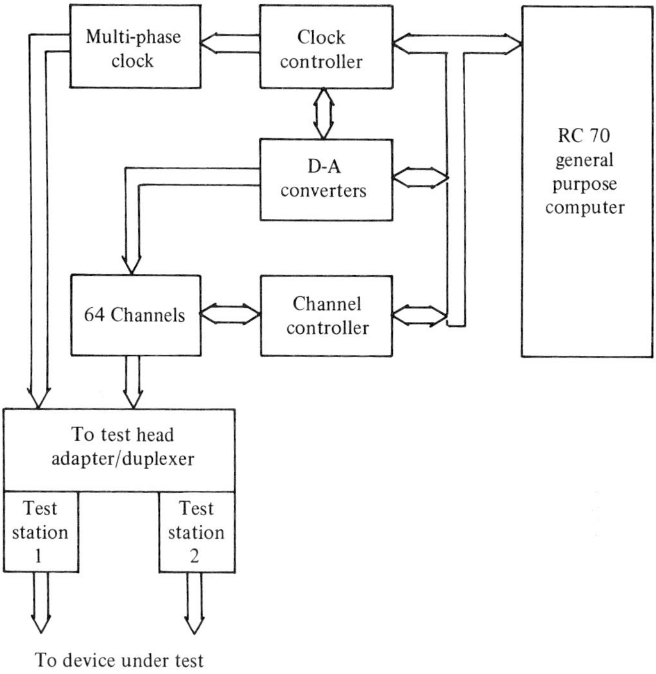

Figure 6–23 PAFT II system block diagram.

Figure 6-24 is a simplified block diagram showing the interfaces with the array under test. Shift registers, which are loaded through the channel controller, hold the input test patterns and predicted (or correct) output patterns. As the registers recirculate, these patterns are applied to the array inputs and to comparator circuits which check the actual array outputs against the predicted patterns and programmed output amplitudes. Since the system is controlled by a general-purpose computer, substantial modifications in the test language, testing techniques, and operator communications may be achieved by modifications to the controlling software.

One technique of testing is termed "cyclic." In this mode of operation, test patterns are limited to some maximum length. If the register length is exactly 400 bits, the test patterns must also be 400 bits long and will be repeated each time the register recirculates. Tests shorter than the maximum are easily padded out with continuous 1s or 0s to fill the register.

After the registers are loaded and the test is initiated, the first input test sequence is applied without activating the comparators. This gives one complete cycle during which the inputs can place all delay elements or

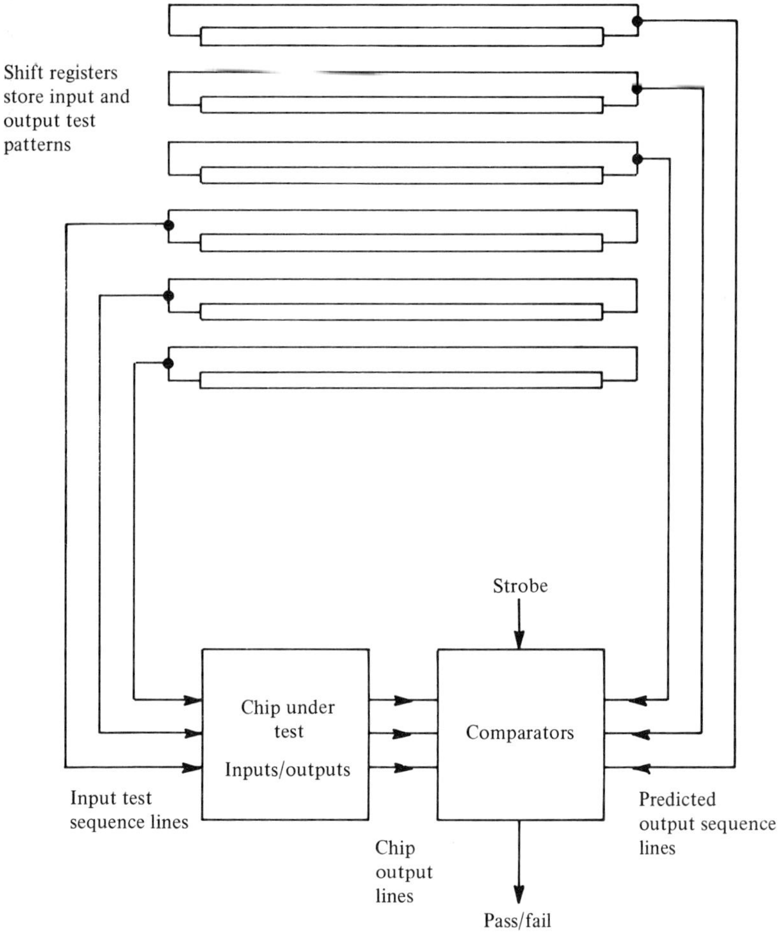

Figure 6–24 Simplified block diagram of PAFT II.

flip-flops on the chip to a known state. The comparators are then enabled during the next revolution.

The cyclic test has the advantage that "don't care" output test pattern bits are not required when the test starts. On the other hand, it has the disadvantage that the test must be written in cyclic form, i.e., so that the state of the chip at the end of the test is identical to the state of the chip at the beginning of the test.

A number of MOS LSI compatible test systems are currently available. Most of these systems feature computer control and associated software packages. Among these are the Fairchild Sentry 400, the Teradyne J283, and the Macrodata 200.

The Fairchild 400 test system, shown in Figure 6-25, functions much like a general-purpose computer. A central processor interfaces with disc, tape, teletype, card reader, and as many as four multiplexed test stations. Each test station may function as an automatic handler or wafer probe station.

A minimum configuration will perform static and dynamic tests on one 30-pin combinatorial or sequential network at a time. Expansion modules can be added in 30-pin increments to a total of 480 pins at four independently operating 120-pin test stations. Logic and functional tests can range up to 286,000 tests per second. Absolute dc parameter measurements, such as input pin leakages and saturation voltages, are made at a rate of 250 tests per second at each station. A software system that includes a compiler and disc operating programs provides a special test oriented language to the user.

The Teradyne J283 test system, shown in Figure 6-26, consists of a general-purpose computer with its input/output equipment, a main frame, and a satellite station that services up to four test stations. The standard test station configurations allow testing of arrays with up to 40 or 64 pins, with possible expansion to 120 pins. The functional testing rate is from 20 to 50 kHz, with a range of 20 to 30 kHz typical for LSI arrays. dc parameter tests are also possible. Again, a software system is provided which controls all operations, including data logging and reporting.

The Macrodata series 200 test system shown in Figure 6-27, incorporates many of the features of the other test systems. The modular design

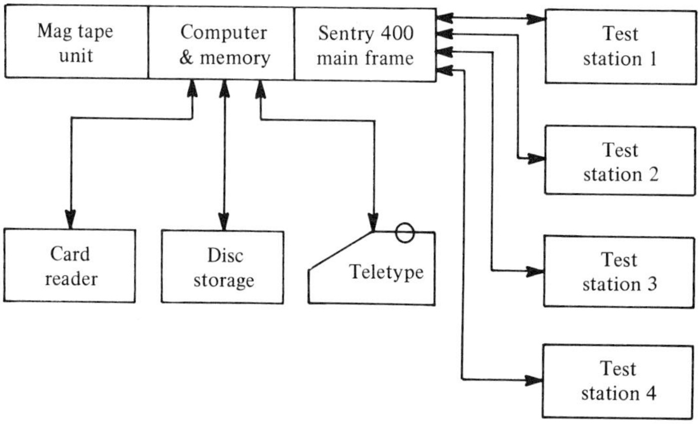

Figure 6–25 Fairchild sentry 400 test system.

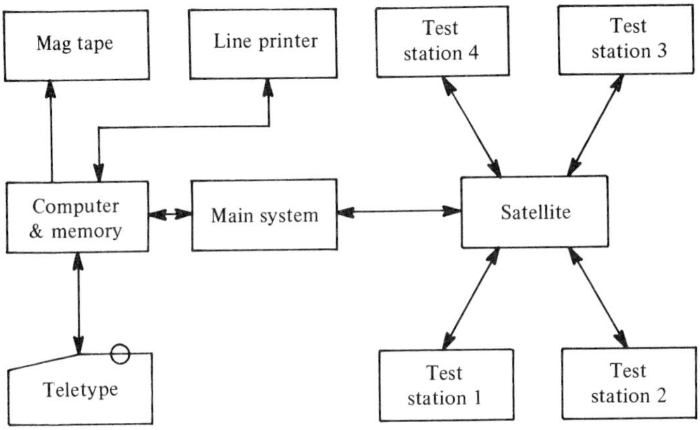

Figure 6–26 Teradyne test system configuration.

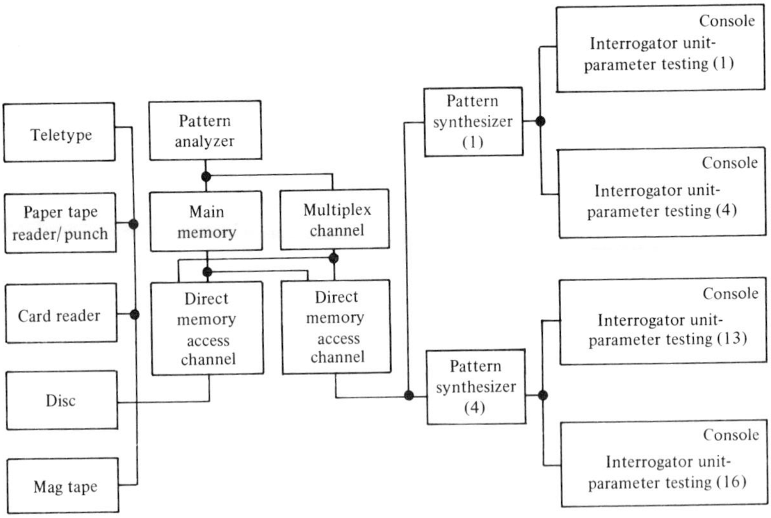

Figure 6–27 Macrodata test system.

can accommodate up to 16 test stations per computer, plus the addition of a number of peripheral units. This system is capable of performing digital functional and parameter tests. A unique feature is the location of the driver/receiver/comparator circuits within inches of the device under test, thus enabling on-wafer testing at speeds of up to 2 MHz. This allows circuits to be tested at designed frequencies, thus eliminating the subsequent cost of assembly and handling of speed sensitive dice. The minimum configuration provides 16 pins and is modularly expandable to 64 pins. The system also automatically controls all mechanical functions required at the wafer probe station such as probe down-sensing, off-wafer detection, and inking of bad dice.

GENERAL REFERENCES

D. P. Lindorff, *Theory of Sampled Data Control Systems,* John Wiley & Sons, New York, 1965.

G. E. Penisten, "Impact of LSI Technology on the Electronics Market," paper presented at WESCON, San Francisco, August 1969.

G. F. Watson, "LSI and Systems/The Changing Interface," *Electronics,* **42,** 78 (March 1969).

W. A. Notz et al., "Benefitting the System Designers," *Electronics,* **40,** 130–133 (February 20, 1967).

L. A. Taylor, "Mechanization of Filter Algorithms" and "Real Time Digital Filtering and Spectrum Analysis," *Proc. Nat'l. Elec. Conf. Seminar,* St. Charles, Ill., June 1969.

L. C. Hobbs, "Effects on Large Arrays on Machine Organization and Hardware/Software Tradeoffs," *Proc. Fall Joint Computer Conf.,* 89–96 (1966).

7

MOS Memory Products

7.1 INTRODUCTION

Memory circuits are an essential part of any digital system. In digital systems, memory circuits are often repetitive within the system, and common blocks are found in many different systems. Because of this, memory circuits can be considered as large volume, standard product subsystems. This is considerably different from the logic circuits that generally cannot be defined as common arrays of gates that are uniform from system to system.

In attempting to generate standard memory products, three functional product classifications have been established:

1. serial memory (shift register)
2. read write, random access memory (RAM)
3. read only, random access memory (ROM)

These functional blocks consist of ordered circuit arrays which provide a higher circuit density than non-ordered configurations. The large volume production of these standard product blocks dramatically lowers the cost per equivalent gate.

The advantage of these functional blocks can best be demonstrated by comparison of array complexity for various types of digital arrays. Figure 7-1 shows a graphic representation of several types of arrays and their average number of equivalent gates* per package. The chart shows

*An equivalent gate is arbitrarily defined as 1 bit of delay in the case of serial memory and 8 gate inputs (memory bits and decode gate inputs) in the case of read only memories.

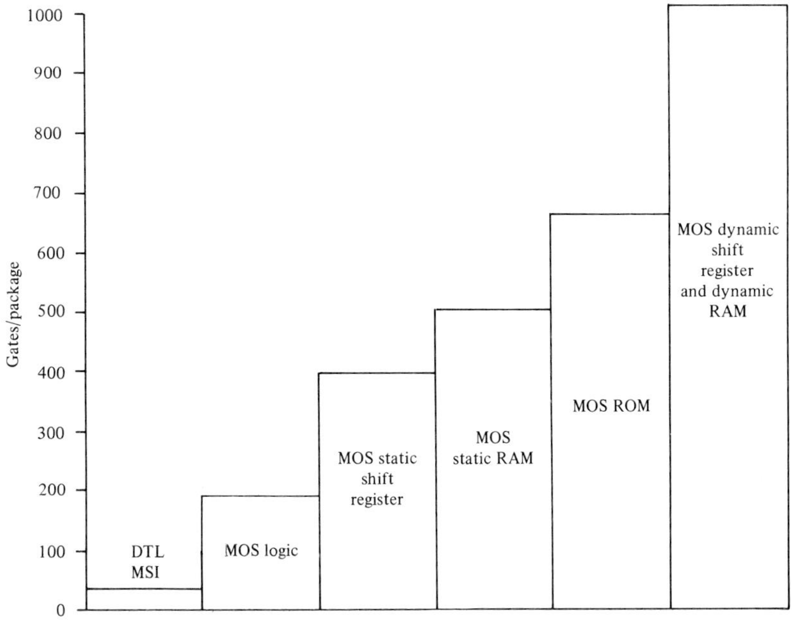

Figure 7–1 Comparison of circuit complexity of LSI arrays.

minimum improvement of 400 percent in gate count for structured MOS arrays compared to random logic arrays. The chart also demonstrates that substantially more logic complexity can be achieved with MOS as compared to the highest level of integration that is feasible with bipolar technology.

7.2 ALTERABLE RANDOM ACCESS MEMORY

A computer system is comprised of four interacting subsystems: (1) an arithmetic-logic unit; (2) a memory; (3) a control unit; and (4) an input/output unit. The memory subsystem has the ability to store and retrieve information (arranged in words) on command from the control unit. In random access memory (RAM) operation, a desired memory location can be accessed during one processor clock period, as opposed to sequential memories that require multiple bit time delays to access a desired memory location.

The RAM subsystem is divided into four parts: (1) decode, (2) storage, (3) control, and (4) data input/output buffering as shown in Figure 7-2. A memory operation is initiated by presenting an address to the decode inputs, and giving a fetch or store command to the input/output control.

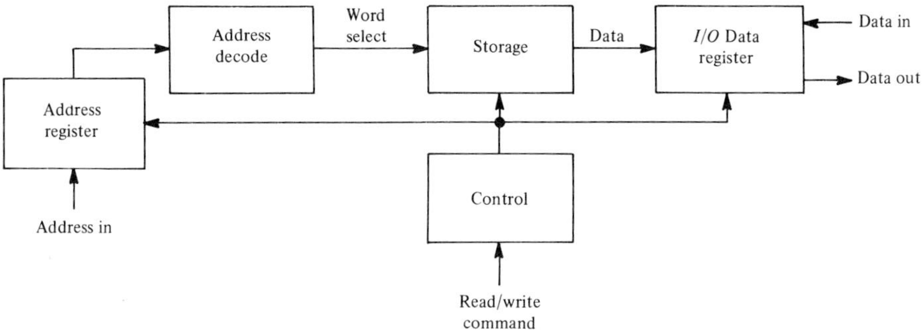

Figure 7–2 Random access memory.

The address is decoded and a memory location selected. The selected memory word contents are then read or altered as required.

The memory function can be asynchronous or synchronous with respect to processor operation. In synchronous operation, the flow of address words and storage information to and from the memory is controlled by the same command clock as the remainder of the system. In asynchronous operation, the memory must communicate with the processor to determine a mutually agreeable time for data transfer to take place. Most memory systems operate synchronously, which eliminates the need for external data transfer control.

By adding logic capability to the basic storage capability, the memory in Figure 7-2 can be converted to a content addressable memory (CAM). A CAM is useful in performing search operations on large data banks. The content addressable feature is incorporated in Figure 7-2 by adding a mask register and a flag register. The CAM will accept a data word and a mask word as inputs and will search all address locations simultaneously for a match between the unmasked bits of the data word and all stored words. Address locations that contain a stored word that matches the input data word in the unmasked bit positions provide an output to the flag register.

7.3 MOS MECHANIZATION

The memory function can be mechanized using various technologies. The storage array can be implemented with magnetic cores, plated wire, arrays of electronic elements, etc. An MOS implementation of the entire memory function usually consists of an array of circuits interconnected as shown in Figure 7-3. The MOS array typically contains the decode, storage, and data input/output steering functions.

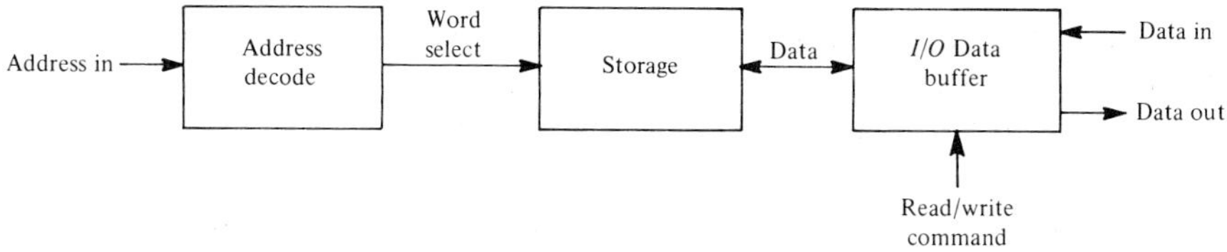

Figure 7–3 MOS memory subsystem.

7.3.1 Static Memory Cell

Several types of static MOS storage elements have been proposed; however, all these elements work on the flip-flop feedback principle. The simplest storage cell is constructed as shown in Figure 7-4. This cell employs the basic flip-flop and requires five or six devices per cell, depending on whether a single or dual data bus is desired. The advantages of this cell are its noise immunity and wide electrical operating range. Noise immunity is important in an integrated array because of the possibility of signal coupling from cell to cell.

The major disadvantage of this cell is the large amount of silicon area required for implementation. The inverter device must be large with respect to the load device because of the nature of the cell. The area required to construct six devices, of which only four can be of minimum size, is comparatively large. Using conventional *P*-MOS technology, a 256-bit array of cells requires a die 130 mils on a side.

The static nature of circuit operation causes average power dissipation to be quite large. Power dissipation is multiplied by the number of cells in the array, and can be enormous when many arrays are interconnected to form a memory system. This drawback can be eliminated by relying on

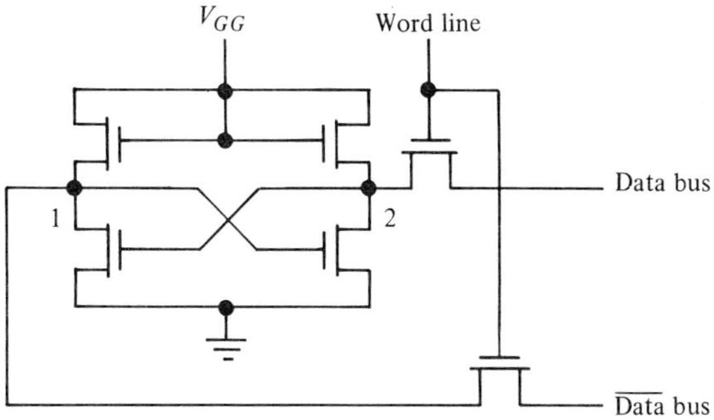

Figure 7–4 Static memory cell.

the capacitance at nodes 1 and 2 to store information, and turning on the load devices periodically to recharge the capacitor.

Several variations on the static memory cell have been proposed. A separate data bus may be employed to write into the cell. This change requires additional area for an extra interconnect line and two MOS devices. The advantage gained by adding these elements is the elimination of the long read recovery time after a write operation.

A refinement of the clocked load technique employs the data and the data interconnect lines as a means of providing "refresh" current to the storage nodes. This results in the elimination of the two load devices from the cell structure shown in Figure 7-4, but places somewhat of a restriction on the device sizes used to implement the cell.

7.3.2 Dynamic Memory Cell

Several types of cells that employ dynamic ratioless circuits can also be used to implement the storage function. The basic configuration of the dynamic memory bit is shown in Figure 7-5.

The operation of the circuit in this cell is discussed in detail in Chapter 5. The information contained in the cell is continuously clocked from the ϕ_1 inverter gate to the ϕ_2 inverter gate and back to the ϕ_1 gate. The advantage of this cell is low power dissipation and small cell size. Although seven devices are required, all can be of minimum size. Also, the signal and clock node interconnections can be arranged to minimize interconnect area. This cell requires only a single data bus for proper operation.

The low noise immunity of this cell is a serious disadvantage. This is complicated by the fact that two clock signals are fed to the cell array.

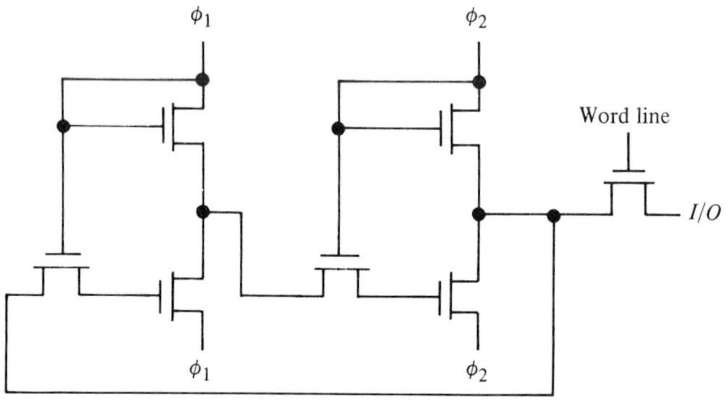

Figure 7–5 Dynamic memory cell.

Careful circuit layout is required to isolate the storage nodes from false information transfer through capacitive coupling to other circuit nodes.

A cell based on ratioless circuits that contain no internal feedback is illustrated in Figure 7-6. The figure shows one feedback stage with many storage cells. A single storage cell is also shown enclosed in dotted lines in Figure 7-6.

Since only three MOS devices and five interconnect lines are required, the resultant cell area is quite small. Arrays of 1024 bits have been fabricated in a reasonable area. This circuit also exhibits the disadvantage of low noise immunity of the previously discussed dynamic cell.

Data are written into the cell through Node B by enabling the write line (W_3) during the ϕ_1 to ϕ_2 dead time. Data are read by enabling the read line (R_3) during the ϕ_2 to ϕ_1 dead time and observing Node B.

This cell imposes a restriction on the timing of the memory system since periodic information refreshing is required. The information storage occurs on Node 1 and this node must be refreshed periodically by turning on the read line (R_3) during the ϕ_2 to ϕ_1 dead time. This causes the stored data to be inverted and shifted to the inverter gate (Node A) of the feedback stage. During the ϕ_1 to ϕ_2 dead time the data are inverted again and shifted to Node B. The data are rewritten back into the cell (Node 1) by enabling the cell write line (W_3). This refresh operation must be repeated for all the cells serviced by the feedback stage.

Several timing schemes can be used to implement refresh. One technique is to use a periodic interrupt for a number of bit times to sequence through all the storage cells that are connected to an inverter stage. However, the system interrupt requirement presents a serious disadvantage. An

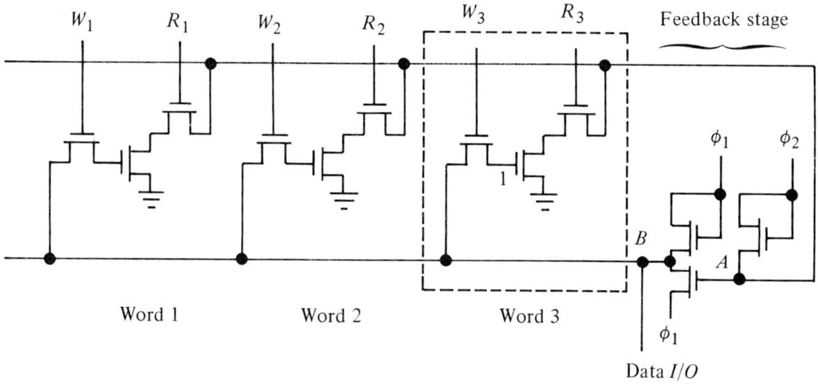

Figure 7–6 Three-device dynamic memory cell.

alternate technique which eliminates this disadvantage is to accomplish a normal store or fetch operation during the first half of a memory cycle. During the last half cycle, a cell is refreshed. The cell to be refreshed is determined by a counter that sequences through all address locations that are connected to a common feedback stage. Although this technique doubles the memory cycle time, computer architects have argued that this is not a system restriction, because of arithmetic and control unit limitations. They point out that since the processor requires time after memory access to process data, that time can be used for memory refreshing at no loss of computer cycle time.

7.3.3 Content Addressable Memory Cell

Figure 7-7 shows the basic static flip-flop shown in Figure 7-4 with the required extra logic to convert it for use as a memory cell in a CAM or associative memory. The cell functions in the content addressable mode when the word line is held at ground and the Address/Flag Control line is negative. A ground on both D_S and $\bar{D}_S$ "masks" or disables the cell match operation. A match operation is initiated by establishing a negative level on D_S or $\bar{D}_S$. A negative level on D_S initiates a "match 1" operation. A ground on D_C (negative on $\bar{D}_C$ is a mismatch and causes current to flow from (V) to the word line. A negative on $\bar{D}_S$ initiates a "match 0" operation. A ground on $\bar{D}_C$ causes mismatch current to flow. Any unmasked bit

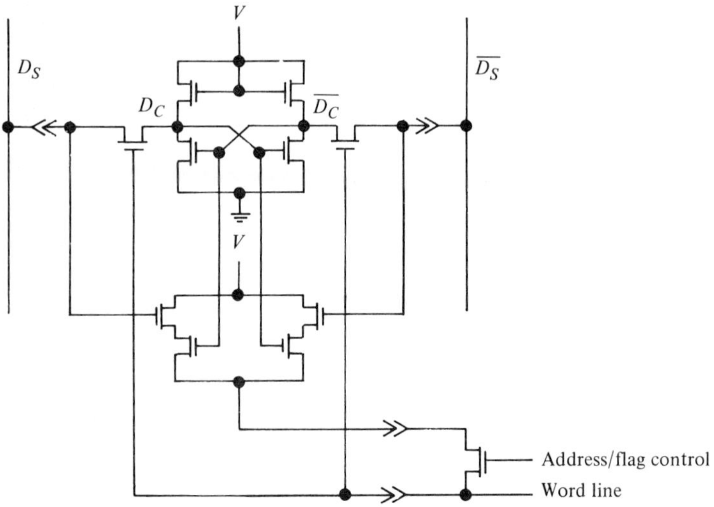

Figure 7–7 CAM memory cell.

in a stored word can contribute mismatch current to the word line. The same cell can function as a normal memory element with the control line held at ground. A negative voltage on the word line connects the cell storage nodes (D_C, $\bar{D}_C$) to the data bus (D_S, $\bar{D}_S$) for a read or write operation.

7.3.4 Decode Logic

Two techniques of MOS implementation of the decode function are illustrated in Figure 7-8. Figure 7-8a shows the one-dimensional decode of a simple 4-bit address and resultant decode states. This is the simplest implementation of the decode function. An alternate approach that reduces the total gate count employs a two-dimensional decode that specifies a memory location by the intersection of an X decode true and a Y decode true condition. This type of decode is demonstrated in Figure 7-8b.

Several circuit considerations influence the choice between a one-dimensional or a two-dimensional mechanization of the decode function. The predominant design problems are associated with the high fanout of the decode gates. The decode gate is required to drive the gate capacitance of all bit lines in the output memory word which it interrogates. This drive affects the power dissipation of the decode gate. Array designs are especially sensitive to power dissipation, because the power level chosen for a decode gate is multiplied by the number of decode gates to arrive at total power consumption.

The one-dimensional decode has an advantage with respect to fanout. Figure 7-8a shows that the decode gate has a fanout of one word location. The disadvantage is that more gates are required for a given number of

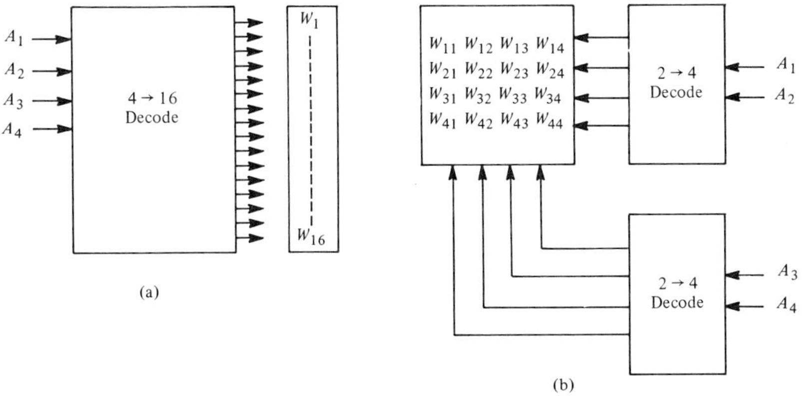

Figure 7–8 Decode logic implementation.

address locations. This results in a larger usage of silicon area and a greater total power dissipation. The two-dimensional mechanization, on the other hand, requires fewer gates, resulting in a lower power dissipation and smaller circuit size. Notice in Figure 7-8b that only eight decode outputs (4 columns) provide the same number of decode states as the one-dimensional decoder. However, the two-dimensional decode is inferior with respect to fanout. Each decode output of Figure 7-8b must drive four times the number of memory locations as the one-dimensional mechanization.

The decoder design substantially affects the speed of the memory system. Large capacitive loading of decode outputs causes considerable propagation delay. Since the speed power product of dynamic circuits is superior to that of static circuits, high-speed operation dictates the use of dynamic decode logic.

7.3.5 Input/Output Buffers

Input/output buffering can be accomplished in several different ways. Most implementations contain series MOS devices in the data lines to provide steering for bilateral data flow, as shown in the dual bus example (Figure 7-9).

A major design consideration is the trade-off between good drive capability and minimum read access time. Large buffer devices are required to achieve the drive capability required by bipolar current sinking logic. The small current available from the memory cell requires considerable charge

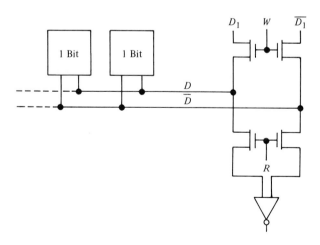

Figure 7–9 Data input/output buffering.

time to establish a logic signal on the large gate capacitance of this type of driver. An alternate approach employs a differential amplifier external to the circuit array to sense the small currents of the memory cell. This technique results in faster read access if the data output lines can be successfully protected from noise signals.

7.4 MEMORY SYSTEM DESIGN TRADE-OFFS

The design of a memory system using MOS arrays requires familiarization with a new set of design relationships. The first and simplest consideration in memory system design involves array organization. Arrays should be organized to contain a maximum number of words and a minimum number of bits. This minimizes array pin-count and allows for maximum flexibility of stored word length.

The large packing density achievable with the dynamic memory cell reduces the system cost and size significantly. With dynamic storage, approximately twice as many storage bits can be fabricated on a chip of reasonable size than with static storage. On the other hand, refreshing is a constraint associated with dynamic storage. As mentioned previously, refreshing can be accomplished with minimum extra hardware, if the memory cycle time is not critical. If a large cycle time is not desirable, a system interrupt must be provided for refreshing. This is a definite constraint on both memory and total system design.

System size also influences the decision on system voltage and logic levels. The bipolar system and MOS memory arrays have different optimum operating points. If the memory subsystem is small, the interface and power supplies should be chosen to minimize the memory interface hardware requirements and provide compatibility with the available bipolar supplies. When MOS is interfaced directly with bipolar current sinking logic, the system component cost is reduced, but at the expense of MOS array size, power dissipation, and operating speed.

7.4.1 Speed

Optimization of either a large or small system with respect to speed involves the choice of interface circuits and system operating voltages. The input interface should be selected to provide good MOS operating characteristics at the expense of interface hardware. The system operating voltages should be chosen to maximize the MOS array speed since this is the slowest component in the system. The output interface should be capable of operating directly from the small current signals available from the MOS array, since a considerable amount of time is required to buffer the signals. The system should be organized to prevent large fan-in and fanout on memory data buses that feed the memory arrays. It may be desirable to provide a means on the MOS array to disconnect the array from the system data output bus. This will reduce bus loading when the array is not being read.

In summary, it can be said that although minimum cost dictates dynamic logic in large systems, it requires minimum interface and support hardware in small systems. Maximum speed calls for extensive interface and support circuits in any size system.

7.5 READ ONLY MEMORY CONCEPTS

A read only memory (ROM) functional block was first proposed in an attempt to arrange the architecture of computers to use the capabilities of random access read/write storage more efficiently. Initially, it was observed that a significant amount of the storage capability of a computer was assigned the task of storing micro-instructions for program execution. The storage content of these instructions remained relatively fixed in most cases. For greater efficiency, the micro-instruction function was assigned to a separate nonalterable memory that could be mechanized more effectively than read/write storage. Thus the initial concept of ROMs closely parallels the concept of alterable storage.

As in the case of alterable storage, the ROM subsystem consists of three subsystem functions: (1) an array of memory cells, (2) a decoder to randomly select one of the memory words, and (3) an output buffer stage to amplify the selected memory output signals. Figure 7-10 illustrates the combination of these functions.

7.6 ROM MECHANIZATION

The decode function of the ROM block diagram is implemented using the same techniques as those employed for an alterable memory. The differences in mechanization occur in the memory array and memory output buffer, since storage is permanent and buffering need only be unilateral in the case of a ROM.

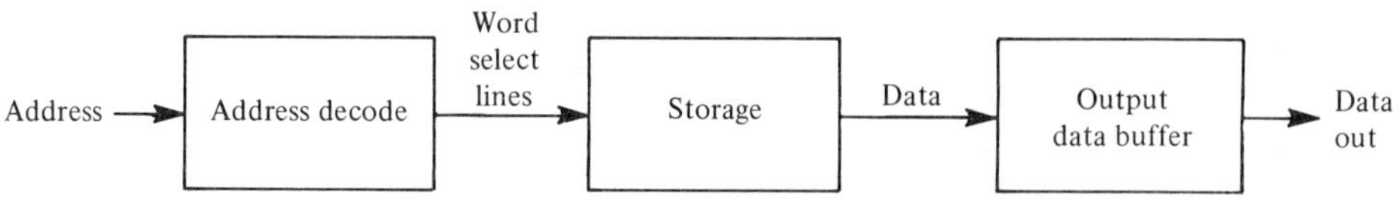

Figure 7–10 ROM block diagram.

7.6.1 Memory Array

The largest portion of a ROM is the memory array. The logical structure of a memory array is shown in Figure 7-11. The memory array performs a NOR-ing function (negative true logic) on all decode output word drive lines connected to the NOR gate input. The dotted lines represent selective connections of decode outputs to the NOR gate. This selection is accomplished by the presence or absence of a thin oxide gate over the MOS device in the memory array. The presence of a gate on the decoded word line input turns the memory device on and causes a ground voltage to appear at the memory bit line output. If two-level decode is employed, the *Y* decode input device is used to select one of several columns of memory devices.

As in the decode mechanization, circuit design trade-offs greatly influence the exact structure of the memory array. The physical implementation of Figure 7-11 employs two long *P*-type diffusions to connect the drain and source terminals of the memory devices in a column. The large junction capacitance and distributed resistance associated with these *P*-regions cause slow signal propagation through this logic level.

The area of the *P*-region can be reduced if the memory storage is arranged in multiple columns of shorter length. This is accomplished by

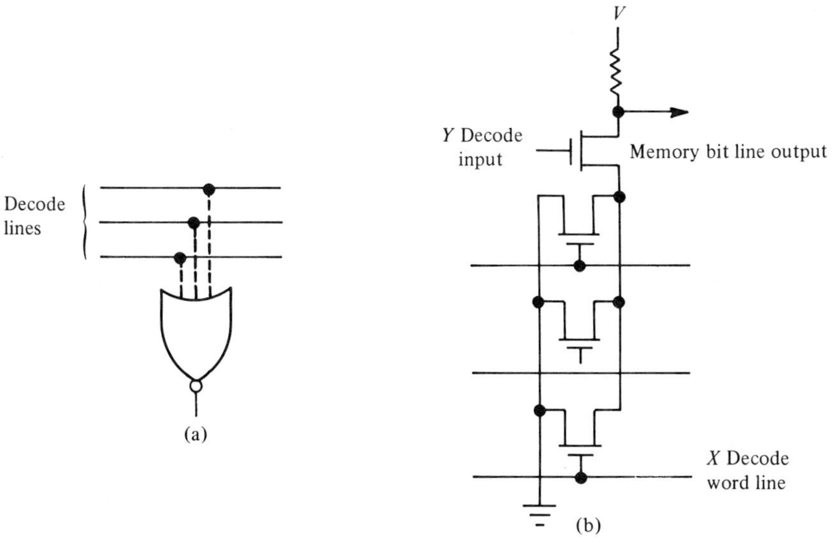

Figure 7–11 ROM memory array implementation.

using the two-level decode technique. However, multiple columns affect circuit layout. When the memory columns are shortened by doubling the number of columns, there is a corresponding doubling effect on the layout dimension required in the direction perpendicular to the columns. Since square arrays are desirable, this technique of improving the capacitive loading can be carried only so far.

7.6.2 Input/Output Buffering

Some type of buffering circuit is usually required on the decode input as well as on the memory bit line outputs. Input buffering is required for the conversion from bipolar to MOS voltage levels. If MOS levels are provided, no input buffering is required on the chip. The main problem at the bit line outputs is that the output of the memory array is incapable of driving the high capacitance associated with the package interface. A trade-off is involved in deciding on a suitable output buffer for driving an external current sinking bipolar logic gate. A buffer that is compatible with bipolar circuits requires a large area and dissipates considerable power. The actual number of outputs required tends to multiply these disadvantages. The alternative to this type of buffer is an external discrete buffer circuit that requires a smaller drive current. The external buffers are an added cost that must be considered. The factors of ROM cost, as reflected in array area, number of outputs, power dissipation, and cost of external buffers, should be examined in an effort to optimize the ROM output interface.

7.7 SYSTEM IMPLEMENTATION USING ROMs

The ROM is thought of as a memory when used in microprogramming, table look-up, and code conversion applications. As an illustration of a table look-up application, consider the function $f(X) = X^2$. All possible binary values of the independent variable X are represented by the list of input addresses in the left block of Table 7-1. The memory word in the right block is programmed to contain the binary square of its associated

Table 7–1 ROM Truth Table

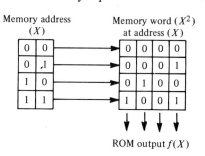

Memory address (X)		→	Memory word (X^2) at address (X)			
0	0	→	0	0	0	0
0	1	→	0	0	0	1
1	0	→	0	1	0	0
1	1	→	1	0	0	1

↓ ↓ ↓ ↓

ROM output $f(X)$

memory address in the left block. $f(X)$ is generated by applying X in the form of an address to the address inputs. The ROM decodes the address and selects the memory word stored in the address location (whose value is the binary square of the address) and presents this memory word to the ROM outputs. A similar operation occurs when the ROM is used as a code converter. The left side of the table would represent a listing of the word set to be converted. The right side of the table would contain a list of the word set in the desired output code.

A ROM could also be considered as a logic function generator. Use of the ROM as a function generator requires an understanding of the basic Boolean algebra sum-of-products theorem, which states that any output function of a set of independent variables may be generated by the AND-OR, two-level combination of the independent variables. The simple Boolean function in Figure 7-12a and its two-level equivalent (7-12b) demonstrate this theorem.

Notice in Figure 7-12b, the first gate levels generate the product terms of the input variables. This is the function the ROM address decode gates perform. Notice also the equivalence of the second level OR gate to the summing function of the ROM memory matrix. The truth table in Figure 7-12c shows how the memory matrix of an 8 word $X1$ output ROM would be programmed to generate the required Boolean function. One bit line of the memory word generates one function of the independent input variables. The number of independent output functions possible is equal to the number of bits in the memory word.

When the ROM is viewed as a function generator, Table 7-1 takes on a new meaning. Instead of concentrating on horizontal rows that contain address-memory word associations, the vertical columns on the right side

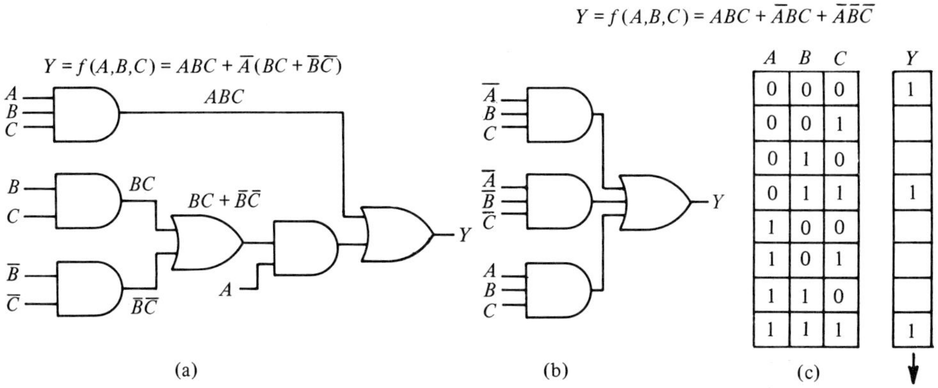

A	B	C	Y
0	0	0	1
0	0	1	
0	1	0	
0	1	1	1
1	0	0	
1	0	1	
1	1	0	
1	1	1	1

Figure 7–12 ROM logic implementation.

Table 7–2 Barker Code Conversion

Barker word									Subset A $A_1,A_2,A_3, = f\ (B_5 - - - - B_9)$				Subset C $C_1,C_2,C_3, = f\ (B_1 - - - - B_4)$					ASCII word						
B_1	B_2	B_3	B_4	B_5	B_6	B_7	B_8	B_9	A_1	A_2	A_3	*Member*	C_1	C_2	C_3	C_4	*Member*	D_1	D_2	D_3	D_4	D_5	D_6	D_7
1	1	0	0	0	0	0	0	0	0	0	0	0	0	0	0	0	0	0	0	0	0	0	0	1
1	0	1	0	0	0	0	0	0				0	0	0	0	1	1	0	0	0	0	0	1	0
1	0	0	1	0	0	0	0	0				0	0	0	1	0	2	0	0	0	0	1	0	1
1	0	0	0	1	0	0	0	0	0	0	1	1	0	0	1	1	3	0	0	0	0	1	1	0
1	0	0	0	0	1	0	0	0	0	1	0	2					3	0	0	0	1	0	0	1
1	0	0	0	0	0	1	0	0	0	1	1	3					3	0	0	0	1	0	1	0
1	0	0	0	0	0	0	1	0	1	0	0	4					3	0	0	0	1	1	0	1
1	0	0	0	0	0	0	0	1	1	0	1	5					3	0	0	0	1	1	1	0
0	1	1	0	0	0	0	0	0				0	0	1	0	0	4	0	0	1	0	0	0	1
0	1	0	1	0	0	0	0	0				0	0	1	0	1	5	0	0	1	0	0	1	0
0	1	0	0	1	0	0	0	0				1	0	1	1	0	6	0	1	1	0	1	0	1
0	1	0	0	0	1	0	0	0				2					6	1	0	0	0	1	1	0
0	1	0	0	0	0	1	0	0				3					6	0	0	1	1	0	0	1
0	1	0	0	0	0	0	1	0				4					6	0	0	1	1	0	1	0
0	1	0	0	0	0	0	0	1				5					6	0	0	1	1	1	0	1
0	0	1	1	0	0	0	0	0				0	0	1	1	1	7	0	0	1	1	1	1	0

that contain output functions of the independent input variables become the focal point. A particular column contains programmed ones at row intersections that correspond to input product terms contained in the sum of products equation defining the desired output function.

7.8 ROM TECHNIQUES

Now that the internal circuits that comprise a ROM and the two alternate functional concepts of ROM operation have been reviewed, let us consider an example to demonstrate several techniques of logic implementation using ROMs. Suppose a logic network must be created to translate from a 9-bit Barker code (refer to Table 7-2) to the 7-bit ASCII code. The word set described by the Barker code contains 16 members. This code conversion could be carried out in a straightforward manner using an ROM as described previously. The resulting ROM which requires 3548 bits is shown in Figure 7-13a.

7.8.1 Cascading ROMs

An alternate method of implementation can be employed to minimize the number of ROM bits required. Consider for example the multiple ROM shown in Figure 7-13b. This implementation makes use of the set of intermediate state assignments described in Table 7-2, Subset *A* and *C*. The 16-member Barker code word set can be described by products of words from the two subsets of 5 and 7 members, respectively. These subsets can then be translated into the ASCII code using a smaller number of ROM bits than originally required. For the Barker to ASCII translation using a two-level conversion scheme, 592 bits are required.

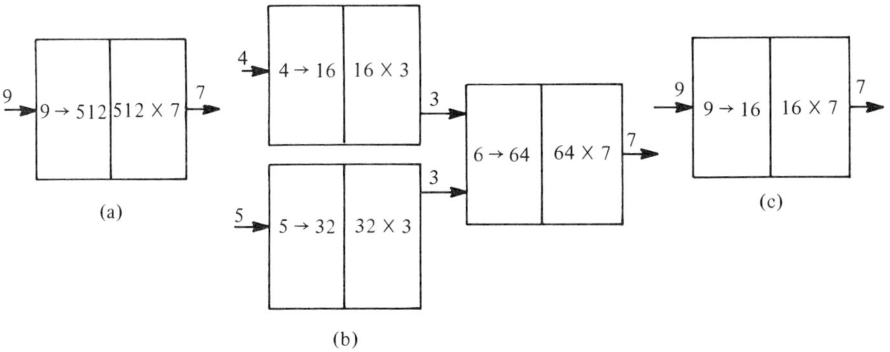

Figure 7–13 Barker code to ASCII code converter ROMs.

7.8.2 ROMs with Partial Decode

Another approach to the code conversion problem is shown in Figure 7-13c. Here the constraint that all possible input product terms be generated is removed. Implementation under this relaxation is best visualized by using the function generator concept of operation. Each output can be described by the OR-ing of input combinations that produce a one on the output. The 16 input combinations of codes to be converted represent the set of product terms that are capable of generating a 1 on the output. Therefore, a decoder (product term generator) that decodes 9 to 16 is sufficient. The resulting memory matrix contains 16 words of 7 bits or a total of 112 bits. Programming of both memory and decode matrices is required with this technique.

Although the third technique represents an optimum solution to the decoding problem from a standpoint of bit requirement, it should be pointed out that this approach is implemented with a ROM custom-organized in the most efficient organization for the particular requirement. Conversely, the ROMs of the second implementation do not necessarily represent a most efficient solution. It has been observed that most general logic functions can be implemented by including enough decode gates to generate approximately 25 percent of the possible input product terms. If the ROM proposed in Figure 7-13c were built as a standard product to this criteria, it would contain a 9 to 128 decoder which would result in an 896-bit ROM instead of the 112-bit ROM described.

The reduction in the size of the memory achieved by lifting the constraint on product terms, as demonstrated in the third mechanization example of the code translator, has applications that extend to the general problem of random logic implementation. The straight ROM approach to logic implementation usually suffers from comparison with the custom array approach. The custom array does not contain all the unused product terms and associated memory storage of the standard ROM and, therefore, is a more efficient mechanization of the desired logic. ROM construction that allows for partial generation of product terms, as demonstrated in the previous example, can in some cases, result in a more efficient mechanization than custom logic. This necessitates the choice of ROM vs. custom logic on an individual basis as the most economical implementation technique.

7.8.3 Synchronous ROM

To demonstrate further the logic capabilities of ROMs, consider the addition of a bit of delay to the functional block diagram of a ROM. With

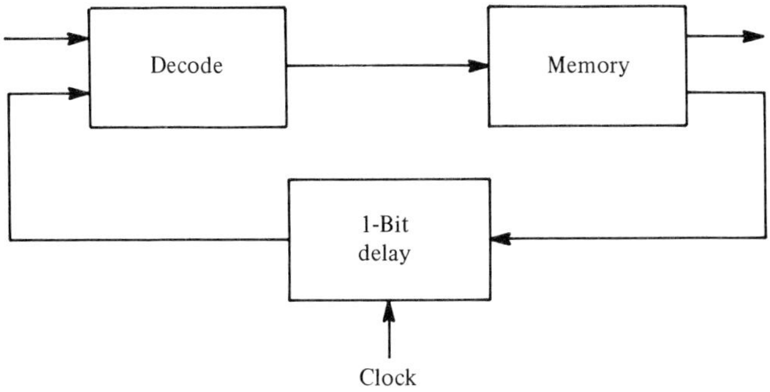

Figure 7–14 Sequential ROM.

this feature the ROM is upgraded from the class of asynchronous logic networks to the class of synchronous systems. A ROM with synchronous delay can be used to implement sequential machines by providing feedback from ROM outputs to inputs as shown in Figure 7-14. The capability of a ROM is greatly enhanced by including this feature.

It is shown by Taylor Booth[1] that Mealy and Moore machines, constructed as shown in Figure 7-15, can be used to implement any sequential or combinatorial logic network conceivable. In both types of logic networks, the present state of the machine depends on the inputs and the previous state of the machine. The Mealy machine is possibly more flexible since the present value of the machine outputs depend on the present inputs as well as the past state of the machine. This flexibility allows the outputs to be asynchronously affected by the present inputs.

As a practical example, consider the implementation of an up-down counter using this technique. Figure 7-16 shows a 256 × 10 ROM and a flow chart that describes its memory program. The memory is programmed to contain a binary number that is one bit greater than the binary address assigned to the memory location. A count sequence is initiated by setting

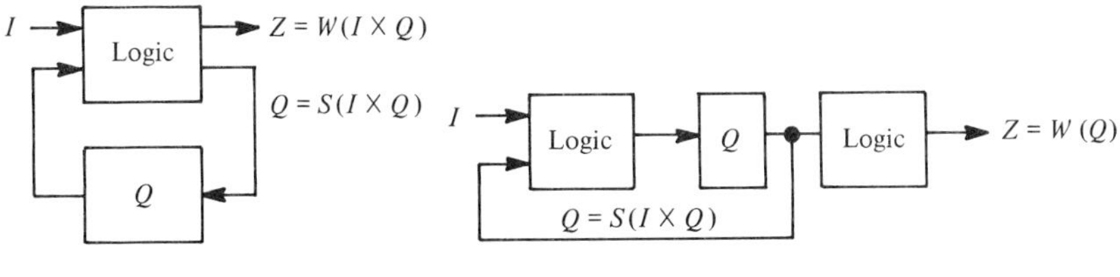

(a) Mealy machine (b) Moore machine

Figure 7–15 Block diagram representation of sequential machines.

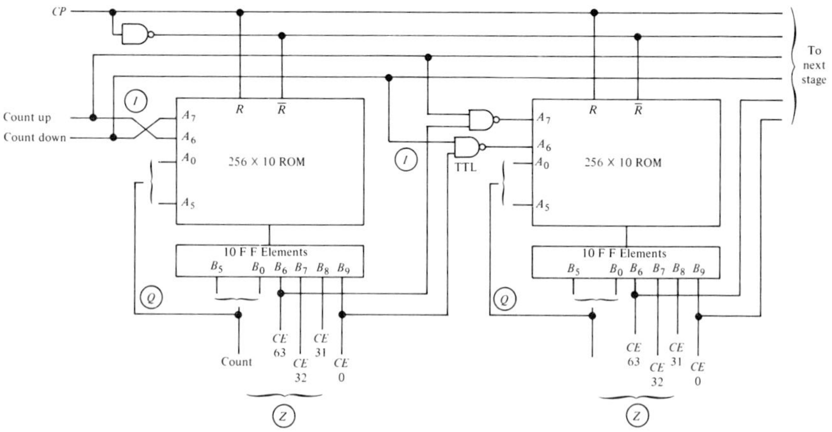

Figure 7–16a Block diagram of a 256 × 10 ROM as a sequential machine.

the inputs at the zero count. This input calls the binary one count from the memory to the outputs during bit time one. At bit time two, the one count on the outputs is fed back to the inputs. This input calls the binary two count from the memory during bit time two. During bit time three the two count on the output is fed back to the input and the process is repeated as long as the sequencing clock is maintained. Since this program uses only a fourth of the ROM storage, a down count and a no count algorithm can also be programmed into the array.

Notice the similarity to a Moore machine. With the storage element on the ROM input instead of the ROM outputs, the counter would convert to a Mealy machine capable of asynchronous signal propagation from inputs to outputs.

7.9 SERIAL MEMORY

Serial memory storage is required in almost every digital data processing system. The length of the storage register can vary from 1 bit to very long lengths. Serial memories implemented with MOS appear in small lengths intermixed with random logic in some system. They are used both for delay elements and data storage and manipulation. Some applications such as drum storage and CRT display storage require bit lengths that make long register arrays feasible. MOS serial memories offer variable data rates that cannot be achieved with other serial memory technologies.

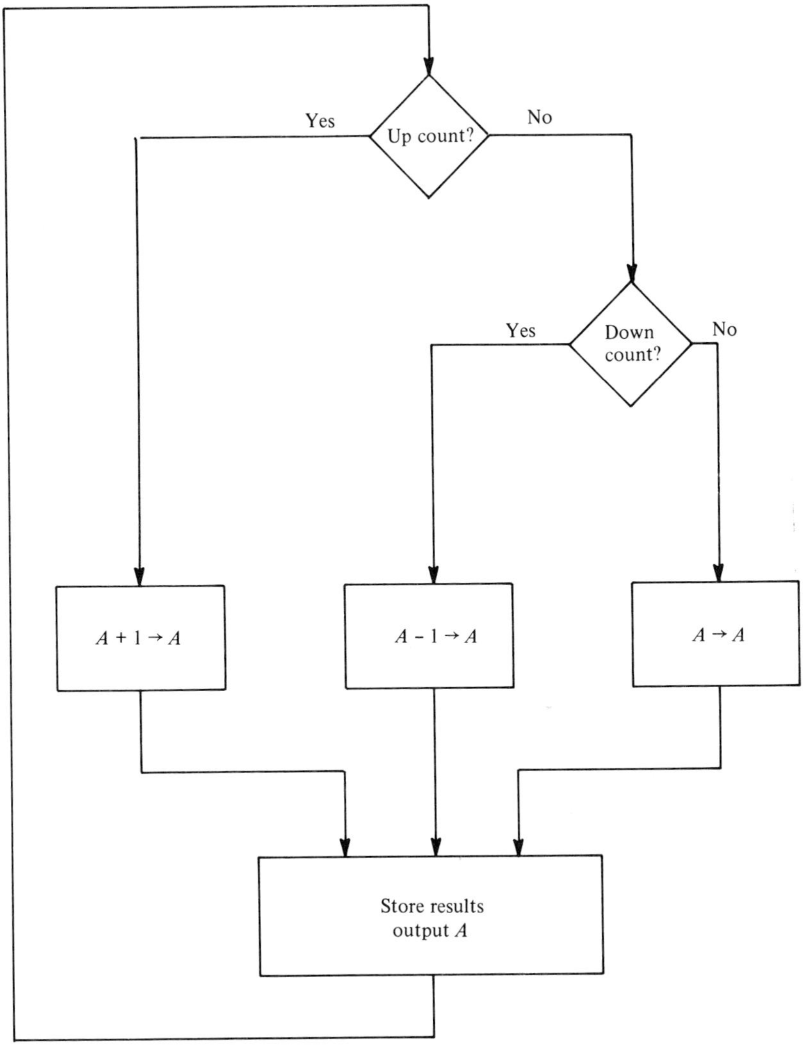

Figure 7–16b Flow chart of a 256 × 10 ROM as a sequential machine.

7.10 SERIAL MEMORY MECHANIZATION

A serial memory subsystem is mechanized with internal storage cells connected in series and an output buffer to convert the low level MOS signals to a level capable of driving an external load. A block diagram representation of a serial memory is shown in Figure 7-17. The basic internal bit can be either static or dynamic; however, both types require a 2-phase clock.

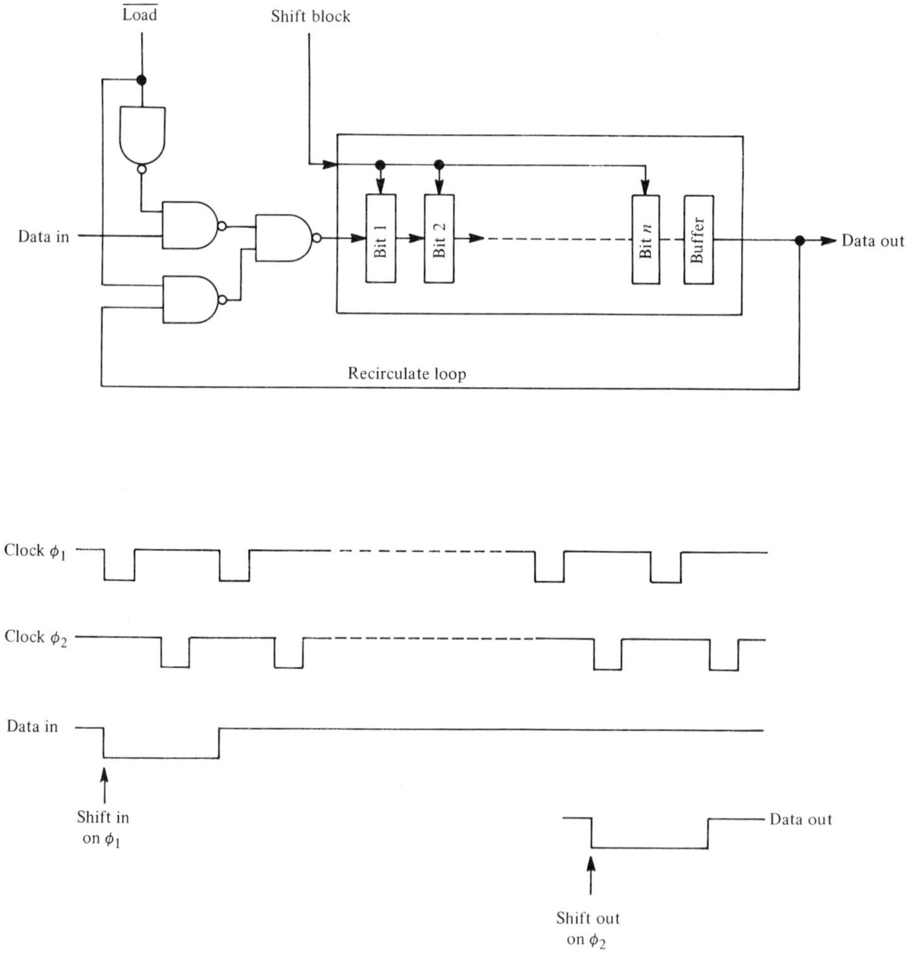

Figure 7–17 Serial memory subsystem.

7.10.1 Static Bit

The primary advantage of the static cell is its ability to store data permanently during the absence of clock pulsing. The primary disadvantage is the large area required to construct the cell. This results in a restriction on the maximum number of bits per package. Implementation of this type of cell also requires a substantial amount of dc power.

7.10.2 Dynamic Bits

A dynamic circuit implementation of the storage bit relies on capacitance storage of information. A number of circuit configurations of this type of storage cell are shown in Chapter 5. The advantage of this type of storage cell is a low area requirement due to the size of the devices and the lack of a ground bus. Serial memory subsystems that are implemented with this cell have approximately twice as many bits per package as those which use static bits. Furthermore, the dc power dissipation for this cell is also greatly reduced. The fact that dynamic cells cannot store information indefinitely is unimportant in most applications of serial memories.

The timing of this memory is shown in Figure 7-17. Data on the input of a serial register is sampled when the phase 1 clock is at a negative level. The sampled information appears on the output when the phase 2 clock goes negative n-bit times later. The output remains valid until the next negative-going edge of the phase 2 clock. It is evident that the actual delay in this register is $(n + \frac{1}{2})$ bits. Output information is normally sampled during phase 1 of bit time $n + 1$ for driving into logic or another serial memory input.

One common mode of operation is to have a serial memory recirculate the information stored. The gates that accomplish this are shown in Figure 7-17. The steering logic at the input of the memory selects between a recirculating and a new data entry mode of operation. In same cases, this control function is made a part of the internal MOS array.

The maximum speed of MOS serial memory is limited to approximately 10 MHz with present P-MOS technologies. If higher speed operation is desired, a multiplexing scheme, as shown in Figure 7-18, can be employed. By multiplexing two shift registers, the maximum data rate can be doubled for a given clock rate. This is usually accomplished at the disadvantage of multiplying the number of input and output pins of the MOS device by two and adding some external gating. If a lower maximum data rate (less than 10 MHz) is satisfactory, the multiplexing function can be performed internally, that is, on the same chip that contains the shift registers, thereby

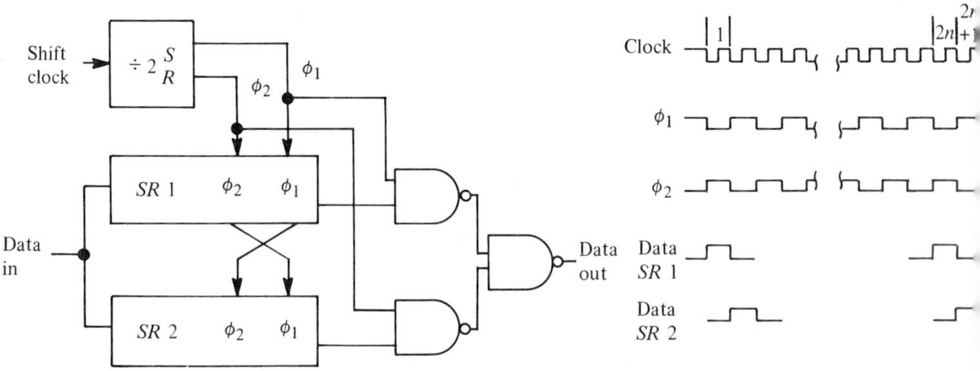

Figure 7–18 Multiplexed serial memory

eliminating the need for additional pin connections. The registers are both n-bits long. Note that the phase 1 and phase 2 connections are reversed on *SR* 2. This allows data to be shifted in and out at twice the maximum allowable register clock rate. During phase 1, input data is sampled by *SR* 1, and during phase 2, the input data is sampled by *SR* 2. The data shifted in during phase 1 is available at the *SR* 1 output after the register has received n phase 1 clock pulses. This is equivalent to $2n$ shift clock pulses. Therefore, the multiplexed registers appear to be $2n$ bits long. At the $2n$ shift clock pulse the data is fed to the outputs from *SR* 1 through the enable gate. During the $2n + 1$ shift clock pulse, phase 2 goes negative and the data previously sampled on *SR* 2 is enabled and fed to the output. This technique may be used to multiply the data rate until the maximum frequency of the output steering logic is reached.

7.10.3 Applications

Serial memories are readily adaptable to computer peripheral equipment, radar signal processors, desk calculators, and where slow access buffer storage is required.

A block diagram of a CRT display system is shown in Figure 7-19. In this system a shift register is used to convert data from serial to parallel and as a recirculating storage buffer. The recirculating store is implemented as shown in Figure 7-17. A page of data is entered in 6 bit serial (one character) segments. The recirculating memory can store 10 lines of display with 40 characters in each line. Once placed in the memory the data are recirculated and the CRT is continually refreshed. The serial to parallel function is implemented by a 6-bit shift register with each bit connected to an output pin through an output buffer as shown in Figure 7-20. Six bits

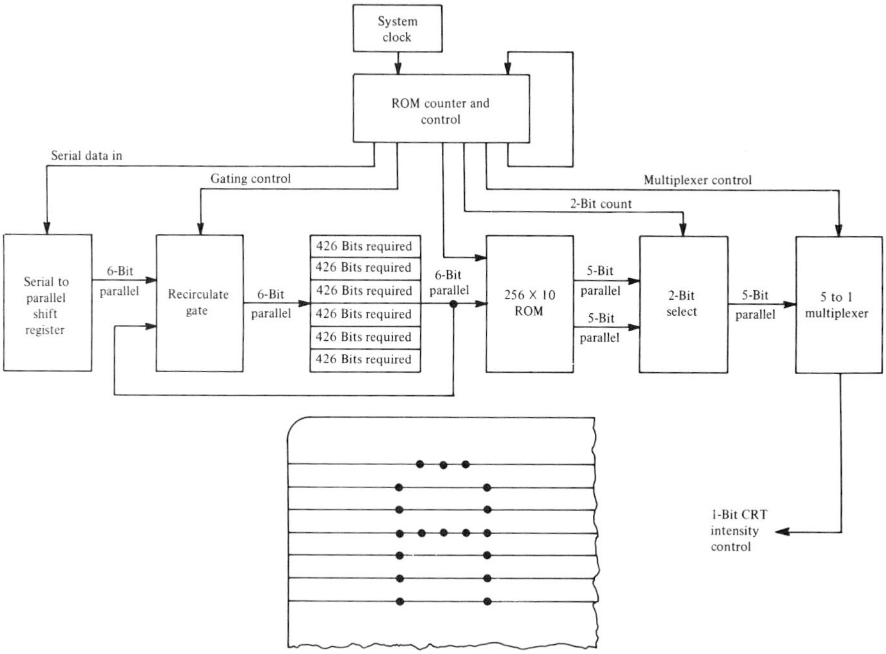

Figure 7–19 Display system using shift registers and ROMs.

are shifted in serially and then shifted out through the buffers in parallel.

The serial to parallel register can be converted into a 12-state Johnson counter by the addition of an inverting feedback loop, as illustrated in Figure 7-21. The counter starts with all zeros stored. The register is first filled with ones during six clock periods and then filled with zeros during

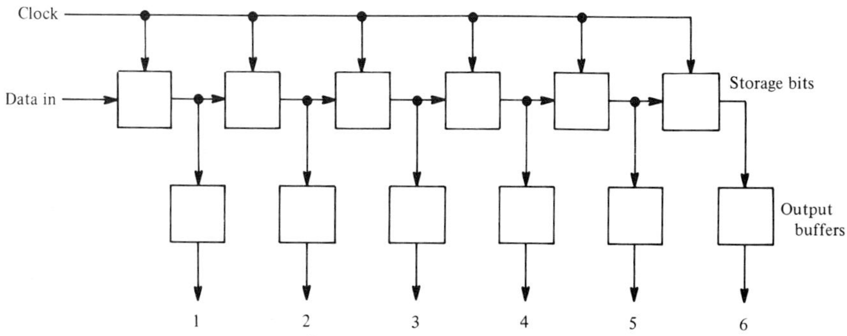

Figure 7–20 Serial to parallel conversion.

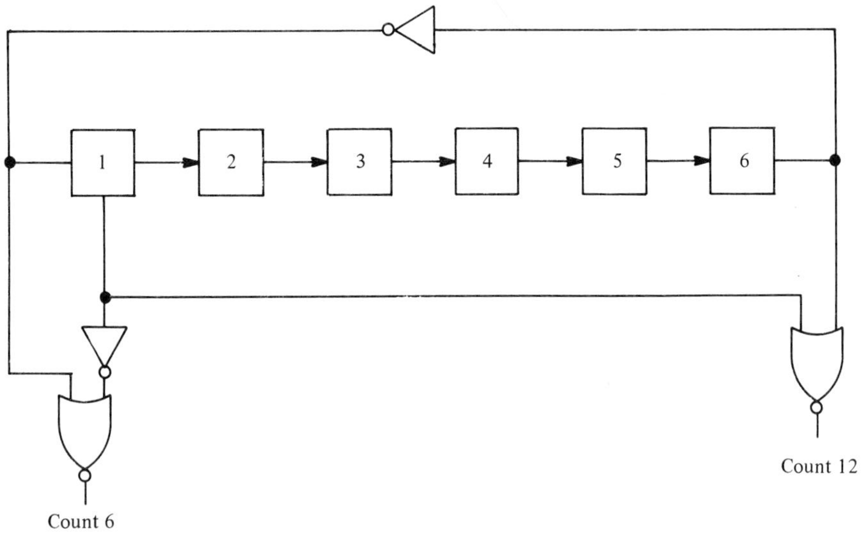

Figure 7–21 Johnson counter.

the next six clock periods. The 6 count and 12 count decode logic is shown to illustrate how the state of the counter can be decoded.

A recirculating serial memory may also be used as a word generator in test equipment designed to check out digital systems. This type of testing is required on airborne equipment and integrated circuits. A 100-bit word generator might be implemented as shown in Figure 7-22.

The generator provides a data pattern at the output continuously recirculating the serial memory contents. Data are loaded into the memory by

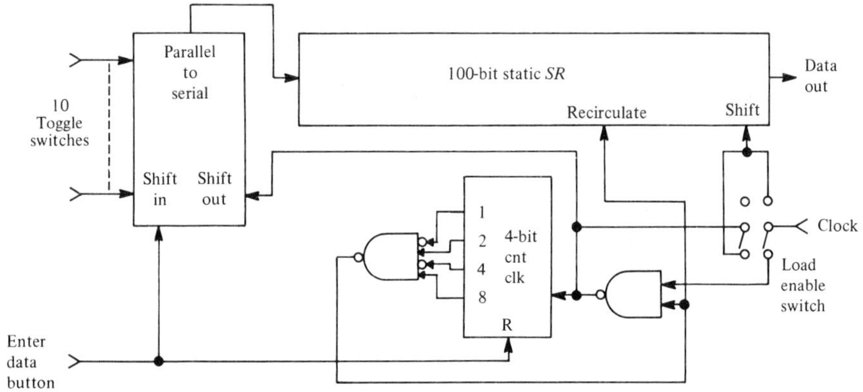

Figure 7–22 100-bit word generator.

disconnecting the clock from the shift register. This stops the recirculation of the register. The first 10 bits of the 100-bit word are placed on the toggle switches. When the enter data button is pressed, the 4-bit counter counts 10 clock pulses and then cuts off the clock pulse train. The ten clock pulses shift the data on the toggle switches into the 100-bit register. This procedure is repeated 10 times to load 100 bits. After the register is loaded, the load enable switch is repositioned to connect the clock to the register and the new word pattern is recirculated.

REFERENCES

1. T. L. Booth, *Sequential Machines and Automation Theory,* p. 69, John Wiley & Sons, New York, 1967.

GENERAL REFERENCES

J. Karp, "Dynamic Refresh Memories," *IEEE Conven. Digest,* 36–37 (1971).

W. S. Sander, "Semiconductor Memory Circuits and Technology," *Fall Joint Computer Conf. Digest,* 1205–1211 (1968).

R. H. Crawford, *MOS FET in Circuit Design,* McGraw-Hill, New York, 1967.

F. Kvamme, "Standard Read Only Memories Simplify Complex Logic Design," *Electronics,* **43,** 88–95 (January 1970).

L. Boysel, "Random Access Memory Packs More Bits Per Chip," *Electronics,* **43,** 109–115 (February 1970).

8

Topology-Array Layout

8.1 INTRODUCTION

In this chapter, layout and the steps that lead to final artwork and mask preparation are examined. Some of the basic layout considerations that are important in MOS LSI design are also discussed. The development of an MOS LSI array normally follows a path similar to that shown in Figure 8-1. The steps that lead up to the layout of an array have been discussed in detail in Chapters 4, 5, and 6. The final steps in the development sequence, i.e., fabrication and testing, have also been discussed in earlier chapters.

8.2 GENERAL LAYOUT PHILOSOPHY

In the layout of an MOS LSI array, a number of important factors must be considered. The primary objective is usually efficient utilization of area; however, consideration must be given to all aspects of circuit performance. Many times trade-offs must be made between optimum area usage, performance, and reproducibility. Such trade-offs may involve a careful examination of the circuit design and device parameters and usually require close coordination between the circuit design engineer and the layout designer. Particular attention must be devoted to such performance factors as circuit transient response, noise coupling, storage node leakage, and deterioration of voltage levels due to accumulated diffused region resistance.

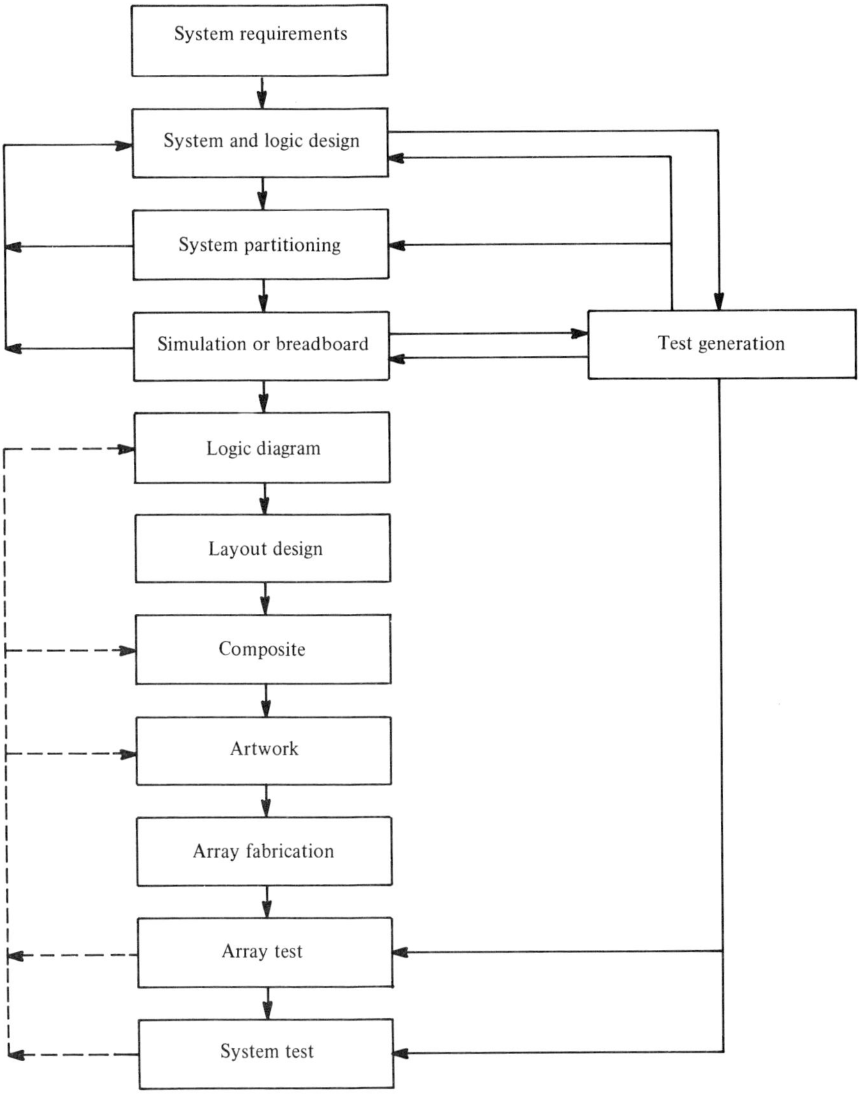

Figure 8–1 Array development flow chart.

In addition, a very important consideration is layout simplicity. In order to reduce the possibility of error in mask fabrication as well as in the calculation of circuit performance, the layout should be as simple as possible to accomplish the desired function. Complex layout configurations should be avoided except where specific characteristics are desired.

8.3 COMPOSITE DRAWING

A composite drawing is the basic tool used to translate an array into graphic form. It is a single drawing made on grid paper or other material, such as Mylar, which shows the photomasking levels that are used in device fabrication. The composite is drawn to scale, usually 500×, since at this magnification even the smallest device geometries can be drawn accurately and with little difficulty.

Using the English system of measurement, the scale of 500× allows increments of 0.1 mil (final size) to be used with generally available grid sizes. Unique symbology is required for each photomask so that each level can be separated when photomasks are prepared.

The photomasking levels that are usually identified on a composite drawing are:

Metal areas
Diffused regions
Gate or thin oxide regions
Connections between metal and diffused areas (cutouts)

Basic symbols used to identify these are shown in Figure 8-2. These symbols are not standardized throughout the industry, although those shown here are a representative set utilized by a number of MOS manufacturers.

A portion of an array composite drawing is shown in Figure 8-3 with the symbols defined in Figure 8-2. A standard output inverter circuit is shown with important areas identified. Scale has not been preserved in this illustration.

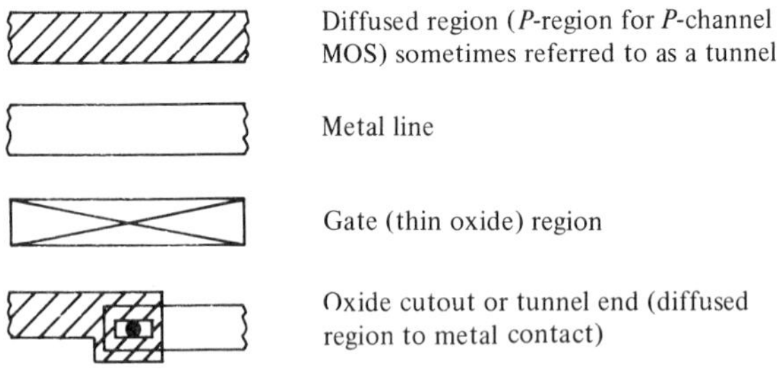

Figure 8–2 Typical symbols used on composite drawings.

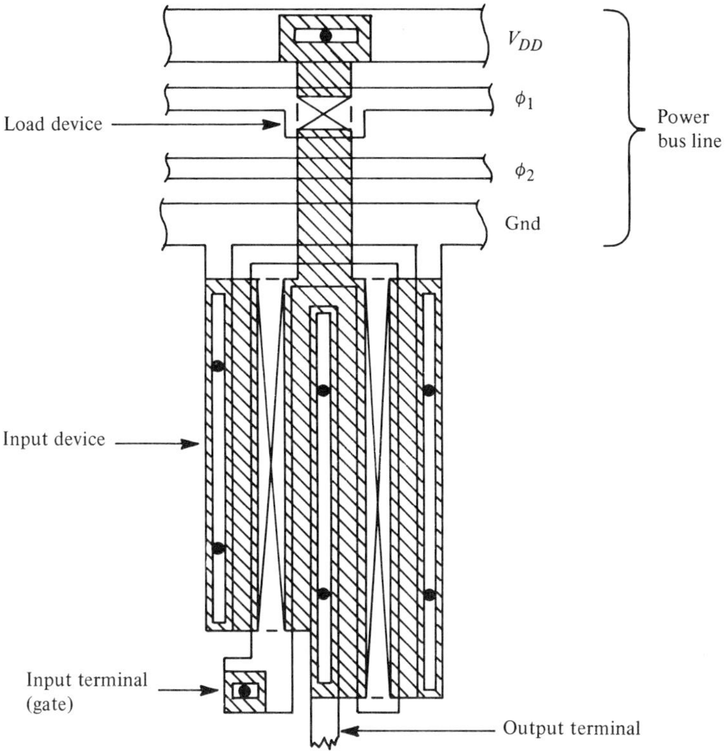

Figure 8–3 Composite drawing of a standard inverter cell.

8.4 BASIC CIRCUIT CATEGORIES AND LAYOUT CONSIDERATIONS

LSI arrays may be divided into two categories from layout considerations. One of these includes logic circuits or any circuit containing a significant number of interconnections between logic functions or cells within the array. The second category has a minimum of interconnections and displays a regular organizational pattern. This category consists mainly of memories and registers. Since the approach to the topological design of each of these array categories is different, the categories will be examined separately.

8.4.1 Logic Arrays

Layouts of this type may be prepared manually or with the aid of a computer. Where an array is expected to go into high volume production, the primary concern is reducing die size to a minimum, provided no compromises are made in performance. In order to accomplish this, careful planning of the array must be done and a considerable amount of engineering effort must be directed toward area reduction. For optimum area utilization or if performance is a critical factor, the array is drawn manually.

To start a manual array layout, a plan is prepared by the layout designer and analyzed carefully to ensure proper electrical performance. The planning of a logic array must be coordinated closely with the circuit design engineer. For some arrays, it may be necessary to verify the electrical characteristics of every internal circuit node as it is located to ensure that the complete array will meet the performance specifications. When the layout plan is completed and performance verified, the array composite is drawn in final form.

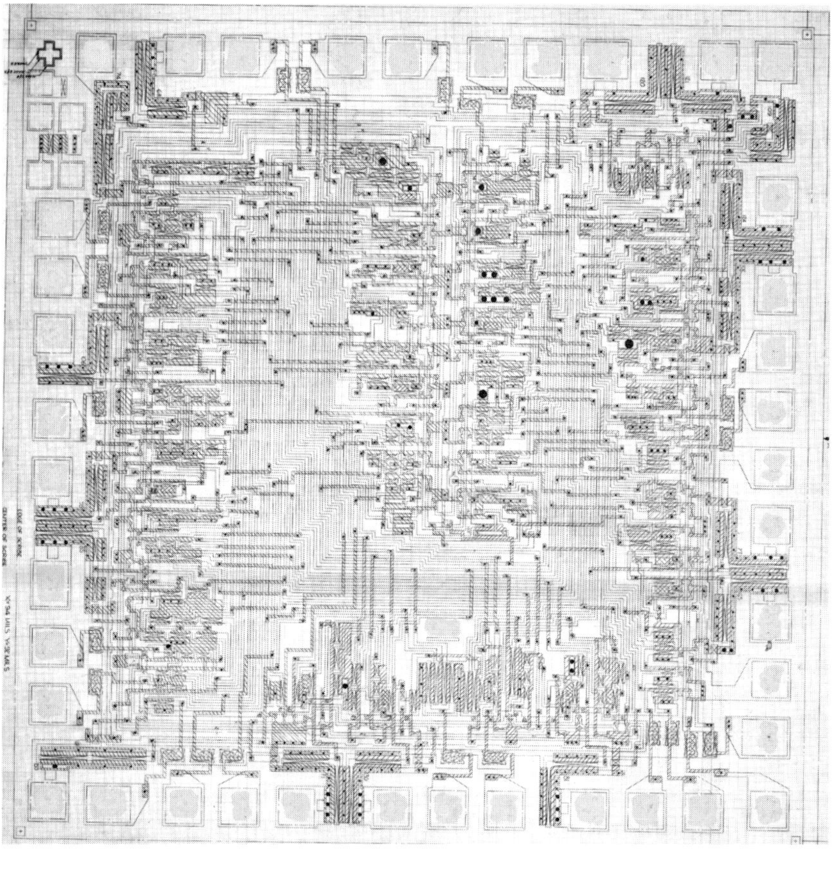

Figure 8–4 Manually drawn composite.

Figure 8-4 is an example of a composite drawing of an array layout done manually to achieve performance characteristics within a narrowly specified range of speed and power. The array shown in this figure is a general combinational logic circuit implemented with 2-phase ratioed techniques. This layout clearly demonstrates efficient area utilization. By carefully arranging the position of devices and allowing devices to be placed between pads and interconnections, the overall size of the array has been kept to a minimum. It should be noted that there are a number of different device sizes utilized throughout the array indicating that individual node capacitances were calculated and the appropriate device sizes were specified for each node. This layout is an example of a completely custom design approach.

Some other significant observations can be made by examining this layout. As pointed out in Chapters 4 and 6, outputs from an array occupy a significant area when compared to internal logic elements. The outputs in the array of Figure 8-4 can be identified by the close proximity of the bonding pads to the large geometry output driver circuits, and it can be seen that the driver circuit and pad occupy an area equivalent to several internal logic elements.

The interconnections between logic elements on this array occupy a very significant portion of the total area. Therefore, in minimizing silicon area for general logic arrays, maximum effort should be devoted to arranging the location of the logic elements to minimize this interconnection area. In order to accomplish this, various specialized techniques have been developed, some of which utilize computer aids. However, a skilled layout designer can frequently produce a very compact and efficient layout relying primarily on his talent and experience.

Following the preparation of the composite drawing, a number of checks are made to ensure circuit continuity, electrical performance, and proper layout techniques. The composite is usually photoreduced 2½ times to produce a photocomposite 200× final size. The reduction is necessary for large arrays so that the composite will be sufficiently small to fit on a coordinatograph table for rubylith cutting. Rubylith is a clear Mylar with a ruby-colored overlay that can be peeled. The ruby surface is cut with a knife, which is controlled by the coordinatograph. Five different rubies are cut, each ruby being used to generate a subsequent mask. After cutting is complete, the areas that are to be clear are cut completely around their periphery and the ruby-colored overlay material is peeled off. The rubies are then checked to assure their accuracy.

If large volume production is not anticipated or if the development cycle time is short, the layout may be accomplished with the aid of standard cells rather than custom designing each logic element. The arrangement of these standard cells to minimize the total chip size is the primary task in the design of a given logic array, as mentioned in Chapter 6.

A standard cell family suitable for use in general logic arrays usually consists of from 30 to 100 individual cell types, each of which has been carefully designed and characterized. Performance of each cell is ensured by maintaining the same device geometries and layout every time the cell is used. Figure 8-5 shows the schematic diagram of a typical standard cell together with a combination logic diagram and block composite that indicates the actual size of this cell. Standard cells offer an advantage to the designer in that various computer aids can be easily utilized to simplify the array design and layout tasks, thereby improving turnaround time and reducing the possibility of error. The penalty paid in using standard cells is some compromise in area utilization.

A number of computer programs exist to aid in optimizing standard cell placement, interconnection routing, and artwork generation. These programs are usually designed to be used with automated hardware capable of either directly generating the various photomasks required by the process or cutting rubylith masks at an enlarged scale.

A simplified flow diagram of a typical placement, routing, and artwork generation program is shown in Figure 8-6. The particular program illustrated in the flow chart of Figure 8-6 is designed for use with an automated plotting/cutting table which is designed to generate 200× rubylith masks. After the logic diagram has been completed and partitioned into a standard cell compatible format, the logic is entered into the program. Topological features and electrical parameters for each member of the standard cell family to be used are stored in the computer memory. Each time a specific cell type is placed by the computer program, it is reproduced with exactly the same device locations and dimensions internal to the cell, thereby assuring uniform performance characteristics.

An intermediate output, such as a block composite showing the cell locations and interconnections, is frequently used as a means of verifying proper cell placement and circuit continuity for a given array. At this point, the designer can make any manual adjustments necessary to improve performance or layout efficiency. Following a block composite, a multilevel composite is frequently plotted as a final check of the array design. This document may also serve as a record for future reference and checking purposes. Finally, after all checks have been made on the composite, the circuit is committed to the artwork stage where either rubylith masks are cut or photographic plates are made directly.

Figure 8-7 shows an enlarged photograph of a complex logic array designed using standard cells which were placed and interconnected by com-

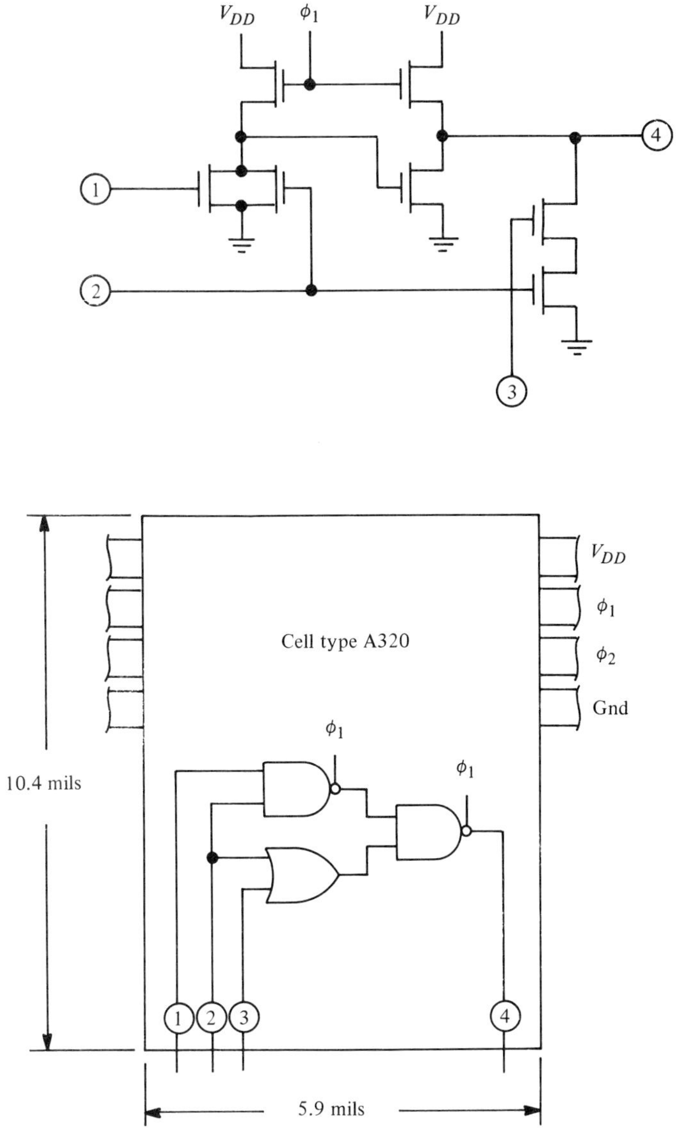

Figure 8–5 Typical standard cell logic and block composite drawing.

puter program with a minimum of manual editing. The actual size of this array is 155 × 160 mils and contains more than 100 logic elements or cells.

It can be seen in this photograph that the area occupied by the inter-

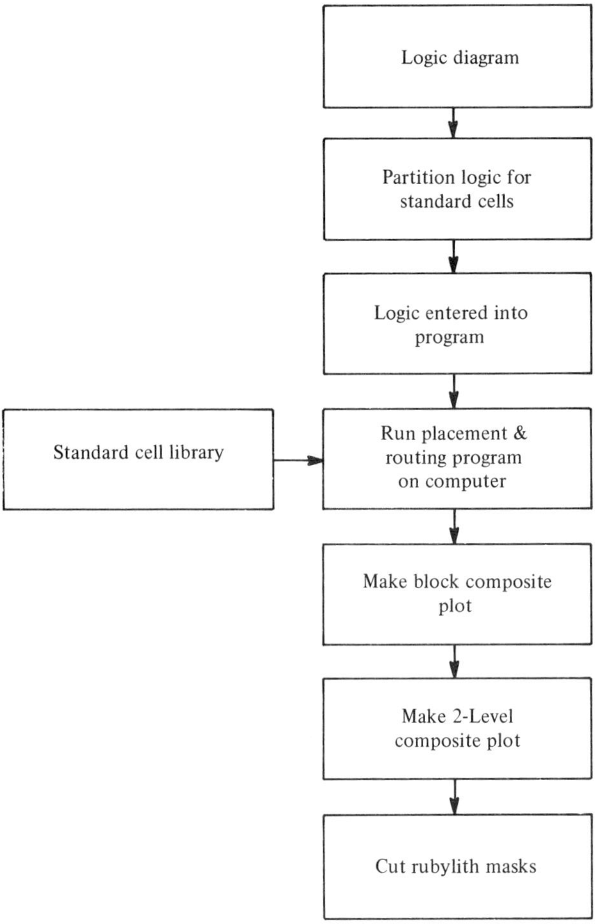

Figure 8–6 Simplified flow diagram of a typical placement and routing program with standard cells.

connections between the standard cells is greater than 50 percent of the total array area. In addition, it is interesting to note the contrast in packing density between the custom designed array of Figure 8-3 and this automated layout. Again it is important to remember that the primary importance of a computer aided layout is rapid turnaround and lower development cost.

Typically, a placement and routing program such as the one described here can be arranged to provide electrical parameter printouts such as propagation delays or loading capacitances for the internal elements on

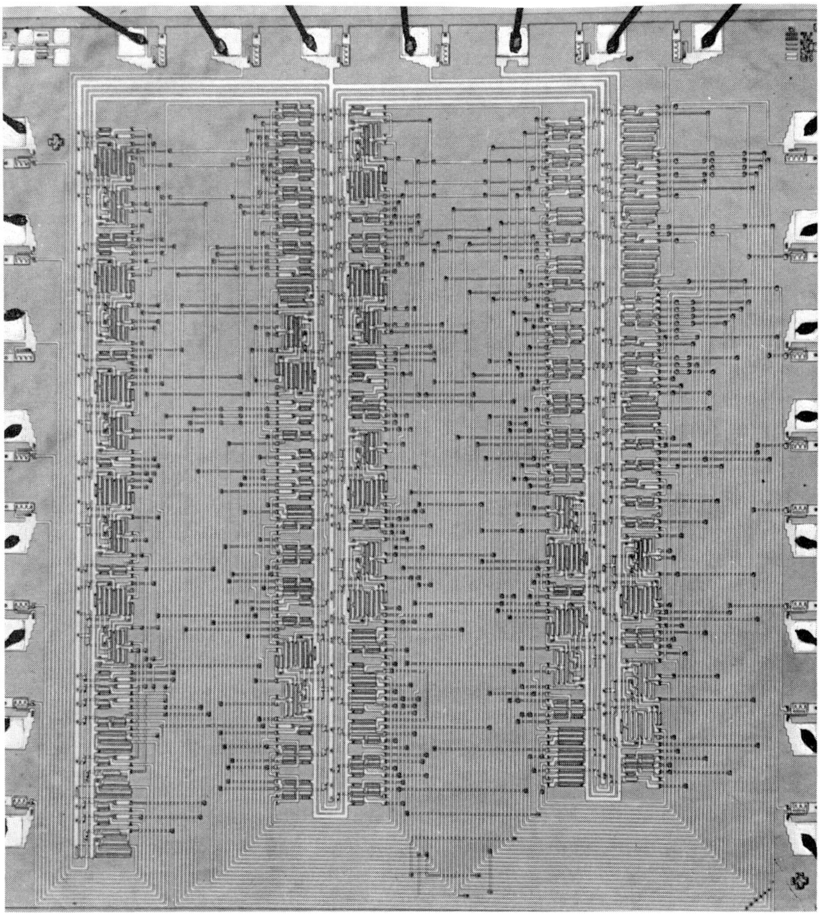

Figure 8–7 Logic array designed with standard cells.

the array. Information such as this can then be used to verify that performance is within the array specification limits.

8.4.2 Ordered Arrays

This category of circuits includes all types of registers and memories and is characterized by an ordered arrangement of logic elements or cells that require a minimum of interconnections external to the cells. The efficient layout of an array of this type depends primarily on a carefully designed cell or group of cells together with a placement plan that not only utilizes area efficiently but does not compromise cell performance characteristics.

The cells used in an ordered array are usually designed and characterized in advance of a given array requirement. The cells are then reproduced either manually or with the aid of a computer program and placed according to the array plan. The design of a cell may be quite involved, depending on the speed or special performance characteristics desired, and it may involve a number of iterations of the layout before the designer is satisfied that electrical performance and layout are effectively merged. Frequently, test circuits are designed and fabricated to verify the cell design before application in an actual array. Ratioless cells are among the most critical from a performance/layout standpoint and require special attention to assure good cell design.

The arrangement of cells for an ordered array is usually approached by making a plan of the layout which indicates the position of each cell, routing of power and clock lines, and pad locations. From this plan the engineer can verify the performance characteristics and make any necessary adjustments to device sizes or cell locations prior to entering the final drafting or artwork stages.

The array plan, once complete, may be used either in conjunction with a design automation system or a composite drawing and rubylith masks may be prepared manually. In using the automated approach, the topology details of the cells to be used are stored in computer memory and reproduced at designated locations according to the array plan. Depending on the specific hardware associated with the computer aided design (CAD) system, various input and output forms may be utilized.

Figure 8-8 shows a typical intermediate output from a CAD system. This output is a multilevel composite drawing showing the location of cells and interconnections for a quad 40-bit dynamic shift register. This type of output is used to verify proper cell placement prior to mask production.

8.5 DESIGN RULES

The factors that govern topology may be grouped into two general categories: graphical and electrical. Graphical factors are related only to process resolutions and tolerances. Electrical factors are concerned with limitations such as drain to source breakdown characteristics or metal line current density. Both of these topology governing factors are usually defined explicitly as a group of "design rules." Design rules differ from one MOS LSI manufacturer to another, but they basically define the minimum sizes and maximum ratings dictated by each process consistent with reliable, high volume reproducibility.

Presented here are a number of basic design rules typical of those used

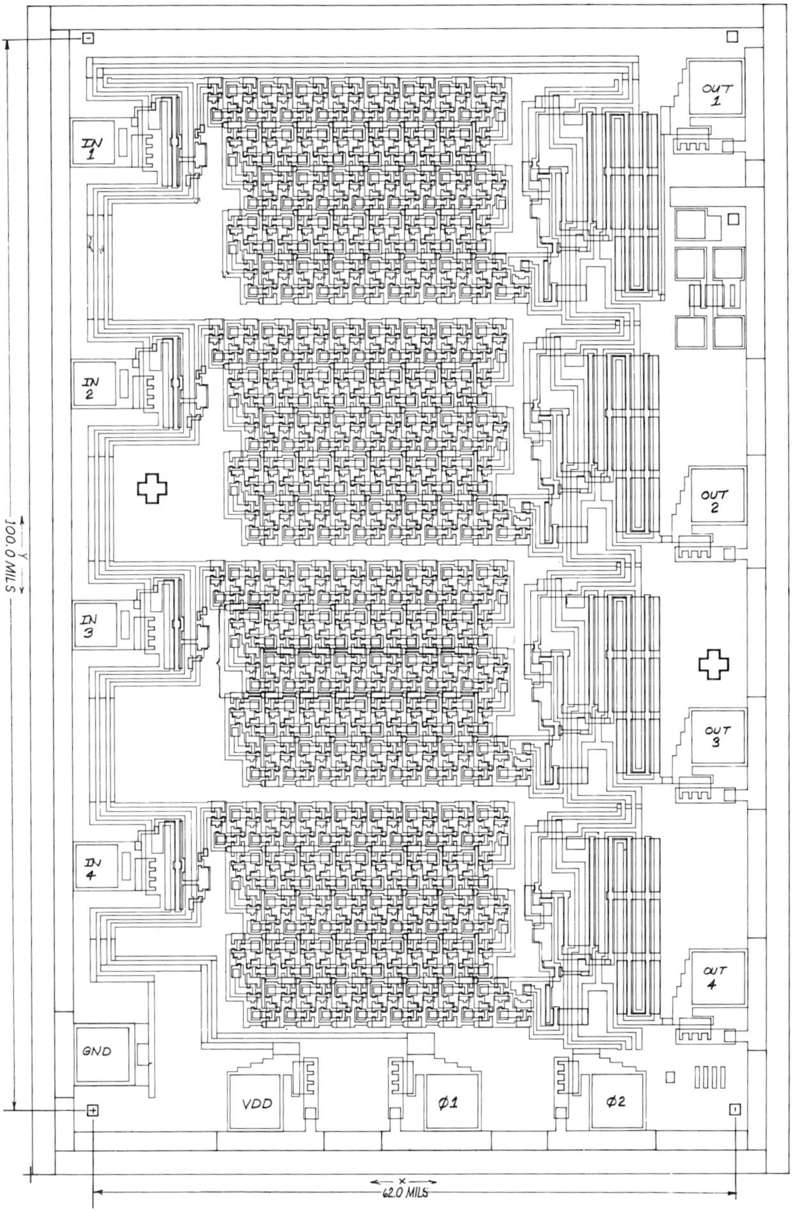

Figure 8–8 CAD prepared composite of quad 40 bit shift register.

throughout the industry. A range of minimum dimensions is presented for each rule based on a cross section of values from various MOS LSI manufacturers. The dimensions given here are in English units (mils).

Figure 8-9 shows a portion of a composite drawing of an MOS LSI array which is intended to illustrate the basic design rules given in this chapter. The numbers shown in the figures refer to the applicable rules.

It is important to note that only the basic rules are listed here. Many

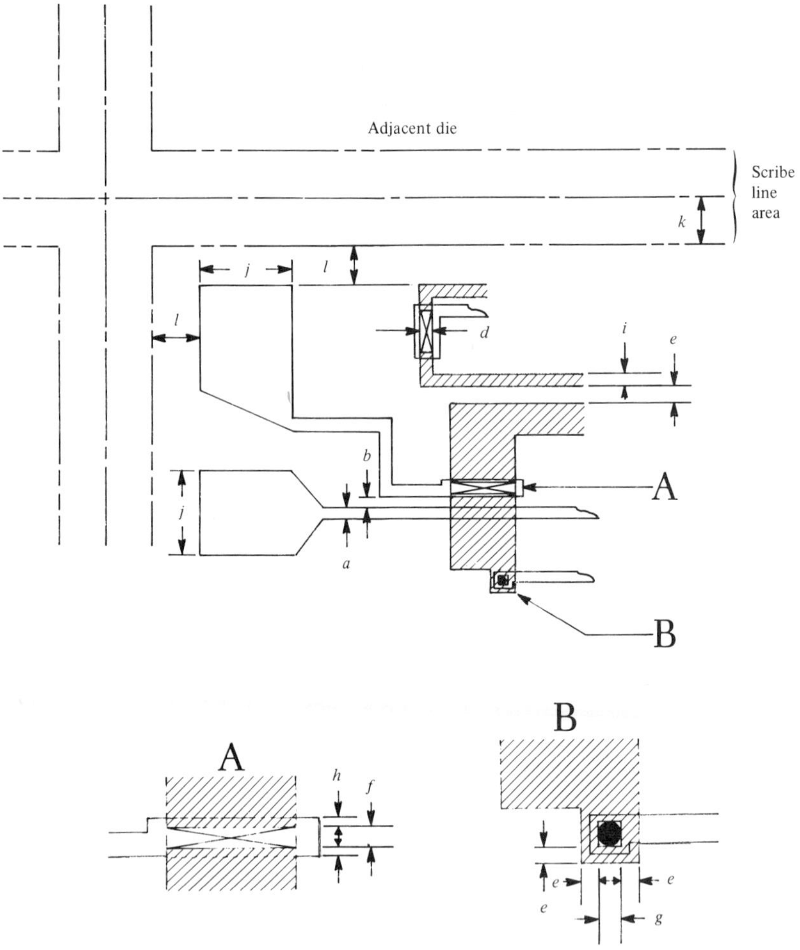

Figure 8–9 Portion of an MOS LSI array composite showing use of the basic design rules.

additional rules may be necessary, depending on the particular process characteristics. Many of these rules are carefully guarded by each device manufacturer, since they may reveal specific process steps or limits considered proprietary. The topology considerations that have been discussed in this chapter are intended to familiarize the reader with the type of constraints that are placed on the array designer and to provide some guidelines to aid in system partitioning and array complexity estimation.

8.5.1 Graphic Design Rules

The following limiting dimensions are based on a number of factors imposed by tolerances and size limitations in both photomask and wafer fabrication and are referenced to the indicated positions on Figure 8-9.

Minimum metal line width	0.2 to 0.5 mill (a)
Minimum metal line spacing	0.3 to 0.5 mil (b)
Minimum diffused region width	0.2 to 0.5 mil (c)
Minimum gate (thin oxide) width	0.2 to 0.4 mil (d)
Minimum oxide cutout (contact) dimension	0.2 to 0.4 mil (e)
Minimum metal border around gate (thin oxide) region	0.0 to 0.2 mil (f)
Minimum diffused region border around contact area	0.1 to 0.3 mil (g)

8.5.2 Electrical Design Rules

The following rules are based upon the electrical breakdown and leakage characteristics of two adjacent diffused regions:

Minimum device channel length (source to drain spacings)	0.4 to 0.6 mil (h)
Minimum diffused region spacing (other than device channel spacing)	0.4 to 0.7 mil (i)

The following rule ensures maximum reliability by keeping the current density in the metal lines well below levels that are likely to cause failure. In pulsed operation, the rule applies to average current.

maximum current rating for a metal (aluminum) line	20 to 30 mA/mil

8.5.3 Special Design Rules

These rules are required for the processing of arrays beyond the wafer state:

Bonding pad minimum dimension	4.0 to 5.0 mils sq. (j)

This minimum dimension assures adequate area to accommodate ball or ultrasonic bonding as well as test probing of the pad.

Minimum dimension from edge to center of scribe line	1.7 to 2.5 mils (k)

This defines the separation between arrays on a wafer and ensures adequate area for scribing.

Minimum dimension from edge of scribe line to pads, metal lines, or diffused areas	1.5 to 2.0 mils (l)

This ensures that no damage will result from scribing and separation of the individual arrays.

GENERAL REFERENCES

Engineering staff, *Polycell LSI,* Motorola Semiconductor Products Division, July 1970.

Lester H. Hazlett, "I am an Integrated Circuit Design Engineer," *EDN* (June 1, 1969).

Lawrence Curran, "Computers Make a Big Difference in MOS Designs," *Electronics* (October 13, 1969).

Engineering staff, *Micromosaic Arrays—MOS Approach to Custom LSI,* Fairchild Semiconductor, 1969.

Joseph Mittleman, "Computer Aided Design, Act Two: Admission Price Exceeds Forecasts," *Electronics* (June 9, 1969).

Engineering staff, *Building Block Handbook,* Philco-Ford Corporation, Microelectronics Division, October 1967.

Engineering staff, *Electrical Performance of AMI Standard Cells,* American Microsystems, Inc., September 1969.

9
Reliability of MOS Devices and Arrays

9.1 INTRODUCTION

The increasing complexity of MOS devices and arrays has made it more reasonable to consider the end reliability of the system and to place far less emphasis on the reliability of the individual part. Although it is quite unreasonable to expect a highly complex semiconductor to be as reliable as a simple one, it is completely reasonable to expect the ultimate system reliability to improve with devices that have many functions on one chip, as will be shown in this chapter along with the real causes of failure.

Also discussed in this chapter are the significant failure modes and symptoms, circuit design techniques for improving reliability, failure rates of complex MOS arrays, screening techniques for improving failure rates, a test evaluation device for measuring reliability, and failure analysis techniques.

9.2 SIGNIFICANT FAILURE MODES AND SYMPTOMS

In the semiconductor industry, failure may mean anything from a catastrophic inoperable condition after some period of use to some minor cosmetic defect that bothers an individual who is concerned with the overall quality of the product he is buying. The discussion of failures will be confined to problems that cause inoperability in system use, or dramatic changes in performance during high-stress testing.

Over the past years there has

been much concern over parametric shifts in the bipolar semiconductor products that have been available. A bipolar integrated circuit must be capable of driving a specified number of low impedance circuits. In order to do this, it must produce sufficient current to fanout into a useful number of such devices. This critical parameter, commonly called $I_{\text{available}}$, can shift slowly and eventually cause the part in question to provide too little current to perform its function. High-stress testing quite often resulted in this type of parametric shift. The MOS large-scale array, however, is normally expected to work into very high impedances and the basic parameter of interest is output voltage in either the logic 0 or 1 state. The inherent design of these devices makes drift at the output unlikely. In general, a large-scale MOS array is more likely to show symptoms of catastrophic failure rather than parametric drift. The underlying problem could be parametric drift within the circuit, but the ultimate symptom of failure is an improper output pulse train pattern. There are few useful parameters that are readable at the terminals of such a complex array. An evaluation device specifically designed to provide useful parameters at the terminals will be discussed later in this chapter.

In discussing failure modes, little emphasis, if any, will be placed on those mechanical modes that are common to both MOS and bipolar devices. Mechanically, a large-scale MOS array does not differ from a bipolar integrated circuit. MOS ICs are produced from silicon slices with no epitaxial layer. Bonding and die attach techniques are those that have been used commonly for ten years. In general, the dice are larger than those made in the past, and the number of leads to the outside world have been increased by a factor of about three. This would merely indicate a higher probability of failure due to any problems associated with bonding or die attaching. The package count in an MOS system is reduced dramatically enough to enhance system reliability, despite the increased possibility of a mechanical failure in a given circuit. Failure modes associated with packageing are, in general, common between the two technologies, although some failure modes are brought on by the gross size of LSI packaging.

9.2.1 Open Circuits

Many of the failures discussed here are related to complexity and the number of functions on each chip. However, at this point in time, most complex chips are produced through MOS technology and these modes of failure are applicable to MOS. Regardless of complexity, the basic failures occur within small cells of individual MOS transistors. The normal digital circuit is composed exclusively of MOS transistors and MOS loads, with an occasional diffused resistor or MOS capacitor. In general, resistance and capacitance are distributed throughout the circuit. Consider, then, an individual MOS transistor with a basic cross section as shown in Figure 9-1.

This single transistor is a high input impedance switch with a limited number of possibilities for failure. The switch is expected to turn on within a specified range of voltages applied to the gate and is expected to turn off within another range of voltages. In the off condition, the resistance is expected to be extremely high; while in the on condition, the resistance is expected to be as low as possible commensurate with other parametric requirements. Only those defective conditions that directly change these requirements have any real importance. Most real defects exhibit themselves as resistors varying from 0 Ω to large values inserted into the theoretical model of an MOS transistor.

The most obvious and important mode of failure is an open circuit within the MOS device or in the conductive interconnect that leads to the device. Assuming that a device is totally operable at the time of shipment, an open circuit can only be caused by excessive current density or a break caused by thermal or mechanical shock. Mechanical shock is more likely to cause an open where the die is joined to the package by fine wires bonded by thermal-compression or ultrasonic techniques. Much has been written about the various ways in which lead bonds can fail. However, a point not commonly considered is that, theoretically, resonance of such internal wire interconnects occurs in the range between 18 and 60 kHz. The

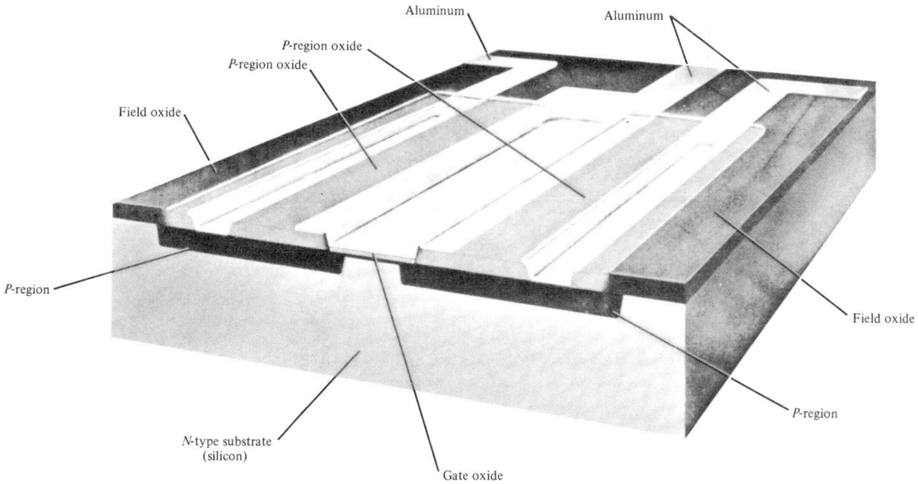

Figure 9–1 Basic structure of elemental MOS transistor.

internal wires are not very vulnerable to shock or vibration levels under 10,000 g's with short pulse durations. They are, however, very vulnerable to resonance.

One of the most damaging environments is ultrasonic cleaning commonly used for flux removal on printed circuit boards. Ultrasonic cleaners inherently generate a broad spectrum of frequencies with high energy content in the resonant range. Exposure to ultrasonics is most damaging when the entire semiconductor device is submerged in the cleaning bath. External leads subjected to this condition will not transmit enough energy into the device structure to cause harm. For this reason, many designs call for semiconductors to be mounted on one side of a board only so that the board may be cleaned ultrasonically by the flotation method. The entire board can be immersed except for the actual semiconductor package.

Of concern in the monolithic chip itself are interconnects generally formed by deposition of aluminum to a thickness of approximately 6000 Å. If the circuit is designed conservatively to limit current densities within the known safe range, deposited aluminum interconnects are highly reliable on the flat surface of a chip. Thickness control is easily achieved and the surface is vulnerable only to mechanical abuse during assembly. Any scratch may reduce the cross section of the aluminum and cause the current density to increase. However, it is relatively simple to see such defects at a magnification of 100×. On the other hand, there are subtle ways by which the cross section may be reduced, and these defects are not easily seen through a microscope.

By design, the surface of the circuit is irregular because of the need for different oxide thicknesses. As shown in Figure 9-1, the oxide varies in thickness between 16,000 and 1600 Å. The metal must be deposited as uniformly as possible over the tens of thousands of steps in present designs. Several years ago, it was common practice to vacuum deposit aluminum in a bell jar by placing aluminum clips on a tungsten wire with the silicon wafers placed below the wire. Through resistance heating, the aluminum slugs were evaporated and metal atoms traveled in all directions. Thickness control was difficult, but there was considerable assurance that the coating over the sides of each oxide step was sufficiently thick to keep current densities within tolerances. This method of metalizing, however, was extremely "dirty" since one of the first by-products of this type of metalization is sodium ions. As will be shown later, sodium ions are totally unacceptable in an MOS structure.

Today, the electron beam evaporation technique is being used to achieve clean aluminum deposition and control the thickness of the metal. In this method, an electron beam strikes a crucible of pure aluminum, creating a point source of molten aluminum. As shown in Figure 9-2, the point source

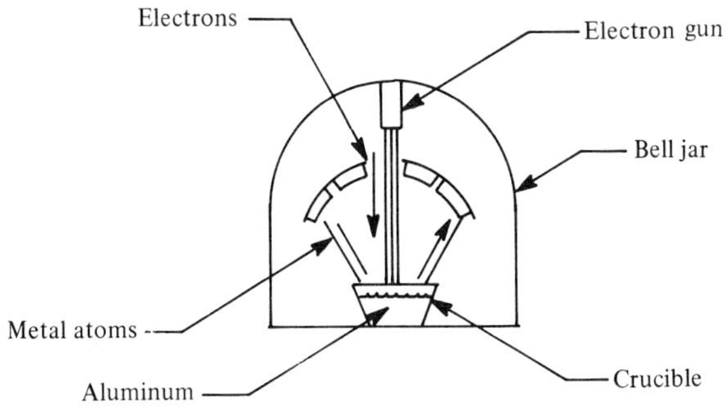

Figure 9–2 Electron beam evaporation.

causes the metal to be evaporated normal to the surface of the silicon wafers.

At the same time, the thickness of metal deposited on the essentially vertical walls of the oxide steps is less controllable. The best analogy is that of snow falling when there is no wind. In Figure 9-3, consider the depth of the snow deposited on a flat-roofed building and assume that the snow is very wet.

The snow deposited on the top will be of uniform thickness. The wetness of the snow will cause some adherence on the side of the building. The thickness will vary according to the moisture content of the snow. Extremely small changes in the angle of snow fall will also cause some deposit to accumulate on the side of the building. A bulge will form on the corner of the roof because the snow is more likely to adhere to snow than to the side of the building. This bulge will then cause a shadow effect which

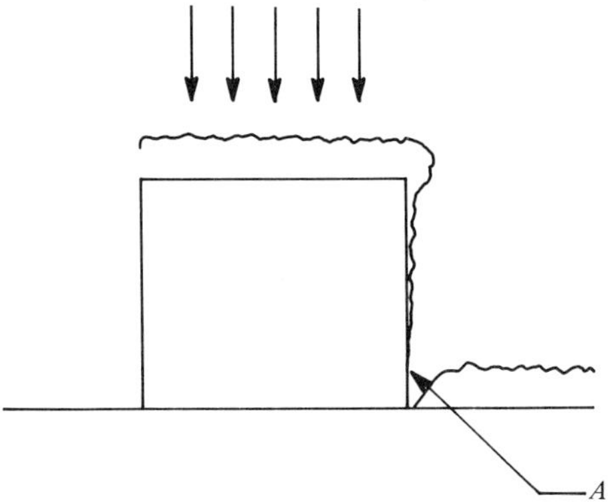

Figure 9–3 Metal deposition analogy.

will cause the drop-off in snow depth at position *A*. Some distance from the edge of the building, there will be the same snow depth as that on the roof. This is precisely what occurs during electron beam evaporation when a single source of metal atoms is used.

In the case of the devices, it is easy to see that the aluminum interconnects will provide very low current density on the flat surfaces, but the current density can rise astronomically as this current passes over the uncontrollable step. Figure 9-4 is a photograph taken at a magnification of 10,000× which shows such a step and dramatically points out the discontinuities in metal over the nearly vertical oxide wall. Fortunately, there are at least two approaches to solving this problem.

In keeping with our snow analogy, one merely has to provide the effect of a strong wind to provide continuity down the side of the oxide step. By using multiple point sources or providing an appropriate motion to the

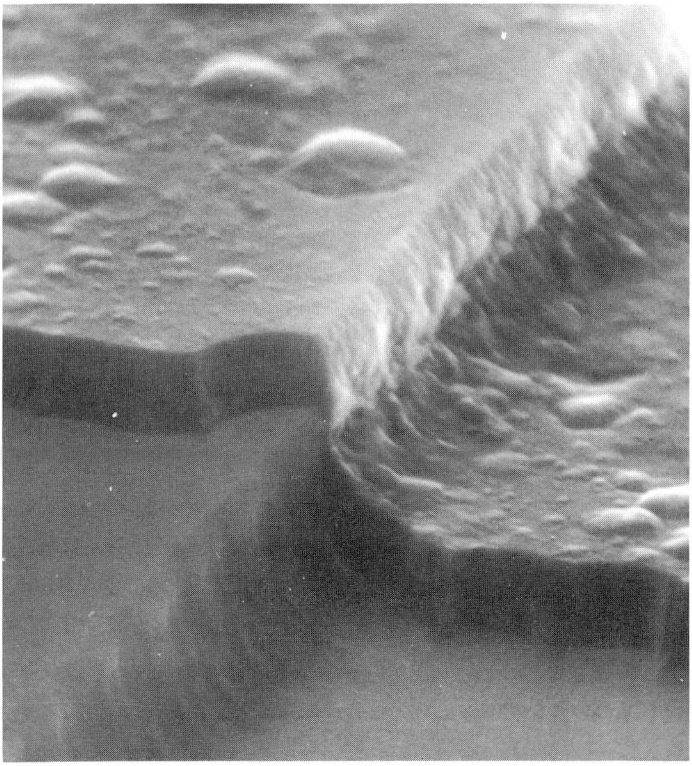

Figure 9–4 Basic interconnect failure mode.

wafers that receive the aluminum deposition, one can provide better deposition on these steps. Several techniques have been used successfully to minimize field failures to levels that are acceptable to users of these circuits. Figure 9-5 is an example of the results of deposition at many angles. One cannot be sure that the thickness is the same as that on the surface but, qualitatively, the metal is continuous and should provide much lower current density. Since the basic problem is high current density rather than just thin metal, there are also design techniques to enhance the process techniques. These are discussed in Section 9.3, "Design Aspects for Reliability."

The failure mode discussed above is amplified by externally supplied heat as well as the heat generated where high resistances are encountered in the aluminum interconnects. Compatible with good general practice, the

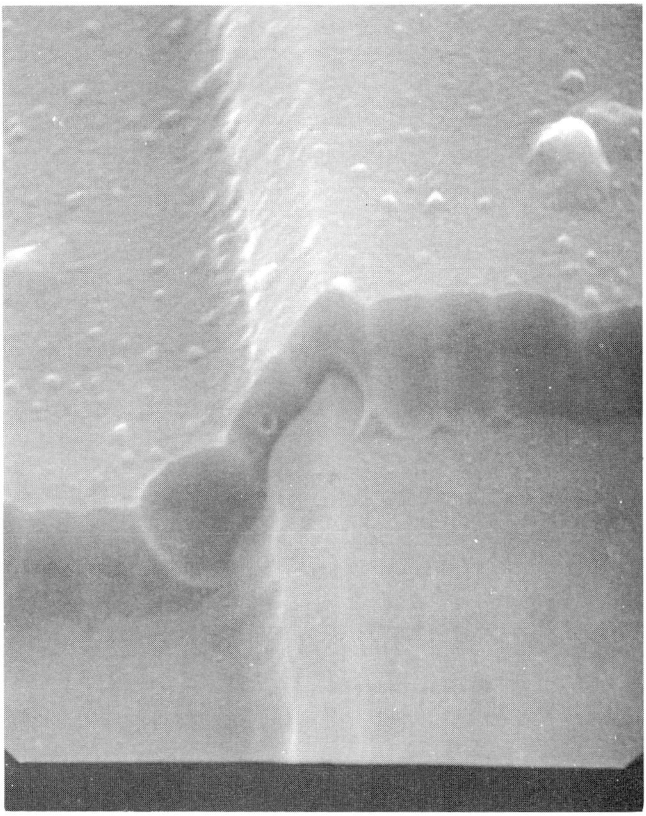

Figure 9–5 Satisfactory metal interconnect.

effect of the mode should be minimized by operating in an environment as cold as possible, with the least current possible.

Thermal shock has been mentioned as the other stress that can cause a circuit to open. Since the coefficients of expansion of aluminum and silicon are different, metal interconnects may crack. The most likely set of conditions that would worsen this mode would be thick surface metal accompanied by very thin metal deposition over steps. This can be further magnified by poor adhesion of the deposited aluminum caused by surface contamination. It should be fairly obvious that long stretches of deposited metal that have no adhesion would be subject to fracture, more from thermal shock than from mechanical shock.

Also worth mentioning is the bond-associated failure mode which was common in earlier days of integrated circuit manufacture. The reader will find many references are made to intermetallic phase formations, most of which are discussed in connection with lead bond failures. An even more insidious failure mode is the growth of intermetallics from a bonding pad out along a narrow conductive interconnect. Inevitably, the deposited interconnect will fail if a thermal compression bond using gold wire is placed directly onto the narrow aluminum interconnect. It can take thousands of hours for this mode to develop, and there are no useful screens to remove it. Therefore, the problem must be corrected by design and by proper placement of thermal compression bonds on bonding pads. Figure 9-6 is an illustration of proper and improper bonding techniques as related to this failure mode.

9.2.2 Short Circuits

The short circuit is the second obvious symptom that is caused by innumerable failure modes. There are more types of short circuits or partial short circuits than there are types of open circuits. The simpler types of short circuits are ones that occur above the oxide layers. First and most significant is the short circuit that is caused by contaminants between two conductive areas of different potential. Small pieces of lint and skin have been the most common causes. These appear after the dice have been inspected and considered clean. To keep things in perspective, it must be remembered that spacing between conductive areas may be on the order of 0.1 mil. Potentially conductive particles of that size are not detectable except at high magnification. There is no known method for detecting them after a device is encapsulated. Therefore, one either eliminates the source of particles or provides protection against the symptoms. The latter will be discussed in Section 9.3, "Design Aspects for Reliability."

Another significant mode is related to metal deposition. Line intercon-

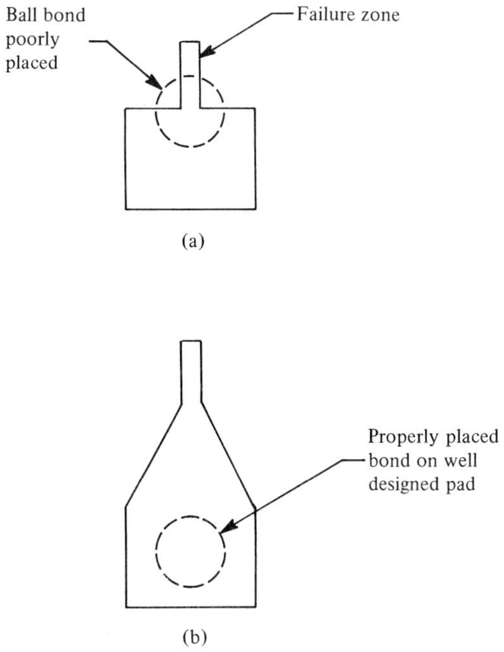

Figure 9–6 Proper bonding pad design and bond position.

nects are defined by photoresist. Certain faults in the photoresist or in the actual metal-defining mask may cause a bridging which can be detected optically and should not be a significant field failure mode. However, near-bridging does reduce the space between interconnects and could decrease the resistance. Such an anomaly in the metal in combination with contaminants provides a path for a high-resistance short.

A likely time-dependent mode of failure stems from liquid contaminants left on the surface as a result of poor cleaning procedures. There are infinite occasions during a manufacturing operation in which the surface can be contaminated by residual salts, organic contaminants, silicones, skin oils, cosmetics, etc. All provide potentially conductive paths between purposely isolated circuit elements. In performing an analysis on this kind of failure, one must remember that contaminants of this nature are just as harmful on the surface of the device package as they are on the die itself. The package should be investigated as a source of problem before the die.

Surface short circuits may also be caused by metallic particles. Common pre-form material used for soldering the die to the package is a source of small pieces of conductive material. Extraneous bonding wires are similarly dangerous and are not easily detectable after the package is sealed.

Solder-sealed packages, while inherently very reliable, provide a source of molten tin-gold eutectic which can splatter onto the die surface during sealing operations. In gold-plated packages plating can flake off and cause circuit failure. This particular mode of failure is most dangerous in the zero gravity environment where particles are free to float. Silicon particles from the die itself are also present inside the package, but the conductivity of the silicon is low enough that this mode is quite insignificant.

Within the structure, there are some interesting and significant failure modes. The degree to which surface metal is alloyed into base silicon is a subject of much discussion. It is well known that over-alloying is extremely dangerous. During the course of normal integrated circuit manufacturing, the aluminum is alloyed into all *P*-region contact cutouts in a closely controlled furnace environment. However, further alloying does occur in several subsequent operations, and this additional alloying may not be as well controlled as the initial step which is introduced purposely to produce the alloying action. The temperatures commonly used in die attaching and sealing will cause significant alloying. This is one reason why thermal compression bonding should be avoided. The time that a die is subjected to temperatures about 400°C on a heater block varies widely depending on the number of leads that must be attached. Hence the amount of alloying varies widely. At some point during this alloying, the aluminum preferentially alloys along crystal boundaries as shown in Figure 9-7. The total effect of this alloying is unknown, but it is obvious that when alloying proceeds from two isolated *P*-regions and meet at a common point, a short circuit will occur. These triangles appear to be on the same plane as the metal; however, a triangle may actually extend out from a *P*-region all the way to a metal interconnect line without causing a short circuit. It is only when two triangles appear to meet that a short circuit results. There is no reason to believe that this alloying continues at any significant pace during normal operation of the final circuit. Therefore, it is not believed that this phenomenon causes a reliability problem. However, it may have significant effects on initial performance.

Although anomalies in *P*-regions are not normally considered to cause failure modes in the field (since they cannot migrate), there is one type of anomaly that could cause the formation of a spurious transistor. Figure 9-8 shows a plan view of this anomaly.

In the figure, a spurious transistor might not even exist due to the *P*-region to *P*-region spacing designed into the part, but the protrusion on the left-hand *P*-region minimizes the spacing and sets up a condition where the spurious transistor could arise.

The most insidious short circuit failure mode is the ruptured oxide that causes a short circuit between surface metalization and the substrate material. Referring to Figure 9-1 again, note the varying thickness of oxide in

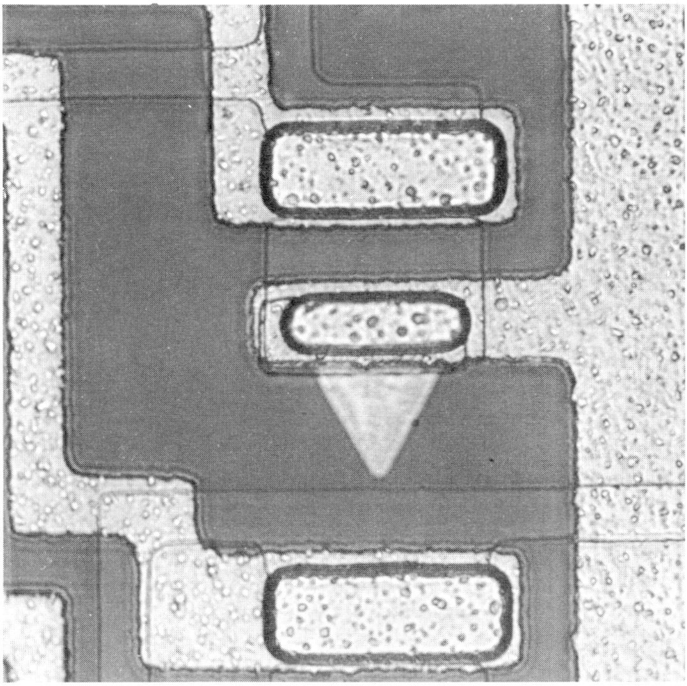

Figure 9–7 Preferential alloying of surface interconnect.

a typical MOS array. The oxide is at its thinnest dimension where a gate overlaps a *P*-region. All other areas of the chip have an oxide layer composed of two separately grown layers. This is done to discourage propagation of pinholes completely through the oxide. In the case of the gate overlap, the oxide is not only at its thinnest, but is also a single layer of oxide. To compound the problem, there is inevitably an oxide step in the same region which causes a point of electrical stress. It is here that oxides are most likely to rupture under undue stress, and it is also the most likely place for anomalies to appear in the oxide. Oxide anomalies are normally so small that they cannot be viewed with optical microscopes. It is impossible to see most of these defects even with modern scanning electron microscopes. These anomalies can be minimized by assuring a defect-free oxide and by assuring that the oxide is thick enough at the thinnest point. The oxide thickness is so closely related to parametric performance that one is not completely free to adjust oxide thickness for the sake of reliability.

Results of failure analyses on finished devices show that most ruptured

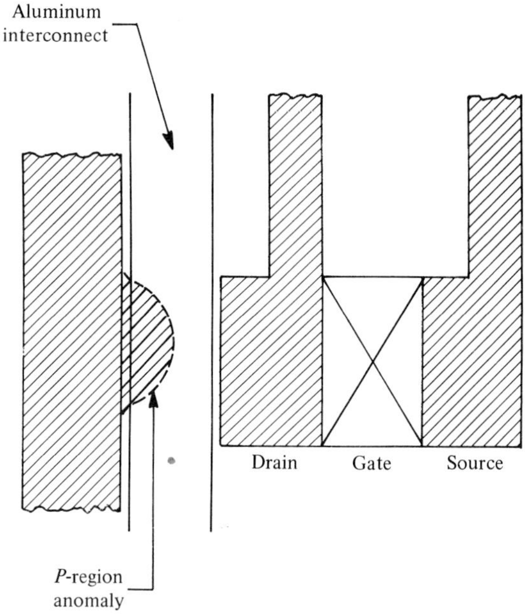

Figure 9–8 Spurious transistors.

gates occur due to mishandling. A high-quality MOS array has a gate breakdown voltage of more than 100 V. However, the devices have an extremely high input impedance, making them sensitive to accumulation of static charges. Once the rupture occurs, there is no effective way to determine whether an oxide anomaly existed. A typical oxide rupture is shown in Figure 9-9.

The best known preventive is an effective protective device at each terminal. The protective device clamps any incoming voltage to a level below the gate breakdown voltage. Details of gate protection devices are included in Chapter 2. This failure mode is by far the most significant from the standpoint of MOS. The other modes discussed are functions of complexity rather than MOS technology.

9.2.3 Threshold Drift

Another MOS failure mode is the threshold drift phenomenon caused by the presence of ions under gate structures. This phenomenon is similar to channeling in *PNP* or *NPN* transistors, but the symptoms are different. Figure 9-10 shows a typical MOS transistor structure with positive ion contamination in the gate area. Figure 9-11 shows a similar situation with

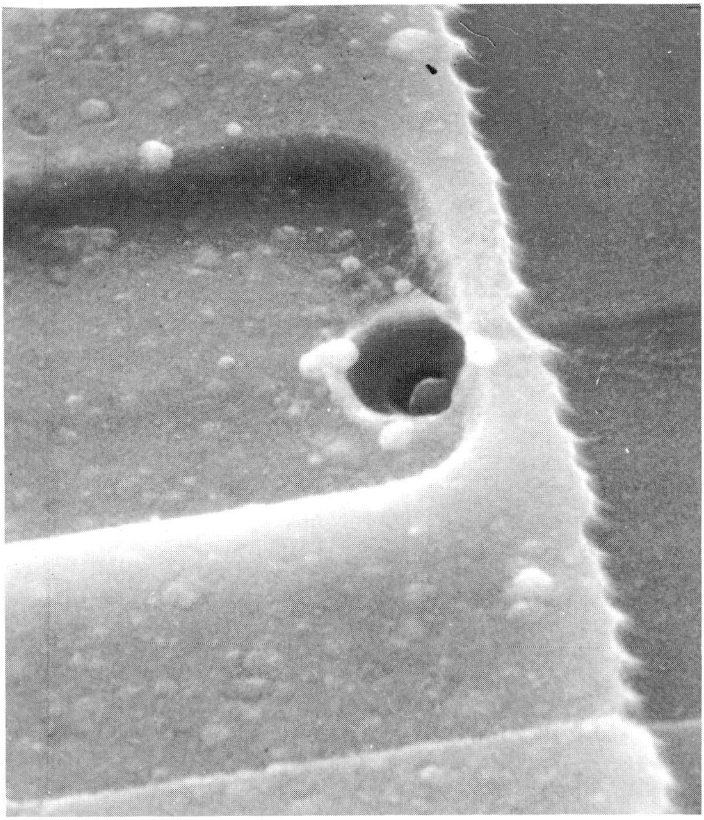

Figure 9–9 Result of voltage overstress.

negative ions. It is more likely that positive ions will exist, since sodium ions are common in nature.

The positive ion condition is best described with electrical biases as shown in Figure 9-10. These biases are abnormal, but are used as exaggerated case for purposes of evaluation. Assume that the device is biased as shown at some high temperature greater than 100°C. The positive gate voltage will drive positive ions toward the substrate. Since equilibrium must be achieved, negative charges must arrive at the interface to offset the positive charge. Once the temperature is lowered to room temperature and the bias removed, a semi-permanent condition exists where the threshold voltage of the device will have risen significantly because of the need to overcome negative charges in the channel area.

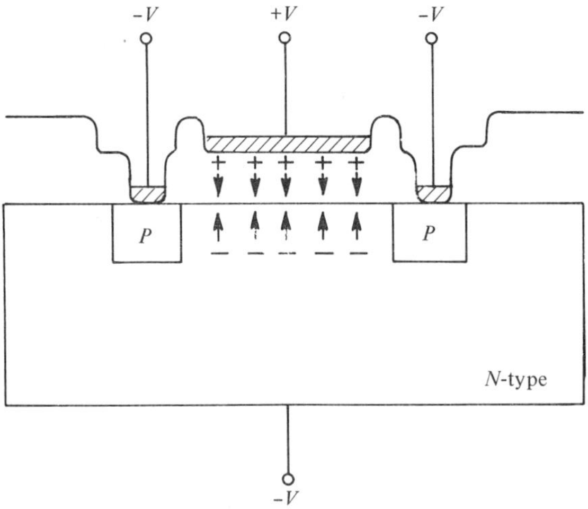

Figure 9–10 Threshold voltage drift mechanism.

In normal operation, of course, the gate is negative with respect to substrate and the opposite effect will take place. Positive ions in the oxide will be driven from the channel area, causing the threshold voltage to be lowered. In general, this is a fairly good condition for digital applications. Analog applications are more sensitive since a change in any direction may affect circuit function. In a sense, an MOS device is self-cleaning during

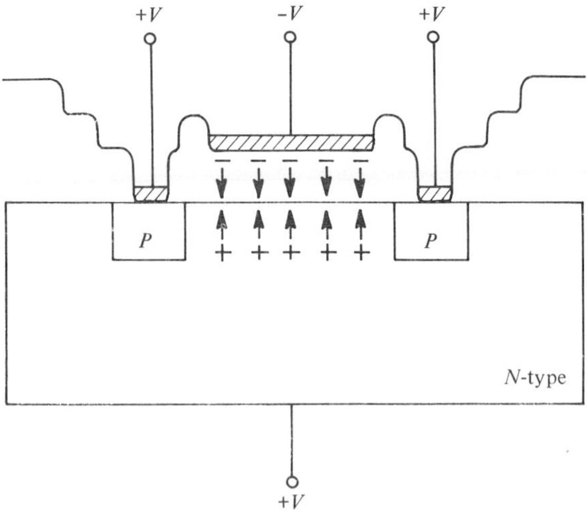

Figure 9–11 Second threshold voltage drift mechanism.

normal operation. However, the presence of sodium (Na^+) or other positive alkaline contaminants can cause very serious yield problems because they increase the threshold voltages beyond specified limits.

Figure 9-11 which shows negative ion contamination is somewhat hypothetical, since no mobile negative ion has been identified in silicon dioxide. The biases required to exaggerate the problem are reversed from Figure 9-10.

Another problem related to ionic contamination is surface contamination, which may cause leakage between source and drain. Looking at the device from the outside, it would appear to require less gate voltage to achieve a certain (although probably low) drain current. Therefore, the threshold voltage may appear to be low, which is likely to seriously reduce the noise immunity of the device.

From a practical point of view, the preceding discussions have covered the most important failure modes of MOS arrays, since they are the most likely to occur. Other modes, such as open bonds, hermetic seal failure, etc., have been covered extensively in the literature; therfore, they will not be discussed here.

9.3 DESIGN ASPECTS FOR RELIABILITY

The failure modes of MOS arrays have been categorized as varying degrees of short circuits, open circuits, and threshold drift. At this point, a discussion of design considerations related to these failure modes should be useful to a potential designer. Failures in the open circuit mode were limited to open metal interconnects on the surface of the die and open connections from the die to the chip.

With regard to surface metalization, it is important at the design stage to assure that current densities in the interconnects are low enough that metal migration never occurs. Since metal migration occurs at current densities greater than 10^6 A/sq cm, an adequate safety margin can be built into all circuits. While the industry tends to allow one order of magnitude safety factor, it is very practical in MOS circuit design to allow at least two orders of magnitude, or 10^4 A/sq cm. This safety factor may appear to be extravagant, but it must be remembered that metal interconnects that pass over sharp steps in the oxide may not be as thick as indicated by surface measurements. In fact, the thickness is rather uncontrollable and it is best to assume an extreme worst-case condition. One might reasonably assume that this metal is only 10 percent as thick as the surface metal. In the process of designing a circuit then, the weakest point is the minimum cross section of the metal as it drops into a contact aperture.

An early design error was to define minimum size contact apertures that have the same width as the minimum metal interconnect. This meant that, with perfect alignment, all current carried in a particular line would have to pass over oxide steps. Since perfect alignment was rarely possible, some of the metal bypassed the contact aperture and the current density close to the aperture was low. An example of this earlier faulty design rule is shown in Figure 9-12. The effect of misalignment is also depicted. This is the only case in which misalignment was beneficial.

An obvious answer to the problem is to totally surround the contact aperture with metal, to provide not only a good bypass but also a large cross-sectional area for current into the contact area. In general, this design consideration is most important in ground and power lines, although it is safest to apply good design practices throughout the entire circuit. One must remember that high current transients are always possible, even on a clock or input line. The small added area due to this type of metal design rule pays for itself in yield alone. This is a classic example of where yield and reliability are directly proportional.

Another design consideration for preventing opens involves the shape of the bonding pads. While thermal compression bonds are becoming less common, the advent of ultrasonic ball bonding with gold wire should bring back gold-aluminum metal systems. Much has been said about the problems caused by intermetallic phases, the least of which is the failure of a bond between the gold wire and an aluminum bonding pad. A major problem arises when an operator places a gold ball bond on a very narrow metal interconnect. Figure 9-6 shows a case in which failure is bound to occur after sufficient hours of operation at high temperature. In a manufacturing environment, the design must be such that it is easy for an operator to do what is right, even if he is undisciplined. One cannot always allocate large areas to bonding pads; but the pad can be tapered into a lead-away in such a manner that the operator has sufficient area in which to make the bond.

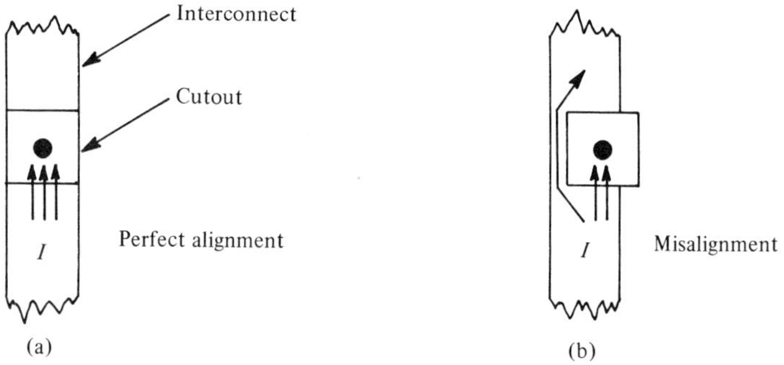

Figure 9–12 Poor oxide cutout design.

Device performance must not be totally forgotten in the attempt to improve reliability. A general rule should be to make the bonding pad as large as possible without any regard to uniformity of appearance, providing that the additional area does not increase the die size or increase the capacitance beyond useful limits. It is quite common today to define a minimum bonding pad as 4 mils square when 1 mil wire is used for connection to the package leads. Positioning of the lead bond is not nearly so important with an aluminum-aluminum system; it is far more critical with a gold-aluminum system. With the latter system, intermetallics form where the gold wire is bonded to the aluminum and can grow very rapidly along a very narrow aluminum stripe at temperatures above 250°C during the manufacturing process. The resultant intermetallic is quite brittle and is very likely to open as the stripe expands and contracts during temperature changes.

Short circuits present many difficult problems from a design viewpoint. Short circuits may be caused by conductive particles, faulty oxides, chemical contaminants, spurious transistor action, metal etching problems, and placement of bonding wires. The first and most important consideration is the particle failure mode. It is common to passivate the entire surface of a die with materials such as silane and silicon nitride. Such passivation, if done well, provides near-perfect protection against surface short circuits from particles or chemical contaminants. Even full surface passivation, of course, does not prevent short circuits or high resistance paths between adjacent bonding wires. Those areas, however, are small compared to the overall surface of the die.

Whenever one considers the addition of a process step, one must consider the effects of unexpected contaminants being introduced by the step itself. Many corrective actions result in unanticipated problems. As an example, consider overall passivation of a die by silane deposition. The purpose of the silane coating is to prevent external contaminants from causing resistive short circuits. One must remember that deposition of "dirty" silane would provide a built-in resistive short circuit instead of protecting against this mode of failure. If contaminants are present before silane is deposited, then leakage can occur between adjacent circuit elements. If one assumes that particle contamination is far more common than chemical contamination, it is logical to deposit the passivation layer only over metal interconnects. A method has been developed to do this without ever allowing the silane or its by-products to touch thermally grown oxides between metallic stripes. Such a technique provides protection against particles without introducing extraneous leakage when faulty deposition occurs. It may be assumed that even dirty silane on top of unipotential surfaces will be harmless since no biases exist to cause ions to move about and cause problems. This approach to design is far more practical than attempting to purge the environment of all contaminants or to closely control contaminant deposition along with silane or other passivating materials.

The spurious transistor was mentioned in Section 9.2.2 as a possible source of current leakage. To avoid this problem, the spacing between unrelated diffused *P*-regions should be maximized commensurate with space considerations. In an MOS transistor, the spacing between source and drain is critical to device performance. The spacing referred to here is that between the source or drain of a transistor and some unrelated diffused interconnect. Close spacing of *P*-regions with improper location of a metal clock line could form a spurious MOS device if the field threshold degraded below a certain level. Although field thresholds may not change downward, it is much more practical to design the circuit to continue to function even with severe degradation in threshold voltage. Such considerations must be given in every phase of design.

Another important design consideration is the protective device which should clamp spurious high voltage pulses and accumulation of static charges at some level which will not damage high impedance MOS gates. Here again the protective device must be made as tolerant as possible to heavy stress. The protective device should truly protect the circuitry against as high a voltage as is practical. This protective device must react very rapidly compared to the rate at which an input gate capacitance will charge.

In summary then, design considerations are not dictated by circuit performance alone. The circuit must be producible in volume with high yield. This would indicate a trade-off study between a large die with wide metal to metal spacings, wide diffused region spacings, large bonding pads, and large protective devices, and a smaller die with all spacings and elements reduced to the ultimate small size.

The cost of producing a single wafer full of arrays is fixed. The cost of each circuit sold is a function of how many good dice are produced from each wafer. Very small dice provide more potential salable units. Large dice having similar but "safe" circuitry will have a higher yield per hundred patterns. The object is to produce the most salable devices from a single wafer.

9.4 FAILURE RATES

The average aerospace user has become accustomed to using failure rate figures because his ultimate customer demands an analysis of the mean time between failures in his system. In view of the variety of MOS circuits and their varying complexity, it is difficult to establish absolute failure rates. On the other hand, there is some validity to relative failure rates. An example would best explain this point of view. If one assumes a failure rate for a given component and uses this number in a system calculation, one can very easily compare relative failure rates for two different approaches to building a system. There is little inherent error if a consistent failure rate is used for comparison. However, one should not take seriously the mean time between failure calculations based on a ficticious failure rate for the components involved. Ten years ago it was proven that single silicon transistors could be produced with a failure rate of approximately 0.001 percent per thousand hours. Since that time, little information has been generated because of the expense involved. One can justify, however, that higher levels of integration produce better system failure rates than can be generated by using single transistors or simple integrated circuits. Consider a system which uses one thousand common integrated circuits in 14-lead packages. For the sake of discussion, assume that each of the leads is functional. If performance characteristics allow this system to be changed over to MOS LSI, it is quite likely that it would result in a system having 25 components with 40-leads each. If we further assume that any component failure causes system failure, it could be said that any one of the MOS LSI devices could have a failure rate 40 times greater than any one of the integrated circuits used in the original design. With this assumption, the component contribution to system failure rate would be equal. It is difficult to believe that a single LSI element would have a failure rate that high. Take any one failure mode, such as lead bond failure, and look at the situation in the system. The original system had 28,000 lead bonds in 1000 integrated circuits. The LSI version has 200-lead-bonds in the entire system. This is a very significant reduction in the probability of lead bond failure. Similarly, the number of solder joints is reduced by the same proportion. It is quite likely that the number of printed circuit cards in the system would be reduced by a factor of ten. Along with this reduction in printed circuit cards is the reduction in the number of connectors. Furthermore, the MOS LSI version will consume far less power and generate far less heat. Semiconductors are affected by heat and are less prone to fail in cool environments. So even if one assumes a failure rate 40 times that established for a common integrated circuit, there would be reliability benefits derived from the lesser quantity of circuit boards, connectors, solder joints, etc. It is recommended that when quantitative system failure rates are necessary, a very high failure rate number be assumed for any particular LSI unit in computing system MTBF. If the number is satisfactory, then there is no need to justify a ficticious, low-component failure rate number which cannot be supported. It is already well-established that MOS array failure rates in system usage are under 0.1 percent per thousand hours. The most realistic range to use in calculations is between 0.01

and 0.1 percent per thousand hours. The choice of the exact number would depend on complexity. Presently, 0.1 percent should be used for circuits that contain approximately 6000 individual transistors on a single chip, and 0.01 percent would be most applicable to a simple circuit that contains about 200 transistors on a single chip. Both failure rates are large enough to be acceptable to most knowledgeable users.

More appropriate to this modern technology is a knowledge of process stability. If one has confidence in the parametric stability of a given process, it is easy to justify that improvements in system level reliability can be achieved readily by using large-scale arrays produced by the simple MOS process. Caution must be taken in the selection of processes because of the sensitivity of MOS devices to contamination. This sensitivity is advantageous, since there are certain screening techniques that will locate faults at an early stage. This will be discussed in the section on screening.

9.5 USEFUL SCREENS FOR IMPROVING DEVICE RELIABILITY

MIL-STD-883 provides a listing of common screens that are useful or have become traditional in the semiconductor industry. In that document, the screens are classified into three lists that are related to the end reliability needed for the system being designed. That list will be used as a basis for discussion.

It has been said that one must design reliability into a product, instead of attempting to inspect quality into the product. In general this is true, especially in a very high-yield product line. Unfortunately, overall yield of any semiconductor product is low, and inspection becomes a very important means for assuring the integrity of the parts. The screening methods described in the MIL-STD-883 are aimed at removing devices that have inherent reliability problems not readily detected by electrical testing alone. Screening is accomplished by optical means prior to final seal and by high mechanical, electrical, or thermal stressing after final seal. Properly manufactured semiconductors will withstand thermal environments from cryogenic temperatures up to approximately 350°C. Mechanically, the average device should withstand acceleration stresses on the order of 150,000 g's in the constant acceleration mode and 15,000 to 20,000 g's in the shock mode where shock pulse lengths are on the order of 0.1 ms. Electrically, allowable stresses are related to breakdown voltages as viewed at the inputs and current carrying capability of internal conductors. In general, the industry promotes the idea of excessive stress testing where the philosophy is to remove all of those devices that are different from the normal population. This philosophy differs from most industries where an attempt is made merely to assure that products will meet the stresses expected in normal use.

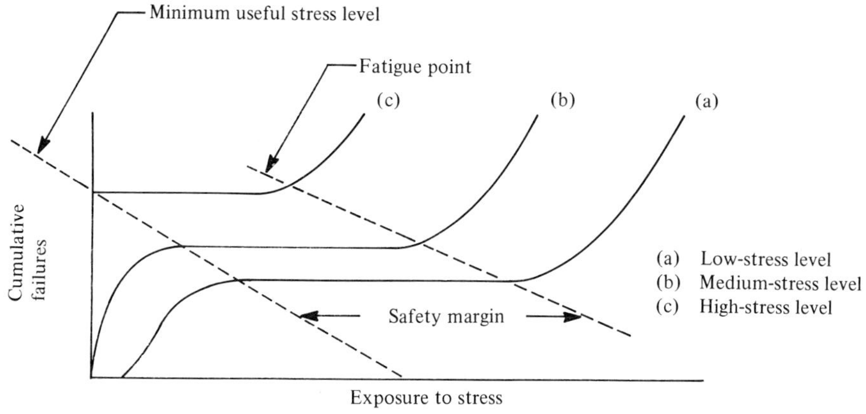

Figure 9–13 Generalized stress vs. failure curve.

By definition, 100 percent screening must be nondestructive. When a new and useful screen is being investigated, one must define the term nondestructive. Figure 9-13 shows generalized curves of cumulative failures vs. the number of exposures to stress. Close to the axis we see failures that result from substandard material or processes. Out to the right we see failures due to excessive fatigue or excessive stress levels. If there is a flat portion of sufficient length in the curve one could define a screen as being nondestructive. As a specific example, consider the effect of shocking each component as a means of removing defective lead bonds. Then the curve would have the dimensions shown in Figure 9-14. This curve is a factual one. The flat portion indicates a fairly efficient level of screening after one exposure to shock and shows clearly that the part would survive up to 25 exposures to the same shock without any further degradation. This is a

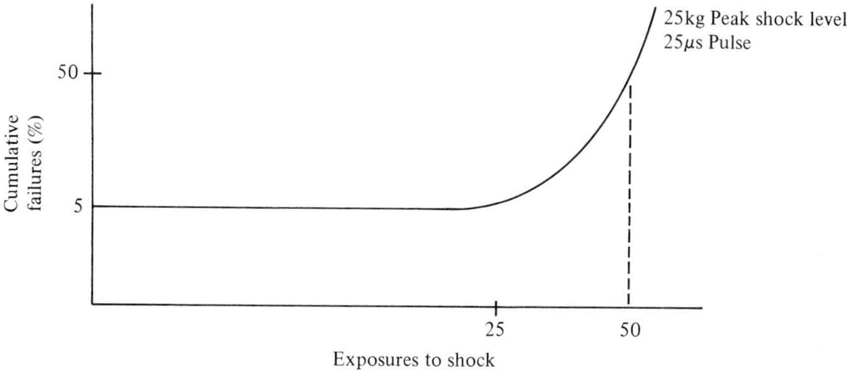

Figure 9–14 Specific stress vs. failure curve.

clear indication that the test is nondestructive and is safe for production screening. Any other type of stress can be evaluated and presented in this fashion, although some of the common screens will not necessarily cause the fatiguing shown in the generalized curve.

9.5.1 Optical Screening

The most effective screen for assuring semiconductor reliability is detailed and efficient optical inspection of the die and its package. Historically, field failures have been due to poor inspection of the device immediately before final seal. In previous sections of this chapter, failures were categorized as varying degrees of short circuits or open circuits. Figures 9-15 and 9-16 are examples of these problems in complex integrated circuits. Certain criteria can be established that are clear-cut, though perhaps arbitrary in nature. Open circuits that can be seen at practical magnification (100×) will be confined to the metalized pattern and the lead bonds. The detection

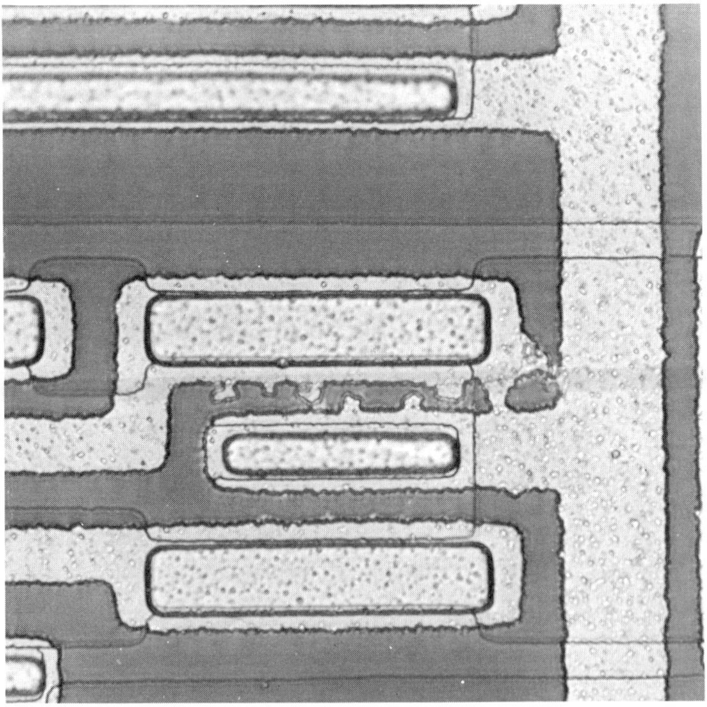

Figure 9–15 Short-circuit mechanism caused by bridged metal interconnect.

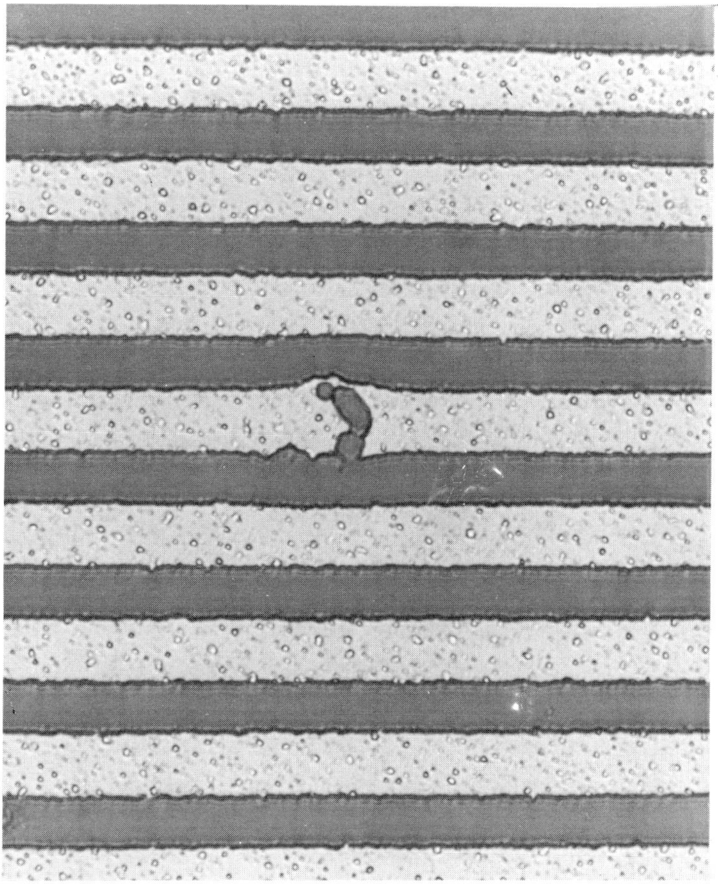

Figure 9–16 Potential open circuit.

of actual open circuits before final seal is an economic measure. More important from the reliability viewpoint is finding conditions that would *lead* to open circuits at some point after the circuit is put in operation. Therefore, optical inspection should be aimed at finding points where the cross section of metal interconnects has been reduced significantly. Today it is commonly accepted that a cross section should not be reduced to less than 50 percent. This assumes that the circuit is properly designed to limit the current density, as discussed in Section 9.3.

Judgment on lead bonding quality is much more subtle. Most integrated circuits are connected to the outside world with ultrasonic wedge

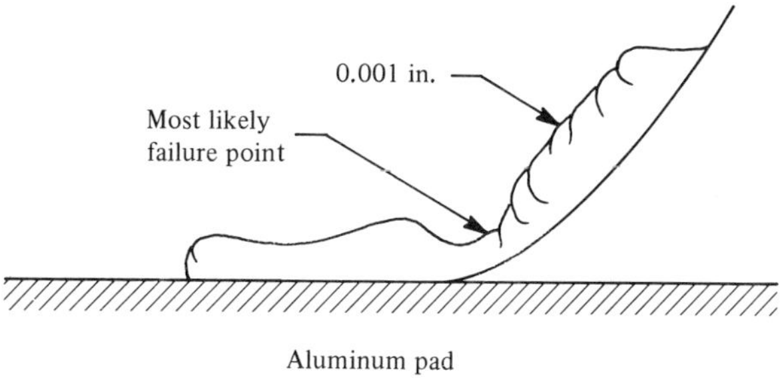

Figure 9–17 Ultrasonic wedge bond failure mechanism.

bonds. The shape of the bond is as shown in Figure 9-17. One cannot tell whether a bond is adhering well, but a fair judgment can be made on the thickness of the connecting wire at the heel. If this cross section is too narrow, subsequent vibration or shock may cause failure. There is little standardization in the industry on the appearance of a lead bond. A good description of a bond appears in MIL-STD-883, Test Method 2010.

In the area of short circuits, we again see some clear-cut problems and some subtle ones. There are many sources of extraneous metal that can cause short circuits. It is absolutely necessary that the enclosed package not contain any loose conductive particles that could possibly bridge between exposed metal areas. In today's very complex circuits, such particles may be fractions of a mil long and be totally invisible to any technique such as X-ray which might be used after sealing. A more subtle problem arises from chemical contaminants. It would appear economically justifiable to reject any device that has visible liquid residue of unknown origin on its surface. Subsequent exposure to heat may cause the contaminant to migrate and cause internal circuit leakage. If the contaminant appears to be a solid, it is likely to have no effect on a passivated circuit if it cannot bridge between bare metal areas.

There are many other aspects to visual screening, but the significant ones have been covered.

9.5.2 Mechanical Screening

Mechanical screening is generally done so that poor bonds separate from the circuit, causing it to fail during final electrical test. This screening is also effective for causing the die to separate from the package in the event that die bonding is ineffective. Screening methods should be designed to keep the package from being damaged. The most common mechanical screen is constant acceleration. A common method for many years was to centrifuge parts at 20,000 g's in the orientation that would tend to pull the die off the package or pull the bonds off the die. The very low mass of an aluminum wedge bond makes it difficult to apply enough force to be effective. A good ultrasonic bond with 1 mil aluminum wire will have a pull strength in excess of 4 grams. It is impossible to achieve such a pull by centrifuging. However, the centrifuge remains a useful tool for removing the poorest bonds. Centrifuging should be done at the highest level possible with safety considerations in mind. At this time 40,000 g's is a common requirement. Although, it is highly unlikely that any *system* could withstand 40,000 g's.

A more effective method for removing defective lead bonds is a shock test. Empirically, we have seen that a very short-pulse, high-g shock test will cause a higher failure rate than constant acceleration at much higher g-levels. For many package configurations 100 percent shock testing is extremely costly and very slow. For the common TO-5 or TO-18 configurations, equipment is available for shock testing at approximately 25,000 g's with a pulse width close to 25 μs. Commercially available equipment allows this type of testing at a rate of 2000 pieces per hour. With larger packages, expensive jigging is required, and commercial shock test equipment that will impart enough energy to be useful is not readily available. Shock testing as a screen for larger packages is uneconomical, even though the test itself might be useful in removing latent failures.

9.5.3 Thermal Screens

Thermal screens fall in two categories. The first and most useful is thermal cycling or thermal shock aimed at exaggerating the effects of differential expansion. Repeated exposure to hot and cold temperature extremes can cause fracturing of metal at the interfaces between two materials. Repeated temperature cycling is useful in detecting defective surface metalization. In the section on failure modes (Section 9.2.1), it was shown that metal can be extremely thin over oxide steps. If a scanning electron microscope is not available for examining metal, repeated thermal cycling can indicate disastrous metal conditions. Common military requirements today call for temperature cycling between −65°C and +200°C. In general, ten complete temperature cycles are specified. It is generally considered that 50 or more cycles are required to cause failure in marginal metalization. Although the efficiency of such a screen is in doubt, it is not costly. Where the ultimate in reliability is required, it is a useful tool.

A second class of thermal stressing is simple storage at high temperatures. This is occasionally specified as an inexpensive substitute for operational burn-in. High temperatures will cause contaminants to become mobile. If a device is found satisfactory after electrical testing, subsequent exposure to high temperature may cause mobile contaminants to move to a more critical area. Electrical testing after high temperature exposure may disclose some defective parts. A great difficulty is that the screen may also make marginally acceptable devices look extremely good. If a contaminant is concentrated in critical circuit areas, high-temperature exposure may cause the material to migrate to a less critical area, making the circuit look better. Therefore, high-temperature bake is not recommended unless electrical biases are provided while the parts are exposed to high temperature. In any case, high-temperature bake causes an artificial redistribution of contaminants and the distribution immediately changes upon application of bias.

9.5.4 Operational Burn-In

Operational burn-in can be performed in several ways. The method used is related more to funds available than any other single factor. To be effective, an operational burn-in should be performed at high temperature to increase the mobility of contaminants. A full operational burn-in can be performed that tends to reject devices with excessive contamination and those with poor quality metal interconnects. In the discussion of functional testing, it was seen that digital test words usually could be applied to devices in such a way that all transistors were turned on and off on command. If these same test words were applied to a complex circuit for a long period of time under high temperature, excess contamination would migrate toward junctions and channel areas causing the circuits to become inoperable. At the same time, current would flow through the conductors at a level that approximates operating conditions. A high-ambient temperature will accelerate the failures.

In the case of MOS structures, migration of contaminants will occur very rapidly, causing significant threshold shifts and junction degradation within 100 hours. This type of screening is less effective as far as metal problems are concerned. Whereas a number of metal defects may possibly exist which could cause metal interconnects to fail in a very few hours, it may in fact take thousands of hours before failure occurs. Therefore, the effectiveness of the screen is very high for contamination failures and is unknown for metal failures. The expense of full operational burn-in is probably not justified because of this unknown factor. A common substitute is called high-temperature reverse bias testing (HTRB). In this test, the burn-in circuit is very simple. One provides a high-voltage bias to most diffused regions within a circuit while providing a high-ambient temperature. This causes all contamination to move toward junction or channel areas. Thus, it is possible to quickly eliminate parts that contain excess contamination without the risk of reducing the useful life of the device.

In summary, it is economically more feasible to perform the high-temperature reverse bias burn-in for a reasonable period of time. One week is a general guideline for the length of this type of burn-in. These expenditures for screening are related to the overall economics of assembling an electronics system. In many cases it is possible to achieve 95 percent of the benefits with only 50 percent of the cost. As an example, one might consider the ultimate operational burn-in requirement to call for 1000 hours of operation with reading and recording of data. It is quite probable that 95 percent of the latent failures could be removed by the more economical alternative of using HTRB and 100 percent test of significant parameters after the completion of burn-in.

9.5.5 X-Ray Screening

For years, it has been common to X-ray packaged semiconductor devices in one or two orientations. In the days of simple transistor structures with thermal compression bonds, the X-ray method provided a good screen for gross workmanship errors such as large metal particles and badly damaged internal lead wires. The construction of early transistors was simple and, in general, welded enclosures were used. Present day MOS large-scale integrated circuits have aluminum ultrasonic bonds and solder-sealed enclosures. X-ray photographs do not show the bond or the die but do show voids in the die attach material and excess solder flow from the solder seal. Voids in the die attach area are almost meaningless since MOS circuits require a very minimal contact to ground. Generally speaking, the dice are connected to ground. Bipolar devices may require a very high-quality contact between the die and ground due to saturation voltage requirements. This is not true in MOS circuits. The solder seals tend to cause confusion when X-rayed. Solder flows toward the hottest, cleanest areas on the lid of the package and an X-ray technician cannot effectively tell the difference between this harmless solder flow and extraneous solder that might appear inside the package. Thus, internal lead damage cannot be seen and excess metal particles are hardly definable from excess solder flow caused during the sealing cycle. Therefore, this screen becomes of doubtful value.

9.6 RELIABILITY EVALUATION

The advent of large-scale MOS arrays has led to large-scale usage of custom circuits. These custom circuits each replace as many as 50 common integrated circuits or 6000 individual transistors. This very fact means that each circuit is more expensive per chip and the production volume of individual circuits is lower. This environment does not encourage large-scale reliability testing of each and every circuit as a means of establishing failure rates. However, the custom circuit aspect of the business does encourage standardization of processing. Otherwise, each custom circuit becomes a new adventure. This processing standardization makes it possible for the first time to use a specially designed reliability evaluation device to judge the reliability of a multitude of custom circuits. MOS processing is far more amenable to standardization than bipolar processing. It would appear valid then to start judging the process rather than the individual device. For a given manufacturer using standardization philosophies, it might be said that all custom circuits are members of one product family. A device designed to evaluate process control does not necessarily assure the user that 6000 transistors on one chip will interface long enough to make the circuit useful. Therefore, it is also necessary to support this type of testing with occasional large-scale tests of complex devices to give total assurance. Most large-scale arrays can only be evaluated from the viewpoint of operability with known input conditions. The reliability evaluation device can be designed to provide parametric measurements of performance over a period of time.

The wafer fabrication process is most likely to suffer mainly from contamination or defective metal deposition. There are many other factors that might cause poor yield, but there are no other significant factors that would affect reliability. Thus, the reliability test vehicle is useful from a wafer processing point of view, but is not useful for checking the quality of workmanship in an assembly line. By nature, the test vehicle is a special part and is not necessarily assembled by the same personnel who build working circuits.

Figure 9-18 shows a useful reliability evaluation device that contains

1. A discrete inverter and MOS load device
2. An MOS field threshold device
3. A large *P-N* junction with enhancement gate
4. A straight MOS capacitor identical in size to the enhancing gate of Item 3.
5. A large *P-N* junction area identical in size to Item 3 but without gate enhancement

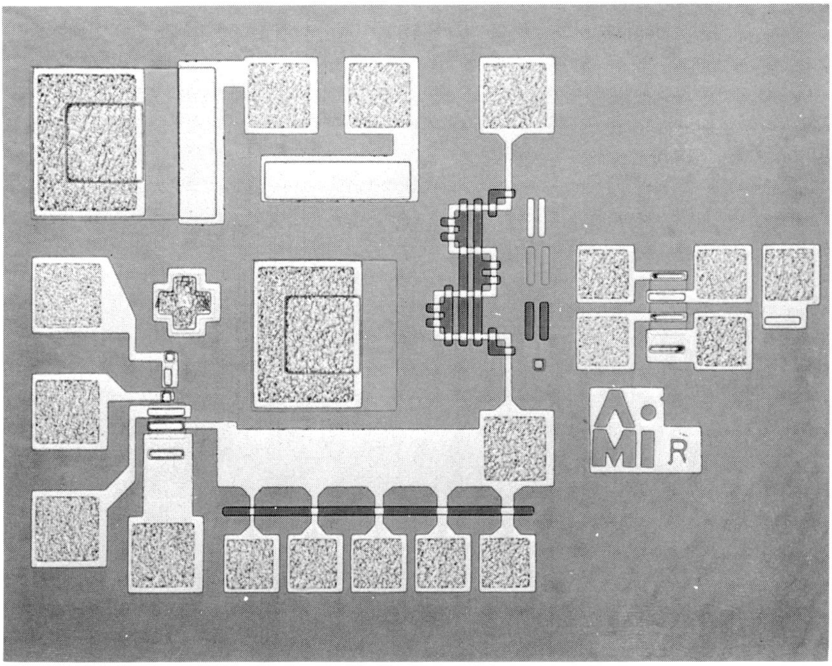

Figure 9–18 Reliability evaluation device.

6. A long metal stripe of minimum design rule width that crosses a large number of etched cutouts
7. A series of metal stripes that cross a deeply etched area

The metal deposition aspect of processing is covered in two ways. The small stripes of metal that pass through a deep cutout can be used for process control of metalization since the burn-out characteristics of these stripes are indicative of metal thickness. While it is simple, quantitatively, to measure metal thickness on the surface of a die, it is nearly impossible to measure metal thickness as the interconnect passes over an oxide step. The current at which this step burns out should be indicative of metal thickness at these steps. To be consistent in measurements of burn-out current, a current generator should be used that increases current at a constant rate with respect to time. A rate of 25 mA/s has been found to be a satisfactory level. Under these conditions the burn-out current appears to form a normal distribution. This makes it easy to apply classical control chart methods for evaluating metal immediately after the alloying operation.

The long metal stripe represents a minimum design width for a specified metal thickness. One can force a known current through this stripe for a period of time, assuming a current density based on surface metal thickness. Failures during the course of a life test would be indicative of metal migration, implying either a faulty design or faulty metal deposition over steps.

The inverter load and field threshold devices can be operated in many different modes simultaneously. For example, it is possible to operate the inverter at a fixed frequency, while operating the load device under high-temperature reverse bias conditions with either positive or negative gate bias. Positive gate bias is very useful for exposing the presence of positive ions in the gate area, while negative gate bias would be more indicative of performance under actual use. Sufficient testing of small elements as this would provide failure rate information for individual MOS transistors. Knowing the elemental failure rates, one might make a good estimate of the relative circuit failure rates of very simple circuits and very complex circuits.

Also on the chip is a large *P*-region enhanced by a gate. This allows evaluation of a very large area device which would have a high probability of junction defects. Separate from this is another *P*-region of like size and an isolated gate structure identical to the one enhancing the large *P*-region. Evaluation of these two elements would help to locate a source of contamination that caused difficulties in the gate-enhanced *P*-region.

We have then a device that can be evaluated in many ways at the same time. If such a device is included with production circuits, it is valid to conclude that good performance of the reliability test vehicle implies good working circuits. There is merit in being able to evaluate a standard circuit on a daily basis rather than to randomly evaluate different types of circuits where data comparison is meaningless.

9.7 USEFUL TOOLS AND TECHNIQUES IN ANALYSIS

In failure analysis, there are two characteristic defect situations: those devices that have defects that are readily visible, and those that are not easily discernible. Examples of the defects in the first category would be a cracked die, broken lead bonds, visually ruptured gates, vaporized or burned metalization. This type of characteristic would manifest itself initially in dc parameter testing and would be optically visible. On the other hand, there are usually a large number of parts in which there is no visual cause for rejection under optical microscopy; the device may have passed dc parameter testing, but failed functionally. To determine the true cause for failure, more sophisticated techniques and instrumentation are required, such as electrolytic plating and micro-cross-sectioning along with a scanning electron microscope (SEM), microprobe, and phase contrast interferometer instrumentation. At times, there are pinholes in gates that are not easily seen or recognized even when viewed with a SEM at 10,000 power. A technique that has been used successfully in detecting oxide defects is the plating technique. Figure 9-19 shows a setup where a wafer is placed in a conduc tive beaker that contains a suitable electrolyte, such as acetic acid, or alcohol, or even water. The wafer must have the oxide stripped from the back side in order to make contact with the conductive beaker. In this configuration, the copper will plate onto any exposed silicon area. As the plating action takes place, gases from the bare silicon evolve and expand, and a bubbling action occurs in the presence of an oxide defect. If left under bias long enough, the defects are decorated with copper salts that can be counted easily under a microscope. In this manner, the basic quality of an oxide can be determined. It is generally conceded that most oxide anomalies occur because of residual contaminants on the silicon wafer at the beginning of the oxidation operation. A likely contaminant is the lapping compound used to polish the wafer prior to oxidation. By reviewing the oxide defect on an historical basis, the number of defects contributed after each oxidation step can be established.

The capability of cross-sectioning a device or die has proven itself to be a highly useful tool. It allows one actually to see and measure oxide thicknesses, metalization thickness, and any deformation in the metal, oxide, and/or *P*-region when staining is applied.

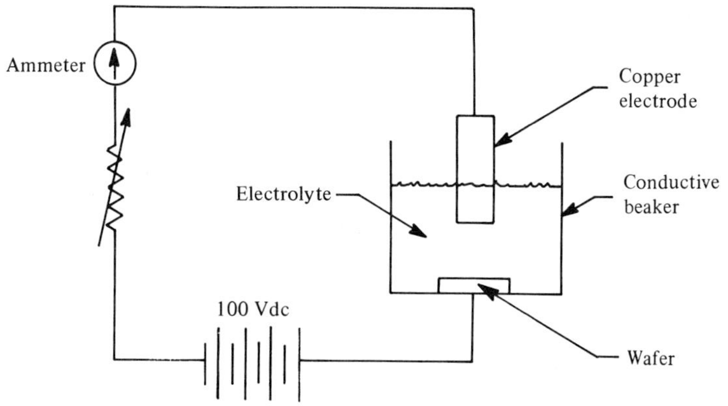

Figure 9–19 Detection of oxide pinholes.

The SEM is an absolute necessity in scanning devices for open metallization cracks or micro-cracks that can only be seen at 10,000 power or greater. Although the optical microscope itself if a useful tool, it has a high magnification level of approximately 2000 power, while its maximum useful range is less than 1000 power. The optimum resolution is on the order of 3500 Å. The microscope is limited in its ability to view three-dimensional objects by shallow depth of field, whereas the SEM can surpass this in every respect and also provide the three-dimensional capability necessary. Perhaps one of the greatest advantages of the SEM is its ability to perform voltage contrast tests that indicate voltage potential through visual illumination. The microprobe, although not widely used, provides a particle analysis that gives an elemental breakdown of the specimen itself. It has been employed on foreign material in and on the surface of specimens.

The phase contrast interferometer provides an accurate means for determining junction depth, such as on a wafer that has been grooved, as well as many other measurement applications.

The combined use of all the instruments and techniques just described can achieve highly satisfying results. As an example, in troubleshooting a large-scale array after having covered all the preliminary stages (such as the initial testing of the device, optical inspection that revealed no cause for failure, and functional probe that actually isolated the failure area to one discrete device within the array), it was noted that a *P*-region contact was not making contact to the *P*-region itself and that further probing of the device was useless. It could have been concluded at this time that the device was inoperative and/or the cutout window was not making contact with the *P*-region. However, the device was evaluated with a SEM. During the examination, it was discovered that the metal line and associated cutout region were intact with no open or cracked metal. It was then decided that a voltage contrast test be performed, again using the SEM. Here the device was put into functional operation, and it was seen by voltage contrast, through *P*-region illumination, that the device was truly operational and that the *P*-region did indeed have a voltage on it, but the cutout area associated with the *P*-region was not illuminated. This meant that the cutout itself was not coming into contact with the *P*-region or a contaminant had formed between the *P*-region and the cutout area. The cutout region was then cross-sectioned and the *P*-region stained. Examination showed a contaminant above the *P*-region area and below the *P*-region cutout. The contaminant was then subjected to microprobe analysis and it was discovered that the oxide had not been entirely cleaned from the cutout area prior to metalization.

Failure analysis is the single most important means of providing useful feedback, since in the ultimate sense, returned devices that do not operate, as well as those that fail final test, are the real failures of concern. Many characteristics of semiconductors are called failures arbitrarily to provide a safe margin for error. But the feedback of information from returned and failed devices in final test are the most valuable data in improving future designs and cutting costs on present designs without compromising reliability.

Appendix

Some Concepts of Surface Physics Applied to the Analysis of FET Characteristics*

In Chapter 2, the pertinent surface physics of MOS** structures was given and then used to develop the theory of operation of field effect transistors. In this appendix, the relationship of the channel conductance (G_0) to the conductance of the surface ($\Delta\sigma$) under all conditions of surface potential, i.e., accumulation, depletion, and inversion, is defined. An understanding of the latter relationship is essential for correct interpretation of the $G_0 - V_g$ characteristic as a function of the surface charge array.

1. Channel Conductance, Surface Conductance and Surface Charge Structure

As has been indicated in Chapter 2, the slope and voltage axis offset of G_0 (with V_D approaching zero) is controlled by the net charge at the surface and the work function difference and is a function of the minority carrier mobility.

The surface potential (u_s)*** at

*Part of work done by G. P. Walker, P. Lin and W. Ford (Reference 2) at Raytheon Co., Mountain View, California.

**A better term would be CIS for conductor insulator semiconductor; however, MOS will be used here. The basic arguments, of course, apply to most of the gate electrode and gate dielectric materials.

***The notation and sign convention of Many, Goldstein and Grover[1] is used in this section, except for some transistor terms.

$$u = q\phi/kT;\ q\phi \equiv E_F - E_i$$

where q is the electronic charge, ϕ the potential in volts, and T the absolute temperature. u, then, is the dimensionless potential with subscript b for the bulk and s for the surface.

which G_0 becomes zero, has been commonly taken to be approximately equal to $-u_b$, the bulk potential (a more exact value will be given later). Thus, the Fermi level, at turn-on is about halfway ($u_s \simeq -u_b = -11.5$ for 3.5 Ωcm material, for example) between mid-gap and the valence band, for an N-type body. The surface states that lie more than 3 (units of u) above this point will have no influence on the slope of the $G_0 - V_g$ trace, as V_g is increased beyond turn-on. The charge in these states will, of course, enter into the net charge and affect the voltage axis offset at every point.

The P-regions that form ohmic contacts to the channel when the surface is inverted, block the possibility of observing any charge changes in the surface states for voltages less than the turn-on voltage (V_T). If extra ohmic contacts to the N-bulk are provided, then the conductance change may also be followed in the depletion and accumulation regions.

Figure 1[2] shows, among other features, the surface conductance ($\Delta\sigma$) (change with respect to the bulk value) vs. V_g characteristic of a special device having contacts that are ohmic in all ranges of surface potential. As will be seen, this classical field effect display is closely related to the $G_0 - V_g$ characteristic of a transistor having the same bulk and surface properties. This relationship is well worth appreciating, since it enables one to interpret the $G_0 - V_g$ characteristic with an awareness of the contribution of the charge centers not traversed when V_g is varied over the range beyond turn-on.

The surface conductance ($\Delta\sigma$) analogous to the bulk conductance is,

$$\Delta\sigma = q\mu_{ns}\Delta N + q\mu_{ps}\Delta P \tag{1}$$

Here, μ_s is the surface mobility for electrons or holes and ΔN, ΔP the change in electron or hole density from bulk values. Convenient numerical solutions are given by Many, Goldstein and Grover[1] (and others, given in their references) for the surface excesses of electrons and holes,

$$\Delta N = \int_0^\infty (n - n_b)\, dZ \tag{2}$$

$$\Delta P = \int_0^\infty (p - p_b)\, dZ \tag{3}$$

All of the space charge (Q_{sc}) is included in the sum of ΔN and ΔP,

$$Q_{sc} = q(\Delta P - \Delta N) \tag{4}$$

For the theoretical case, i.e., no surface state charge, Q_{ss} (ϕ_s), no oxide charge (Q_{ox}) and zero work function difference (ϕ_{MS}) then, V_g (the voltage applied to the gate) drops across the oxide and the space charge region.

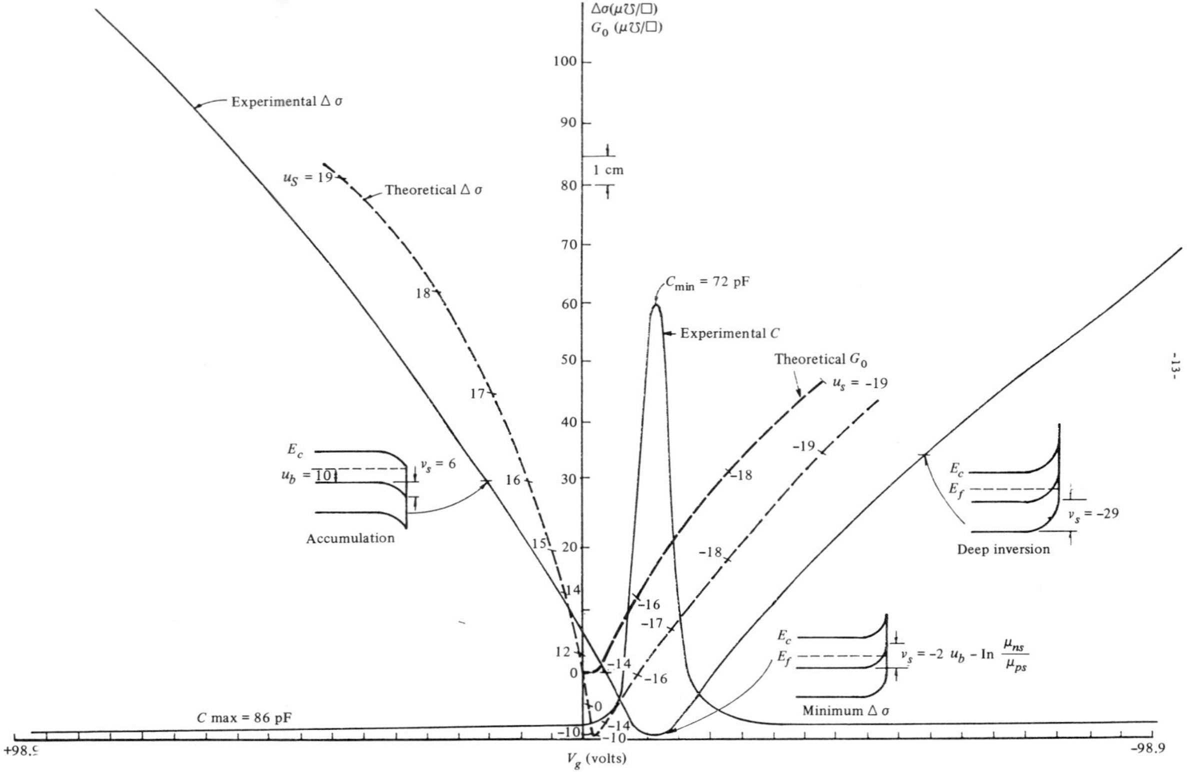

Figure 1 Surface conductance and capacity vs. gate voltage for 17 Ω cm *N*-Si at 297°K. Typical high temperature deposited oxide with FAN etch-DI quench-HF dip-DI rinse pretreatment. Oxide thickness ≈ 8000 Å.

C_{max} = 86 pF, scale factor = 1 pF/cm Horizontal scale factor = 5.2 V/cm

FP area = 2.23×10^{-2} cm^2 (111) surface

Sample number T382B

Note: Numbers on ordinate for $\Delta\sigma$ or G_0 only. Use scale factor for capacity.

$$V_g = V_s + V_{ox}$$

Since for the theoretical case,

$$V_{ox} = \frac{-Q_{sc}}{C_{ox}}$$

then

$$V_g = V_s - \frac{Q_{sc}}{C_{ox}} \tag{5}$$

Here, $V_s = \phi_s - \phi_b$ is the band bending, in volts, and C_{ox} is the geometrical capacity of the gate oxide per unit area.

It should be noted that whereas the interface state and oxide charge were treated as an operational or lumped constant in Chapter 2, here, Q_{ss} (ϕ_s) is a variable with surface potential and Q_{ox} represents the charge in the oxide.

$$Q_{ss}(\phi_s) = qN_t f(E_t^f) \tag{6}$$

N_t is the density of states and the Fermi function $f\ (E_t^f)$ represents f_p (E_t^f) for acceptor like states and $f_n\ (E_t^f)$ for donor like states.[1] Q_{ox} is the net effect (on the space charge) of all the charges in the oxide.[3]

Equations 1 and 5 can be used to plot a theoretical $\Delta\sigma$ vs. V_g relation such as that shown in Figure 1. For this figure, surface mobility values determined by Walker, Lin, and Ford[2] have been used; however, other values can be inserted in the equation equally well. The correctness of any detailed charge analysis depends, of course, on the chosen mobilities $[\mu_s = f\ (\phi_s)]$ being the correct ones. From the theoretical curve, surface potential points can be transferred along lines of constant conductance. The conductance for the experimental case is only a function of the mobile charge in the space charge region. On Figure 1, band diagrams indicate the surface condition at various points.

Provided that all of the conditions of a good, equilibrium measurement have been met, the experimental $\Delta\sigma$ curve of Figure 1 can be meaningfully analyzed to yield the energy levels and densities of the interface states and the magnitude of the oxide charge. The horizontal difference, at every point, of the experimental and theoretical curves is shown below.

$$-\phi_{MS} + \frac{Q_{ss}\phi_s}{C_{ox}} + \frac{Q_{ox}}{C_{ox}}$$

The work function can be obtained from Deal, Snow, and Mead[4] and the charge terms plotted vs. u_s. Many et al.[1] indicate the manner in which this plot is fitted with an appropriate form of $f\ (E_t^f)$.

This analysis is a major goal of the surface worker. The purpose here, however, is to demonstrate how an appreciation of the broader (in u_s) display of surface changes leads to a more useful interpretation of readily observable transistor characteristics such as the $G_0 - V_g$ and $C_g - V_g$ (gate capacity vs. gate voltage). Also, it can be noted, that considerable information about the surface charge can be obtained by just a visual interpretation of Figure 1 and of the associated $G_0 - V_g$ characteristic, now to be discussed.

2. $G_0 - V_g$ CHARACTERISTIC

Returning to Equation 1, but considering now a theoretical (no surface states) transistor with the bulk parameters and oxide thickness of the field effect device of Figure 1*; but, with contacts only to the channel region during inversion (conventional *P*-diffused contacts as for the usual transistor construction). Under these conditions, it will not be possible to observe changes in ΔN due to the *P-N* junction. Expression 1 reduces to:

$$\Delta\sigma_{ch} = q\mu_{ps}\Delta P \tag{7}$$

This is the expression for the channel conductance of a transistor except that the change in conductance is with respect to the surface conductance at flat bands. However, as Garrett and Brattain[5] have pointed out, the change of ΔP from flat bands to the point where the channel has a signi- cant minority carrier concentration is small enough that it is a good approximation to consider $G_0 = \Delta\sigma_{ch}$, i.e., the theoretical channel conductance will go approximately to zero at the turn-off point. This can be seen from Figure 1 also, where the channel conductance has been plotted. Surface potential points on this curve lie along the same gate voltage line as for the previously developed theoretical curve, since the band bending is the same for the same voltage for both theoretical cases.

The channel conductance (G_0) is offset from the $\Delta\sigma$ curve by the fact that ΔN is absent from the G_0 expression. ΔN increases (negatively) only slightly in this region. The band bending at turn-on can now be seen to be equal to that at the minimum of the $\Delta\sigma - V_g$ curve. The value is obtained[1] by setting the derivative of $q\ (\mu_{ns}\ \Delta N + \mu_{ps}\ \Delta P) = 0$ and is

$$V_{sm} \simeq -2u_b - \frac{\ell n\ \mu_{ns}}{\mu_{ps}} \tag{8}$$

*The bulk resistivity and oxide thickness used here are more appropriate for detailed surface state analysis than for useful transistors; however, the experimental and theoretical behavior will be quite analogous when other resistivities and oxide thicknesses are substituted.

where V_{sm} is the potential barrier (band bending) at the surface at the minimum. V_{sm} is dimensionless and equals qV_{sm}/kT, where k is the Boltzman constant.

The expressions that were used to compute the theoretical $G_0 - V_g$ relation are

$$G_0 = q\mu_{ps}\Delta P$$

$$V_g = V_s - \frac{Q_{sc}}{C_{ox}}$$

Q_{sc} is obtained from Equation 4.* G_0 is in mhos per square in this equation.

The $G_0 - V_g$ characteristic of a real (experimental) transistor, plotted on Figure 1 would have the same relationship to the experimental trace that the two theories, just discussed, have to each other. That is, it would be offset (from the theory) and modified in slope by the interface and oxide charges and work function difference and would go to zero along a horizontal line passing through the turn-off point of the theoretical G_0 curve.

3. CHARGE ANALYSIS FROM THE *G-V* AND *C-V* CHARACTERISTICS

As for the *C-V* characteristic, also shown (experimental) in Figure 1, horizontal shifts of the $G_0 - V_g$ trace with respect to the theoretical can be taken to be equal to the net surface state and oxide charge plus that due to the work function difference. This is particularly tempting if, for example, a voltage-temperature stress has resulted in a parallel shift of the curve. The pitfall in such a procedure is vividly displayed in Figure 2.[2] Here, although some of the change in surface potential sensitive charge centers has occurred near the minimum, a major cause for the shift, of this unstable oxide is due to a charge center well outside the range of both the $G_0 - V_g$ and the *C-V* characteristics.** Also, note that until a determination of the net charge in all such interface centers has been made, nothing can be said about the possible presence of charge in the oxide or beyond the surface potential range of even this more extensive measurement.

Apart from the matter of limited surface potential range common to both the *G-V* and *C-V* methods, the *C-V* measurement can give misleading results because of an unfortunate choice of signal frequency. In

*Note: Q_{sc} is not equal to just $q\Delta P$.

**Detailed analysis gives $N_t = 3.5 \times 10^{11}$ cm^{-1} for this trap at $(E_t - E_i)/kT = 16$.

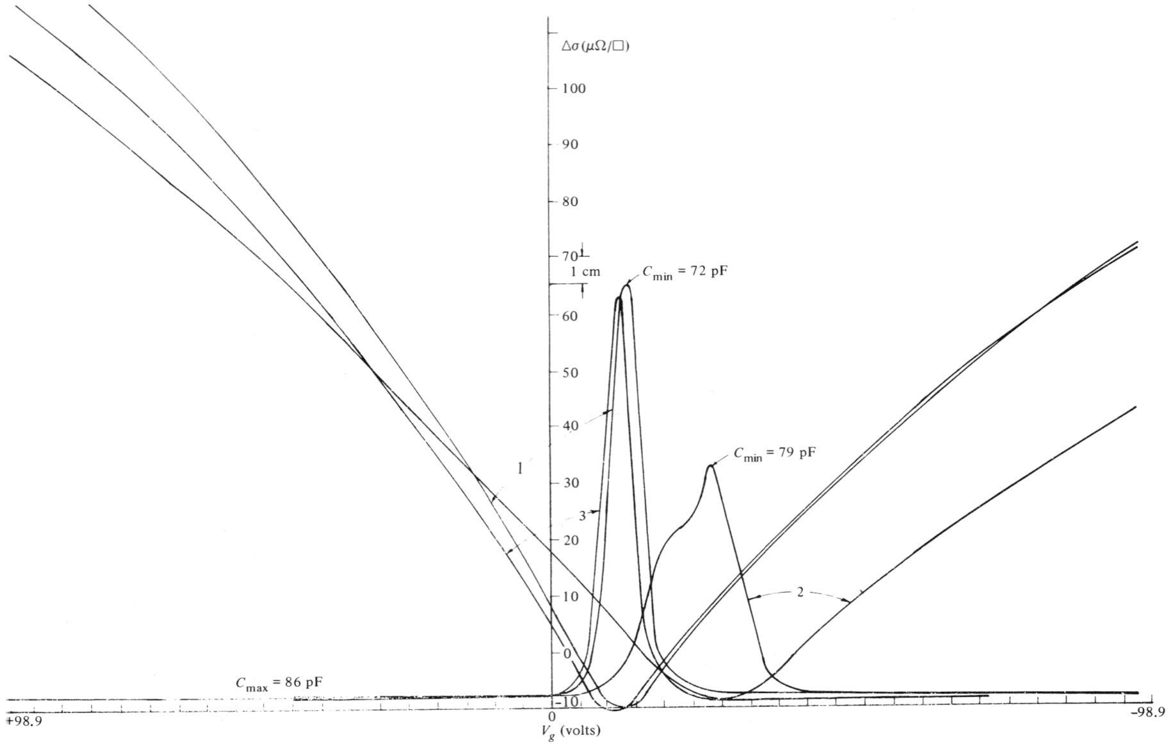

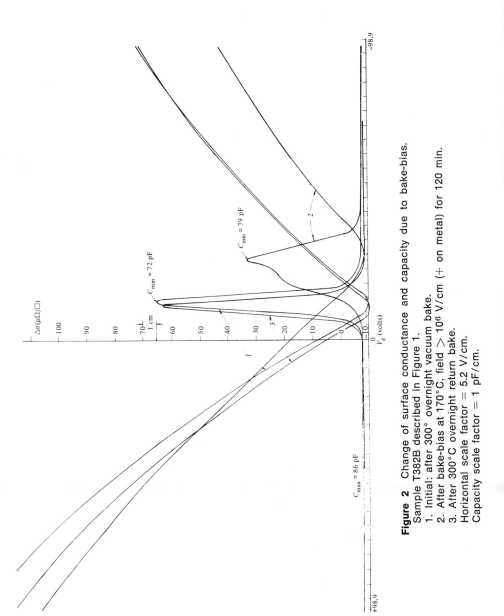

Figure 2 Change of surface conductance and capacity due to bake-bias.
Sample T382B described in Figure 1.

1. Initial: after 300° overnight vacuum bake.
2. After bake-bias at 170°C, field $> 10^6$ V/cm (+ on metal) for 120 min.
3. After 300°C overnight return bake.

Horizontal scale factor = 5.2 V/cm.
Capacity scale factor = 1 pF/cm.

order to be able to fit the experiment to theory (an equilibrium theory), it is necessary that the signal frequency be high enough to eliminate the response of the interface states,

$$\frac{1}{C} = \frac{1}{C_{ox}} + \frac{1}{C_{sc} + C_{ss}} \tag{9}$$

where C_{sc} is the capacity of the space charge region and C_{ss} the capacity of the surface states. The limitations and advantages of this measurement have been extensively discussed.[7-10, 12] Grove[12] also discusses the G-V relation with regard to the surface charge structure. An analysis of the electronic structure from G-V measurements at various temperatures is presented in Reference 11. It should be recalled that the G-V measurement, while free of the surface state equilibrium problem of the C-V measurement, needs a correct value of the surface mobility vs. ϕ_s for the theoretical curve to be correct.

With the correct set of values for the surface mobility, a G-V curve can be analyzed over the range of sensitivity to yield the energy levels and densities of discrete interface states. The same analysis can be applied to any C-V experimental curve providing that the signal frequency is high enough.* Many and Gerlich[13] have pointed out that the slope of the Q_{ss} (ϕ_s) vs. u_s curve at $u_s = (E_t - E_i)/kT$ is $N_t/4$. This result came from taking the derivative of the expression for the charge in a set of states of a single energy level

$$Q_{ss} = -\frac{qN_t}{1 + e^{(E_t - E_i/kT - u_s)}} \tag{10}$$

for acceptor like states with an analogous expression of positive sign for donor like states.

A good procedure is to measure the $G_0 - V_g$ and $C_g - V_g$ characteristics for the same large test transistor. The device should be large enough such that geometrical and diffusion factors will not give too much error. The overlap capacity should be made relatively small or accounted for. These can be recorder traces on the same graph. The theoretical $G_0 - V_g$ relation is plotted on the same graph using Equations 5 and 7. As previously discussed, the horizontal offset at every point between the theoretical and experimental is

*In principle, the frequency must be somewhat greater than 50 MHz to eliminate the response of all the fast states; however, if the sensitive range of the experimental C-V relation can be caused (by raising the signal frequency) to correspond to the theoretical C-V plot, except for voltage offset, then, in this range the trace can be meaningfully analyzed.

$$-\phi_{MS} + \frac{Q_{ss}(\phi_s)}{C_{ox}} + \frac{Q_{ox}}{C_{ox}} \tag{11}$$

As before, the work function difference is determined for the particular materials involved (Equation 4); the C_{ox} is obtained from the deep accumulation region of the *C-V* experimental curve; and Q_{ss} $(\phi_s) + Q_{ox}$ is plotted vs. u_s. Now, before examining this, consider the *C-V* plot.

If the frequency were high enough, Equation 9 reduces to

$$\frac{1}{C} = \frac{1}{C_{ox}} + \frac{1}{C_{sc}} \tag{12}$$

From Frankl[14] the expression for C_{sc} is,

$$C_{sc} = \frac{dQ_{sc}}{dV_s} = \frac{\kappa\epsilon_0}{L_D}\frac{\partial F_s}{\partial u_s} \tag{13}$$

$$= \frac{\kappa_1\epsilon_0}{L_D}\frac{\sinh u_s - \sinh u_b}{F(u_s \cdot u_b)}$$

This can be used to generate the *C-V* curve (high frequency case) and convenient tables are supplied in Reference 15.* However, Goetzberger[6] has supplied a very adequate set of theortical *C-V* curves which can be used. As for the $G_0 - V_g$ case, the difference in the theoretical and experimental *C-V* curves is Equation 11. Hence, Q_{ss} $(\phi_s) + Q_{ox}$ from this can be plotted versus u_s on the same graph as for G_0.

This also nicely extends the range of u_s covered. Now, if Q_{ox} is believed to be a fixed quantity, the slope at any point of inflection can be set equal to $N_t/4$, as previously discussed. The procedure is to fit the experimental curve with the distribution function as carried out by Many and Gerlich.[13] It should be recalled that this procedure is limited to just the surface states in the energy range of sensitivity of the G_0 and C traces.

One should also be aware that the usual theory assumes that the impurity redistribution is negligible. For deposited oxides, put down on surfaces that have no pile-up or depletion of bulk impurities, the results should be correct. For surfaces that have been subjected to thermal oxidation, the error in the results will be larger as a function of the degree of pile-up or depletion. The relative location of the *C-V* minimum to the theoretical can be used as a means of assessing the extent of this; providing that the $G_0 - V_g$ curve indicates a low Q_{ss} (ϕ_s) near the turn-on point. Where extensive redistribution has occurred, the results will be questionable, at best, but may still be a useful aid in process development.

*The tables there are also very useful for determining $\triangle P$ vs. u_s that is required for the $G_0 - V_g$ theory.

REFERENCES

1. A. Many, Y. Goldstein, and N. B. Grover, chap. 4, p. 4 of chap 5. *Semiconductor Surfaces,* American Elsevier, New York, 1965.
2. G. Walker, P. Lin, and W. Ford, "Coincident Field Effect Conductance and Capacity Measurements on Silicon," Tech. Report No. 49, Proj. 1231–00009, Raytheon Co., Mountain View, Calif., March 1967.
3. E. H. Snow et al., "Ion Transport Phenomena in Insulating Films," *J. Appl. Phys.,* **36,** 1664 (1965).
4. B. E. Deal, E. H. Snow, and C. A. Mead, "Barrier Energies in Metal-Silicon Dioxide-Silicon Structures," *J. Phys. Chem. Solids,* **27,** 1873–1879, Pergamon Press, London, 1966.
5. C. G. B. Garrett and W. H. Brattain, "Physical Theory of Semiconductor Surfaces," *Phys. Rev.,* **99,** 376 (1955).
6. A. Goetzberger, "Ideal MOS Curves for Silicon," *Bell Sys. Tech. J.,* **45,** no. 7, 1097 (September 1966).
7. K. H. Zaininger and G. Warfield, "Limitation of the MOS Capacitance Method for the Determination of Semiconductor Surface Properties," *IEEE Trans. Electron Devices,* **ED-12,** 179 (April 1965).
8. A. S. Grove et al., "Investigation of Thermally Oxidized Silicon Surfaces Using Metal-Oxide-Semiconductor Structures," *Solid State Elec.,* **8,** 145 (1965).
9. A. G. Revesz and K. H. Zaininger, "The Si-SiO_2 Solid-Solid Interface System," *RCA Rev.,* **29,** no. 1 (March 1968).
10. E. Kooi, *The Surface Properties of Oxidized Silicon,* Springer-Verlag, New York, 1967
11. E. Arnold, "Surface Charges and Surface Potential in Silicon Surface Inversion Layers," *IEEE Trans. Electron Devices,* **15,** 1003 (1968).
12. A. S. Grove, *Physics and Technology of Semiconductor Devices,* John Wiley & Sons, New York, 1967.
13. A. Many and D. Gerlich, "Distribution and Cross Sections of Fast States on Germanium Surfaces in Various Gaseous Ambients," *Phys. Rev.,* **107,** 404 (1957).
14. D. R. Frankl, "Some Effects of Material Parameters on the Design of Surface Space Charge Varactors," *Solid State Elec.,* **2,** 71–76 (1961).
15. D. R. Frankl, *Electrical Properties of Semiconductor Surfaces,* Pergamon Press, London, 1967.

GENERAL REFERENCES

R. M. Warner, Jr. and J. N. Fordemwalt, eds., *Integrated Circuits,* McGraw-Hill New York, 1965. Bulk and junction properties.

C. T. Sah, "Characteristics of the Metal-Oxide-Semiconductor Transistor," *IEEE Trans. Electron Devices,* 324–345 (July 1964).

C. T. Sah and H. C. Pao, "The Effects of Fixed Bulk Charge on the Characteristics of Metal-Oxide-Semiconductor Transistors," *IEEE Trans. Electron Devices,* 393–409 (April 1966).

M. M. Atalla, Tannenbaum, and E. J. Scheibner, "Stabilization of Silicon Surfaces by Thermally Grown Oxides," *Bell Sys. Tech. J.,* **38,** 749 (1959). Field effect measurements on thick oxides.

B. E. Deal, E. L. McKenna, and P. L. Castro, "Characteristics of Fast Surface States Associated with SiO_2-Si and Si_2N_4-SiO_2-Si Structures," *J. Electrochem. Soc.,* **116,** 997 (1969). Recent work and references.

R. Lindner, "Semiconductor Surface Varactor," *Bell Sys. Tech. J.,* **41,** 903 (1962). Basic concepts of surface capacity.

P. V. Gray, "The Silicon-Silicon Dioxide System," *Proc. IEEE,* **57,** no. 9, 1543–1551 (September 1969). Recent work and references.

Index

INDEX